McGraw-Hill Ryerson

Mathematics of Data Management

Authors

Barbara J. Canton
B.A.(Hons.), B.Ed., M.Ed.
Limestone District School Board

Wayne Erdman
B.Math., B.Ed.
Toronto District School Board

Jeff Irvine
B.Math., B.Ed., M.A., M.B.A., C.F.P.
Peel District School Board

Louis Lim
B.Sc.(Hons.), B.Ed., M.Ed.
York Region District School Board

Fran McLaren
B.Sc., B.Ed.
Upper Grand District School Board

Roland W. Meisel
B.Sc., B.Ed., M.Sc.
Port Colborne, Ontario

David Tallach Miller
B.Sc., B.Ed.
District School Board of Niagara

Jacob Speijer
B.Eng., M.Sc.Ed., P.Eng.
District School Board of Niagara

Technology/Web Consultants

Roland W. Meisel
B.Sc., B.Ed., M.Sc.

David Tallach Miller
B.Sc., B.Ed.

Assessment Consultant

Fran McLaren
B.Sc., B.Ed.

Pedagogical Consultant

Cynthia Ballheim
B.Sc., M.A.

**McGraw-Hill
Ryerson**

Toronto   Montréal   Boston   Burr Ridge, IL   Dubuque, IA   Madison, WI   New York
San Francisco   St. Louis   Bangkok   Bogotá   Caracas   Kuala Lumpur   Lisbon   London
Madrid   Mexico City   Milan   New Delhi   Santiago   Seoul   Singapore   Sydney   Taipei

**McGraw-Hill
Ryerson Limited**

A Subsidiary of The **McGraw·Hill** Companies

*COPIES OF THIS BOOK
MAY BE OBTAINED BY
CONTACTING:*

McGraw-Hill Ryerson Ltd.

WEB SITE:
http://www.mcgrawhill.ca

E-MAIL:
orders@mcgrawhill.ca

TOLL-FREE FAX:
1-800-463-5885

TOLL-FREE CALL:
1-800-565-5758

OR BY MAILING YOUR
ORDER TO:
McGraw-Hill Ryerson
Order Department
300 Water Street
Whitby, ON L1N 9B6

Please quote the ISBN and
title when placing your
order.

Student text ISBN:
0-07-552912-2

Student e-book ISBN:
0-07-090758-7

**McGraw-Hill Ryerson
Mathematics of Data Management**

ISBN 0-07-552912-2/0-07-091714-0

http://www.mcgrawhill.ca

2 3 4 5 6 7 8 9 0 TRI 0 9 8 7 6 5 4 3 2

Printed and bound in Canada

Care has been taken to trace ownership of copyright material contained in this text. The publishers will gladly take any information that will enable them to rectify any reference or credit in subsequent printings.

Corel® Quattro® Pro 8 and Quattro® Pro 9 are trademarks or registered trademarks of Corel Corporation or Corel Corporation Limited in Canada, the United States, and/or other countries.

Microsoft® Excel are either registered trademarks or trademarks of Microsoft Corporation in the United States and/or other countries.

Fathom Dynamic Statistics™ Software and *The Geometer's Sketchpad*®, Key Curriculum Press, 1150 65th Street, Emeryville, CA 94608, 1-800-995-MATH.

National Library of Canada Cataloguing in Publication Data

Main entry under title:
    McGraw-Hill Ryerson mathematics of data management

Includes index.
ISBN 0-07-552912-2

    1. Probabilities. 2. Mathematical statistics. 3. Permutations.
4. Combinations. I. Canton, Barbara II. Title: Mathematics of data management.

QA273.M344 2002   519   C2001-904162-4

PUBLISHER: Diane Wyman
DEVELOPMENTAL EDITORS: David Peebles, Jean Ford, Tom Gamblin
SENIOR SUPERVISING EDITOR: Carol Altilia
CONTRIBUTING EDITOR: Jennifer Burnell
COPY EDITOR: Rosina Daillie
PERMISSIONS EDITORS: Maria DeCambra, Linda Tanaka
EDITORIAL ASSISTANT: Erin Parton
JUNIOR EDITORS: Christopher Cappadocia, Cheryl Stallabrass
ASSISTANT PROJECT COORDINATORS: Melissa Nippard, Janie Reeson
PRODUCTION SUPERVISOR: Yolanda Pigden
PRODUCTION COORDINATOR: Paula Brown
COVER DESIGN: Dianna Little
INTERIOR DESIGN: Pronk&Associates
ART DIRECTION: Tom Dart/First Folio Resource Group, Inc.
ELECTRONIC PAGE MAKE-UP: Tom Dart, Greg Duhaney, Alana Lai, Claire Milne/First Folio Resource Group, Inc.
COVER IMAGE: John Warden/Getty Images/Stone. Chart: From "The Demographic Population Viability of Algonquin Wolves," by John Vucetich and Paul Paquet, prepared for the Algonquin Wolf Advisory Committee.

Acknowledgements

Reviewers of *McGraw-Hill Ryerson Mathematics of Data Management*

The authors and editors of McGraw-Hill Ryerson Mathematics of Data Management wish to thank the reviewers listed below for their thoughtful comments and suggestions. Their input has been invaluable in ensuring that this text meets the needs of the students and teachers taking this course.

C. Ann Clark
District School Board of Niagara

Chris Dearling
Burlington, Ontario

Eric Forshaw
Greater Essex County District School Board

Mary-Beth Fortune
Peel District School Board

Marilyn Hurrell
Thunder Bay, Ontario

Gwyn Jackson
District School Board of Niagara

Ann Kajander
Lakehead University and Lakehead Public Schools

David Kay
Peel District School Board

Sabina Knight
District School Board of Niagara

Anastasia Liebster
Toronto District School Board

Doug McMillan
Toronto District School Board

Peter Saarimaki
Scarborough, Ontario

Silvana Simone
Toronto District School Board

Al Smith
Kawartha Pine Ridge District School Board

Bob Smith
Rainbow District School Board

Charles Stewart
Toronto District School Board

Accuracy Reviewer
Tom Gamblin, M.A. (Mathematics),
Ph.D. (Combinatorics)
Toronto, Ontario

Contents

CHAPTER 1
Tools for Data Management

CHAPTER 2
Statistics of One Variable

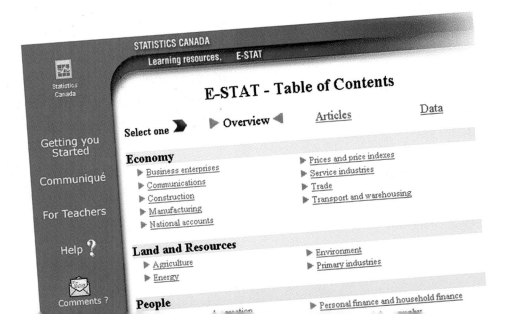

CHAPTER 6

Introduction to Probability

CHAPTER 7

Probability Distributions

CHAPTER 8

The Normal Distribution

CHAPTER 9

Culminating Project: Integration of the Techniques of Data Management

Preface

McGraw-Hill Ryerson Mathematics of Data Management gives students a solid background in both the mathematical theory and the practical techniques for managing and analysing data. This course provides a valuable foundation for studies in fields involving the collection and interpretation of data, including business, engineering, sociology, and all of the natural sciences.

Text Organization

This text gives students immediate opportunities to apply the mathematical principles they learn to real-life situations. Chapter 1 introduces a variety of powerful mathematical and technology tools. In Chapters 2 and 3, students apply these tools to the statistics of one and two variables. Then, Chapters 4, 5, and 6 explain the concepts and techniques of combinatorial analysis and probability theory. Chapters 7 and 8 apply these concepts and techniques to probability distributions and statistical calculations.

Chapter 9 gives students a framework for creatively integrating the concepts and analytical tools from the entire course in an open-ended **Culminating Project**. This project is a key part of the curriculum. The text is structured to help students plan their projects early in the course and work on them throughout the school year. **Timeline charts** at key points in the book provide guidance for keeping projects on track. Four smaller **Projects** are woven through the book to give students a chance to apply what they have learned and develop the skills they will need for the culminating project. **Project Prep** boxes identify concepts and tools that students can apply to their projects. **Research Skills** and **Oral Presentation Skills Appendices** provide further support.

The book concludes with a convenient summary of **Key Equations**.

Chapter Features

The first eight chapters open with a **Chapter Problem** providing a practical context for the mathematics, followed by a **Review of Prerequisite Skills** that covers concepts learned in previous courses or other chapters. Descriptors guide students to the **Review of Prerequisite Skills Appendix,** which has additional material on the various topics.

Almost all sections start with an **Investigate & Inquire,** an inquiry-based activity that encourages students to explore the mathematical concepts taught in the section. The concepts are demonstrated in numerous examples with complete solutions using both paper-and-pencil and technology methods. A summary of **Key Concepts** and a set of **Communicate Your Understanding** questions provide a helpful study guide and reinforce learning of the material in the section.

Numerous **Practise** and **Apply, Solve, Communicate** questions let students practise and apply their learning to real-life problems while covering all of the achievement categories.

The chapters conclude with a section-by-section **Review of Key Concepts** and a model **Chapter Test** to help students prepare for tests and examinations.

Assessment

Each section includes an extensive set of review and exercise questions. Descriptors indicate questions that are particularly well-suited for assessment of thinking/inquiry/problem solving, communication, and application skills. In addition, each chapter has **Achievement Checks** for major concepts. These specially prepared questions are accessible to all students and provide opportunities for assessing performance in all achievement-chart categories. The **Projects** and the **Culminating Project** also serve as open-ended assessment tasks. The **Chapter Tests**, **Cumulative Reviews**, and **Course Review** all simulate real test-taking situations.

Technology

This text makes extensive use of graphing calculators and spreadsheets. Where appropriate, worked examples also include an alternative solution using Fathom™, a ministry-licensed statistical software package specially designed for use in schools. However, pencil-and-paper worked examples introduce the concepts so that students will understand the mathematics that the technology uses. This approach also ensures that students who do not have ready access to computers or graphing calculators can follow all of the material in the course.

The worked examples using technology have numerous screen captures, and the **Technology Appendix** details the functions and keystrokes for TI-83 calculators, spreadsheets, and Fathom™. Data sets in the worked examples and the exercises have deliberately been kept to a modest size so that students can, if necessary, do most calculations with an inexpensive business or scientific calculator. The **Student e-book** enclosed with this textbook includes larger data sets in a variety of downloadable formats.

The **Student e-book** contains the complete text, plus a number of enhancements. In addition to the larger data sets, these enhancements include templates, interactive Java™ applets demonstrating difficult concepts, and an answer section with full graphics. Throughout the printed text, the e-book content icon indicates topics with supplementary material on the CD.

Web Connections give details on the links from the McGraw-Hill Ryerson web
site to interesting sites that extend the content in the text. These links are kept
up-to-date and include a direct connection to **E-STAT,** Statistics Canada's
remarkable interactive educational resource.

From start to finish, *McGraw-Hill Ryerson Mathematics of Data Management* is
designed to present a challenging curriculum in an exciting and accessible format.

Rating Universities and Colleges

Background

During this school year, most of you will be selecting a university or college program. *Maclean's* magazine annually ranks universities and community colleges across Canada. This project will assist you in making an informed choice as to which university or community college to attend.

Your Task

You will research and analyse current data on university and community college rankings and customize them to your own or a friend's situation. You will use spreadsheets, the Internet, matrices, and graph theory to access, manipulate, and display data and to develop a plan for visiting selected universities or community colleges.

Developing an Action Plan

You will need to find data on colleges and universities, develop a ratings system, and plan a visit to your top five choices.

CHAPTER 1

Tools for Data Management

Specific Expectations	Section
Locate data to answer questions of significance or personal interest, by searching well-organized databases.	1.3
Use the Internet effectively as a source for databases.	1.3
Create database or spreadsheet templates that facilitate the manipulation and retrieval of data from large bodies of information that have a variety of characteristics.	1.2, 1.3, 1.4
Represent simple iterative processes, using diagrams that involve branches and loops.	1.1
Represent complex tasks or issues, using diagrams.	1.1, 1.5
Solve network problems, using introductory graph theory.	1.5
Represent numerical data, using matrices, and demonstrate an understanding of terminology and notation related to matrices.	1.6, 1.7
Demonstrate proficiency in matrix operations, including addition, scalar multiplication, matrix multiplication, the calculation of row sums, and the calculation of column sums, as necessary to solve problems, with and without the aid of technology.	1.6, 1.7
Solve problems drawn from a variety of applications, using matrix methods.	1.6, 1.7

Chapter Problem

VIA Rail Routes

When travelling by bus, train, or airplane, you usually want to reach your destination without any stops or transfers. However, it is not always possible to reach your destination by a non-stop route. The following map shows the VIA Rail routes for eight major cities. The arrows represent routes on which you do not have to change trains.

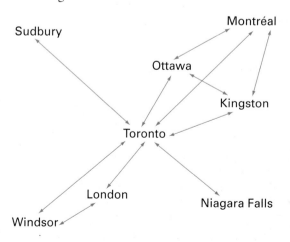

1. a) List several routes you have travelled where you were able to reach your destination directly.

   b) List a route where you had to change vehicles exactly once before reaching your destination.

2. a) List all the possible routes from Montréal to Toronto by VIA Rail.

   b) Which route would you take to get from Montréal to Toronto in the least amount of time? Explain your reasoning.

3. a) List all the possible routes from Kingston to London.

   b) Give a possible reason why VIA Rail chooses not to have a direct train from Kingston to London.

This chapter introduces graph theory, matrices, and technology that you can use to model networks like the one shown. You will learn techniques for determining the number of direct and indirect routes from one city to another. The chapter also discusses useful data-management tools including iterative processes, databases, software, and simulations.

Review of Prerequisite Skills

If you need help with any of the skills listed in **purple** below, refer to Appendix A.

1. **Order of operations** Evaluate each expression.

   a) $(-4)(5) + (2)(-3)$

   b) $(-2)(3) + (5)(-3) + (8)(7)$

   c) $(1)(0) + (1)(1) + (0)(0) + (0)(1)$

   d) $(2)(4) + \dfrac{12}{3} - (3)^2$

2. **Substituting into equations** Given $f(x) = 3x^2 - 5x + 2$ and $g(x) = 2x - 1$, evaluate each expression.

   a) $f(2)$

   b) $g(2)$

   c) $f(g(-1))$

   d) $f(g(1))$

   e) $f(f(2))$

   f) $g(f(2))$

3. **Solving equations** Solve for x.

   a) $2x - 3 = 7$

   b) $5x + 2 = -8$

   c) $\dfrac{x}{2} - 5 = 5$

   d) $4x - 3 = 2x - 1$

   e) $x^2 = 25$

   f) $x^3 = 125$

   g) $3(x + 1) = 2(x - 1)$

   h) $\dfrac{2x - 5}{2} = \dfrac{3x - 1}{4}$

4. **Graphing data** In a sample of 1000 Canadians, 46% have type O blood, 43% have type A, 8% have type B, and 3% have type AB. Represent these data with a fully-labelled circle graph.

5. **Graphing data** Organize the following set of data using a fully-labelled double-bar graph.

City	Snowfall (cm)	Total Precipitation (cm)
St. John's	322.1	148.2
Charlottetown	338.7	120.1
Halifax	261.4	147.4
Fredericton	294.5	113.1
Québec City	337.0	120.8
Montréal	214.2	94.0
Ottawa	221.5	91.1
Toronto	135.0	81.9
Winnipeg	114.8	50.4
Regina	107.4	36.4
Edmonton	129.6	46.1
Calgary	135.4	39.9
Vancouver	54.9	116.7
Victoria	46.9	85.8
Whitehorse	145.2	26.9
Yellowknife	143.9	26.7

6. **Graphing data** The following table lists the average annual full-time earnings for males and females. Illustrate these data using a fully-labelled double-line graph.

Year	Women ($)	Men ($)
1989	28 219	42 767
1990	29 050	42 913
1991	29 654	42 575
1992	30 903	42 984
1993	30 466	42 161
1994	30 274	43 362
1995	30 959	42 338
1996	30 606	41 897
1997	30 484	43 804
1998	32 553	45 070

7. **Using spreadsheets** Refer to the spreadsheet section of Appendix B, if necessary.

   a) Describe how to refer to a specific cell.

   b) Describe how to refer to a range of cells in the same row.

   c) Describe how to copy data into another cell.

   d) Describe how to move data from one column to another.

   e) Describe how to expand the width of a column.

   f) Describe how to add another column.

   g) What symbol must precede a mathematical expression?

8. **Similar triangles** Determine which of the following triangles are similar. Explain your reasoning.

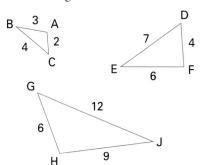

9. **Number patterns** Describe each of the following patterns. Show the next three terms.

   a) 65, 62, 59, …

   b) 100, 50, 25, …

   c) $1, -\dfrac{1}{2}, \dfrac{1}{4}, -\dfrac{1}{8}, \ldots$

   d) a, b, aa, bb, aaa, bbbb, aaaa, bbbbbbbb, …

10. **Ratios of areas** Draw two squares on a sheet of grid paper, making the dimensions of the second square half those of the first.

   a) Use algebra to calculate the ratio of the areas of the two squares.

   b) Confirm this ratio by counting the number of grid units contained in each square.

   c) If you have access to *The Geometer's Sketchpad*® or similar software, confirm the area ratio by drawing a square, dilating it by a factor of 0.5, and measuring the areas of the two squares. Refer to the help menu in the software, if necessary.

11. **Simplifying expressions** Expand and simplify each expression.

   a) $(x - 1)^2$

   b) $(2x + 1)(x - 4)$

   c) $-5x(x - 2y)$

   d) $3x(x - y)^2$

   e) $(x - y)(3x)^2$

   f) $(a + b)(c - d)$

12. **Fractions, percents, decimals** Express as a decimal.

   a) $\dfrac{5}{20}$ **b)** $\dfrac{23}{50}$ **c)** $\dfrac{2}{3}$

   d) $\dfrac{138}{12}$ **e)** $\dfrac{6}{7}$ **f)** 73%

13. **Fractions, percents, decimals** Express as a percent.

   a) 0.46   **b)** $\dfrac{4}{5}$ **c)** $\dfrac{1}{30}$

   d) 2.25   **e)** $\dfrac{11}{8}$

The Iterative Process

If you look carefully at the branches of a tree, you can see the same pattern repeated over and over, but getting smaller toward the end of each branch. A nautilus shell repeats the same shape on a larger and larger scale from its tip to its opening. You yourself repeat many activities each day. These three examples all involve an iterative process. **Iteration** is a process of repeating the same procedure over and over. The following activities demonstrate this process.

INVESTIGATE & INQUIRE: Developing a Sort Algorithm

Often you need to sort data using one or more criteria, such as alphabetical or numerical order. Work with a partner to develop an algorithm to sort the members of your class in order of their birthdays. An **algorithm** is a procedure or set of rules for solving a problem.

1. Select two people and compare their birthdays.

2. Rank the person with the later birthday second.

3. Now, compare the next person's birthday with the last ranked birthday. Rank the later birthday of those two last.

4. Describe the continuing process you will use to find the classmate with the latest birthday.

5. Describe the process you would use to find the person with the second latest birthday. With whom do you stop comparing?

6. Describe a process to rank all the remaining members of your class by their birthdays.

7. Illustrate your process with a diagram.

The process you described is an iterative process because it involves repeating the same set of steps throughout the algorithm. Computers can easily be programmed to sort data using this process.

Method 1: Pencil and Paper

1. Using isometric dot paper, draw a large equilateral triangle with side lengths of 32 units.

2. Divide this equilateral triangle into four smaller equilateral triangles.

3. Shade the middle triangle. What fraction of the original triangle is shaded?

4. For each of the unshaded triangles, repeat this process. What fraction of the original triangle is shaded?

5. For each of the unshaded triangles, repeat this process again. What fraction of the original triangle is shaded now?

6. Predict the fraction of the original triangle that would be shaded for the fourth and fifth steps in this iterative process.

7. Predict the fraction of the original triangle that would be shaded if this iterative process continued indefinitely.

Method 2: *The Geometer's Sketchpad®*

1. Open a **new sketch** and a **new script**.

2. Position both windows side by side.

3. Click on **REC** in the script window.

4. In the sketch window, construct a triangle. Shift-click on each side of the triangle. Under the **Construct** menu, choose **Point at Midpoint** and then **Polygon Interior** of the midpoints.

5. Shift-click on one vertex and the two adjacent midpoints. Choose **Loop** in your script.

6. Repeat step 5 for the other two vertices.

7. Shift-click on the three midpoints. From the **Display** menu, choose **Hide Midpoints**.

8. **Stop** your script.

9. Open a new sketch. Construct a new triangle. Mark the three vertices. **Play** your script at a recursion depth of at least 3. You may increase the speed by clicking on **Fast**.

10. **a)** What fraction of the original triangle is shaded

    i) after one recursion?

    ii) after two recursions?

    iii) after three recursions?

   b) Predict what fraction would be shaded after four and five recursions.

   c) Predict the fraction of the original triangle that would be shaded if this iterative (recursion) process continued indefinitely.

11. Experiment with recursion scripts to design patterns with repeating shapes.

The Sierpinski triangle is named after the Polish mathematician, Waclaw Sierpinski (1882–1924). It is an example of a **fractal**, a geometric figure that is generally created using an iterative process. One part of the process is that fractals are made of **self-similar** shapes. As the shapes become smaller and smaller, they keep the same geometrical characteristics as the original larger shape. Fractal geometry is a very rich area of study. Fractals can be used to model plants, trees, economies, or the honeycomb pattern in human bones.

WEB CONNECTION
www.mcgrawhill.ca/links/MDM12

Visit the above web site and follow the links to learn more about the Sierpinski triangle and fractals. Choose an interesting fractal and describe how it is self-similar.

Example 1 Modelling With a Fractal

Fractals can model the branching of a tree. Describe the algorithm used to model the tree shown.

Solution

Begin with a 1-unit segment. Branch off at 60° with two segments, each one half the length of the previous branch. Repeat this process for a total of three iterations.

Arrow diagrams can illustrate iterations. Such diagrams show the sequence of steps in the process.

Example 2 The Water Cycle

Illustrate the water cycle using an arrow diagram.

Solution

The water, or hydrologic, cycle is an iterative process. Although the timing of the precipitation can vary, the cycle will repeat itself indefinitely.

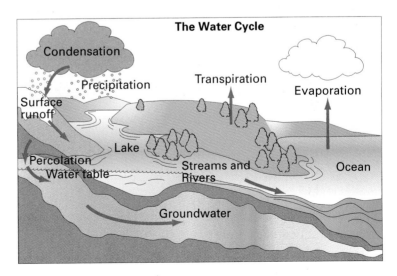

The Water Cycle

Example 3 Tree Diagram

a) Illustrate the results of a best-of-five hockey playoff series between Ottawa and Toronto using a tree diagram.

b) How many different outcomes of the series are possible?

Solution

a) For each game, the tree diagram has two branches, one representing a win by Ottawa (O) and the other a win by Toronto (T). Each set of branches represents a new game in the playoff round. As soon as one team wins three games, the playoff round ends, so the branch representing that sequence also stops.

b) By counting the endpoints of the branches, you can determine that there are 20 possible outcomes for this series.

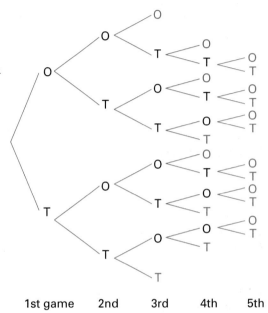

1st game   2nd   3rd   4th   5th

Example 4 Recursive Formula

The **recursive formula** $t_n = 3t_{n-1} - t_{n-2}$ defines a sequence of numbers. Find the next five terms in the sequence given that the **initial** or **seed values** are $t_1 = 1$ and $t_2 = 3$.

Solution

$$t_3 = 3t_2 - t_1$$
$$= 3(3) - 1$$
$$= 8$$

$$t_4 = 3t_3 - t_2$$
$$= 3(8) - 3$$
$$= 21$$

$$t_5 = 3t_4 - t_3$$
$$= 3(21) - 8$$
$$= 55$$

$$t_6 = 3t_5 - t_4$$
$$= 3(55) - 21$$
$$= 144$$

$$t_7 = 3t_6 - t_5$$
$$= 3(144) - 55$$
$$= 377$$

The next five terms are 8, 21, 55, 144, and 377.

Communicate Your Understanding

1. Describe how fractals have been used to model the fern leaf shown on the right.

2. Describe your daily routine as an iterative process.

Practise

1. Which of the following involve an iterative process?

   a) the cycle of a washing machine

   b) your reflections in two mirrors that face each other

   c) the placement of the dials on an automobile dashboard

   d) a chart of sunrise and sunset times

   e) substituting a value for the variable in a quadratic equation

   f) a tessellating pattern, such as paving bricks that fit together without gaps

2. The diagram below illustrates the carbon-oxygen cycle. Draw arrows to show the gains and losses of carbon dioxide.

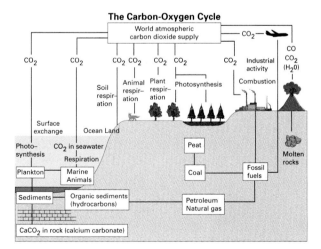

The Carbon-Oxygen Cycle

3. Draw a tree diagram representing the playoffs of eight players in a singles tennis tournament. The tree diagram should show the winner of each game continuing to the next round until a champion is decided.

Apply, Solve, Communicate

4. Draw a diagram to represent the food chain.

5. **Communication** Describe how the tracing of heartbeats on a cardiac monitor or electrocardiogram is iterative. Illustrate your description with a sketch.

6. In the first investigation, on page 6, you developed a sort algorithm in which new data were compared to the lowest ranked birthday until the latest birthday was found. Then, the second latest, third latest, and so on were found in the same manner.

    a) Write a sort algorithm in which this process is reversed so that the highest ranked item is found instead of the lowest.

    b) Write a sort algorithm in which you compare the first two data, then the second and third, then the third and fourth, and so on, interchanging the order of the data in any pair where the second item is ranked higher.

7. **Application** Sierpinski's carpet is similar to Sierpinski's triangle, except that it begins with a square. This square is divided into nine smaller squares and the middle one is shaded. Use paper and pencil or a drawing program to construct Sierpinski's carpet to at least three stages. Predict what fraction of the original square will be shaded after n stages.

8. **Application** In 1904 the Swedish mathematician Helge von Koch (1870–1924) developed a fractal based on an equilateral triangle. Using either paper and pencil or a drawing program, such as *The Geometer's Sketchpad®*, draw a large equilateral triangle and trisect each side. Replace each middle segment with two segments the same length as the middle segment, forming an equilateral triangle with the base removed, as shown below.

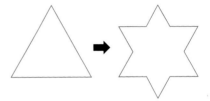

Repeat the process of trisection and replacement on each of the 12 smaller segments. If you are using a computer program, continue this process for at least two more iterations.

    a) How many segments are there after three iterations?

    b) How many segments are there after four iterations?

    c) What pattern can you use to predict the number of segments after n iterations?

9. The first two terms of a sequence are given as $t_1 = 2$ and $t_2 = 4$. The recursion formula is $t_n = (t_{n-1})^2 - 3t_{n-2}$. Determine the next four terms in the sequence.

10. Each of the following fractal trees has a different algorithm. Assume that each tree begins with a segment 1 unit long.

  a) Illustrate or describe the algorithm for each fractal tree.

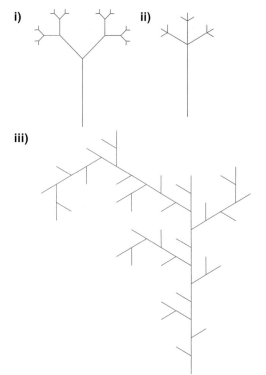

  i)   **ii)**

  iii)

  b) What is the total length of the branches in each tree?

  c) An interesting shape on a fractal tree is a spiral, which you can trace by tracing a branch to its extremity. Are all spirals within a tree self-similar?

  d) Write your own set of rules for a fractal tree. Draw the tree using paper and pencil or a drawing program.

11. Inquiry/Problem Solving Related to fractals is the mathematical study of chaos, in which no accurate prediction of an outcome can be made. A random walk can illustrate such "chaotic" outcomes.

  a) Select a starting point near the centre of a sheet of grid paper. Assign the numbers 1 to 4 to the directions north, south, east, or west in any order. Now, generate random whole numbers between 1 and 4 using a die, coin, or graphing calculator. Draw successive line segments one unit long in the directions corresponding to the random numbers until you reach an edge of the paper.

  b) How would a random walk be affected if it were self-avoiding, that is, not allowed to intersect itself? Repeat part a) using this extra rule.

  c) Design your own random walk with a different set of rules. After completing the walk, trade drawings with a classmate and see if you can deduce the rules for each other's walk.

WEB CONNECTION

www.mcgrawhill.ca/links/MDM12

To learn more about chaos theory, visit the above web site and follow the links. Describe an aspect of chaos theory that interests you.

12. Use the given values for t_1 to find the successive terms of the following recursive formulas. Continue until a pattern appears. Describe the pattern and make a prediction for the value of the nth term.

  a) $t_n = 2^{-t_{n-1}}$; $t_1 = 0$

  b) $t_n = \sqrt{t_{n-1}}$; $t_1 = 256$

  c) $t_n = \dfrac{1}{t_{n-1}}$; $t_1 = 2$

13. a) Given $t_1 = 1$, list the next five terms for the recursion formula $t_n = n \times t_{n-1}$.

b) In this sequence, t_k is a factorial number, often written as $k!$ Show that

$$t_k = k!$$
$$= k(k-1)(k-2)\ldots(2)(1).$$

c) Explain what $8!$ means. Evaluate $8!$

d) Explain why factorial numbers can be considered an iterative process.

e) Note that

$$(2^5)(5!)$$
$$= (2 \times 2 \times 2 \times 2 \times 2)(5 \times 4 \times 3 \times 2 \times 1)$$
$$= (2 \times 5)(2 \times 4)(2 \times 3)(2 \times 2)(2 \times 1)$$
$$= 10 \times 8 \times 6 \times 4 \times 2$$

which is the product of the first five even positive integers. Write a formula for the product of the first n even positive integers. Explain why your formula is correct.

f) Write $\dfrac{10!}{(2^5)(5!)}$ as a product of consecutive odd integers.

g) Write a factorial formula for the product of

i) the first six odd positive integers

ii) the first ten odd positive integers

iii) the first n odd positive integers

14. Inquiry/Problem Solving Recycling can be considered an iterative process. Research the recycling process for a material such as newspaper, aluminum, or glass and illustrate the process with an arrow diagram.

15. Inquiry/Problem Solving The infinite series $S = \cos\theta + \cos^2\theta + \cos^3\theta + \ldots$ can be illustrated by drawing a circle centred at the origin, with radius of 1. Draw an angle θ and, on the x-axis, label the point $(\cos\theta, 0)$ as P_1. Draw a new circle, centred at P_1, with radius of $\cos\theta$. Continue this iterative process. Predict the length of the line segment defined by the infinite series $S = \cos\theta + \cos^2\theta + \cos^3\theta + \ldots$.

16. Communication Music can be written using fractal patterns. Look up this type of music in a library or on the Internet. What characteristics does fractal music have?

17. Computers use binary (base 2) code to represent numbers as a series of ones and zeros.

Base 10	Binary
0	0
1	1
2	10
3	11
4	100
⋮	⋮

a) Describe an algorithm for converting integers from base 10 to binary.

b) Write each of the following numbers in binary.

i) 16   **ii)** 21

iii) 37   **iv)** 130

c) Convert the following binary numbers to base 10.

i) 1010   **ii)** 100000

iii) 111010   **iv)** 111111111

1.2 Data Management Software

INVESTIGATE & INQUIRE: Software Tools

1. List every computer program you can think of that can be used to manage data.

2. Sort the programs into categories, such as word-processors and spreadsheets.

3. Indicate the types of data each category of software would be best suited to handle.

4. List the advantages and disadvantages of each category of software.

5. Decide which of the programs on your list would be best for storing and accessing the lists you have just made.

Most office and business software manage data of some kind. Schedulers and organizers manage lists of appointments and contacts. E-mail programs allow you to store, access, and sort your messages. Word-processors help you manage your documents and often have sort and outline functions for organizing data within a document. Although designed primarily for managing financial information, spreadsheets can perform calculations related to the management and analysis of a wide variety of data. Most of these programs can easily transfer data to other applications.

Database programs, such as Microsoft® Access and Corel® Paradox®, are powerful tools for handling large numbers of records. These programs produce **relational databases**, ones in which different sets of records can be linked and sorted in complex ways based on the data contained in the records. For example, many organizations use a relational database to generate a monthly mailing of reminder letters to people whose memberships are about to expire. However, these complex relational database programs are difficult to learn and can be frustrating to use until you are thoroughly familiar with how they work. Partly for this reason, there are thousands of simpler database programs designed for specific types of data, such as book indexes or family trees.

Of particular interest for this course are programs that can do statistical analysis of data. Such programs range from modest but useful freeware to major data-analysis packages costing thousands of dollars. The more commonly used programs include MINITAB™, SAS, and SST (Statistical

Software Tools). To demonstrate statistical software, some examples in this book have alternative solutions that use Fathom™, a statistical software package specifically designed for use in schools.

Data management programs can perform complex calculations and link, search, sort, and graph data. The examples in this section use a spreadsheet to illustrate these operations. A spreadsheet is software that arranges data in rows and columns. For basic spreadsheet instructions, please refer to the spreadsheet section of Appendix B. If you are not already familiar with spreadsheets, you may find it helpful to try each of the examples yourself before answering the Practise questions at the end of the section. The two most commonly used spreadsheets are Corel® Quattro® Pro and Microsoft® Excel.

Formulas and Functions

A formula entered in a spreadsheet cell can perform calculations based on values or formulas contained in other cells. Formulas retrieve data from other cells by using cell references to indicate the rows and columns where the data are located. In the formulas C2*0.05 and D5+E5, each reference is to an individual cell. In both Microsoft® Excel and Corel® Quattro® Pro, it is good practice to begin a formula with an equals sign. Although not always necessary, the equals sign ensures that a formula is calculated rather than being interpreted as text.

Built-in formulas are called **functions**. Many functions, such as the SUM function or MAX function use references to a range of cells. In Corel® Quattro® Pro, precede a function with an @ symbol. For example, to find the total of the values in cells A2 through A6, you would enter

Corel® Quattro® Pro: @SUM(A2..A6)   Microsoft® Excel: SUM(A2:A6)

Similarly, to find the total for a block of cells from A2 through B6, enter

Corel® Quattro® Pro: @SUM(A2..B6)   Microsoft® Excel: SUM(A2:B6)

A list of formulas is available in the Insert menu by selecting Function…. You may select from a list of functions in categories such as Financial, Math & Trig, and Database.

Example 1 Using Formulas and Functions

The first three columns of the spreadsheet on the right list a student's marks on tests and assignments for the first half of a course. Determine the percent mark for each test or assignment and calculate an overall midterm mark.

	A	B	C	D	E	F
1	NAME	MARK	OUT OF			
2	Quiz 1	7	10			
3	Assign 1	12	12			
4	Test 1	45	50			
5	Assign 2A	10	10			
6	Assign 2B	5	9			
7	Test 2	32	45			
8	Quiz 3A	13.5	15			
9	Quiz 3B	8	10			
10	Test 3	28	40			

Solution

In column D, enter formulas with cell referencing to find the percent for each individual mark. For example, in cell D2, you could use the formula B2/C2*100.

Use the SUM function to find totals for columns B and C, and then convert to percent in cell D12 to find the midterm mark.

	A	B	C	D	E	F	G	H	I	J	K	L
1	NAME	MARK	OUT OF	PERCENT								
2	Quiz 1	7	10	70								
3	Assign 1	12	12	100								
4	Test 1	45	50	90								
5	Assign 2A	10	10	100								
6	Assign 2B	5	9	55.55556								
7	Test 2	32	45	71.11111								
8	Quiz 3A	13.5	15	90								
9	Quiz 3B	8	10	80								
10	Test 3	28	40	70								
11												
12	MIDTERM	160.5	201	79.85075								
13												

Relative and Absolute Cell References

Spreadsheets automatically adjust cell references whenever cells are copied, moved, or sorted. For example, if you copy a SUM function, used to calculate the sum of cells A3 to E3, from cell F3 to cell F4, the spreadsheet will change the cell references in the copy to A4 and E4. Thus, the value in cell F4 will be the sum of those in cells A4 to E4, rather than being the same as the value in F3.

Because the cell references are relative to a location, this automatic adjustment is known as relative cell referencing. If the formula references need to be kept exactly as written, use absolute cell referencing. Enter dollar signs before the row and column references to block automatic adjustment of the references.

Fill and Series Features

When a formula or function is to be copied to several adjoining cells, as for the percent calculations in Example 1, you can use the Fill feature instead of Copy. Click once on the cell to be copied, then click and drag across or down through the range of cells into which the formula is to be copied.

To create a sequence of numbers, enter the first two values in adjoining cells, then select Edit/Fill/Series to continue the sequence.

Example 2 Using the Fill Feature

The relationship between Celsius and Fahrenheit temperatures is given by the formula Fahrenheit = 1.8 × Celsius + 32. Use a spreadsheet to produce a conversion table for temperatures from 1°C to 15°C.

Solution

Enter 1 into cell **E2** and 2 into cell **E3**. Use the Fill feature to put the numbers 3 through 15 into cells **E4** to **E16**. Enter the conversion formula E2*1.8+32 into cell **F2**. Then, use the Fill feature to copy the formula into cells **F3** through **F16**. Note that the values in these cells show that the cell references in the formulas did change when copied. These changes are an example of relative cell referencing.

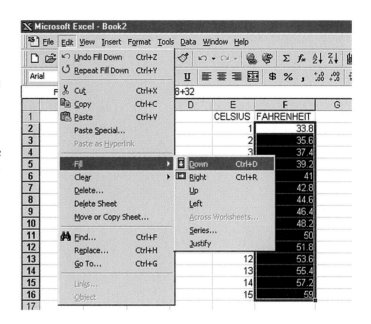

Charting

Another important feature of spreadsheets is the ability to display numerical data in the form of charts or graphs, thereby making the data easier to understand. The first step is to select the range of cells to be graphed. For non-adjoining fields, hold down the **Ctrl** key while highlighting the cells. Then, use the Chart feature to specify how you want the chart to appear.

You can produce legends and a title for your graph as well as labels for the axes. Various two- and three-dimensional versions of bar, line, and circle graphs are available in the menus.

Example 3 Charting

The results and standings of a hockey league are listed in this spreadsheet. Produce a two-dimensional bar chart using the **TEAM** and **POINTS** columns.

	A	B	C	D	E	F	G	H
1	TEAM	WINS	LOSSES	TIES	GF	GA	POINTS	
2	Blades	12	3	5	70	56	29	
3	Ice Hogs	6	8	3	46	46	15	
4	Leaves	7	5	8	55	51	22	
5	Nice Guys	10	5	2	64	42	22	
6	Rinkrats	6	7	3	58	63	15	
7	Slapshots	2	15	1	24	71	5	
8								

Solution

Holding down the Ctrl key, highlight cells A1 to A7 and then G1 to G7. Use the Chart feature and follow the on-screen instructions to customize your graph. You will see a version of the bar graph as shown here.

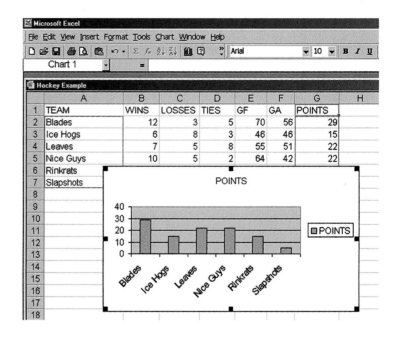

Sorting

Spreadsheets have the capability to sort data alphabetically, numerically, by date, and so on. The sort can use multiple criteria in sequence. Cell references will adjust to the new locations of the sorted data. To sort, select the range of cells to be sorted. Then, use the Sort feature.

Select the criteria under which the data are to be sorted. A sort may be made in ascending or descending order based on the data in any given column. A sort with multiple criteria can include a primary sort, a secondary sort within it, and a tertiary sort within the secondary sort.

Example 4 Sorting

Rank the hockey teams in Example 3, counting points first (in descending order), then wins (in descending order), and finally losses (in ascending order).

Solution

When you select the Sort feature, the pop-up window asks if there is a header row. Confirming that there is a header row excludes the column headings from the sort so that they are left in place. Next, set up a three-stage sort:
• a primary sort in descending order, using the points column
• then, a secondary sort in descending order, using the wins column
• finally, a tertiary sort in ascending order, using the losses column

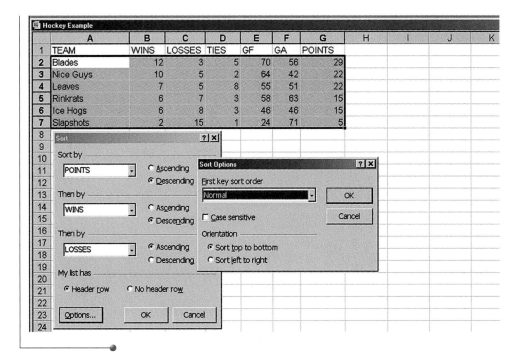

Searching

To search for data in individual cells, select **Find and Replace**.

Then, in the dialogue box, enter the data and the criteria under which you are searching. You have the option to search or to search and replace.

A filtered search allows you to search for rows containing the data for which you are searching.

Arrows will appear at the top of each column containing data. Clicking on an arrow opens a pull-down menu where you can select the data you wish to find. The filter will then display only the rows containing these data. You can filter for a specific value or select **custom…** to use criteria such as **greater than, begins with,** and **does not contain.** To specify multiple criteria, click the **And** or **Or** options. You can set different filter criteria for each column.

Example 5 Filtered Search

In the hockey-league spreadsheet from Example 3, use a filtered search to list only those teams with fewer than 16 points.

Solution

In Microsoft® Excel, select Data/Filter/Autofilter to begin the filter process. Click on the arrow in the POINTS column and select custom... In the dialogue window, select is less than and key in 16.

In Corel® Quattro® Pro, you use Tools/Quickfilter/custom....

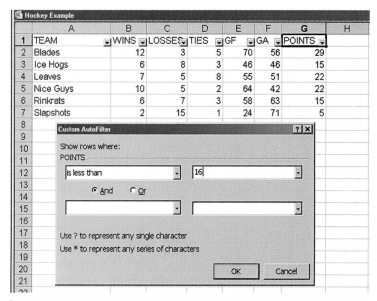

Now, the filter shows only the rows representing teams with fewer than 16 points.

Adding and Referencing Worksheets

To add worksheets within your spreadsheet file, click on one of the sheet tabs at the bottom of the data area. You can enter data onto the additional worksheet using any of the methods described above or you can copy and paste data from the first worksheet or from another file.

To reference data from cells in another worksheet, preface the cell reference with the worksheet number for the cells.

Such references allow data entered in sheet A or sheet 1 to be manipulated in another sheet without changing the values or order of the original data. Data edited in the original sheet will be automatically updated in the other sheets that refer to it. Any sort performed in the original sheet will carry through to any references in other sheets, but any other data in the secondary sheets will not be affected. Therefore, it is usually best to either reference all the data in the secondary sheets or to sort the data only in the secondary sheets.

Project Prep

The calculation, sorting, and charting capabilities of spreadsheets could be particularly useful for your tools for data management project.

Example 6 Sheet Referencing

Reference the goals for (GF) and goals against (GA) for the hockey teams in Example 3 on a separate sheet and rank the teams by their goals scored.

Solution

Sheet 2 needs only the data in the columns titled **GF** and **GA** in sheet 1. Notice that cell **C2** contains a cell reference to sheet 1. This reference ensures the data in cell **F2** of sheet 1 will carry through to cell **C2** of sheet 2 even if the data in sheet 1 is edited. Although the referenced and sorted data on sheet 2 appear as shown, the order of the teams on sheet 1 is unchanged.

| C2 | | = | =Sheet1!F2 |

Hockey Example

	A	B	C	D
1	TEAM	GF	GA	
2	Blades	70	56	
3	Nice Guys	64	42	
4	Rinkrats	58	63	
5	Leaves	55	51	
6	Ice Hogs	46	46	
7	Slapshots	24	71	

Key Concepts

- Thousands of computer programs are available for managing data. These programs range from general-purpose software, such as word-processors and spreadsheets, to highly specialized applications for specific types of data.

- A spreadsheet is a software application that is used to enter, display, and manipulate data in rows and columns. Spreadsheet formulas perform calculations based on values or formulas contained in other cells.

- Spreadsheets normally use relative cell referencing, which automatically adjusts cell references whenever cells are copied, moved, or sorted. Absolute cell referencing keeps formula references exactly as written.

- Spreadsheets can produce a wide variety of charts and perform sophisticated sorts and searches of data.

- You can add additional worksheets to a file and reference these sheets to cells in another sheet.

Communicate Your Understanding

1. Explain how you could use a word-processor as a data management tool.

2. Describe the advantages and drawbacks of relational database programs.

3. Explain what software you would choose if you wanted to determine whether there was a relationship between class size and subject in your school. Would you choose different software if you were going to look at class sizes in all the schools in Ontario?

4. Give an example of a situation requiring relative cell referencing and one requiring absolute cell referencing.

5. Briefly describe three advantages that spreadsheets have over hand-written tables for storing and manipulating data.

Practise

1. **Application** Set up a spreadsheet page in which you have entered the following lists of data. For the appropriate functions, look under the **Statistical** category in the **Function** list.

   Student marks:
   65, 88, 56, 76, 74, 99, 43, 56, 72, 81, 80, 30, 92

   Dentist appointment times in minutes:
   45, 30, 40, 32, 60, 38, 41, 45, 40, 45

   a) Sort each set of data from smallest to greatest.

   b) Calculate the mean (average) value for each set of data.

   c) Determine the median (middle) value for each set of data.

   d) Determine the mode (most frequent) value for each set of data.

2. Using the formula features of the spreadsheet available in your school, write a formula for each of the following:

   a) the sum of the numbers stored in cells A1 to A9

   b) the largest number stored in cells F3 to K3

   c) the smallest number in the block from A1 to K4

   d) the sum of the cells A2, B5, C7, and D9

   e) the mean, median, and mode of the numbers stored in the cells F5 to M5

   f) the square root of the number in cell A3

   g) the cube of the number in cell B6

   h) the number in cell D2 rounded off to four decimal places

   i) the number of cells between cells D3 and M9 that contain data

   j) the product of the values in cells A1, B3, and C5 to C10

   k) the value of π

Apply, Solve, Communicate

3. Set up a spreadsheet page that converts angles in degrees to radians using the formula Radians = $\pi \times$ Degrees/180, for angles from 0° to 360° in steps of 5°. Use the series capabilities to build the data in the Degrees column. Use π as defined by the spreadsheet. Calculations should be rounded to the nearest hundredth.

4. The first set of data below represents the number of sales of three brands of CD players at two branches of Mad Dog Music. Enter the data into a spreadsheet using two rows and three columns.

Branch	Brand A	Brand B	Brand C
Store P	12	4	8
Store Q	9	15	6

   The second set of data represents the prices for these CD players. Enter the data using one column into a second

Brand	Price
A	$102
B	$89
C	$145

   sheet of the same spreadsheet workbook. Set up a third sheet of the spreadsheet workbook to reference the first two sets of data and calculate the total revenue from CD player sales at each Mad Dog Music store.

5. **Application** In section 1.1, question 12, you predicted the value of the nth term of the recursion formulas listed below. Verify your predictions by using a spreadsheet to calculate the first ten terms for each formula.

   a) $t_n = 2^{-t_{n-1}}; t_1 = 0$

   b) $t_n = \sqrt{t_{n-1}}; t_1 = 256$

   c) $t_n = \dfrac{1}{t_{n-1}}; t_1 = 2$

6. a) Enter the data shown in the table below into a spreadsheet and set up a second sheet with relative cell references to the Name, Fat, and Fibre cells in the original sheet.

Nutritional Content of 14 Breakfast Cereals (amounts in grams)							
Name	Protein	Fat	Sugars	Starch	Fibre	Other	TOTALS
Alphabits	2.4	1.1	12.0	12.0	0.9	1.6	
Bran Flakes	4.4	1.2	6.3	4.7	11.0	2.4	
Cheerios	4.0	2.3	0.8	18.7	2.2	2.0	
Crispix	2.2	0.3	3.2	22.0	0.5	1.8	
Froot Loops	1.3	0.8	14.0	12.0	0.5	1.4	
Frosted Flakes	1.4	0.2	12.0	15.0	0.5	0.9	
Just Right	2.2	0.8	6.6	17.0	1.4	2.0	
Lucky Charms	2.1	1.0	13.0	11.0	1.4	1.5	
Nuts 'n Crunch	2.3	1.6	7.1	16.5	0.7	1.8	
Rice Krispies	2.1	0.4	2.9	22.0	0.3	2.3	
Shreddies	2.9	0.6	5.0	16.0	3.5	2.0	
Special K	5.1	0.4	2.5	20.0	0.4	1.6	
Sugar Crisp	2.0	0.7	14.0	11.0	1.1	1.2	
Trix	0.9	1.6	13.0	12.0	1.1	1.4	
AVERAGES							
MAXIMUM							
MINIMUM							

b) On the first sheet, calculate the values for the TOTALS column and AVERAGES row.

c) Determine the maximum and minimum values in each column.

d) Rank the cereals using fibre content in decreasing order as a primary criterion, protein content in decreasing order as a secondary criterion, and sugar content in increasing order as a tertiary criterion.

e) Make three circle graphs or pie charts: one for the averages row in part b), one for the cereal at the top of the list in part d), and one for the cereal at the bottom of the list in part d).

f) Perform a search in the second sheet to find the cereals containing less than 1 g of fat and more than 1.5 g of fibre. Make a three-dimensional bar graph of the results.

C

7. In section 1.1, question 10, you described the algorithm used to draw each fractal tree below. Assuming the initial segment is 4 cm in each tree, use a spreadsheet to determine the total length of a spiral in each tree, calculated to 12 iterative stages.

a)

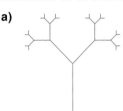

b)

8. Communication Describe how to lock column and row headings in your spreadsheet software so that they remain visible when you scroll through a spreadsheet.

9. Inquiry/Problem Solving Outline a spreadsheet algorithm to calculate $n \times (n - 1) \times (n - 2) \ldots 3 \times 2 \times 1$ for any natural number n without using the built-in factorial function.

Introduction to Fathom™

Fathom™ is a statistics software package that offers a variety of powerful data-analysis tools in an easy-to-use format. This section introduces the most basic features of Fathom™: entering, displaying, sorting, and filtering data. A complete guide is available on the Fathom™ CD. The real power of this software will be demonstrated in later chapters with examples that apply its sophisticated tools to statistical analysis and simulations.

Appendix B includes details on all the Fathom™ functions used in this text.

When you enter data into Fathom™, it creates a **collection**, an object that contains the data. Fathom™ can then use the data from the collection to produce other objects, such as a graph, table, or statistical test. These secondary objects display and analyse the data from the collection, but they do not actually contain the data themselves. If you delete a graph, table, or statistical test, the data still remains in the collection.

Fathom™ displays a collection as a rectangular window with gold balls in it. The gold balls of the collection represent the original or "raw" data. Each of the gold balls represents a **case**. Each case in a collection can have a number of **attributes**. For example the cases in a collection of medical records could have attributes such as the patient's name, age, sex, height, weight, blood pressure, and so on. There are two basic types of attributes, **categorical** (such as male/female) and **continuous** (such as height or mass). The **case table** feature displays the cases in a collection in a format similar to a spreadsheet, with a row for each case and a column for each attribute. You can add, modify, and delete cases using a case table.

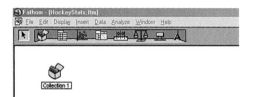

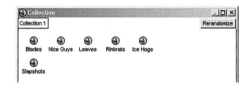

• **Example 1 Tables and Graphs**

a) Set up a collection for the hockey league standings from section 1.2, Example 3 on page 17.
b) Graph the Team and Points attributes.

Solution

a) To enter the data, start Fathom™ and drag the **case table** icon 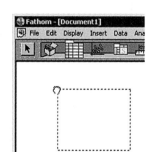 from the menu bar down onto the work area.

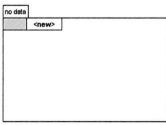

Click on the attribute <new>, type the heading **Team**, and press **Enter**. Fathom™ will automatically create a blank cell for data under the heading and start a new column to the right of the first. Enter the heading **Wins** at the top of the new column, and continue this process to enter the rest of the headings. You can type entries into the cells under the headings in much the same way as you would enter data into the cells of a spreadsheet.

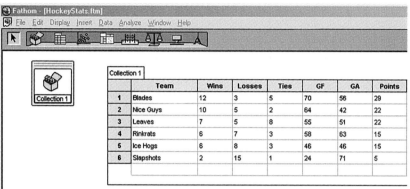

	Team	Wins	Losses	Ties	GF	GA	Points
1	Blades	12	3	5	70	56	29
2	Nice Guys	10	5	2	64	42	22
3	Leaves	7	5	8	55	51	22
4	Rinkrats	6	7	3	58	63	15
5	Ice Hogs	6	8	3	46	46	15
6	Slapshots	2	15	1	24	71	5

Note that Fathom™ has stored your data as **Collection 1**, which will remain intact even if you delete the **case table** used to enter the data. To give the collection a more descriptive name, double-click on **Collection 1** and type in **HockeyStats**.

b) Drag the **graph icon** 📊 onto the work area. Now, drag the **Team** attribute from the **case table** to the *x*-axis of the graph and the **Points** attribute to the *y*-axis of the graph.

Your graph should look like this:

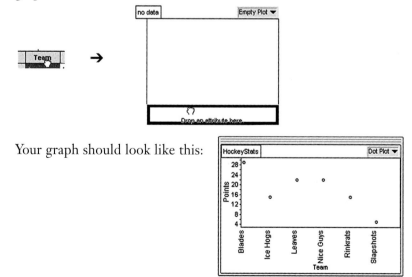

Fathom™ can easily sort or filter data using the various attributes.

Example 2 Sorting and Filtering

a) Rank the hockey teams in Example 1 by points first, then by wins if two teams have the same number of points, and finally by losses if two teams have the same number of points and wins.

b) List only those teams with fewer than 16 points.

c) Set up a separate table showing only the goals for (GF) and goals against (GA) data for the teams and rank the teams by their goals scored.

Solution

a) To **Sort** the data, right-click on the **Points** attribute and choose **Sort Descending**. Fathom™ will list the team with the most points first, with the others following in descending order by their point totals. To set the secondary sort, right-click on the **Wins** attribute and choose **Sort Descending**. Similarly, right-click on the **Losses** attribute and choose **Sort Ascending** for the final sort, giving the result below.

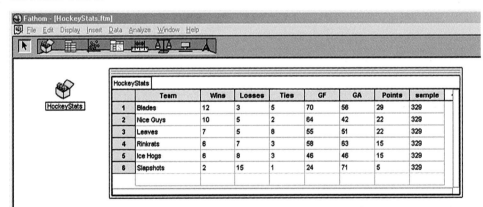

b) To **Filter** the data, from the **Data** menu, choose **Add Filter**. Click on the plus sign beside **Attributes**.

Now, double-click on the **Points attribute**, choose the **less-than** button $<$, and type 16. Click the **Apply** button and then **OK**.

The results should look like this:

	Team	Wins	Losses	Ties	GF	GA	Points	sample	<n
1	Rinkrats	6	7	3	58	63	15	329	
2	Ice Hogs	6	8	3	46	46	15	329	
3	Slapshots	2	15	1	24	71	5	329	

The **Filter** is listed at the bottom as Points < 16.

c) Click on HockeyStats, and then drag a new table onto the work area. Click on the Wins attribute. From the **Display** menu, choose **Hide Attribute**. Use the same method to hide the **Losses**, **Ties**, and **Points** attributes. Right-click the GF attribute and use **Sort Descending** to rank the teams.

	Team	GF	GA
1	Blades	70	56
2	Nice Guys	64	42
3	Rinkrats	58	63
4	Leaves	55	51
5	Ice Hogs	46	46
6	Slapshots	24	71

HockeyStats

1. Enter the data from Example 1 into Fathom™. Use the built-in functions in Fathom™ to find the following.

   a) the mean of goals against (GA)

   b) the largest value of goals for (GF)

   c) the smallest value of GF

   d) the sum of GA

   e) the sum of GA and GF for each case

For details on functions in Fathom™, see the Fathom™ section of Appendix B or consult the Fathom™ Help screen or manual.

2. a) Set up a new collection with the following student marks:
      65, 88, 56, 76, 74, 99, 43, 56, 72, 81, 80, 30, 92

   b) Sort the marks from lowest to highest.

   c) Calculate the mean mark.

   d) Determine the median (middle) mark.

3. Explain how you would create a graph of class size versus subjects in your school using Fathom™.

4. Briefly compare the advantages and disadvantages of using Fathom™ and spreadsheets for storing and manipulating data.

WEB CONNECTION

www.mcgrawhill.ca/links/MDM12

For more examples, data, and information on how to use Fathom™, visit the above web site and follow the links.

1.3 Databases

A **database** is an organized store of records. Databases may contain information about almost any subject—incomes, shopping habits, demographics, features of cars, and so on.

INVESTIGATE & INQUIRE: Databases in a Library

In your school or local public library, log on to the library catalogue.

1. Describe the types of fields under which a search can be conducted (e.g., subject).

2. Conduct a search for a specific topic of your choice.

3. Describe the results of your search. How is the information presented to the user?

INVESTIGATE & INQUIRE: The E-STAT Database

1. Connect to the Statistics Canada web site and go to the E-STAT database. Your school may have a direct link to this database. If not, you can follow the Web Connection links shown here. You may need to get a password from your teacher to log in.

2. Locate the database showing the educational attainment data for Canada by following these steps:

   a) Click on <u>Data</u>.

   b) Under the heading *People*, click on <u>Education</u>.

   c) Click on <u>Educational Attainment</u>, then under the heading *Census databases*, select <u>Educational Attainment</u> again.

   d) Select <u>Education, Mobility and Migration</u> for the latest census.

WEB CONNECTION

www.mcgrawhill.ca/links/MDM12

To connect to E-STAT visit the above web site and follow the links.

---

3. Scroll down to the heading *University, pop. 15 years and over by highest level of schooling*, hold down the Ctrl key, and select all four subcategories under this heading. View the data in each of the following formats:

   a) table   **b)** bar graph   **c)** map

4. Describe how the data are presented in each instance. What are the advantages and disadvantages of each format? Which format do you think is the most effective for displaying this data? Explain why.

5. Compare the data for the different provinces and territories. What conclusions could you draw from this data?

A database **record** is a set of data that is treated as a unit. A record is usually divided into **fields** that are reserved for specific types of information. For example, the record for each person in a telephone book has four fields: last name, first name or initial, address, and telephone number. This database is sorted in alphabetical order using the data in the first two fields. You search this database by finding the page with the initial letters of a person's name and then simply reading down the list.

A music store will likely keep its inventory records on a computerized database. The record for each different CD could have fields for information, such as title, artist, publisher, music type, price, number in stock, and a product code (for example, the bar code number). The computer can search such databases for matches in any of the data fields. The staff of the music store would be able to quickly check if a particular CD was in stock and tell the customer the price and whether the store had any other CDs by the same artist.

Databases in a Library

A library catalogue is a database. In the past, library databases were accessed through a card catalogue. Most libraries are now computerized, with books listed by title, author, publisher, subject, a Dewey Decimal or Library of Congress catalogue number, and an international standard book number (ISBN). Records can be sorted and searched using the information in any of the fields.

Such catalogues are examples of a well-organized database because they are easy to access using keywords and searches in multiple fields, many of which are cross-referenced. Often, school libraries are linked to other libraries. Students have access to a variety of print and online databases in the library. One powerful online database is Electric Library Canada, a database of books, newspapers, magazines, and television and radio transcripts. Your school probably has access to it or a similar library database. Your local public library may also have online access to Electric Library Canada.

Project Prep

Skills in researching library and on-line databases will help you find the information needed for your tools for data management project.

Statistics Canada

Statistics Canada is the federal government department responsible for collecting, summarizing, analysing, and storing data relevant to Canadian demographics, education, health, and so on. Statistics Canada maintains a number of large databases using data collected from a variety of sources including its own research and a nation-wide census. One such database is **CANSIM II** (the updated version of the Canadian Socio-economic Information Management System), which profiles the Canadian people, economy, and industries. Although Statistics Canada charges a fee for access to some of its data, a variety of CANSIM II data is available to the public for free on Statistics Canada's web site.

Statistics Canada also has a free educational database, called **E-STAT**. It gives access to many of Statistics Canada's extensive, well-organized databases, including CANSIM II. E-STAT can display data in a variety of formats and allows students to download data into a spreadsheet or statistical software program.

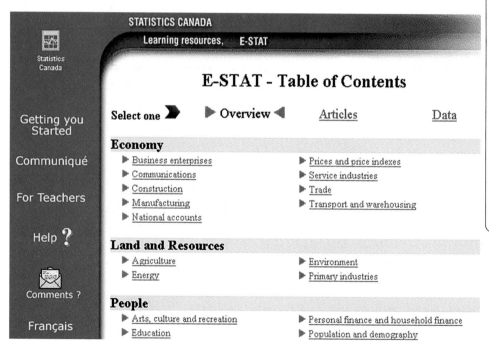

Method 1: Pencil and Paper

1. Using isometric dot paper, draw a large equilateral triangle with side lengths of 32 units.

2. Divide this equilateral triangle into four smaller equilateral triangles.

3. Shade the middle triangle. What fraction of the original triangle is shaded?

4. For each of the unshaded triangles, repeat this process. What fraction of the original triangle is shaded?

5. For each of the unshaded triangles, repeat this process again. What fraction of the original triangle is shaded now?

6. Predict the fraction of the original triangle that would be shaded for the fourth and fifth steps in this iterative process.

7. Predict the fraction of the original triangle that would be shaded if this iterative process continued indefinitely.

Method 2: *The Geometer's Sketchpad®*

1. Open a new sketch and a new script.

2. Position both windows side by side.

3. Click on REC in the script window.

4. In the sketch window, construct a triangle. Shift-click on each side of the triangle. Under the Construct menu, choose Point at Midpoint and then Polygon Interior of the midpoints.

5. Shift-click on one vertex and the two adjacent midpoints. Choose Loop in your script.

6. Repeat step 5 for the other two vertices.

7. Shift-click on the three midpoints. From the Display menu, choose Hide Midpoints.

8. Stop your script.

9. Open a new sketch. Construct a new triangle. Mark the three vertices. Play your script at a recursion depth of at least 3. You may increase the speed by clicking on Fast.

10. a) What fraction of the original triangle is shaded
    i)   after one recursion?
    ii) after two recursions?
    iii) after three recursions?

    b) Predict what fraction would be shaded after four and five recursions.

    c) Predict the fraction of the original triangle that would be shaded if this iterative (recursion) process continued indefinitely.

11. Experiment with recursion scripts to design patterns with repeating shapes.

The Sierpinski triangle is named after the Polish mathematician, Waclaw Sierpinski (1882–1924). It is an example of a **fractal**, a geometric figure that is generally created using an iterative process. One part of the process is that fractals are made of **self-similar** shapes. As the shapes become smaller and smaller, they keep the same geometrical characteristics as the original larger shape. Fractal geometry is a very rich area of study. Fractals can be used to model plants, trees, economies, or the honeycomb pattern in human bones.

WEB CONNECTION

www.mcgrawhill.ca/links/MDM12

Visit the above web site and follow the links to learn more about the Sierpinski triangle and fractals. Choose an interesting fractal and describe how it is self-similar.

Example 1   Modelling With a Fractal

Fractals can model the branching of a tree. Describe the algorithm used to model the tree shown.

Solution

Begin with a 1-unit segment. Branch off at 60° with two segments, each one half the length of the previous branch. Repeat this process for a total of three iterations.

Arrow diagrams can illustrate iterations. Such diagrams show the sequence of steps in the process.

Example 2   The Water Cycle

Illustrate the water cycle using an arrow diagram.

Solution

The water, or hydrologic, cycle is an iterative process. Although the timing of the precipitation can vary, the cycle will repeat itself indefinitely.

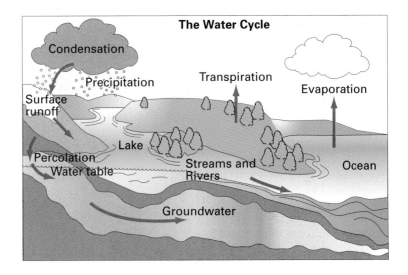

Communicate Your Understanding

1. For a typical textbook, describe how the table of contents and the index are sorted. Why are they sorted differently?

2. Describe the steps you need to take in order to access the 1860–61 census results through E-STAT.

Practise

1. Which of the following would be considered databases? Explain your reasoning.
   a) a dictionary
   b) stock-market listings
   c) a catalogue of automobile specifications and prices
   d) credit card records of customers' spending habits
   e) an essay on Shakespeare's *Macbeth*
   f) a teacher's mark book
   g) the *Guinness World Records* book
   h) a list of books on your bookshelf

Apply, Solve, Communicate

2. Describe each field you would include in a database of
   a) a person's CD collection
   b) a computer store's software inventory
   c) a school's textbook inventory
   d) the backgrounds of the students in a school
   e) a business's employee records

3. a) Describe how you would locate a database showing the ethnic makeup of your municipality. List several possible sources.

   b) If you have Internet access, log onto E-STAT and go to the data on ethnic origins of people in large urban centres:
      i) Select <u>Data</u> on the Table of Contents page.
      ii) Select <u>Population and Demography</u>.
      iii) Under *Census*, select <u>Ethnic Origin</u>.
      iv) Select <u>Ethnic Origin and Visible Minorities</u> for the latest census in large urban centres.
      v) Enter a postal code for an urban area and select two or more ethnic origins while holding down the Ctrl key.
      vi) View table, bar graph, and map in turn and describe how the data are presented in each instance.

   c) Compare these results with the data you get if you leave the postal code section line blank. What conclusions could you draw from the two sets of data?

4. Application

a) Describe how you could find data to compare employment for males and females. List several possible sources.

b) If you have Internet access, log onto E-STAT and go to the data on employment and work activity:

   i) Under the *People* heading, select Labour.

   ii) Under the *Census databases* heading, select Salaries and Wages.

   iii) Select Sources of Income (Latest census, Provinces, Census Divisions, Municipalities).

   iv) While holding down the Ctrl key, click on All persons with employment income by work activity, Males with employment income by work activity, and Females with employment income by work activity.

   v) Download this data as a spreadsheet file. Record the path and file name for the downloaded data.

c) Open the data file with a spreadsheet. You may have to convert the format to match your spreadsheet software. Use your spreadsheet to

   i) calculate the percentage difference between male and female employment

   ii) display all fields as a bar graph

5. Communication Go to the reference area of your school or local library and find a published database in print form.

a) Briefly describe how the database is organized.

b) Describe how to search the database.

c) Make a list of five books that are set up as databases. Explain why they would be considered databases.

6. Application The Internet is a link between many databases. Search engines, such as Yahoo Canada, Lycos, Google, and Canoe, are large databases of web sites. Each search engine organizes its database differently.

a) Use three different search engines to conduct a search using the keyword *automobile*. Describe how each search engine presents its data.

b) Compare the results of searches with three different search engines using the following keywords:

   i) computer monitors

   ii) computer+monitors

   iii) computer or monitors

   iv) "computer monitors"

7. Use the Internet to check whether the map of VIA Rail routes at the start of this chapter is up-to-date. Are there still no trains that go from Montréal or Kingston right through to Windsor?

8. Communication Log on to the Electric Library Canada web site or a similar database available in your school library. Enter your school's username and password. Perform a search for magazine articles, newspaper articles, and radio transcripts about the "brain drain" or another issue of interest to you. Describe the results of your search. How many articles are listed? How are the articles described? What other information is provided?

1.4 Simulations

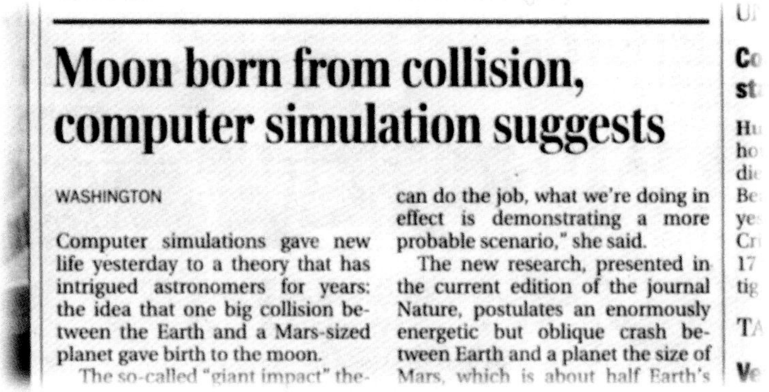

A **simulation** is an experiment, model, or activity that imitates real or hypothetical conditions. The newspaper article shown here describes how astrophysicists used computers to simulate a collision between Earth and a planet the size of Mars, an event that would

Moon born from collision, computer simulation suggests

WASHINGTON

Computer simulations gave new life yesterday to a theory that has intrigued astronomers for years: the idea that one big collision between the Earth and a Mars-sized planet gave birth to the moon.

The so-called "giant impact" the-

can do the job, what we're doing in effect is demonstrating a more probable scenario," she said.

The new research, presented in the current edition of the journal Nature, postulates an enormously energetic but oblique crash between Earth and a planet the size of Mars, which is about half Earth's

be impossible to measure directly. The simulation showed that such a collision could have caused both the formation of the moon and the rotation of Earth, strengthening an astronomical theory put forward in the 1970s.

INVESTIGATE & INQUIRE: Simulations

For each of the following, describe what is being simulated, the advantages of using a simulation, and any drawbacks.

a) crash test dummies **b)** aircraft simulators

c) wind tunnels **d)** zero-gravity simulator

e) 3-D movies **f)** paint-ball games

g) movie stunt actors **h)** grow lights

i) architectural scale models

In some situations, especially those with many variables, it can be difficult to calculate an exact value for a quantity. In such cases, simulations often can provide a good estimate. Simulations can also help verify theoretical calculations.

Example 1 Simulating a Multiple-Choice Test

When writing a multiple-choice test, you may have wondered "What are my chances of passing just by guessing?" Suppose that you make random guesses on a test with 20 questions, each having a choice of 5 answers. Intuitively, you would assume that your mark will be somewhere around 4 out of 20 since there is a 1 in 5 chance of guessing right on each question. However, it is possible that you could get any number of the questions right—anywhere from zero to a perfect score.

a) Devise a simulation for making guesses on the multiple-choice test.

b) Run the simulation 100 times and use the results to estimate the mark you are likely to get, on average.

c) Would it be practical to run your simulation 1000 times or more?

Solution 1   Using Pencil and Paper

a) Select any five cards from a deck of cards. Designate one of these cards to represent guessing the correct answer on a question. Shuffle the five cards and choose one at random. If it is the designated card, then you got the first question right. If one of the other four cards is chosen, then you got the question wrong.

Put the chosen card back with the others and repeat the process 19 times to simulate answering the rest of the questions on the test. Keep track of the number of right answers you obtained.

b) You could run 100 simulations by repeating the process in part a) over and over. However, you would have to choose a card 2000 times, which would be quite tedious. Instead, form a group with some of your classmates and pool your results, so that each student has to run only 10 to 20 repetitions of the simulation.

Make a table of the scores on the 100 simulated tests and calculate the mean score. You will *usually* find that this average is fairly close to the intuitive estimate of a score around 4 out of 20. However, a mean does not tell the whole story. Tally up the number of times each score appears in your table. Now, construct a bar graph showing the frequency for each score. Your graph will look something like the one shown.

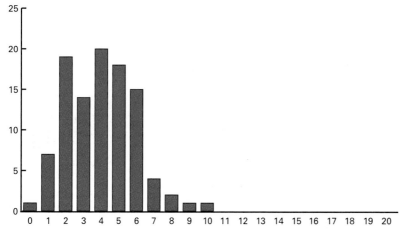

This graph gives a much more detailed picture of the results you could expect. Although 4 is the most likely score, there is also a good chance of getting 2, 3, 5, or 6, but the chance of guessing all 20 questions correctly is quite small.

c) Running the simulation 1000 times would require shuffling the five cards and picking one 20 000 times—possible, but not practical.

Solution 2   Using a Graphing Calculator

See Appendix B for more details on how to use the graphing calculator and software functions in Solutions 2 to 4.

a) You can use random numbers as the basis for a simulation. If you generate random integers from 1 to 5, you can have 1 correspond to a correct guess and 2 through 5 correspond to wrong answers.

Use the STAT EDIT menu to view lists L1 and L2. Make sure both lists are empty. Scroll to the top of L1 and enter the randInt function from the MATH PRB menu. This function produces random integers.

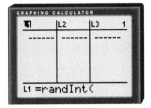

Enter 1 for the lower limit, 5 for the upper limit, and 20 for the number of trials. L1 will now contain 20 random integers between 1 and 5. Next, sort the list with the SortA function on the LIST OPS menu. Press 2nd 1 to enter L1 into the sort function. When you return to L1, the numbers in it will appear in ascending order. Now, you can easily scroll down the list to determine how many correct answers there were in this simulation.

b) The simplest way to simulate 100 tests is to repeat the procedure in part a) and keep track of the results by entering the number of correct answers in L2. Again, you may want to pool results with your classmates to reduce the number of times you have to enter the same formula over and over. If you know how to program your calculator, you can set it to re-enter the formulas for you automatically. However, unless you are experienced in programming the calculator, it will probably be faster for you to just re-key the formulas.

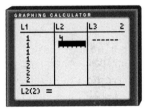

Once you have the scores from 100 simulations in L2, calculate the average using the mean function on the LIST MATH menu. To see which scores occur most frequently, plot L2 using STAT PLOT.

- **i)** Turn off all plots except Plot1.
- **ii)** For Type, choose the bar-graph icon and enter L2 for Xlist. Freq should be 1, the default value.
- **iii)** Use ZOOM/ZoomStat to set the window for the data. Press WINDOW to check the window settings. Set Xscl to 1 so that the bars correspond to integers.
- **iv)** Press GRAPH to display the bar graph.

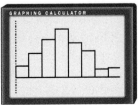

c) It is possible to program the calculator to run a large number of simulations automatically. However, the maximum list length on the TI-83 Plus is 999, so you would have to use at least two lists to run the simulation a 1000 times or more.

Solution 3 Using a Spreadsheet

a) Spreadsheets have built-in functions that you can use to generate and *count* the random numbers for the simulation.

The RAND() function produces a random real number that is equal to or greater than zero and less than one. The INT function rounds a real number down to the nearest integer. Combine these functions to generate a random integer between 1 and 5.

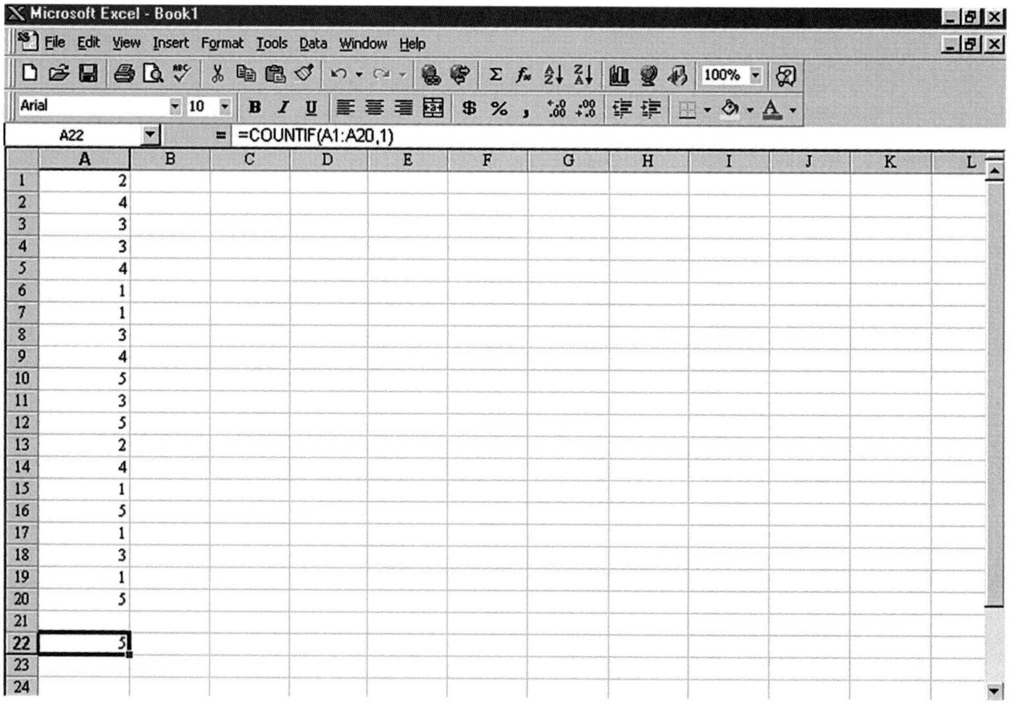

In Microsoft® Excel, you can use RANDBETWEEN only if you have the Analysis Toolpak installed.

Enter the formula INT(RAND( )*5)+1 or RANDBETWEEN(1,5) in A1 and copy it down to A20. Next, use the COUNTIF function to count the number of *1*s in column A. Record this score in cell A22.

Microsoft Excel - Book1
File Edit View Insert Format Tools Data Window Help
Arial
A22
=COUNTIF(A1:A20,1)
A B C D E F G H I J K L
1 2
2 4
3 3
4 3
5 4
6 1
7 1
8 3
9 4
10 5
11 3
12 5
13 2
14 4
15 1
16 5
17 1
18 3
19 1
20 5
21
22 5
23
24

b) To run 100 simulations, copy A1:A22 into columns B through CV using the Fill feature. Then, use the average function to find the mean score for the 100 simulated tests. Record this average in cell B23.

Next, use the COUNTIF function to find the number of times each possible score occurs in cells A22 to CV22. Enter the headings SUMMARY, Score, and Frequency in cells A25, A26, and A27, respectively. Then, enter 0 in cell B26 and highlight cells B26 through V26. Use the Fill feature to enter the integers 0 through 20 in cells B26 through V26. In B27, enter the formula for the number of zero scores; in C27, the number of *1*s; in D27, the number of *2*s; and so on, finishing with V27 having the number of perfect

scores. Note that by using absolute cell referencing you can simply copy the COUNTIF function from B27 to the other 20 cells.

Finally, use the Chart feature to plot frequency versus score. Highlight cells A26 through V27, then select Insert/Chart/XY.

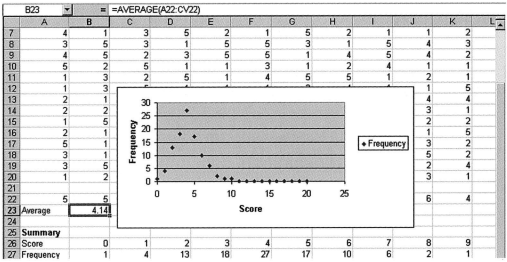

	B23	▼	=	=AVERAGE(A22:CV22)							
	A	B	C	D	E	F	G	H	I	J	K	L
7	4	1	3	5	2	1	5	2	1	1	2	
8	3	5	3	1	5	5	3	1	5	4	3	
9	4	5	2	3	5	5	1	4	5	4	2	
10	5	2	5	1	1	3	1	2	4	1	1	
11	1	3	2	5	1	4	5	5	1	2	1	
12	1	3								1	5	
13	2	1								4	4	
14	2	2								3	1	
15	1	5								2	2	
16	2	1								1	5	
17	5	1								3	2	
18	3	1								5	2	
19	3	5								2	4	
20	1	2								3	1	
21												
22	5	5								6	4	
23	Average	4.14										
24												
25	Summary											
26	Score	0	1	2	3	4	5	6	7	8	9	
27	Frequency	1	4	13	18	27	17	10	6	2	1	

c) The method in part b) can easily handle 1000 simulations or more.

Solution 4   Using Fathom™

a) Fathom™ also has built-in functions to generate random numbers and count the scores in the simulations.

Launch Fathom™ and open a new document if necessary. Drag a new **collection** box to the document and rename it MCTest. Right-click on the box and create 20 new cases.

Drag a **case table** to the work area. You should see your 20 cases listed. Expand the table if you cannot see them all on the screen.

Rename the <new> column Guess. Right-click on Guess and select Edit Formula, Expand Functions, then Random Numbers. Enter 1,5 into the **randomInteger() function** and click OK to fill the Guess column with random integers between 1 and 5. Scroll down the column to see how many correct guesses there are in this simulation.

MCTest		
	Guess	**<new>**
1	5	
2	2	
3	2	
4	5	
5	5	
6	3	
7	5	
8	4	
9	4	
10	3	
11	1	
12	1	
13	1	
14	3	
15	2	
16	3	
17	5	
18	4	
19	4	
20	4	

b) You can run a new simulation just by pressing Ctrl-Y, which will fill the Guess column with a new set of random numbers. Better still, you can set Fathom™ to automatically repeat the simulation 100 times automatically and keep track of the number of correct guesses.

First, set up the count function. Right–click on the **collection** box and select Inspect Collection. Select the Measures tab and rename the <new> column Score. Then, right-click the column below Formula and select Edit Formula, Functions, Statistical, then One Attribute. Select count, enter Guess = 1 between the brackets, and click OK to count the number of correct guesses in your case table.

Click on the MCTest **collection** box. Now, select Analyse, Collect Measures from the main menu bar, which creates a new collection box called Measures from MCTest. Click on this box and drag a new **case table** to the document. Fathom™ will automatically run five simulations of the multiple-choice test and show the results in this **case table** .

To simulate 100 tests, right-click on the Measures from MCTest **collection** box and select Inspect Collection. Turn off the animation in order to speed up the simulation. Change the number of measures to 100. Then, click on the Collect More Measures button. You should now have 100 measures in the **case table** for Measures from MCTest.

Next, use the **mean function** to find the average score for these simulations. Go back to the Inspect Measures from MCTest **collection** box and change the column heading <new> to Average. Right-click on Formula and select Edit Formula, Functions, Statistical, then One Attribute. Select mean, enter Score between the brackets, and select OK to display the mean mark on the 100 tests.

Finally, plot a histogram of the scores from the simulations. Drag the **graph icon** onto the work area. Then, drag the Score column from the Measures from MCTest **case table** to the horizontal axis of the graph. Fathom™ then automatically produces a dot plot of your data. To display a histogram instead, simply click the menu in the upper right hand corner of the graph and choose Histogram.

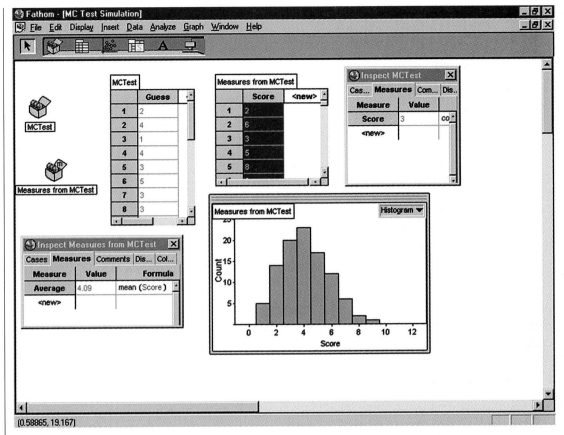

c) Fathom™ can easily run this simulation 1000 times or more. All you have to do is change the number of measures.

Key Concepts

- Simulations can be useful tools for estimating quantities that are difficult to calculate and for verifying theoretical calculations.

- A variety of simulation methods are available, ranging from simple manual models to advanced technology that makes large-scale simulations feasible.

Communicate Your Understanding

1. Make a table summarizing the pros and cons of the four simulation methods used in Example 1.

2. A manufacturer of electric motors has a failure rate of 0.2% on one of its products. A quality-control inspector needs to know the range of the number of failures likely to occur in a batch of 1000 of these motors. Which tool would you use to simulate this situation? Give reasons for your choice.

Practise

1. Write a graphing calculator formula for

a) generating 100 random integers between 1 and 25

b) generating 24 random integers between −20 and 20

2. Write a spreadsheet formula for

a) generating 100 random numbers between 1 and 25

b) generating 100 random integers between 1 and 25

c) generating 16 random integers between −40 and 40

d) counting the number of entries that equal 42.5 in the range C10 to V40

Apply, Solve, Communicate

3. Communication Identify two simulations you use in everyday life and list the advantages of using each simulation.

4. Describe three other manual methods you could use to simulate the multiple-choice test in Example 1.

5. Communication

a) Describe a calculation or mechanical process you could use to produce random integers.

b) Could you use a telephone book to generate random numbers? Explain why or why not.

6. Application A brother and sister each tell the truth two thirds of the time. The brother stated that he owned the car he was driving. The sister said he was telling the truth. Develop a simulation to show whether you should believe them.

7. Inquiry/Problem Solving Consider a random walk in which a coin toss determines the direction of each step. On the odd-numbered tosses, walk one step north for heads and one step south for tails. On even-numbered tosses, walk one step east for heads and one step west for tails.

a) Beginning at position (0, 0) on a Cartesian graph, simulate this random walk for 100 steps. Note the coordinates where you finish.

b) Repeat your simulation 10 times and record the results.

c) Use these results to formulate a hypothesis about the endpoints for this random walk.

d) Change the rules of the random walk and investigate the effect on the end points.

Knowledge/ Understanding	Thinking/Inquiry/ Problem Solving	Communication	Application

8. a) Use technology to simulate rolling two dice 100 times and record the sum of the two dice each time. Make a histogram of the sums.

b) Which sum occurs most often? Explain why this sum is likely to occur more often than the other sums.

c) Which sum or sums occur least often? Explain this result.

d) Suppose three dice are rolled 100 times and the sums are recorded. What sums would you expect to be the most frequent and least frequent? Give reasons for your answers.

9. Communication Describe a quantity that would be difficult to calculate or to measure in real life. Outline a simulation procedure you could use to determine this quantity.

1.5 Graph Theory

Graph theory is a branch of mathematics in which graphs or networks are used to solve problems in many fields. Graph theory has many applications, such as
- setting examination timetables
- colouring maps
- modelling chemical compounds
- designing circuit boards
- building computer, communication, or transportation networks
- determining optimal paths

In graph theory, a graph is unlike the traditional Cartesian graph used for graphing functions and relations. A **graph** (also known as a **network**) is a collection of line segments and **nodes**. Mathematicians usually call the nodes **vertices** and the line segments **edges**. Networks can illustrate the relationships among a great variety of objects or sets.

This network is an illustration of the subway system in Toronto. In order to show the connections between subway stations, this map is not to scale. In fact, networks are rarely drawn to scale.

INVESTIGATE & INQUIRE: Map Colouring

In each of the following diagrams the lines represent borders between countries. Countries joined by a line segment are considered **neighbours**, but countries joining at only a single point are not.

1. Determine the smallest number of colours needed for each map such that all neighbouring countries have different colours.

   a)

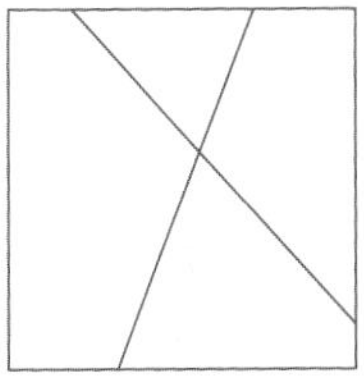

   b)

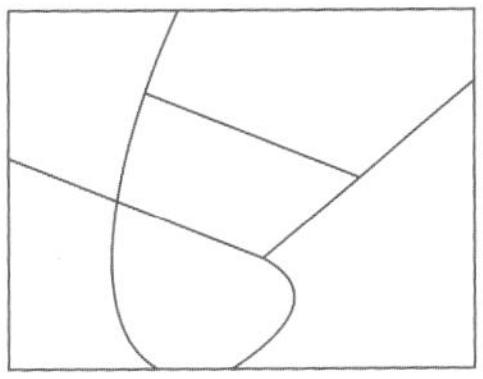

c) 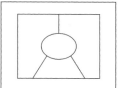   **d)** 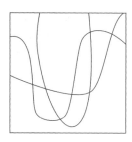   **e)**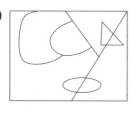

2. Make a conjecture regarding the maximum number of colours needed to colour a map. Why do you think your conjecture is correct?

Although the above activity is based on maps, it is very mathematical. It is about solving problems involving **connectivity**. Each country could be represented as a node or **vertex**. Each border could be represented by a segment or **edge**.

Example 1 Representing Maps With Networks

Represent each of the following maps with a network.

a)   **b)**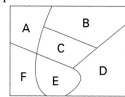

Solution

a) Let A, B, C, and D be vertices representing countries A, B, C, and D, respectively. A shares a border with both B and D but not with C, so A should be connected by edges to B and D only. Similarly, B is connected to only A and C; C, to only B and D; and D, to only A and C.

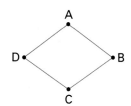

b) Let A, B, C, D, E, and F be vertices representing countries A, B, C, D, E, and F, respectively. Note that the positions of the vertices are not important, but their interconnections are. A shares borders with B, C, and F, but not with D or E. Connect A with edges to B, C, and F only. Use the same process to draw the rest of the edges.

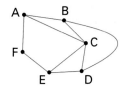

As components of networks, edges could represent connections such as roads, wires, pipes, or air lanes, while vertices could represent cities, switches, airports, computers, or pumping stations. The networks could be used to carry vehicles, power, messages, fluid, airplanes, and so on.

If two vertices are connected by an edge, they are considered to be **adjacent**. In the network on the right, A and B are adjacent, as are B and C. A and C are not adjacent.

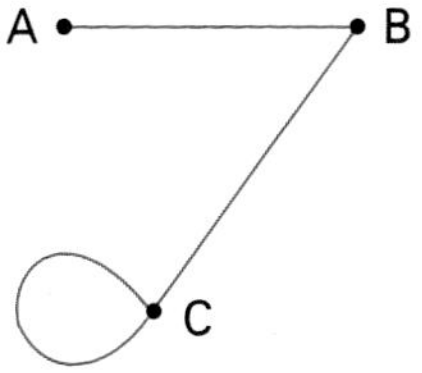

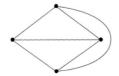

The number of edges that begin or end at a vertex is called the **degree** of the vertex. In the network, A has degree 1, B has degree 2, and C has degree 3. The loop counts as both an edge beginning at C and an edge ending at C.

Any connected sequence of vertices is called a **path**. If the path begins and ends at the same vertex, the path is called a **circuit**. A circuit is independent of the starting point. Instead, the circuit depends on the route taken.

Example 2 Circuits

Determine if each path is a circuit.

a) 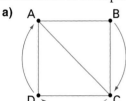 b) 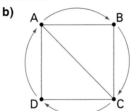 c)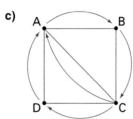

Solution

a) Path: BC to CD to DA
    Since this path begins at B and ends at A, it is not a circuit.

b) Path: BC to CD to DA to AB
    This path begins at B and ends at B, so it is a circuit.

c) Path: CA to AB to BC to CD to DA
    Since this path begins at C and ends at A, it is not a circuit.

A network is **connected** if and only if there is at least one path connecting each pair of vertices. A **complete** network is a network with an edge between every pair of vertices.

Connected but not complete: Not all vertices are joined directly.   *Connected and complete: All vertices are joined to each other by edges.*   *Neither connected nor complete: Not all vertices are joined.*

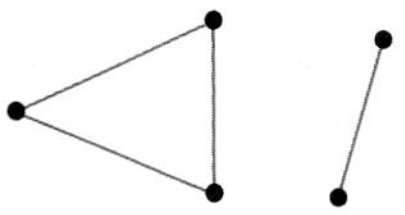

In a **traceable** network all the vertices are connected to at least one other vertex and all the edges can be travelled exactly once in a continuous path.

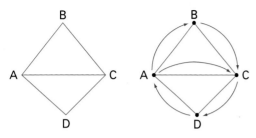
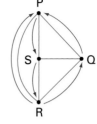

Traceable: All vertices are connected to at least one other vertex, and the path from A to B to C to D to A to C includes all the edges without repeating any of them.

Non-traceable: No continuous path can travel all the edges only once.

Example 3   The Seven Bridges of Koenigsberg

The eighteenth-century German town of Koenigsberg (now the Russian city of Kaliningrad) was situated on two islands and the banks of the Pregel River. Koenigsberg had seven bridges as shown in the map.

People of the town believed—but could not prove—that it was impossible to tour the town, crossing each bridge exactly once, regardless of where the tour started or finished. Were they right?

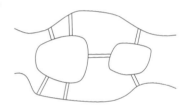

Solution

Reduce the map to a simple network of vertices and edges. Let vertices A and C represent the mainland, with B and D representing the islands. Each edge represents a bridge joining two parts of the town.

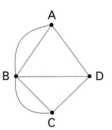

If, for example, you begin at vertex D, you will leave and eventually return but, because D has a degree of 3, you will have to leave again.

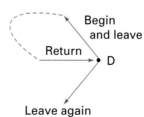
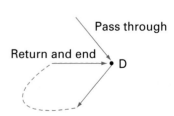

Conversely, if you begin elsewhere, you will pass through vertex D at some point, entering by one edge and leaving by another. But, because D has degree 3, you must return in order to trace the third edge and, therefore,

must end at D. So, your path must either begin or end at vertex D. Because all the vertices are of odd degree, the same argument applies to all the other vertices. Since you cannot begin or end at more than two vertices, the network is non-traceable. Therefore, it is indeed impossible to traverse all the town's bridges without crossing one twice.

Leonhard Euler developed this proof of Example 3 in 1735. He laid the foundations for the branch of mathematics now called graph theory. Among other discoveries, Euler found the following general conditions about the traceability of networks.

- A network is traceable if it has only vertices of even degree (even vertices) or exactly two vertices of odd degree (odd vertices).

- If the network has two vertices of odd degree, the tracing path must begin at one vertex of odd degree and end at the other vertex of odd degree.

Example 4 Traceability and Degree

For each of the following networks,
a)   list the number of vertices with odd degree and with even degree
b)   determine if the network is traceable

i)    ii)    iii) 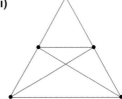   iv)

Solution

i) **a)** 3 even vertices 0 odd vertices **b)** traceable	ii) **a)** 0 even vertices 4 odd vertices **b)** non-traceable	iii) **a)** 3 even vertices 2 odd vertices **b)** traceable	iv) **a)** 1 even vertex 4 odd vertices **b)** non-traceable

If it is possible for a network to be drawn on a two-dimensional surface so that the edges do not cross anywhere except at vertices, it is **planar.**

Example 5 Planar Networks

Determine whether each of the following networks is planar.

a)   **b)** 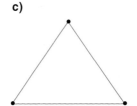   **c)**    **d)** 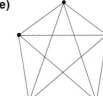   **e)**

Solution

a) Planar

b) Planar

c) Planar

d)

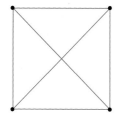 can be redrawn as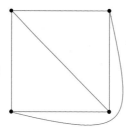

Therefore, the network is planar.

e)

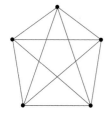 cannot be redrawn as a planar network: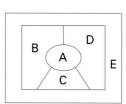

Therefore, the network is non-planar.

Example 6 Map Colouring (The Four-Colour Problem)

A graphic designer is working on a logo representing the different tourist regions in Ontario. What is the minimum number of colours required for the design shown on the right to have all adjacent areas coloured differently?

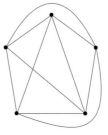

Solution

Because the logo is two-dimensional, you can redraw it as a planar network as shown on the right. This network diagram can help you see the relationships between the regions. The vertices represent the regions and the edges show which regions are adjacent. Vertices A and E both connect to the three other vertices but not to each other. Therefore, A and E can have the same colour, but it must be different from the colours for B, C, and D. Vertices B, C, and D all connect to each other, so they require three different colours. Thus, a minimum of four colours is necessary for the logo.

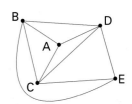

This example is a specific case of a famous problem in graph theory called the four-colour problem. As you probably conjectured in the investigation at the start of this section, the maximum number of colours required in *any* planar map is four. This fact had been suspected for centuries but was not proven until 1976. The proof by Wolfgang Haken and Kenneth Appel at the University of Illinois required a supercomputer to break the proof down into cases and many years of verification by other mathematicians. Non-planar maps can require more colours.

WEB CONNECTION
www.mcgrawhill.ca/links/MDM12

Visit the above web site and follow the links to find out more about the four-colour problem. Write a short report on the history of the four-colour problem.

Example 7 Scheduling

The mathematics department has five committees. Each of these committees meets once a month. Membership on these committees is as follows:

Committee A: Szczachor, Large, Ellis
Committee B: Ellis, Wegrynowski, Ho, Khan
Committee C: Wegrynowski, Large
Committee D: Andrew, Large, Szczachor
Committee E: Bates, Card, Khan, Szczachor

What are the minimum number of time slots needed to schedule the committee meetings with no conflicts?

Solution

Draw the schedule as a network, with each vertex representing a different committee and each edge representing a potential conflict between committees (a person on two or more committees). Analyse the network as if you were colouring a map.

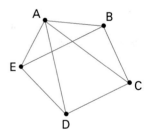

The network can be drawn as a planar graph. Therefore, a maximum of four time slots is necessary to "colour" this graph. Because Committee A is connected to the four other committees (degree 4), at least two time slots are necessary: one for committee A and at least one for all the other committees. Because each of the other nodes has degree 3, at least one more time slot is necessary. In fact, three time slots are sufficient since B is not connected to D and C is not connected to E.

Time Slot	Committees
1	A
2	B, D
3	C, E

Project Prep

Graph theory provides problem-solving techniques that will be useful in your tools for data management project.

- In graph theory, a graph is also known as a network and is a collection of line segments (edges) and nodes (vertices).

- If two vertices are connected by an edge, they are adjacent. The degree of a vertex is equal to the number of edges that begin or end at the vertex.

- A path is a connected sequence of vertices. A path is a circuit if it begins and ends at the same vertex.

- A connected network has at least one path connecting each pair of vertices. A complete network has an edge connecting every pair of vertices.

- A connected network is traceable if it has only vertices of even degree (even vertices) or exactly two vertices of odd degree (odd vertices). If the network has two vertices of odd degree, the tracing must begin at one of the odd vertices and end at the other.

- A network is planar if its edges do not cross anywhere except at the vertices.

- The maximum number of colours required to colour any planar map is four.

Communicate Your Understanding

1. Describe how to convert a map into a network. Use an example to aid in your description.

2. A network has five vertices of even degree and three vertices of odd degree. Using a diagram, show why this graph cannot be traceable.

3. A modern zoo contains natural habitats for its animals. However, many of the animals are natural enemies and cannot be placed in the same habitat. Describe how to use graph theory to determine the number of different habitats required.

Practise

1. For each network,
   i) find the degree of each vertex
   ii) state whether the network is traceable

   a) 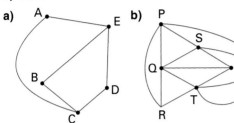 **b)**

2. Draw a network diagram representing the maps in questions 1d) and 1e) of the investigation on pages 41 and 42.

3. **a)** Look at a map of Canada. How many colours are needed to colour the ten provinces and three territories of Canada?

   b) How many colours are needed if the map includes the U.S.A. coloured with a single colour?

Apply, Solve, Communicate

4. The following map is made up of curved lines that cross each other and stop only at the boundary of the map. Draw three other maps using similar lines. Investigate the four maps and make a conjecture of how many colours are needed for this type of map.

   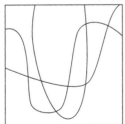

5. Is it possible to add one bridge to the Koenigsberg map to make it traceable? Provide evidence for your answer.

6. **Inquiry/Problem Solving** The following chart indicates the subjects studied by five students.

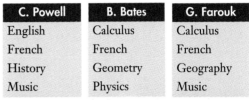

C. Powell	B. Bates	G. Farouk
English	Calculus	Calculus
French	French	French
History	Geometry	Geography
Music	Physics	Music

E. Ho	N. Khan
Calculus	English
English	Geography
Geometry	Mathematics of Data
Mathematics of Data	Management
Management	Physics

   a) Draw a network to illustrate the overlap of subjects these students study.

   b) Use your network to design an examination timetable without conflicts.

   (*Hint*: Consider each subject to be one vertex of a network.)

7. A highway inspector wants to travel each road shown once and only once to inspect for winter damage. Determine whether it is possible to do so for each map shown below.

   a)

   b)

   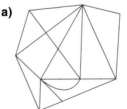

   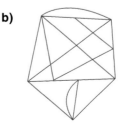

8. Inquiry/Problem Solving

a) Find the degree of each vertex in the network shown.

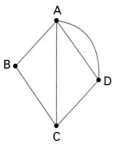

b) Find the sum of the degrees of the vertices.

c) Compare this sum with the number of edges in the network. Investigate other networks and determine the sum of the degrees of their vertices.

d) Make a conjecture from your observations.

9. a) The following network diagram of the main floor of a large house uses vertices to represent rooms and edges to represent doorways. The exterior of the house can be treated as one room. Sketch a floor plan based on this network.

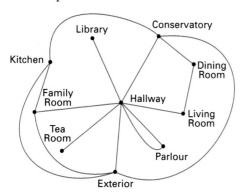

b) Draw a floor plan and a network diagram for your own home.

10. Application

a) Three houses are located at positions A, B, and C, respectively. Water, gas, and electrical utilities are located at positions D, E, and F, respectively. Determine whether the houses can each be connected to all three utilities without any of the connections crossing. Provide evidence for your decision. Is it necessary to reposition any of the utilities? Explain.

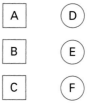

b) Show that a network representing two houses attached to n utilities is planar.

11. The four Anderson sisters live near each other and have connected their houses by a network of paths such that each house has a path leading directly to each of the other three houses. None of these paths intersect. Can their brother Warren add paths from his house to each of his sisters' houses without crossing any of the existing paths?

12. In the diagram below, a sheet of paper with a circular hole cut out partially covers a drawing of a closed figure. Given that point A is inside the closed figure, determine whether point B is inside or outside. Provide reasons for your answer.

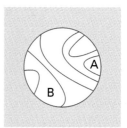

13. Application A communications network between offices of a company needs to provide a back-up link in case one part of a path breaks down. For each network below, determine which links need to be backed up. Describe how to back up the links.

a)

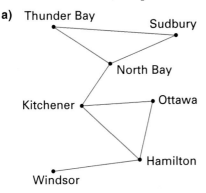

b)

14. During an election campaign, a politician will visit each of the cities on the map below.

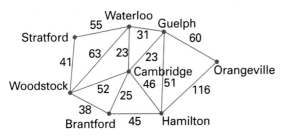

a) Is it possible to visit each city only once?

b) Is it possible to begin and end in the same city?

c) Find the shortest route for visiting all the cities. (*Hint*: You can usually find the shortest paths by considering the shortest edge at each vertex.)

15. In a communications network, the **optimal path** is the one that provides the fastest link. In the network shown, all link times are in seconds.

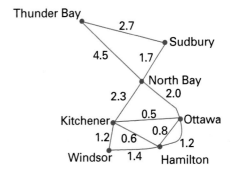

Determine the optimal path from

a) Thunder Bay to Windsor

b) Hamilton to Sudbury

c) Describe the method you used to estimate the optimal path.

16. A salesperson must travel by air to all of the cities shown in the diagram below. The diagram shows the cheapest one-way fare for flights between the cities. Determine the least expensive travel route beginning and ending in Toronto.

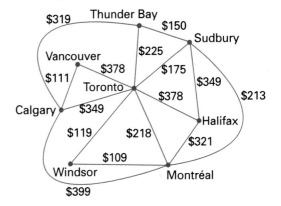

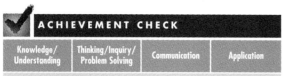
17. The diagram below shows the floor plan of a house.

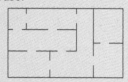

a) Find a route that passes through each doorway of this house exactly once.

b) Use graph theory to explain why such a route is possible.

c) Where could you place two exterior doors so that it is possible to start outside the house, pass through each doorway exactly once, and end up on the exterior again? Explain your reasoning.

d) Is a similar route possible if you add three exterior doors instead of two? Explain your answer.

C

18. a) Six people at a party are seated at a table. No three people at the table know each other. For example, if Aaron knows Carmen and Carmen knows Allison, then Aaron and Allison do not know each other. Show that at least three of the six people seated at the table must be strangers to each other. (*Hint*: Model this situation using a network with six vertices.)

b) Show that, among five people, it is possible that no three all know each other and that no three are all strangers.

19. Inquiry/Problem Solving Use graph theory to determine if it is possible to draw the diagram below using only three strokes of a pencil.

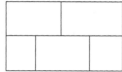

20. Communication

a) Can a connected graph of six vertices be planar? Explain your answer.

b) Can a complete graph of six vertices be planar? Explain.

21. Can the graph below represent a map in two dimensions. Explain.

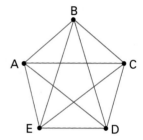

22. Can a network have exactly one vertex with an odd degree? Provide evidence to support your answer.

23. Communication A graph is regular if all its vertices have the same degree. Consider graphs that do not have either loops connecting a vertex back to itself or multiple edges connecting any pair of vertices.

a) Draw the four regular planar graphs that have four vertices.

b) How many regular planar graphs with five vertices are there?

c) Explain the difference between your results in parts a) and b).

1.6 Modelling With Matrices

A **matrix** is a rectangular array of numbers used to manage and organize data, somewhat like a table or a page in a spreadsheet. Matrices are made up of horizontal rows and vertical columns and are usually enclosed in square brackets. Each number appearing in the matrix is called an **entry**. For instance, $A = \begin{bmatrix} 5 & -2 & 3 \\ 2 & 1 & 0 \end{bmatrix}$ is a matrix with two rows and three columns, with entries 5, –2, and 3 in the first row and entries 2, 1, and 0 in the second row. The **dimensions** of this matrix are 2×3. A matrix with m rows and n columns has dimensions of $m \times n$.

INVESTIGATE & INQUIRE: Olympic Medal Winners

At the 1998 Winter Olympic games in Nagano, Japan, Germany won 12 gold, 9 silver, and 8 bronze medals; Norway won 10 gold, 10 silver, and 5 bronze medals; Russia won 9 gold, 6 silver, and 3 bronze medals; Austria won 3 gold, 5 silver, and 9 bronze medals; Canada won 6 gold, 5 silver, and 4 bronze medals; and the United States won 6 gold, 3 silver, and 4 bronze medals.

1. Organize the data using a matrix with a row for each type of medal and a column for each country.

2. State the dimensions of the matrix.

3. **a)** What is the meaning of the entry in row 3, column 1?

   b) What is the meaning of the entry in row 2, column 4?

4. Find the sum of all the entries in the first row of the matrix. What is the significance of this **row sum**? What would the **column sum** represent?

5. Use your matrix to estimate the number of medals each country would win if the number of Olympic events were to be increased by 20%.

6. **a)** Interchange the rows and columns in your matrix by "reflecting" the matrix in the diagonal line beginning at row 1, column 1.

   b) Does this **transpose matrix** provide the same information? What are its dimensions?

7. State one advantage of using matrices to represent data.

In general, use a capital letter as the symbol for a matrix and represent each entry using the corresponding lowercase letter with two indices. For example,

$$A = \begin{bmatrix} a_{11} & a_{12} & a_{13} \\ a_{21} & a_{22} & a_{23} \\ a_{31} & a_{32} & a_{33} \end{bmatrix} \qquad B = \begin{bmatrix} b_{11} & b_{12} \\ b_{21} & b_{22} \\ b_{31} & b_{32} \end{bmatrix} \qquad C = \begin{bmatrix} c_{11} & c_{12} & c_{13} & \cdots & c_{1n} \\ c_{21} & c_{22} & c_{23} & \cdots & c_{2n} \\ c_{31} & c_{32} & c_{33} & \cdots & c_{3n} \\ \vdots & \vdots & \vdots & \vdots & \vdots \\ c_{m1} & c_{m2} & c_{m3} & \cdots & c_{mn} \end{bmatrix}$$

Here, a_{ij}, b_{ij}, and c_{ij} represent the entries in row i and column j of these matrices.

The transpose of a matrix is indicated by a superscript t, so the transpose of A is shown as A^t. A matrix with only one row is called a **row matrix**, and a matrix with only one column is a **column matrix**. A matrix with the same number of rows as columns is called a **square matrix**.

$$[1 \ -2 \ 5 \ -9] \qquad \begin{bmatrix} -3 \\ 0 \\ 5 \end{bmatrix} \qquad \begin{bmatrix} 3 & 4 & 9 \\ -1 & 0 & 2 \\ 5 & -10 & -3 \end{bmatrix}$$

a row matrix   a column matrix   a square matrix

Example 1 Representing Data With a Matrix

The number of seats in the House of Commons won by each party in the federal election in 1988 were Bloc Québécois (BQ), 0; Progressive Conservative Party (PC), 169; Liberal Party (LP), 83; New Democratic Party (NDP), 43; Reform Party (RP), 0; Other, 0. In 1993, the number of seats won were BQ, 54; PC, 2; LP, 177; NDP, 9; RP, 52; Other, 1. In 1997, the number of seats won were BQ, 44; PC, 20; LP, 155; NDP, 21; RP, 60; Other, 1.

a) Organize the data using a matrix S with a row for each political party.

b) What are the dimensions of your matrix?

c) What does the entry s_{43} represent?

d) What entry has the value 52?

e) Write the transpose matrix for S. Does S^t provide the same information as S?

f) The results from the year 2000 federal election were Bloc Québécois, 38; Progressive Conservative, 12; Liberal, 172; New Democratic Party, 13; Canadian Alliance (formerly Reform Party), 66; Other, 0. Update your matrix to include the results from the 2000 federal election.

Solution

a)

$$
S = \begin{array}{c} \\ \\ \\ \\ \\ \\ \\ \end{array}
\begin{array}{ccc} 1988 & 1993 & 1997 \end{array}
$$

$$
S = \left[\begin{array}{ccc}
0 & 54 & 44 \\
169 & 2 & 20 \\
83 & 177 & 155 \\
43 & 9 & 21 \\
0 & 52 & 60 \\
0 & 1 & 1
\end{array}\right]
\begin{array}{l}
\text{BQ} \\
\text{PC} \\
\text{LP} \\
\text{NDP} \\
\text{RP} \\
\text{Other}
\end{array}
$$

Labelling the rows and columns in large matrices can help you keep track of what the entries represent.

b) The dimensions of the matrix are 6×3.

c) The entry s_{43} shows that the NDP won 21 seats in 1997.

d) The entry s_{52} has the value 52.

e) The transpose matrix is

$$
\begin{array}{ccccccc} & \text{BQ} & \text{PC} & \text{LP} & \text{NDP} & \text{RP} & \text{Other} \end{array}
$$

$$
S^t = \left[\begin{array}{cccccc}
0 & 169 & 83 & 43 & 0 & 0 \\
54 & 2 & 177 & 9 & 52 & 1 \\
44 & 20 & 155 & 21 & 60 & 1
\end{array}\right]
\begin{array}{l}
1988 \\
1993 \\
1997
\end{array}
$$

Comparing the entries in the two matrices shows that they do contain exactly the same information.

f)

$$
\begin{array}{cccc} 1988 & 1993 & 1997 & 2000 \end{array}
$$

$$
\left[\begin{array}{cccc}
0 & 54 & 44 & 38 \\
169 & 2 & 20 & 12 \\
83 & 177 & 155 & 172 \\
43 & 9 & 21 & 13 \\
0 & 52 & 60 & 66 \\
0 & 1 & 1 & 0
\end{array}\right]
\begin{array}{l}
\text{BQ} \\
\text{PC} \\
\text{LP} \\
\text{NDP} \\
\text{CA (RP)} \\
\text{Other}
\end{array}
$$

Two matrices are equal only if each entry in one matrix is equal to the corresponding entry in the other.

For example, $\begin{bmatrix} \dfrac{3}{2} & \sqrt{16} & (-2)^3 \\ 5^{-1} & -4 & -(-2) \end{bmatrix}$ and $\begin{bmatrix} 1.5 & 4 & -8 \\ \dfrac{1}{5} & -4 & 2 \end{bmatrix}$ are equal matrices.

Two or more matrices can be added or subtracted, provided that their dimensions are the same. To add or subtract matrices, add or subtract the corresponding entries of each matrix. For example,

$$\begin{bmatrix} 2 & -1 & 5 \\ 0 & 7 & -8 \end{bmatrix} + \begin{bmatrix} 0 & 5 & -3 \\ -2 & 4 & -1 \end{bmatrix} = \begin{bmatrix} 2 & 4 & 2 \\ -2 & 11 & -9 \end{bmatrix}$$

Matrices can be multiplied by a **scalar** or constant. To multiply a matrix by a scalar, multiply each entry of the matrix by the scalar. For example,

$$-3 \begin{bmatrix} 4 & 5 \\ -6 & 0 \\ 3 & -8 \end{bmatrix} = \begin{bmatrix} -12 & -15 \\ 18 & 0 \\ -9 & 24 \end{bmatrix}$$

Example 2 Inventory Problem

The owner of Lou's 'Lectronics Limited has two stores. The manager takes inventory of their top-selling items at the end of the week and notes that at the eastern store, there are 5 video camcorders, 7 digital cameras, 4 CD players, 10 televisions, 3 VCRs, 2 stereo systems, 7 MP3 players, 4 clock radios, and 1 DVD player in stock. At the western store, there are 8 video camcorders, 9 digital cameras, 3 CD players, 8 televisions, 1 VCR, 3 stereo systems, 5 MP3 players, 10 clock radios, and 2 DVD players in stock. During the next week, the eastern store sells 3 video camcorders, 2 digital cameras, 4 CD players, 3 televisions, 3 VCRs, 1 stereo system, 4 MP3 players, 1 clock radio, and no DVD players. During the same week, the western store sells 5 video camcorders, 3 digital cameras, 3 CD players, 8 televisions, no VCRs, 1 stereo system, 2 MP3 players, 7 clock radios, and 1 DVD player. The warehouse then sends each store 4 video camcorders, 3 digital cameras, 4 CD players, 4 televisions, 5 VCRs, 2 stereo systems, 2 MP3 players, 3 clock radios, and 1 DVD player.

a) Use matrices to determine how many of each item is in stock at the stores after receiving the new stock from the warehouse.

b) Immediately after receiving the new stock, the manager phones the head office and requests an additional 25% of the items presently in stock in anticipation of an upcoming one-day sale. How many of each item will be in stock at each store?

Solution 1   Using Pencil and Paper

a) Let matrix A represent the initial inventory, matrix B represent the number of items sold, and matrix C represent the items in the first shipment of new stock.

$$A = \begin{array}{c} \text{E} \quad \text{W} \\ \begin{bmatrix} 5 & 8 \\ 7 & 9 \\ 4 & 3 \\ 10 & 8 \\ 3 & 1 \\ 2 & 3 \\ 7 & 5 \\ 4 & 10 \\ 1 & 2 \end{bmatrix} \end{array} \begin{array}{l} \text{camcorders} \\ \text{cameras} \\ \text{CD players} \\ \text{TVs} \\ \text{VCRs} \\ \text{stereos} \\ \text{MP3 players} \\ \text{clock radios} \\ \text{DVD players} \end{array} \qquad B = \begin{bmatrix} 3 & 5 \\ 2 & 3 \\ 4 & 3 \\ 3 & 8 \\ 3 & 0 \\ 1 & 1 \\ 4 & 2 \\ 1 & 7 \\ 0 & 1 \end{bmatrix} \qquad C = \begin{bmatrix} 4 & 4 \\ 3 & 3 \\ 4 & 4 \\ 4 & 4 \\ 5 & 5 \\ 2 & 2 \\ 2 & 2 \\ 3 & 3 \\ 1 & 1 \end{bmatrix}$$

Since the dimensions of matrices A, B, and C are the same, matrix addition and subtraction can be performed. Then, the stock on hand before the extra shipment is

$$D = A - B + C = \begin{bmatrix} 5 & 8 \\ 7 & 9 \\ 4 & 3 \\ 10 & 8 \\ 3 & 1 \\ 2 & 3 \\ 7 & 5 \\ 4 & 10 \\ 1 & 2 \end{bmatrix} - \begin{bmatrix} 3 & 5 \\ 2 & 3 \\ 4 & 3 \\ 3 & 8 \\ 3 & 0 \\ 1 & 1 \\ 4 & 2 \\ 1 & 7 \\ 0 & 1 \end{bmatrix} + \begin{bmatrix} 4 & 4 \\ 3 & 3 \\ 4 & 4 \\ 4 & 4 \\ 5 & 5 \\ 2 & 2 \\ 2 & 2 \\ 3 & 3 \\ 1 & 1 \end{bmatrix} = \begin{bmatrix} 6 & 7 \\ 8 & 9 \\ 4 & 4 \\ 11 & 4 \\ 5 & 6 \\ 3 & 4 \\ 5 & 5 \\ 6 & 6 \\ 2 & 2 \end{bmatrix}$$

Let E represent the stock in the stores after the extra shipment from the warehouse.

$$E = 125\% \times D = 1.25 \begin{bmatrix} 6 & 7 \\ 8 & 9 \\ 4 & 4 \\ 11 & 4 \\ 5 & 6 \\ 3 & 4 \\ 5 & 5 \\ 6 & 6 \\ 2 & 2 \end{bmatrix} = \begin{bmatrix} 7.5 & 8.75 \\ 10 & 11.25 \\ 5 & 5 \\ 13.75 & 5 \\ 6.25 & 7.5 \\ 3.75 & 5 \\ 6.25 & 6.25 \\ 7.5 & 7.5 \\ 2.5 & 2.5 \end{bmatrix}$$

Assuming the manager rounds to the nearest whole number, the stock at the eastern store will be 8 video camcorders, 10 digital cameras, 5 CD players, 14 televisions, 6 VCRs, 4 stereo systems, 6 MP3 players, 8 clock radios, and 3 DVD players in stock. At the western store, there will be 9 video camcorders, 11 digital cameras, 5 CD players, 5 televisions, 8 VCRs, 5 stereo systems, 6 MP3 players, 8 clock radios, and 3 DVD players in stock.

Solution 2 Using a Graphing Calculator

a) As in the pencil-and-paper solution, let matrix A represent the initial inventory, matrix B the items sold, and matrix C the first shipment of new stock. Use the **MATRX EDIT** menu to store matrices. Press **ENTER** to select a matrix name, then key in the dimensions and the entries. The calculator will store the matrix until it is cleared or overwritten. Matrix names and entries appear in square brackets on the calculator screen.

Use the **MATRX NAMES** menu to copy the matrices into the expression for D, the matrix representing the stock on hand before the extra shipment. Just move the cursor to the matrix you need and press **ENTER**.

b) To find the stock on hand after the extra shipment for the one-day sale, multiply matrix D by 1.25 and store the result in matrix E. Then, you can use the round function in the **MATH NUM** menu to display the closest whole numbers for the entries in matrix E.

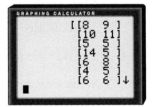

Solution 3 Using a Spreadsheet

a) You can easily perform matrix operations using a spreadsheet. It is also easy to add headings and row labels to keep track of what the entries represent. Enter each matrix using two adjacent columns: matrix A (initial stock) in columns **A** and **B**, matrix B (sales) in columns **C** and **D**, and matrix C (new stock) in columns **E** and **F**.

To find the amount of stock on hand after the first shipment from the warehouse, enter the formula A3−C3+E3 in cell **H3**.

Then, use the Fill feature to copy this formula for the rest of the entries in columns **H** and **I**.

b) Use the Fill feature in a similar way to copy the formula for the entries in matrix E, the stock on hand after the extra shipment from the warehouse. You can use the ROUND function to find the nearest whole number automatically. The formula for cell **J3**, the first entry, is ROUND(1.25*H3,0).

Microsoft Excel - Book2

File Edit View Insert Format Tools Data Window Help

Arial   10   **B** *I* <u>U</u> $ % ,   100%

J3   =   =ROUND(1.25*H3,0)

	A	B	C	D	E	F	G	H	I	J	K	L
1	Matrix A		Matrix B		Matrix C			Matrix D		Matrix E		
2												
3	5	8	3	5	4	4		6	7	8	9	
4	7	9	2	3	3	3		8	9	10	11	
5	4	3	4	3	4	4		4	4	5	5	
6	10	8	3	8	4	4		11	4	14	5	
7	3	1	3	0	5	5		5	6	6	8	
8	2	3	1	1	2	2		3	4	4	5	
9	7	5	4	2	2	2		5	5	6	6	
10	4	10	1	7	3	3		6	6	8	8	
11	1	2	0	1	1	1		2	2	3	3	

Key Concepts

- A matrix is used to manage and organize data.

- A matrix made up of m rows and n columns has dimensions $m \times n$.

- Two matrices are equal if they have the same dimensions and all corresponding entries are equal.

- The transpose matrix is found by interchanging rows with the corresponding columns.

- To add or subtract matrices, add or subtract the corresponding entries of each matrix. The dimensions of the matrices must be the same.

- To multiply a matrix by a scalar, multiply each entry of the matrix by the scalar.

Communicate Your Understanding

1. Describe how to determine the dimensions of any matrix.

2. Describe how you know whether two matrices are equal. Use an example to illustrate your answer.

3. Can transpose matrices ever be equal? Explain.

4. a) Describe how you would add two matrices. Give an example.
   b) Explain why the dimensions of the two matrices need to be the same to add or subtract them.

5. Describe how you would perform scalar multiplication on a matrix. Give an example.

Practise

A

1. State the dimensions of each matrix.

a) $\begin{bmatrix} 4 & 5 & -1 \\ -2 & 3 & 8 \end{bmatrix}$ **b)** $[1 \ \ 0 \ -7]$

c) $\begin{bmatrix} 3 & -9 & -6 \\ 5 & 4 & 7 \\ 1 & 0 & 8 \\ 8 & -1 & 2 \end{bmatrix}$

2. For the matrix $A = \begin{bmatrix} -5 & 3 & 2 \\ 6 & 0 & -1 \\ 4 & 8 & -3 \\ 7 & 1 & -4 \end{bmatrix}$,

a) state the value in entry

   i) a_{21} **ii)** a_{43} **iii)** a_{13}

b) state the entry with value

   i) 4   **ii)** -3 **iii)** 1

3. Let $A = \begin{bmatrix} a & b & c & d & e \\ f & g & h & i & j \\ k & l & m & n & o \\ p & q & r & s & t \end{bmatrix}$

and $B = \begin{bmatrix} u & v \\ w & x \\ y & z \end{bmatrix}$.

For each of the following, replace a_{ij} or b_{ij} with its corresponding entry in the above matrices to reveal a secret message.

a) $a_{33}a_{11}a_{45}a_{43}a_{24}a_{13}a_{15}a_{44}$
    $a_{11}a_{43}a_{15}$ $a_{21}b_{11}a_{34}$

b) a_{24} $a_{32}a_{35}b_{12}a_{15}$ $a_{33}a_{11}a_{45}a_{23}$

c) $b_{21}a_{35}b_{21}$ $a_{45}a_{23}a_{24}a_{44}$
    $a_{24}a_{44}$ $a_{21}b_{11}a_{34}$

4. a) Give two examples of row matrices and two examples of column matrices.

  b) State the dimensions of each matrix in part a).

5. a) Give two examples of square matrices.

  b) State the dimensions of each matrix in part a).

6. a) Write a 3×4 matrix, A, with the property that entry $a_{ij} = i + j$.

  b) Write a 4×4 matrix, B, with the property that entry $b_{ij} = \begin{cases} 3 \text{ if } i = j \\ i \times j \text{ if } i \neq j \end{cases}$

7. Solve for w, x, y, and z.

a) $\begin{bmatrix} x & 4 \\ -2 & 4z - 2 \end{bmatrix} = \begin{bmatrix} 3 & y - 1 \\ w & 6 \end{bmatrix}$

b) $\begin{bmatrix} w^3 & x^2 \\ 2y & 3z \end{bmatrix} = \begin{bmatrix} 8 & 9 \\ 8 - 2y & 2z - 5 \end{bmatrix}$

8. Let $A = \begin{bmatrix} 2 & -1 \\ 3 & 9 \\ 5 & 0 \\ -4 & 1 \end{bmatrix}$, $B = \begin{bmatrix} 3 & 4 \\ -6 & 1 \\ 8 & 2 \\ -1 & -5 \end{bmatrix}$,

and $C = \begin{bmatrix} 3 & -2 & 6 & 5 \\ 1 & 4 & 0 & -8 \end{bmatrix}$.

Calculate, if possible,

a) $A + B$ **b)** $B + A$ **c)** $B - C$

d) $3A$ **e)** $-\dfrac{1}{2}B$ **f)** $2(B - A)$

g) $3A - 2B$

9. Let $A = \begin{bmatrix} 8 & -6 \\ 1 & -2 \\ -4 & 5 \end{bmatrix}$, $B = \begin{bmatrix} 0 & -1 \\ 2 & 4 \\ 9 & -3 \end{bmatrix}$,

and $C = \begin{bmatrix} 2 & 3 \\ 8 & -6 \\ 4 & 1 \end{bmatrix}$.

Show that

a) $A + B = B + A$
   (commutative property)

b) $(A + B) + C = A + (B + C)$
   (associative property)

c) $5(A + B) = 5A + 5B$
   (distributive property)

10. Find the values of w, x, y, and z if

$$\begin{bmatrix} 5 & -1 & 2 \\ 4 & x & -8 \\ 7 & 0 & 3 \end{bmatrix} + 2\begin{bmatrix} 6 & y & 5 \\ -3 & 2 & 1 \\ 2 & -3 & z \end{bmatrix}$$

$$= \frac{1}{2}\begin{bmatrix} 34 & 10 & 24 \\ -4 & 24 & -12 \\ 2w & -12 & 42 \end{bmatrix}$$

11. Solve each equation.

a) $\begin{bmatrix} 3 & 2 & -5 \\ 2 & 0 & 8 \end{bmatrix} + A = \begin{bmatrix} 7 & 0 & 1 \\ -4 & 3 & -2 \end{bmatrix}$

b) $\begin{bmatrix} 5 & 7 \\ 4 & 0 \\ -1 & -3 \end{bmatrix} + y\begin{bmatrix} 1 & 6 \\ 0 & -4 \\ 2 & 5 \end{bmatrix} = \begin{bmatrix} 7 & 19 \\ 4 & -8 \\ 3 & 7 \end{bmatrix}$

Apply, Solve, Communicate

12. **Application** The map below shows driving distances between five cities in Ontario.

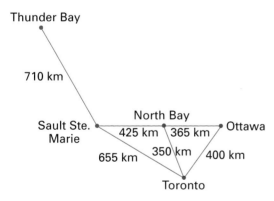

a) Represent the driving distances between each pair of cities with a matrix, A.

b) Find the transpose matrix, A^t.

c) Explain how entry a_{23} in matrix A and entry a_{32} in matrix A^t are related.

13. Nobel prizes are awarded for physics, chemistry, physiology/medicine, literature, peace, and economic sciences. The top five Nobel prize-winning countries are U.S.A. with 67 Nobel prizes in physics, 43 in chemistry, 78 in physiology/medicine, 10 in literature, 18 in peace, and 25 in economic sciences; U.K. with 21 Nobel prizes in physics, 25 in chemistry, 24 in physiology/medicine, 8 in literature, 13 in peace, and 7 in economic sciences; Germany with 20 Nobel prizes in physics, 27 in chemistry, 16 in physiology/medicine, 7 in literature, 4 in peace, and 1 in economic sciences; France with 12 Nobel prizes in physics, 7 in chemistry, 7 in physiology/medicine, 12 in literature, 9 in peace, and 1 in economic sciences; and Sweden with 4 Nobel prizes in physics, 4 in chemistry, 7 in physiology/medicine, 7 in literature, 5 in peace, and 2 in economic sciences.

a) Represent this data as a matrix, N. What are the dimensions of N?

b) Use row or column sums to calculate how many Nobel prizes have been awarded to citizens of each country.

14. The numbers of university qualifications (degrees, certificates, and diplomas) granted in Canada for 1997 are as follows: social sciences, 28 421 males and 38 244 females; education, 8036 males and 19 771 females; humanities, 8034 males and 13 339 females; health professions and occupations, 3460 males and 9613 females; engineering and applied sciences, 10 125 males and 2643 females; agriculture and biological sciences, 4780 males and 6995 females; mathematics and physical sciences, 6749 males and 2989 females; fine and applied arts, 1706 males and 3500 females; arts and sciences, 1730 males and 3802 females.

The numbers for 1998 are as follows: social sciences, 27 993 males and 39 026 females; education, 7565 males and 18 391 females; humanities, 7589 males and 13 227 females; health professions and occupations, 3514 males and 9144 females; engineering and applied sciences, 10 121 males and 2709 females; agriculture and biological sciences, 4779 males and 7430 females;

mathematics and physical sciences, 6876 males and 3116 females; fine and applied arts, 1735 males and 3521 females; arts and sciences, 1777 males and 3563 females.

a) Enter two matrices in a graphing calculator or spreadsheet—one two-column matrix for males and females receiving degrees in 1997 and a second two-column matrix for the number of males and females receiving degrees in 1998.

b) How many degrees were granted to males in 1997 and 1998 for each field of study?

c) How many degrees were granted to females in 1997 and 1998 for each field of study?

d) What is the average number of degrees granted to females in 1997 and 1998 for each field of study?

15. **Application** The table below shows the population of Canada by age and gender in the year 2000.

Age Group	Number of Males	Number of Females
0–4	911 028	866 302
5–9	1 048 247	996 171
10–14	1 051 525	997 615
15–19	1 063 983	1 007 631
20–24	1 063 620	1 017 566
25–29	1 067 870	1 041 900
30–34	1 154 071	1 129 095
35–39	1 359 796	1 335 765
40–44	1 306 705	1 304 538
45–49	1 157 288	1 162 560
50–54	1 019 061	1 026 032
55–59	769 591	785 657
60–64	614 659	641 914
65–69	546 454	590 435
70–74	454 269	544 008
75–79	333 670	470 694
80–84	184 658	309 748
85–89	91 455	190 960
90+	34 959	98 587

a) Create two matrices using the above data, one for males and another for females.

b) What is the total population for each age group?

c) Suppose that Canada's population grows by 1.5% in all age groups. Calculate the anticipated totals for each age group.

16. a) Prepare a matrix showing the connections for the VIA Rail routes shown on page 3. Use a 1 to indicate a direct connection from one city to another city. Use a 0 to indicate no direct connection from one city to another city. Also, use a 0 to indicate no direct connection from a city to itself.

b) What does the entry in row 4, column 3 represent?

c) What does the entry in row 3, column 4 represent?

d) Explain the significance of the relationship between your answers in parts b) and c).

e) Describe what the sum of the entries in the first row represents.

f) Describe what the sum of the entries in the first column represents.

g) Explain why your answers in parts e) and f) are the same.

17. **Inquiry/Problem Solving** Show that for any $m \times n$ matrices, A and B

a) $(A^t)^t = A$ b) $(A + B)^t = A^t + B^t$

18. **Communication** Make a table to compare matrix calculations with graphing calculators and with spreadsheets. What are the advantages, disadvantages, and limitations of these technologies?

19. **Inquiry/Problem Solving** Search the newspaper for data that could be organized in a matrix. What calculations could you perform with these data in matrix form? Is there any advantage to using matrices for these calculations?

1.7 Problem Solving With Matrices

The previous section demonstrated how to use matrices to model, organize, and manipulate data. With multiplication techniques, matrices become a powerful tool in a wide variety of applications.

INVESTIGATE & INQUIRE: Matrix Multiplication

The National Hockey League standings on March 9, 2001 in the Northeast Division are shown below along with the league's point system for a win, loss, tie, or overtime loss (OTL).

Team	Win	Loss	Tie	OTL
Ottawa	39	17	8	3
Buffalo	36	25	5	1
Toronto	31	23	10	5
Boston	28	27	6	7
Montréal	23	36	5	4

Score	Points
Win	2
Loss	0
Tie	1
OTL	1

1. Calculate the number of points for each team in the Northeast Division using the above tables. Explain your method.

2. **a)** Represent the team standings as a 5×4 matrix, A.

   b) Represent the points system as a column matrix, B.

3. Describe a procedure for determining the total points for Ottawa using the entries in row 1 of matrix A and column 1 of matrix B.

4. How could you apply this procedure to find the points totals for the other four teams?

5. Represent the total points for each team as a column matrix, C. How are the dimensions of C related to those of A and B?

6. Would it make sense to define matrix multiplication using a procedure such that $A \times B = C$? Explain your reasoning.

In the above investigation, matrix A has dimensions 5×4 and matrix B has dimensions 4×1. Two matrices can be multiplied when their inner dimensions are equal. The outer dimensions are the dimensions of the resultant matrix when matrices A and B are multiplied.

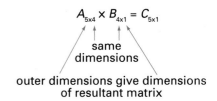

Example 1 Multiplying Matrices

Matrix A represents the proportion of students at a high school who have part-time jobs on Saturdays and the length of their shifts. Matrix B represents the number of students at each grade level.

$$A = \begin{array}{c} \text{Gr 9 \ Gr 10 \ Gr 11 \ Gr 12} \\ \begin{bmatrix} 0.20 & 0.10 & 0.20 & 0.15 \\ 0.25 & 0.30 & 0.25 & 0.45 \\ 0.05 & 0.25 & 0.15 & 0.10 \end{bmatrix} \begin{array}{l} \leq 4\,\text{h} \\ 4.1 - 6\,\text{h} \\ > 6\,\text{h} \end{array} \end{array}
\qquad
B = \begin{array}{c} \text{M \quad F} \\ \begin{bmatrix} 120 & 130 \\ 137 & 155 \\ 103 & 110 \\ 95 & 92 \end{bmatrix} \begin{array}{l} \text{Gr 9} \\ \text{Gr 10} \\ \text{Gr 11} \\ \text{Gr 12} \end{array} \end{array}$$

a) Calculate AB. Interpret what each entry represents.

b) Calculate BA, if possible.

Solution

a) A and B have the same inner dimensions, so multiplication is possible and their product will be a 3×2 matrix: $A_{3\times4} \times B_{4\times2} = C_{3\times2}$

$$AB = \begin{bmatrix} 0.20 & 0.10 & 0.20 & 0.15 \\ 0.25 & 0.30 & 0.25 & 0.45 \\ 0.05 & 0.25 & 0.15 & 0.10 \end{bmatrix} \begin{bmatrix} 120 & 130 \\ 137 & 155 \\ 103 & 110 \\ 95 & 92 \end{bmatrix}$$

$$= \begin{bmatrix} (0.20)(120)+(0.10)(137)+(0.20)(103)+(0.15)(95) & (0.20)(130)+(0.10)(155)+(0.20)(110)+(0.15)(92) \\ (0.25)(120)+(0.30)(137)+(0.25)(103)+(0.45)(95) & (0.25)(130)+(0.30)(155)+(0.25)(110)+(0.45)(92) \\ (0.05)(120)+(0.25)(137)+(0.15)(103)+(0.10)(95) & (0.05)(130)+(0.25)(155)+(0.15)(110)+(0.10)(92) \end{bmatrix}$$

$$\doteq \begin{bmatrix} 73 & 77 \\ 140 & 148 \\ 65 & 71 \end{bmatrix}$$

Approximately 73 males and 77 females work up to 4 h; 140 males and 148 females work 4−6 h, and 65 males and 71 females work more than 6 h on Saturdays.

b) For $B_{4\times2} \times A_{3\times4}$, the inner dimensions are not the same, so BA cannot be calculated.

Technology is an invaluable tool for solving problems that involve large amounts of data.

Example 2 Using Technology to Multiply Matrices

The following table shows the number and gender of full-time students enrolled at each university in Ontario one year.

University	Full-Time Students	Males (%)	Females (%)
Brock	6509	43	57
Carleton	12 376	55	45
Guelph	11 773	38	62
Lakehead	5308	48	52
Laurentian	3999	43	57
McMaster	13 797	46	54
Nipissing	1763	34	66
Ottawa	16 825	42	58
Queen's	13 433	44	56
Ryerson	10 266	47	53
Toronto	40 420	44	56
Trent	3764	36	64
Waterloo	17 568	55	45
Western	21 778	46	54
Wilfred Laurier	6520	45	55
Windsor	9987	46	54
York	27 835	39	61

a) Set up two matrices, one listing the numbers of full-time students at each university and the other the percents of males and females.

b) Determine the total number of full-time male students and the total number of full-time female students enrolled in Ontario universities.

Solution 1 Using a Graphing Calculator

a) Use the **MATRX EDIT** menu to store matrices for a 1×17 matrix for the numbers of full-time students and a 17×2 matrix for the percents of males and females.

b) To multiply matrices, use the **MATRX NAMES** menu. Copy the matrices into an expression such as [A]*[B] or [A][B].

There are 100 299 males and 123 622 females enrolled in Ontario universities.

You can also enter matrices directly into an expression by using the square brackets keys. This method is simpler for small matrices, but does not store the matrix in the MATRX NAMES menu.

Solution 2 Using a Spreadsheet

Enter the number of full-time students at each university as a 17×1 matrix in cells B2 to B18. This placement leaves you the option of putting labels in the first row and column. Enter the proportion of male and female students as a 2×17 matrix in cells D2 to T3.

Both Corel® Quattro® Pro and Microsoft® Excel have built-in functions for multiplying matrices, although the procedures in the two programs differ somewhat.

Corel® Quattro® Pro:
On the Tools menu, select Numeric Tools/Multiply. In the pop-up window, enter the cell ranges for the two matrices you want to multiply and the cell where you want the resulting matrix to start. Note that you must list the 2×17 matrix first.

A:D2		▼ @ { }	0.43					
	A	**B**	**C**	**D**	**E**	**F**	**G**	
1	**University**	**Full-time Students**	**Percent**					
2	Brock	6509	Males	43%	55%	38%	48%	
3	Carleton	12376	Females	57%	45%	62%	52%	
4	Guelph	11773						
5	Lakehead	5308						
6	Laurentian	3999						
7	McMaster	13797						
8	Nipissing	1763						
9	Ottawa	16825						
10	Queen's	13433						
11	Ryerson	10266						
12	Toronto	40420						
13	Trent	3764						
14	Waterloo	17568						
15	Western	21778						
16	Wilfred Laurier	6520						

Matrix Multiply

Matrix 1:
A:D2..T3

Matrix 2:
A:B2..B18

Destination:
A:C20

OK
Cancel
Help

Project Prep

You can apply these techniques for matrix multiplication to the calculations for your tools for data management project.

Microsoft® Excel:
The MMULT(matrix1,matrix2) function will calculate the product of the two matrices but displays only the first entry of the resulting matrix. Use the INDEX function to retrieve the entry for a specific row and column of the matrix.

	A	B	C	D	E	F	G	H	I	J
1	University	Full-time Students	Percent							
2	Brock	6509	Males	43%	55%	38%	48%	43%	46%	34%
3	Carleton	12376	Females	57%	45%	62%	52%	57%	54%	66%
4	Guelph	11773								
5	Lakehead	5308								
6	Laurentian	3999								
7	McMaster	13797								
8	Nipissing	1763								
9	Ottawa	16825								
10	Queen's	13433								
11	Ryerson	10266								
12	Toronto	40420								
13	Trent	3764								
14	Waterloo	17568								
15	Western	21778								
16	Wilfred Laurier	6520								
17	Windsor	9987								
18	York	27835								
19										
20	Totals	Males		100298.7						
21		Females		123622.3						
22										

Identity matrices have the form $I = \begin{bmatrix} 1 & 0 & 0 & 0 & \dots & 0 \\ 0 & 1 & 0 & 0 & \dots & 0 \\ 0 & 0 & 1 & 0 & \dots & 0 \\ 0 & 0 & 0 & 1 & \dots & 0 \\ \vdots & \vdots & \vdots & \vdots & \vdots & \vdots \\ 0 & 0 & 0 & 0 & \dots & 1 \end{bmatrix}$ with entries

of 1 along the main diagonal and zeros for all other entries. The identity matrix with dimensions $n \times n$ is represented by I_n. It can easily be shown that $A_{m \times n} I_n = A_{m \times n}$ for any $m \times n$ matrix A.

For most square matrices, there exists an **inverse matrix** A^{-1} with the property that $AA^{-1} = A^{-1}A = I$. Note that $A^{-1} \neq \dfrac{1}{A}$.

For 2×2 matrices, $AA^{-1} = \begin{bmatrix} a & b \\ c & d \end{bmatrix} \begin{bmatrix} w & x \\ y & z \end{bmatrix} = \begin{bmatrix} 1 & 0 \\ 0 & 1 \end{bmatrix}$

Multiplying the matrices gives four simultaneous equations for w, x, y, and z. Solving these equations yields $A^{-1} = \dfrac{1}{ad - bc} \begin{bmatrix} d & -b \\ -c & a \end{bmatrix}$. You can confirm that $A^{-1}A = I$, also. If $ad = bc$, then A^{-1} does not exist since it would require dividing by zero.

The formulas for the inverses of larger matrices can be determined in the same way as for 2×2 matrices, but the calculations become much more involved. However, it is relatively easy to find the inverses of larger matrices with graphing calculators since they have the necessary formulas built in.

Example 3 Calculating the Inverse Matrix

Calculate, if possible, the inverse of

a) $A = \begin{bmatrix} 3 & 7 \\ 4 & -2 \end{bmatrix}$ **b)** $B = \begin{bmatrix} 6 & 8 \\ 3 & 4 \end{bmatrix}$

Solution 1   Using Pencil and Paper

a) $A^{-1} = \dfrac{1}{ad - bc} \begin{bmatrix} d & -b \\ -c & a \end{bmatrix}$

$\qquad = \dfrac{1}{(3)(-2) - (7)(4)} \begin{bmatrix} -2 & -7 \\ -4 & 3 \end{bmatrix}$

$\qquad = -\dfrac{1}{34} \begin{bmatrix} -2 & -7 \\ -4 & 3 \end{bmatrix}$

$\qquad = \begin{bmatrix} \dfrac{1}{17} & \dfrac{7}{34} \\ \dfrac{2}{17} & -\dfrac{3}{34} \end{bmatrix}$

b) For B, $ad - bc = (6)(4) - (8)(3) = 0$, so B^{-1} does not exist.

Solution 2   Using a Graphing Calculator

a) Use the **MATRX EDIT** menu to store the 2×2 matrix. Retrieve it with the **MATRX NAMES** menu, then use x^{-1} to find the inverse. To verify that the decimal numbers shown are equal to the fractions in the pencil-and-paper solution, use the ▶Frac function from the **MATH NUM** menu.

b) For B, the calculator shows that the inverse cannot be calculated.

Solution 3   Using a Spreadsheet

The spreadsheet functions for inverse matrices are similar to those for matrix multiplication.

a) Enter the matrix in cells A1 to B2.

In Corel® Quattro® Pro, use Tools/Numeric Tools/Invert… to enter the range of cells for the matrix and the cell where you want the inverse matrix to start. Use the Fraction feature to display the entries as fractions rather than decimal numbers.

In Microsoft® Excel, use the MINVERSE function to produce the inverse matrix and the INDEX function to access the entries in it. If you put absolute cell references in the MINVERSE function for the first entry, you can use the Fill feature to generate the formulas for the other entries.
Use the Fraction feature to display the entries as fractions rather than decimal numbers.

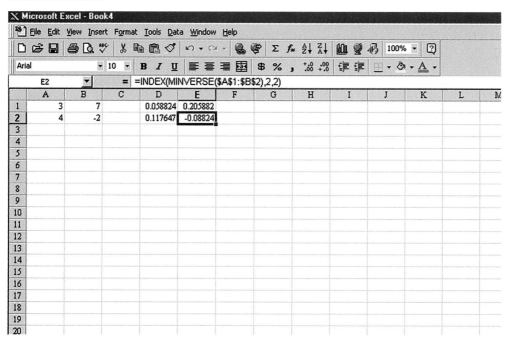

During the 1930s, Lester Hill, an American mathematician, developed methods for using matrices to decode messages. The following example illustrates a simplified version of Hill's technique.

Example 4 Coding a Message Using Matrices

a) Encode the message PHONE ME TONIGHT using 2×2 matrices.

b) Determine the matrix key required to decode the message.

Solution

a) Write the message using 2×2 matrices. Fill in any missing entries with the letter Z.

$$\begin{bmatrix} P & H \\ O & N \end{bmatrix}, \begin{bmatrix} E & M \\ E & T \end{bmatrix}, \begin{bmatrix} O & N \\ I & G \end{bmatrix}, \begin{bmatrix} H & T \\ Z & Z \end{bmatrix}$$

Replace each letter with its corresponding number in the alphabet.

A	B	C	D	E	F	G	H	I	J	K	L	M
1	2	3	4	5	6	7	8	9	10	11	12	13

N	O	P	Q	R	S	T	U	V	W	X	Y	Z
14	15	16	17	18	19	20	21	22	23	24	25	26

$$\begin{bmatrix} 16 & 8 \\ 15 & 14 \end{bmatrix}, \begin{bmatrix} 5 & 13 \\ 5 & 20 \end{bmatrix}, \begin{bmatrix} 15 & 14 \\ 9 & 7 \end{bmatrix}, \begin{bmatrix} 8 & 20 \\ 26 & 26 \end{bmatrix}$$

Now, encode the message by multiplying with a **coding matrix** that only the sender and receiver know. Suppose that you chose $C = \begin{bmatrix} 3 & 1 \\ 5 & 2 \end{bmatrix}$ as your coding matrix.

$$\begin{bmatrix} 3 & 1 \\ 5 & 2 \end{bmatrix}\begin{bmatrix} 16 & 8 \\ 15 & 14 \end{bmatrix} = \begin{bmatrix} 63 & 38 \\ 110 & 68 \end{bmatrix}$$

$$\begin{bmatrix} 3 & 1 \\ 5 & 2 \end{bmatrix}\begin{bmatrix} 5 & 13 \\ 5 & 20 \end{bmatrix} = \begin{bmatrix} 20 & 59 \\ 35 & 105 \end{bmatrix}$$

$$\begin{bmatrix} 3 & 1 \\ 5 & 2 \end{bmatrix}\begin{bmatrix} 15 & 14 \\ 9 & 7 \end{bmatrix} = \begin{bmatrix} 54 & 49 \\ 93 & 84 \end{bmatrix}$$

$$\begin{bmatrix} 3 & 1 \\ 5 & 2 \end{bmatrix}\begin{bmatrix} 8 & 20 \\ 26 & 26 \end{bmatrix} = \begin{bmatrix} 50 & 86 \\ 92 & 152 \end{bmatrix}$$

You would send the message as 63, 38, 110, 68, 20, 59, 35, 105, 54, 49, 93, 84, 50, 86, 92, 152.

b) First, rewrite the coded message as 2×2 matrices.

$$\begin{bmatrix} 63 & 38 \\ 110 & 68 \end{bmatrix}, \begin{bmatrix} 20 & 59 \\ 35 & 105 \end{bmatrix}, \begin{bmatrix} 54 & 49 \\ 93 & 84 \end{bmatrix}, \begin{bmatrix} 50 & 86 \\ 92 & 152 \end{bmatrix}$$

You can decode the message with the inverse matrix for the coding matrix.
$$C^{-1} \times CM = C^{-1}C \times M = IM = M$$
where M is the message matrix and C is the coding matrix.
Thus, the **decoding matrix,** or key, is the inverse matrix of the coding matrix. For the coding matrix used in part a), the key is

$$\begin{bmatrix} 3 & 1 \\ 5 & 2 \end{bmatrix}^{-1} = \frac{1}{(3)(2) - (5)(1)} \begin{bmatrix} 2 & -1 \\ -5 & 3 \end{bmatrix} = \begin{bmatrix} 2 & -1 \\ -5 & 3 \end{bmatrix}$$

Multiplying the coded message by this key gives

$$\begin{bmatrix} 2 & -1 \\ -5 & 3 \end{bmatrix} \begin{bmatrix} 63 & 38 \\ 110 & 68 \end{bmatrix} = \begin{bmatrix} 16 & 8 \\ 15 & 14 \end{bmatrix}$$

$$\begin{bmatrix} 2 & -1 \\ -5 & 3 \end{bmatrix} \begin{bmatrix} 20 & 59 \\ 35 & 105 \end{bmatrix} = \begin{bmatrix} 5 & 13 \\ 5 & 20 \end{bmatrix}$$

$$\begin{bmatrix} 2 & -1 \\ -5 & 3 \end{bmatrix} \begin{bmatrix} 54 & 49 \\ 93 & 84 \end{bmatrix} = \begin{bmatrix} 15 & 14 \\ 9 & 7 \end{bmatrix}$$

$$\begin{bmatrix} 2 & -1 \\ -5 & 3 \end{bmatrix} \begin{bmatrix} 50 & 86 \\ 92 & 152 \end{bmatrix} = \begin{bmatrix} 8 & 20 \\ 26 & 26 \end{bmatrix}$$

The decoded message is 16, 8, 15, 14, 5, 13, 5, 20, 15, 14, 9, 7, 8, 20, 26, 26.
Replacing each number with its corresponding letter in the alphabet gives
PHONEMETONIGHTZZ, the original message with the two Zs as fillers.

Matrix multiplication and inverse matrices are the basis for many computerized encryption systems like those used for electronic transactions between banks and income tax returns filed over the Internet.

Transportation and communication networks can be represented using matrices, called **network matrices**. Such matrices provide information on the number of direct links between two **vertices** or points (such as people or places). The advantage of depicting networks using matrices is that information on indirect routes can be found by performing calculations with the network matrix.

To construct a network matrix, let each vertex (point) be represented as a row and as a column in the matrix. Use 1 to represent a direct link and 0 to represent no direct link. A vertex may be linked to another vertex in one direction or in both directions. Assume that a vertex does not link with itself, so each entry in the main diagonal is 0. Note that the network matrix provides information only on direct links.

Example 5 Using Matrices to Model a Network

Matrixville Airlines offers flights between eight cities as shown on the right.

a) Represent the network using a matrix, A. Organize the matrix so the cities are placed in alphabetical order.

b) Calculate A^2. What information does it contain?

c) How many indirect routes with exactly one change of planes are there from London to Buenos Aires?

d) Calculate $A + A^2$. What information does it contain?

e) Explain what the entry from Vancouver to Paris in $A + A^2$ represents.

f) Calculate A^3. Compare this calculation with the one for A^2.

g) Explain the significance of any entry in matrix A^3.

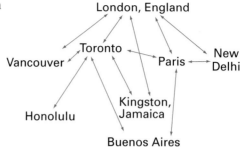

Solution

a)

$$
A = \begin{matrix}
 & \begin{matrix} B & H & K & L & N & P & T & V \end{matrix} & \\
\left[\begin{matrix}
0 & 0 & 0 & 0 & 0 & 1 & 1 & 0 \\
0 & 0 & 0 & 0 & 0 & 0 & 1 & 0 \\
0 & 0 & 0 & 1 & 0 & 0 & 1 & 0 \\
0 & 0 & 1 & 0 & 1 & 1 & 1 & 1 \\
0 & 0 & 0 & 1 & 0 & 1 & 0 & 0 \\
1 & 0 & 0 & 1 & 1 & 0 & 1 & 0 \\
1 & 1 & 1 & 1 & 0 & 1 & 0 & 1 \\
0 & 0 & 0 & 1 & 0 & 0 & 1 & 0 \\
\end{matrix} \right] &
\begin{matrix} B \\ H \\ K \\ L \\ N \\ P \\ T \\ V \end{matrix}
\end{matrix}
$$

b) Since the dimensions of matrix A are 8×8, you may prefer to use a calculator or software for this calculation.

$$
A^2 = \begin{bmatrix}
2 & 1 & 1 & 2 & 1 & 1 & 1 & 1 \\
1 & 1 & 1 & 1 & 0 & 1 & 0 & 1 \\
1 & 1 & 2 & 1 & 1 & 2 & 1 & 2 \\
2 & 1 & 1 & 5 & 1 & 2 & 3 & 1 \\
1 & 0 & 1 & 1 & 2 & 1 & 2 & 1 \\
1 & 1 & 2 & 2 & 1 & 4 & 2 & 2 \\
1 & 0 & 1 & 3 & 2 & 2 & 6 & 2 \\
1 & 1 & 2 & 1 & 1 & 2 & 1 & 2 \\
\end{bmatrix}
$$

The entries in A^2 show the number of indirect routes with exactly one change of planes. A^2 does not contain any information on direct routes.

c) There are two indirect routes with exactly one change of planes from London to Buenos Aires.

London → Paris → Buenos Aires
London → Toronto → Buenos Aires

d) $A + A^2 = \begin{bmatrix} 2 & 1 & 1 & 2 & 1 & 2 & 2 & 1 \\ 1 & 1 & 1 & 1 & 0 & 1 & 1 & 1 \\ 1 & 1 & 2 & 2 & 1 & 2 & 2 & 2 \\ 2 & 1 & 2 & 5 & 2 & 3 & 4 & 2 \\ 1 & 0 & 1 & 2 & 2 & 2 & 2 & 1 \\ 2 & 1 & 2 & 3 & 2 & 4 & 3 & 2 \\ 2 & 1 & 2 & 4 & 2 & 3 & 6 & 2 \\ 1 & 1 & 2 & 2 & 1 & 2 & 2 & 2 \end{bmatrix}$

Since A shows the number of direct routes and A^2 shows the number of routes with one change of planes, $A + A^2$ shows the number of routes with at most one change of planes.

e) The entry in row 8, column 6 of $A + A^2$ shows that there are two routes with a maximum of one change of planes from Vancouver to Paris.

Vancouver → Toronto → Paris
Vancouver → London → Paris

f) $A^3 = \begin{bmatrix} 2 & 1 & 3 & 5 & 3 & 6 & 8 & 3 \\ 1 & 0 & 1 & 3 & 2 & 2 & 6 & 1 \\ 3 & 1 & 2 & 8 & 3 & 4 & 9 & 2 \\ 5 & 3 & 8 & 8 & 7 & 11 & 12 & 8 \\ 3 & 2 & 3 & 7 & 2 & 6 & 5 & 3 \\ 6 & 2 & 4 & 11 & 6 & 6 & 12 & 4 \\ 8 & 6 & 9 & 12 & 5 & 12 & 8 & 9 \\ 3 & 1 & 2 & 8 & 3 & 4 & 9 & 2 \end{bmatrix}$

The calculation of $A^3 = A^2 \times A$ is more laborious than that for $A^2 = A \times A$ since A^2 has substantially fewer zero entries than A does. A calculator or spreadsheet could be useful.

g) The entries in A^3 tell you the number of indirect routes with exactly two changes of planes between each pair of cities.

- To multiply two matrices, their inner dimensions must be the same. The outer dimensions give the dimensions of the resultant matrix: $A_{m \times n} \times B_{n \times p} = C_{m \times p}$. To find the entry with row i and column j of matrix AB, multiply the entries of row i of matrix A with the corresponding entries of column j of matrix B, and then add the resulting products together.

- The inverse of the 2×2 matrix $A = \begin{bmatrix} a & b \\ c & d \end{bmatrix}$ is $A^{-1} = \dfrac{1}{ad - bc} \begin{bmatrix} d & -b \\ -c & a \end{bmatrix}$ provided that $ad \neq bc$. Larger inverse matrices can be found using a graphing calculator or a spreadsheet.

- To represent a network as a matrix, use a 1 to indicate a direct link and a 0 to indicate no direct link. Calculations with the square of a network matrix and its higher powers give information on the various direct and indirect routings possible.

Communicate Your Understanding

1. Explain how multiplying matrices is different from scalar multiplication of matrices.

2. Describe the steps you would take to multiply $\begin{bmatrix} 4 & -2 \\ 1 & 5 \end{bmatrix} \begin{bmatrix} 3 & 6 & 2 & -1 \\ 0 & 4 & 5 & 7 \end{bmatrix}$.

3. Is it possible to find an inverse for a matrix that is not square? Why or why not?

4. Explain why a network matrix must be square.

5. Describe how you would represent the following network as a matrix. How would you find the number of routes with up to three changeovers?

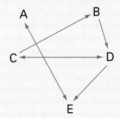

Practise

A

1. Let $A = \begin{bmatrix} 4 & 7 & 0 \\ -3 & -5 & 1 \end{bmatrix}$, $B = \begin{bmatrix} 2 & 9 \\ -7 & 0 \end{bmatrix}$,

$C = \begin{bmatrix} 1 & 5 & 8 \\ 2 & 0 & -4 \\ -3 & -2 & 8 \end{bmatrix}$, $D = \begin{bmatrix} -3 & 1 \\ 2 & 5 \end{bmatrix}$, $E = \begin{bmatrix} 1 \\ 2 \\ -3 \end{bmatrix}$.

Calculate, if possible,

a) BD b) DB c) B^2 d) EA

e) AC f) CE g) DA

2. Given $A = \begin{bmatrix} 4 & 2 \\ 0 & -1 \end{bmatrix}$ and $B = \begin{bmatrix} 0 & 3 \\ -2 & 0 \end{bmatrix}$, show

that $A^2 + 2B^3 = \begin{bmatrix} 16 & -30 \\ 24 & 1 \end{bmatrix}$.

3. If $A = \begin{bmatrix} 0 & 1 \\ 0 & 0 \end{bmatrix}$, show that $A^4 = \begin{bmatrix} 0 & 0 \\ 0 & 0 \end{bmatrix}$, the 2×2 **zero matrix**.

4. Let $A = \begin{bmatrix} 5 & 0 \\ 2 & -1 \end{bmatrix}$, $B = \begin{bmatrix} 3 & 4 \\ -2 & 0 \end{bmatrix}$, $C = \begin{bmatrix} 1 & -3 \\ 0 & 7 \end{bmatrix}$.

Show that

a) $A(B + C) = AB + AC$
(distributive property)

b) $(AB)C = A(BC)$
(associative property)

c) $AB \neq BA$
(not commutative)

5. Find the inverse matrix, if it exists.

a) $\begin{bmatrix} 0 & -1 \\ 2 & 4 \end{bmatrix}$ b) $\begin{bmatrix} 4 & -6 \\ -2 & 3 \end{bmatrix}$ c) $\begin{bmatrix} 3 & 0 \\ -6 & 1 \end{bmatrix}$

d) $\begin{bmatrix} 5 & 3 \\ 4 & 2 \end{bmatrix}$ e) $\begin{bmatrix} 10 & 5 \\ 4 & 2 \end{bmatrix}$

6. Use a graphing calculator or a spreadsheet to calculate the inverse matrix, if it exists.

a) $A = \begin{bmatrix} 1 & -3 & 1 \\ -2 & 1 & 3 \\ 0 & -1 & 0 \end{bmatrix}$

b) $B = \begin{bmatrix} -2 & 0 & 5 \\ 2 & -1 & -1 \\ 3 & 4 & 0 \end{bmatrix}$

c) $C = \begin{bmatrix} 2 & -1 & 1 & 0 \\ 0 & 1 & 0 & 2 \\ -2 & -1 & 0 & 0 \\ 1 & 0 & -1 & 0 \end{bmatrix}$

Apply, Solve, Communicate

B

7. For $A = \begin{bmatrix} 2 & -4 \\ -2 & 5 \end{bmatrix}$ and $B = \begin{bmatrix} 5 & 7 \\ 3 & 4 \end{bmatrix}$, show that

a) $(A^{-1})^{-1} = A$

b) $(AB)^{-1} = B^{-1}A^{-1}$

c) $(A^t)^{-1} = (A^{-1})^t$

8. Application Calculators Galore has three stores in Matrixville. The downtown store sold 12 business calculators, 40 scientific calculators, and 30 graphing calculators during the past week. The northern store sold 8 business calculators, 30 scientific calculators, and 21 graphing calculators during the same week, and the southern store sold 10 business calculators, 25 scientific calculators, and 23 graphing calculators. What were the total weekly sales for each store if the average price of a business calculator is $40, a scientific calculator is $30, and a graphing calculator is $150?

9. Application The manager at Sue's Restaurant prepares the following schedule for the next week.

Employee	Mon.	Tues.	Wed.	Thurs.	Fri.	Sat.	Sun.	Wage Per Hour
Chris	–	8	–	8	8	–	–	$7.00
Lee	4	4	–	–	6.5	4	4	$6.75
Jagjeet	–	4	4	4	4	8	8	$7.75
Pierre	–	3	3	3	3	8	–	$6.75
Ming	8	8	8	8	–	–	–	$11.00
Bobby	–	–	3	5	5	8	–	$8.00
Nicole	3	3	3	3	3	–	–	$7.00
Louis	8	8	8	8	8	–	–	$12.00
Glenda	8	–	–	8	8	8	8	$13.00
Imran	3	4.5	4	3	5	–	–	$7.75

a) Create matrix A to represent the number of hours worked per day for each employee.

b) Create matrix B to represent the hourly wage earned by each employee.

c) Use a graphing calculator or spreadsheet to calculate the earnings of each employee for the coming week.

d) What is the restaurant's total payroll for these employees?

10. According to a 1998 general social survey conducted by Statistics Canada, the ten most popular sports for people at least 15 years old are as follows:

Sport	Total (%)	Male (%)	Female (%)
Golf	7.4	11.1	3.9
Ice Hockey	6.2	12.0	0.5
Baseball	5.5	8.0	3.1
Swimming	4.6	3.6	5.6
Basketball	3.2	4.6	1.9
Volleyball	3.1	3.3	2.8
Soccer	3.0	4.6	1.5
Tennis	2.7	3.6	1.8
Downhill/Alpine Skiing	2.7	2.9	2.6
Cycling	2.5	3.0	2.0

In 1998, about 11 937 000 males and 12 323 000 females in Canada were at least 15 years old. Determine how many males and how many females declared each of the above sports as their favourite. Describe how you used matrices to solve this problem.

11. **Application** A company manufacturing designer T-shirts produces five sizes: extra-small, small, medium, large, and extra-large. The material and labour needed to produce a box of 100 shirts depends on the size of the shirts.

Size	Cloth per shirt (m²)	Labour per 100 shirts (h)
Extra-small	0.8	8
Small	0.9	8.5
Medium	1.2	9
Large	1.5	10
Extra-large	2.0	11

a) How much cloth and labour are required to fill an order for 1200 small, 1500 medium, 2500 large, and 2000 extra-large T-shirts?

b) If the company pays $6.30 per square metre for fabric and $10.70 per hour for labour, find the cost per box for each size of T-shirt.

c) What is the total cost of cloth and labour for filling the order in part a)?

12. Use the coding matrix $\begin{bmatrix} 2 & -3 \\ -2 & 5 \end{bmatrix}$ to encode each message.

a) BIRTHDAY PARTY FRIDAY

b) SEE YOU SATURDAY NIGHT

13. **Application** Use the decoding matrix $\begin{bmatrix} -1 & 2 \\ 2 & -3 \end{bmatrix}$ to decode each message.

a) 64, 69, 38, 45, 54, 68, 31, 44, 5, 115, 3, 70, 40, 83, 25, 49

b) 70, 47, 39, 31, 104, 45, 61, 25, 93, 68, 57, 44, 55, 127, 28, 76

14. a) Create a secret message about 16 to 24 letters long using the coding matrix $\begin{bmatrix} 3 & 5 \\ 1 & 2 \end{bmatrix}$.

b) Trade messages with a classmate and decode each other's messages.

15. Quality education at a school requires open communication among many people.

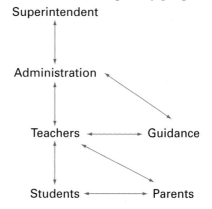

a) Represent this network as a matrix, A.

b) Explain the meaning of any entry, a_{ij}, of matrix A.

c) Describe what the sum of the entries in the third column represents.

d) Calculate A^2.

e) How many indirect links exist with exactly one intermediary between the principal and parents? List these links.

f) Calculate $A + A^2$. Explain what information this matrix provides.

16. Network matrices provide another approach to the Koenigsberg bridges example on page 44.

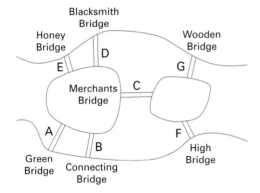

Blacksmith Bridge

Honey Bridge

Wooden Bridge

D

E

G

Merchants Bridge C

A

F

B

Green Bridge

Connecting Bridge

High Bridge

Use network matrices to answer the following questions.

a) How many ways can you get from Honey Bridge to Connecting Bridge by crossing only one of the other bridges? List these routes.

b) How many ways can you get from Blacksmith Bridge to Connecting Bridge without crossing more than one of the other bridges?

c) Is it possible to travel from Wooden Bridge to Green Bridge without crossing at least two other bridges?

17. Use network matrices to find the number of VIA Rail routes from

a) Toronto to Montréal with up to two change-overs

b) Kingston to London with up to three change-overs

18. Inquiry/Problem Solving Create your own network problem, then exchange problems with a classmate. Solve both problems and compare your solutions with those of your classmate. Can you suggest any improvements for either set of solutions?

19. Show how you could use inverse matrices to solve any system of equations in two variables whose matrix of coefficients has an inverse.

20. Communication Research encryption techniques on the Internet. What is meant by 128-bit encryption? How does the system of private and public code keys work?

21. Inquiry/Problem Solving

a) Suppose you receive a coded message like the one in Example 4, but you do not know the coding matrix or its inverse. Describe how you could use a computer to break the code and decipher the message.

b) Describe three methods you could use to make a matrix code harder to break.

22. a) Show that, for any $m \times n$ matrix A and any $n \times p$ matrix B, $(AB)^t = B^t A^t$.

b) Show that, if a square matrix C has an inverse C^{-1}, then C^t also has an inverse, and $(C^t)^{-1} = (C^{-1})^t$.

Review of Key Concepts

1.1 The Iterative Process
Refer to the Key Concepts on page 10.

1. **a)** Draw a tree diagram showing your direct ancestors going back four generations.

   b) How many direct ancestors do you have in four generations?

2. **a)** Describe the algorithm used to build the iteration shown.

   b) Continue the iteration for eight more rows.

   c) Describe the resulting iteration.

```
              MATH
            MATHMATH
        MATH        MATH
       MATHMATHMATHMATH
      MATH              MATH
     MATHMATH          MATHMATH
```

3. **a)** Construct a Pythagoras fractal tree using the following algorithm.
   Step 1: Construct a square.
   Step 2: Construct an isosceles right triangle with the hypotenuse on one side of the square.
   Step 3: Construct a square on each of the other sides of the triangle.
   Repeat this process, with the newly drawn squares to a total of four iterations.

   b) If the edges in the first square are 4 cm, determine the total area of all the squares in the fourth iteration.

   c) Determine the total area of all the squares in the diagram.

4. Design an iterative process using the percent reduction capabilities of a photocopier.

1.2 Data Management Software
Refer to the Key Concepts on page 21.

5. List three types of software that can be used for data management, giving an example of the data analysis you could do with each type.

6. Evaluate each spreadsheet expression.
   a) F2+G7−A12
   where F2=5, G7=−9, and A12=F2+G7
   b) PROD(D3,F9)
   where D3=6 and F9=5
   c) SQRT(B1)
   where B1=144

7. Describe how to reference cells A3 to A10 in one sheet of a spreadsheet into cells B2 to B9 in another sheet.

8. Use a spreadsheet to convert temperatures between −30° C and 30° C to the Fahrenheit scale, using the formula Fahrenheit = 1.8 × Celsius + 32. Describe how you would list temperatures at two-degree intervals in the Celsius column.

1.3 Databases
Refer to the Key Concepts on page 31.

9. Describe the characteristics of a well-organized database.

10. Outline a design for a database of a shoe store's customer list.

11. **a)** Describe the types of data that are available from Statistics Canada's E-STAT database.

    b) What can you do with the data once you have accessed them?

12. What phrase would you enter into a search engine to find

a) the top-selling cookbook in Canada?

b) the first winner of the Fields medal?

c) a list of movies in which bagpipes are played?

1.4 Simulations

Refer to the Key Concepts on page 39.

13. List three commonly used simulations and a reason why each is used.

14. Write out the function to generate a random integer between 18 and 65 using

a) a graphing calculator

b) a spreadsheet

15. A chocolate bar manufacturer prints one of a repeating sequence of 50 brainteasers on the inside of the wrapper for each of its chocolate bars. Describe a manual simulation you could use to estimate the chances of getting two chocolate bars with the same brainteaser if you treat yourself to one of the bars every Friday for five weeks.

16. Outline how you would use technology to run a simulation 500 times for the scenario in question 15.

1.5 Graph Theory

Refer to the Key Concepts on page 48.

17. How many colours are needed to colour each of the following maps?

a)

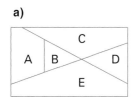

b)

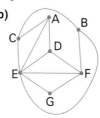

18. State whether each network is

i) connected

ii) traceable

iii) planar

a)

b)

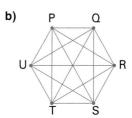

c)

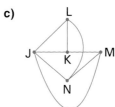

19. For each network in question 18, verify that $V - E + R = 2$, where V is the number of vertices, E is the number of edges, and R is the number of regions in a graph.

20. The following is a listing of viewing requests submitted by patrons of a classic film festival. Use graph theory to set up the shortest viewing schedule that has no conflicts for any of these patrons.
Person A: *Gone With the Wind, Curse of The Mummy, Citizen Kane*
Person B: *Gone With the Wind, Jane Eyre*
Person C: *The Amazon Queen, West Side Story, Citizen Kane*
Person D: *Jane Eyre, Gone With the Wind, West Side Story*
Person E: *The Amazon Queen, Ben Hur*

21. Below is a network showing the relationships among a group of children. The vertices are adjacent if the children are friends.

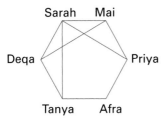

Sarah   Mai

Deqa

Priya

Tanya   Afra

a) Rewrite the network in table form.

b) Are these children all friends with each other?

c) Who has the most friends?

d) Who has the fewest friends?

1.6 Modelling With Matrices

Refer to the Key Concepts on page 59.

22. For the matrix $A = \begin{bmatrix} 2 & -1 & 5 \\ 0 & 4 & 3 \\ 7 & -8 & -6 \\ -2 & 9 & 1 \end{bmatrix}$,

a) state the dimensions

b) state the value of entry

i) a_{32} ii) a_{13} iii) a_{41}

c) list the entry with value

i) 3   ii) 9   iii) −1

23. Write a 4 × 3 matrix, A, with the property that $a_{ij} = i \times j$ for all entries.

24. Given $A = \begin{bmatrix} 3 & 2 & -1 \\ -7 & 0 & 5 \end{bmatrix}$, $B = \begin{bmatrix} 8 & -2 \\ 3 & 4 \\ 2 & 5 \end{bmatrix}$,

$C = \begin{bmatrix} 6 & 1 & -4 \\ -5 & 9 & 0 \end{bmatrix}$, and $D = \begin{bmatrix} 4 & 3 \\ -1 & 7 \\ 6 & 2 \end{bmatrix}$.

Calculate, if possible,

a) $A + C$ b) $C - B$

c) $A + B$ d) $3D$

e) $-\dfrac{1}{2}C$ f) $3(B + D)$

g) $A^t + B$ h) $B^t + C^t$

25. The manager of a sporting goods store takes inventory at the end of the month and finds 15 basketballs, 17 volleyballs, 4 footballs, 15 baseballs, 8 soccer balls, 12 packs of tennis balls, and 10 packs of golf balls. The manager orders and receives a shipment of 10 basketballs, 3 volleyballs, 15 footballs, 20 baseballs, 12 soccer balls, 5 packs of tennis balls, and 15 packs of golf balls. During the next month, the store sells 17 basketballs, 13 volleyballs, 17 footballs, 12 baseballs, 12 soccer balls, 16 packs of tennis balls, and 23 packs of golf balls.

a) Represent the store's stock using three matrices, one each for the inventory, new stock received, and items sold.

b) How many of each item is in stock at the end of the month?

c) At the beginning of the next month, the manager is asked to send 20% of the store's stock to a new branch that is about to open. How many of each item will be left at the manager's store?

26. Outline the procedure you would use to subtract one matrix from another

a) manually

b) using a graphing calculator

c) using a spreadsheet

1.7 Problem Solving With Matrices

Refer to the Key Concepts on page 74.

27. Let $A = \begin{bmatrix} 4 & 5 \\ -6 & 3 \end{bmatrix}$, $B = \begin{bmatrix} 1 & 0 \\ -5 & 7 \end{bmatrix}$,

$C = \begin{bmatrix} 3 & 6 & -1 \\ 2 & 0 & 4 \\ -5 & -2 & 8 \end{bmatrix}$, and $D = \begin{bmatrix} 5 \\ 4 \\ -3 \end{bmatrix}$.

Calculate, if possible,

a) AB **b)** BA **c)** A^2

d) DC **e)** C^2

28. a) Write the transpose of matrices

$A = \begin{bmatrix} 1 & 5 \\ 8 & -2 \end{bmatrix}$ and $B = \begin{bmatrix} 0 & 4 \\ 6 & -1 \end{bmatrix}$.

b) Show whether $(AB)^t = B^t A^t$.

29. A small accounting firm charges $50 per hour for preparing payrolls, $60 per hour for corporate tax returns, and $75 per hour for audited annual statements. The firm did the following work for three of its clients:

XYZ Limited, payrolls 120 hours, tax returns 10 hours, auditing 10 hours

YZX Limited, payrolls 60 hours, tax returns 8 hours, auditing 8 hours

ZXY Limited, payrolls 200 hours, tax returns 15 hours, auditing 20 hours

a) Use matrices to determine how much the accounting firm should bill each client.

b) How can you determine the total billed to the three clients?

30. Suppose you were to encode a message by writing it in matrix form and multiplying by a coding matrix. Would your message be more secure if you then multiplied the resulting matrices by another coding matrix with the same dimensions as the first one? Explain why or why not.

31. a) Write an equation to show the relationship between a matrix and its inverse.

b) Show that $\begin{bmatrix} 1.5 & 0 & -1 \\ 20 & -1.5 & -13 \\ -7.5 & 0.5 & 5 \end{bmatrix}$ is the

inverse of $\begin{bmatrix} 4 & 2 & 6 \\ 10 & 0 & 2 \\ 5 & 3 & 9 \end{bmatrix}$.

c) Find the inverse of $\begin{bmatrix} 4 & 5 \\ 2 & 3 \end{bmatrix}$.

32. The following diagram illustrates the food chains in a pond.

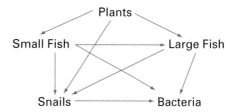

a) Represent these food chains as a network matrix, A.

b) Calculate A^2.

c) How many indirect links with exactly one intermediate step are there from plants to snails?

d) Calculate $A + A^2$. Explain the meaning of any entry in the resulting matrix.

e) Calculate A^3.

f) List all the links with two intermediate steps from plants to bacteria.

Chapter Test

1. a) Describe an iterative process you could use to draw the red path.

 b) Complete the path.

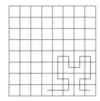

2. Find the first few terms of the recursion formula $t_n = \dfrac{1}{t_{n-1} + 2}$, given $t_1 = 0$.

 Is there a pattern to these terms? If so, describe the pattern.

3. A "fan-out" calling system is frequently used to spread news quickly to a large number of people such as volunteers for disaster relief. The first person calls three people. Each of those people calls an additional three people; each of whom calls an additional three people, and so on.

 a) Use a tree diagram to illustrate a fan-out calling system with sufficient levels to call 50 people.

 b) How many levels would be sufficient to call 500 people?

4. Rewrite each of the following expressions as spreadsheet functions.

 a) C1+C2+C3+C4+C5+C6+C7+C8

 b) The smallest value between cells A5 and G5

 c) $\dfrac{5 - \sqrt{6}}{10 + 15}$

5. Suppose that, on January 10, you borrowed $1000 at 6% per year compounded monthly (0.5% per month). You will be expected to repay $88.88 a month for 1 year. However, the final payment will be less than $88.88. You set up a spreadsheet with the following column headings: MONTH, BALANCE, PAYMENT, INTEREST, PRINCIPAL, NEW BALANCE

 The first row of entries would be:
 MONTH: February
 BALANCE: 1000.00
 PAYMENT: 88.88
 INTEREST: 5.00
 PRINCIPAL: 83.88
 NEW BALANCE: 916.12

 Describe how you would

 a) use the cell referencing formulas and the Fill feature to complete the table

 b) determine the size of the final payment on January 10 of the following year

 c) construct a line graph showing the declining balance

6. Describe how you would design a database of the daily travel logs for a company's salespersons.

7. Describe three different ways to generate random integers between 1 and 50.

8. a) Redraw this map as a network.

 b) How many colours are needed to colour the map? Explain your reasoning.

9. A salesperson must visit each of the towns on the following map.

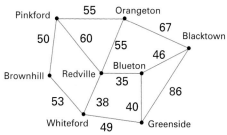

a) Is there a route that goes through each town only once? Explain.

b) Find the shortest route that begins and ends in Pinkford and goes through all the towns. Show that it is the shortest route.

10. The following map shows the bridges of Uniontown, situated on the banks of a river and on three islands. Use graph theory to determine if a continuous path could traverse all the bridges once each.

11. Let $A = \begin{bmatrix} 4 & -2 & 6 \\ -8 & 5 & 9 \\ 0 & 1 & -1 \\ 3 & -7 & -3 \end{bmatrix}$.

a) State the dimensions of matrix A.

b) What is the value of entry a_{23}?

c) Identify the entry of matrix A with value -2.

d) Is it possible to calculate A^2? Explain.

12. Let $A = \begin{bmatrix} 2 \\ 1 \\ 5 \end{bmatrix}$, $B = [7 \ 5 \ 0]$, $C = \begin{bmatrix} 4 & 8 \\ 5 & -3 \end{bmatrix}$,

$D = \begin{bmatrix} 2 & -7 \\ 9 & 1 \end{bmatrix}$, and $E = \begin{bmatrix} 8 & -2 \\ 5 & 0 \\ -4 & 1 \end{bmatrix}$.

Calculate, if possible,

a) $2C + D$ b) $A + B$ c) AD d) EC e) E^t

13. A local drama club staged a variety show for four evenings. The admission for adults was $7.00, for students $4.00, and for children 13 years of age and under $2.00. On Wednesday, 52 adult tickets, 127 student tickets, and 100 child tickets were sold; on Thursday, 67 adult tickets, 139 student tickets, and 115 child tickets were sold; on Friday, 46 adult tickets, 115 student tickets, and 102 child tickets were sold; and on Saturday, 40 adult tickets, 101 student tickets, and 89 child tickets were sold. Use matrices to calculate how much money was collected from admissions.

ACHIEVEMENT CHECK

Knowledge/Understanding	Thinking/Inquiry/Problem Solving	Communication	Application

14. The network diagram below gives the cost of flights between five Canadian cities.

a) Construct a network matrix A for these routes.

b) Calculate A^2 and A^3.

c) How many ways can a person travel from Halifax to Vancouver by changing planes exactly twice? Describe each route. Which route is most economical?

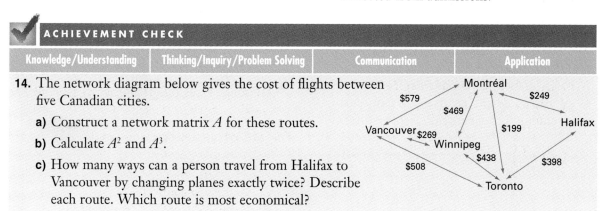

Wrap-Up

Implementing Your Action Plan

1. With your whole class or a small group, brainstorm criteria for ranking universities and community colleges. List the three universities or colleges that you think will most likely be the best choices for you.

2. Have a class discussion on weighting systems.

3. Look up the *Maclean's* university and community college rankings in a library or on the Internet. Note the criteria that *Maclean's* uses.

4. Determine your own set of criteria. These may include those that *Maclean's* uses or others, such as travelling distances, programs offered, size of the city or town where you would be living, and opportunities for part-time work.

5. Choose the ten criteria you consider most important. Research any data you need to rate universities and colleges with these criteria.

6. Assign a weighting factor to each of the ten criteria. For example, living close to home may be worth a weighting of 5 and tuition cost may be worth a weighting of 7.

7. Use a spreadsheet and matrix methods to determine an overall score for each university or community college in Ontario. Then, rank the universities or community colleges on the spreadsheet. Compare your rankings with those in *Maclean's* magazine. Explain the similarities or differences.

8. From your rankings, select the top five universities or community colleges. Draw a diagram of the distances from each university or college to the four others and to your home. Then, use graph theory to determine the most efficient way to visit each of the five universities or community colleges during a five-day period, such as a March break vacation.

9. Based on your project, select your top three choices. Comment on how this selection compares with your original list of top choices.

Suggested Resources

- *Maclean's* magazine rankings of universities and community colleges
- Other publications ranking universities and community colleges
- University and community college calendars
- Guidance counsellors
- Map of Ontario
- Spreadsheets

Refer to section 9.3 for information on implementing an action plan and Appendix C for information on research techniques.

WEB CONNECTION

www.mcgrawhill.ca/links/MDM12

For details of the *Maclean's* rankings of universities and colleges, visit the above web site and follow the links.

Evaluating Your Project

1. Reflect on your weighting formula and whether you believe it fairly ranks the universities and community colleges in Ontario.

2. Compare your rating system to that used by one of your classmates. Can you suggest improvements to either system?

3 What went well in this project?

4. If you were to do the project over again, what would you change? Why?

5. If you had more time, how would you extend this project?

6. What factors could change between now and when you make your final decision about which university or college to attend?

Presentation

Prepare a written report on your findings. Include

- the raw data
- a rationale for your choice of criteria
- a rationale for your weightings
- a printout of your spreadsheet
- a diagram showing the distances between your five highest-ranked universities or community colleges and the route you would use to visit them
- a summary of your findings

Preparing for the Culminating Project

Applying Project Skills

Consider how the data management tools you used on this project could be applied to the culminating project in Chapter 9 to

- access resources
- carry out research
- carry out an action plan
- evaluate your project
- summarize your findings in a written report

Keeping on Track

Now is a good time to draw up a schedule for your culminating project and to investigate methods for selecting a topic. Refer to Chapter 9 for an overview of how to prepare a major project. Section 9.1 suggests methods for choosing a topic. Also, consider how to find the information you will need in order to choose your topic.

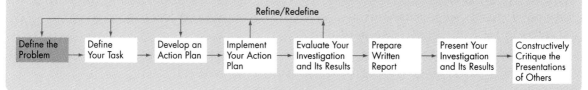

Refine/Redefine

Define the Problem → Define Your Task → Develop an Action Plan → Implement Your Action Plan → Evaluate Your Investigation and Its Results → Prepare Written Report → Present Your Investigation and Its Results → Constructively Critique the Presentations of Others

Cryptographer

In this digital era, information is sent with blinding speed around the world. These transmissions need to be both secure and accurate. Although best known for their work on secret military codes, cryptographers also design and test computerized encryption systems that protect a huge range of sensitive data including telephone conversations among world leaders, business negotiations, data sent by credit-card readers in retail stores, and financial transactions on the Internet. Encrypted passwords protect hackers from reading or disrupting critical databases. Even many everyday devices, such as garage-door openers and TV remote controls, use codes.

Cryptographers also develop error-correcting codes. Adding these special codes to a signal allows a computer receiving it to detect and correct errors that have occurred during transmission. Such codes have numerous applications including CD players, automotive computers, cable TV networks, and pictures sent back to Earth by interplanetary spacecraft.

Modern cryptography is a marriage of mathematics and computers. A cryptographer must have a background in logic, matrices, combinatorics, and computer programming as well as fractal, chaos, number, and graph theory. Cryptographers work for a wide variety of organizations including banks, government offices, the military, software developers, and universities.

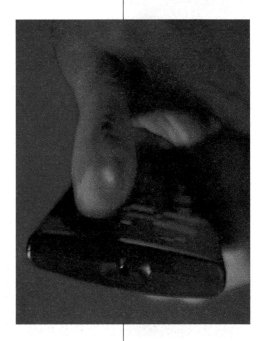

WEB CONNECTION

www.mcgrawhill.ca/links/MDM12

Visit the above web site and follow the links for more information about a career as a cryptographer and about other careers related to mathematics.

Life Expectancies

Background

Do women live longer than men? Do people live longer in warmer climates? Are people living longer today than 50 years ago? Do factors such as education and income affect life expectancy? In this project, you will answer such questions by applying the statistical techniques described in the next two chapters.

Your Task

Research and analyse current data on life expectancies in Canada, and perhaps in other countries. You will use statistical analysis to compare and contrast the data, detect trends, predict future life expectancies, and identify factors that may affect life expectancies.

Developing an Action Plan

You will need to find sources of data on life expectancies and to choose the kinds of comparisons you want to make. You will also have to decide on a method for handling the data and appropriate techniques for analysing them.

CHAPTER 2

Statistics of One Variable

Specific Expectations	Section
Locate data to answer questions of significance or personal interest, by searching well-organized databases.	2.2
Use the Internet effectively as a source for databases.	2.2
Demonstrate an understanding of the purpose and the use of a variety of sampling techniques.	2.3, 2.4
Describe different types of bias that may arise in surveys.	2.4
Illustrate sampling bias and variability by comparing the characteristics of a known population with the characteristics of samples taken repeatedly from that population, using different sampling techniques.	2.4, 2.5, 2.6
Organize and summarize data from secondary sources, using technology.	2.1, 2.2, 2.5, 2.6
Compute, using technology, measures of one-variable statistics (i.e., the mean, median, mode, range, interquartile range, variance, and standard deviation), and demonstrate an understanding of the appropriate use of each measure.	2.5, 2.6
Interpret one-variable statistics to describe characteristics of a data set.	2.5, 2.6
Describe the position of individual observations within a data set, using z-scores and percentiles.	2.6
Explain examples of the use and misuse of statistics in the media.	2.4
Assess the validity of conclusions made on the basis of statistical studies, by analysing possible sources of bias in the studies and by calculating and interpreting additional statistics, where possible.	2.5, 2.6
Explain the meaning and the use in the media of indices based on surveys.	2.2

In earlier times they had no statistics, and so they had to fall back on lies. Hence the huge exaggerations of primitive literature—giants or miracles or wonders! They did it with lies and we do it with statistics; but it is all the same.
—*Stephen Leacock (1869–1944)*

Facts are stubborn, but statistics are more pliable.
—*Mark Twain (1835–1910)*

Chapter Problem

Contract Negotiations

François is a young NHL hockey player whose first major-league contract is up for renewal. His agent wants to bargain for a better salary based on François' strong performance over his first five seasons with the team. Here are some of François' statistics for the past five seasons.

Season	Games	Goals	Assists	Points
1	20	3	4	7
2	45	7	11	18
3	76	19	25	44
4	80	19	37	56
5	82	28	36	64
Total	**303**	**76**	**113**	**189**

1. How could François' agent use these statistics to argue for a substantial pay increase for his client?

2. Are there any trends in the data that the team's manager could use to justify a more modest increase?

As these questions suggest, statistics could be used to argue both for and against a large salary increase for François. However, the statistics themselves are not wrong or contradictory. François' agent and the team's manager will, understandably, each emphasize only the statistics that support their bargaining positions. Such selective use of statistics is one reason why they sometimes receive negative comments such as the quotations above. Also, even well-intentioned researchers sometimes inadvertently use biased methods and produce unreliable results. This chapter explores such sources of error and methods for avoiding them. Properly used, statistical analysis is a powerful tool for detecting trends and drawing conclusions, especially when you have to deal with large sets of data.

Review of Prerequisite Skills

If you need help with any of the skills listed in **purple** below, refer to Appendix A.

1. **Fractions, percents, decimals** The following amounts are the total cost for the items including the 7% goods and services tax (GST) and an 8% provincial sales tax (PST). Determine the price of each item.

   a) watch $90.85

   b) CD $19.54

   c) bicycle $550.85

   d) running shoes $74.39

2. **Fractions, percents, decimals**

   a) How much will Josh make if he receives an 8% increase on his pay of $12.50/h?

   b) What is the net increase in Josh's take-home pay if the payroll deductions total 17%?

3. **Fractions, percents, decimals** What is the percent reduction on a sweater marked down from $50 to $35?

4. **Fractions, percents, decimals** Determine the cost, including taxes, of a VCR sold at a 25% discount from its original price of $219.

5. **Mean, median, mode** Calculate the mean, median, and mode for each set of data.

   a) 22, 26, 28, 27, 26

   b) 11, 19, 14, 23, 16, 26, 30, 29

   c) 10, 18, 30, 43, 18, 13, 10

   d) 70, 30, 25, 52, 12, 70

   e) 370, 260, 155, 102, 126, 440

   f) 24, 32, 37, 24, 32, 38, 32, 36, 35, 42

6. **Graphing data** Consider the following graph, which shows the average price of thingamajigs over time.

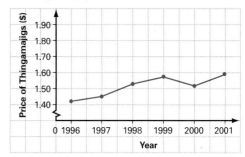

   a) What was the price of thingamajigs in 1996?

   b) In what year did the price first rise above $1.50?

   c) Describe the overall trend over the time period shown.

   d) Estimate the percent increase in the price of thingamajigs from 1996 to 2001.

   e) List the domain and range of these data.

7. **Graphing data** The table below gives the number of CDs sold at a music store on each day of the week for one week.

Day	Number of CDs Sold
Monday	48
Tuesday	52
Wednesday	44
Thursday	65
Friday	122
Saturday	152
Sunday	84

   Display the data on a circle graph.

2.1 Data Analysis With Graphs

Statistics is the gathering, organization, analysis, and presentation of numerical information. You can apply statistical methods to almost any kind of data. Researchers, advertisers, professors, and sports announcers all make use of statistics. Often, researchers gather large quantities of data since larger samples usually give more accurate results. The first step in the analysis of such data is to find ways to organize, analyse, and present the information in an understandable form.

INVESTIGATE & INQUIRE: Using Graphs to Analyse Data

1. Work in groups or as a class to design a fast and efficient way to survey your class about a simple numerical variable, such as the students' heights or the distances they travel to school.

2. Carry out your survey and record all the results in a table.

3. Consider how you could organize these results to look for any trends or patterns. Would it make sense to change the order of the data or to divide them into groups? Prepare an organized table and see if you can detect any patterns in the data. Compare your table to those of your classmates. Which methods work best? Can you suggest improvements to any of the tables?

4. Make a graph that shows how often each value or group of values occurs in your data. Does your graph reveal any patterns in the data? Compare your graph to those drawn by your classmates. Which graph shows the data most clearly? Do any of the graphs have other advantages? Explain which graph you think is the best overall.

5. Design a graph showing the total of the frequencies of all values of the variable up to a given amount. Compare this cumulative-frequency graph to those drawn by your classmates. Again, decide which design works best and look for ways to improve your own graph and those of your classmates.

The unprocessed information collected for a study is called **raw data**. The quantity being measured is the **variable.** A **continuous variable** can have any value within a given range, while a **discrete variable** can have only certain separate values (often integers). For example, the height of students in your school is a continuous variable, but the number in each class is a discrete variable. Often, it is useful to know how frequently the different values of a variable occur in a set of data. **Frequency tables** and **frequency diagrams** can give a convenient overview of the distribution of values of the variable and reveal trends in the data.

A **histogram** is a special form of **bar graph** in which the areas of the bars are proportional to the *frequencies* of the values of the variable. The bars in a histogram are connected and represent a continuous range of values. Histograms are used for variables whose values can be arranged in numerical order, especially continuous variables, such as weight, temperature, or travel time. Bar graphs can represent all kinds of variables, including the frequencies of separate categories that have no set order, such as hair colour or citizenship. A **frequency polygon** can illustrate the same information as a histogram or bar graph. To form a frequency polygon, plot frequencies versus variable values and then join the points with straight lines.

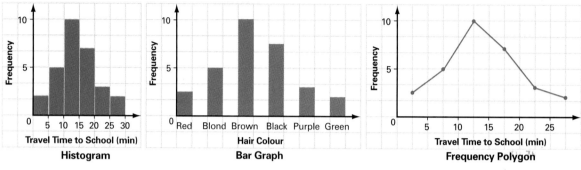

A **cumulative-frequency graph** or **ogive** shows the running total of the frequencies from the lowest value up.

WEB CONNECTION

www.mcgrawhill.ca/links/MDM12

To learn more about histograms, visit the above web site and follow the links. Write a short description of how to construct a histogram.

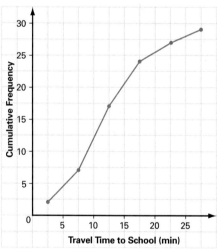

Example 1 Frequency Tables and Diagrams

Here are the sums of the two numbers from 50 rolls of a pair of standard dice.

11	4	4	10	8	7	6	6	5	10	7	9	8	8
4	7	9	11	12	10	3	7	6	9	5	8	6	8
2	6	7	5	11	2	5	5	6	6	5	2	10	9
6	5	5	5	3	9	8	2						

a) Use a frequency table to organize these data.

b) Are any trends or patterns apparent in this table?

c) Use a graph to illustrate the information in the frequency table.

d) Create a cumulative-frequency table and graph for the data.

e) What proportion of the data has a value of 6 or less?

Solution

a) Go through the data and tally the frequency of each value of the variable as shown in the table on the right.

Sum	Tally	Frequency
2	IIII	4
3	II	2
4	III	3
5	⊬⊬⊤ IIII	9
6	⊬⊬⊤ III	8
7	⊬⊬⊤	5
8	⊬⊬⊤ I	6
9	⊬⊬⊤	5
10	IIII	4
11	III	3
12	I	1

b) The table does reveal a pattern that was not obvious from the raw data. From the frequency column, notice that the middle values tend to be the most frequent while the high and low values are much less frequent.

c) The bar graph or frequency polygon makes the pattern in the data more apparent.

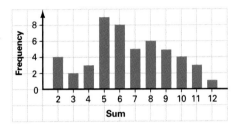
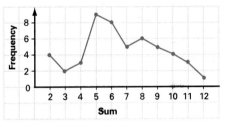

d) Add a column for cumulative frequencies to the table. Each value in this column is the running total of the frequencies of each sum up to and including the one listed in the corresponding row of the sum column. Graph these cumulative frequencies against the values of the variable.

Sum	Tally	Frequency	Cumulative Frequency
2	IIII	4	4
3	II	2	6
4	III	3	9
5	⊬⊬⊤ IIII	9	18
6	⊬⊬⊤ III	8	26
7	⊬⊬⊤	5	31
8	⊬⊬⊤ I	6	37
9	⊬⊬⊤	5	42
10	IIII	4	46
11	III	3	49
12	I	1	50

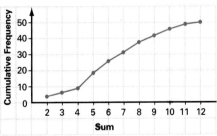

e) From either the cumulative-frequency column or the diagram, you can see that 26 of the 50 outcomes had a value of 6 or less.

When the number of measured values is large, data are usually grouped into **classes** or **intervals**, which make tables and graphs easier to construct and interpret. Generally, it is convenient to use from 5 to 20 equal intervals that cover the entire **range** from the smallest to the largest value of the variable. The interval width should be an even fraction or multiple of the measurement unit for the variable. Technology is particularly helpful when you are working with large sets of data.

Example 2 Working With Grouped Data

This table lists the daily high temperatures in July for a city in southern Ontario.

Day	1	2	3	4	5	6	7	8	9	10	11
Temperature (°C)	27	25	24	30	32	31	29	24	22	19	21

Day	12	13	14	15	16	17	18	19	20	21	22
Temperature (°C)	25	26	31	33	33	30	29	27	28	26	27

Day	23	24	25	26	27	28	29	30	31
Temperature (°C)	22	18	20	25	26	29	32	31	28

a) Group the data and construct a frequency table, a histogram or frequency polygon, and a cumulative-frequency graph.

b) On how many days was the maximum temperature 25°C or less? On how many days did the temperature exceed 30°C?

See Appendix B for more detailed information about technology functions and keystrokes.

Solution 1 Using a Graphing Calculator

a) The range of the data is 33°C − 18°C = 15°C. You could use five 3-degree intervals, but then many of the recorded temperatures would fall on the interval boundaries. You can avoid this problem by using eight 2-degree intervals with the lower limit of the first interval at 17.5°C. The upper limit of the last interval will be 33.5°C.

Use the **STAT EDIT** menu to make sure that lists L1 to L4 are clear, and then enter the temperature data into L1. Use STAT PLOT to turn on **Plot1** and select the histogram icon. Next, adjust the window settings. Set Xmin and Xmax to the lower and upper limits for your intervals and set Xscl to the interval width. Ymin should be 0. Press **GRAPH** to display the histogram, then adjust **Ymax** and **Yscl**, if necessary.

  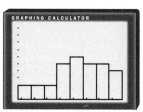

You can now use the TRACE instruction and the arrow keys to determine the tally for each of the intervals. Enter the midpoints of the intervals into L2 and the tallies into L3. Turn off Plot1 and set up Plot2 as an *x-y* line plot of lists L2 and L3 to produce a frequency polygon.

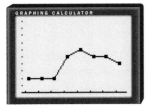

Use the cumSum(function from the LIST OPS menu to find the running totals of the frequencies in L3 and store the totals in L4. Now, an *x-y* line plot of L2 and L4 will produce a cumulative-frequency graph.

b) Since you know that all the temperatures were in whole degrees, you can see from the cumulative frequencies in L4 that there were 11 days on which the maximum temperature was no higher than 25°C. You can also get this information from the cumulative-frequency graph.

You cannot determine the exact number of days with temperatures over 30°C from the grouped data because temperatures from 29.5°C to 31.5°C are in the same interval. However, by interpolating the cumulative-frequency graph, you can see that there were about 6 days on which the maximum temperature was 31°C or higher.

Solution 2   Using a Spreadsheet

a) Enter the temperature data into column A and the midpoints of the intervals into column B. Use the COUNTIF function in column C to tally the cumulative frequency for each interval. If you use absolute cell referencing, you can copy the formula down the column and then change just the upper limit in the counting condition. Next, find the frequency for each interval by finding the difference between its cumulative frequency and the one for the previous interval.

You can then use the Chart feature to produce a frequency polygon by graphing columns B and D. Similarly, charting columns B and C will produce a cumulative-frequency graph.

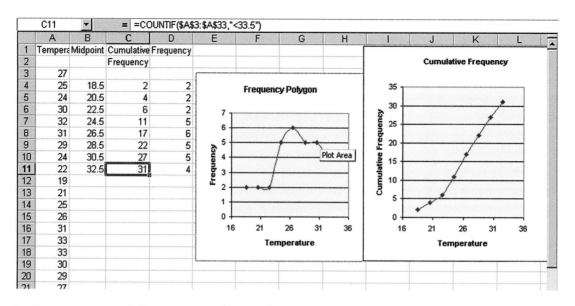

In Corel® Quattro® Pro, you can also use the Histogram tool in the Tools/Numeric Tools/Analysis menu to automatically tally the frequencies and cumulative frequencies.

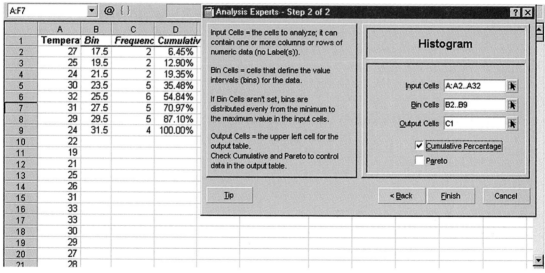

b) As in the solution using a graphing calculator, you can see from the cumulative frequencies that there were 11 days on which the maximum temperature was no higher than 25°C. Also, you can estimate from the cumulative-frequency graph that there were 6 days on which the maximum temperature was 31°C or higher. Note that you could use the COUNTIF function with the raw data to find the exact number of days with temperatures over 30°C.

A **relative-frequency** table or diagram shows the frequency of a data group as a fraction or percent of the whole data set.

Project Prep

You may find frequency-distribution diagrams useful for your statistics project.

● Example 3 Relative-Frequency Distribution

Here are a class' scores obtained on a data-management examination.

78	81	55	60	65	86	44	90
77	71	62	39	80	72	70	64
88	73	61	70	75	96	51	73
59	68	65	81	78	67		

a) Construct a frequency table that includes a column for relative frequency.

b) Construct a histogram and a frequency polygon.

c) Construct a relative-frequency histogram and a relative-frequency polygon.

d) What proportion of the students had marks between 70% and 79%?

Solution

a) The lowest and highest scores are 39% and 96%, which give a range of 57%. An interval width of 5 is convenient, so you could use 13 intervals as shown here. To determine the relative frequencies, divide the frequency by the total number of scores. For example, the relative frequency of the first interval is $\frac{1}{30}$, showing that approximately 3% of the class scored between 34.5% and 39.5%.

Score (%)	Midpoint	Tally	Frequency	Relative Frequency
34.5–39.5	37	I	1	0.033
39.5–44.5	42	I	1	0.033
44.5–49.5	47	–	0	0
49.5–54.5	52	I	1	0.033
54.5–59.5	57	I I	2	0.067
59.5–64.5	62	I I I I	4	0.133
64.5–69.5	67	I I I I	4	0.133
69.5–74.5	72	H H T I	6	0.200
74.5–79.5	77	I I I I	4	0.133
79.5–84.5	82	I I I	3	0.100
84.5–89.5	87	I I	2	0.067
89.5–94.5	92	I	1	0.033
94.5–99.5	97	I	1	0.033

b) The frequency polygon can be superimposed onto the same grid as the histogram.

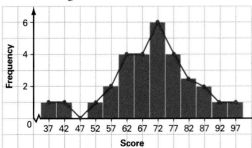

c) Draw the relative-frequency histogram and the relative-frequency polygon using the same procedure as for a regular histogram and frequency polygon. As you can see, the only difference is the scale of the *y*-axis.

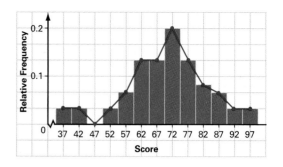

d) To determine the proportion of students with marks in the 70s, add the relative frequencies of the interval from 69.5 to 74.5 and the interval from 74.5 to 79.5:

$0.200 + 0.133 = 0.333$

Thus, 33% of the class had marks between 70% and 79%.

Categorical data are given labels rather than being measured numerically. For example, surveys of blood types, citizenship, or favourite foods all produce categorical data. **Circle graphs** (also known as **pie charts**) and **pictographs** are often used instead of bar graphs to illustrate categorical data.

Example 4 Presenting Categorical Data

The table at the right shows Canadians' primary use of the Internet in 1999.

Illustrate these data with
a) a circle graph
b) a pictograph

Primary Use	Households (%)
E-mail	15.8
Electronic banking	4.2
Purchase of goods and services	3.6
Medical or health information	8.6
Formal education/training	5.8
Government information	7.8
Other specific information	14.7
General browsing	14.2
Playing games	6.7
Chat groups	4.7
Other Internet services	5.8
Obtaining music	5.0
Listening to the radio	3.1

Solution

a)

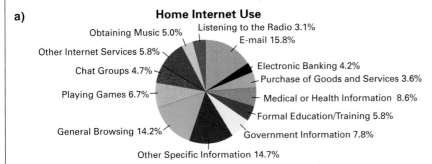

Home Internet Use

Obtaining Music 5.0%
Listening to the Radio 3.1%
E-mail 15.8%
Other Internet Services 5.8%
Chat Groups 4.7%
Electronic Banking 4.2%
Purchase of Goods and Services 3.6%
Playing Games 6.7%
Medical or Health Information 8.6%
Formal Education/Training 5.8%
General Browsing 14.2%
Government Information 7.8%
Other Specific Information 14.7%

b) There are numerous ways to represent the data with a pictograph. The one shown here has the advantages of being simple and visually indicating that the data involve computers.

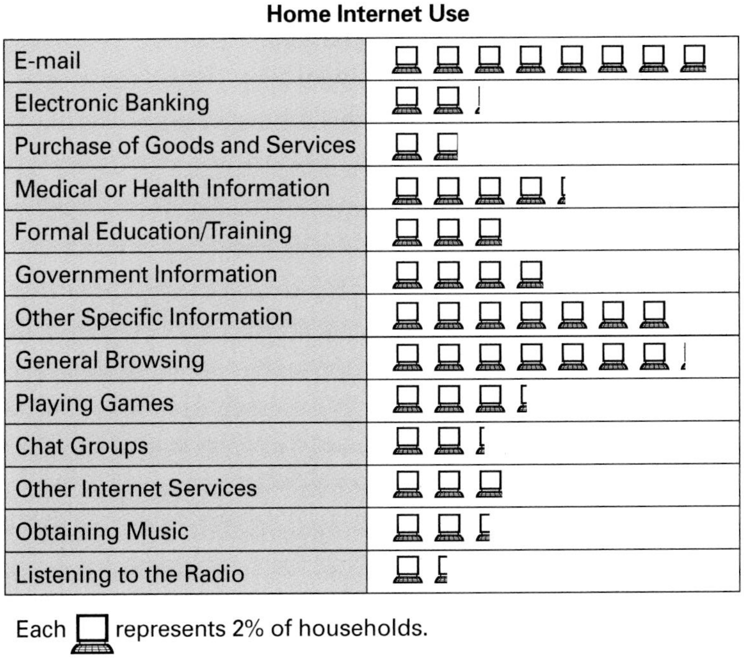

Home Internet Use

E-mail	🖥️🖥️🖥️🖥️🖥️🖥️🖥️🖥️
Electronic Banking	🖥️🖥️╷
Purchase of Goods and Services	🖥️🖥️
Medical or Health Information	🖥️🖥️🖥️🖥️╷
Formal Education/Training	🖥️🖥️🖥️
Government Information	🖥️🖥️🖥️🖥️
Other Specific Information	🖥️🖥️🖥️🖥️🖥️🖥️🖥️
General Browsing	🖥️🖥️🖥️🖥️🖥️🖥️🖥️╷
Playing Games	🖥️🖥️🖥️╷
Chat Groups	🖥️🖥️╷
Other Internet Services	🖥️🖥️🖥️
Obtaining Music	🖥️🖥️╷
Listening to the Radio	🖥️╷

Each 🖥️ represents 2% of households.

You can see from the example above that circle graphs are good for showing the sizes of categories relative to the whole and to each other. Pictographs can use a wide variety of visual elements to clarify the data and make the graph more interesting. However, with both circle graphs and pictographs, the relative frequencies for the categories can be hard to read accurately. While a well-designed pictograph can be a useful tool, you will sometimes see pictographs with distorted or missing scales or confusing graphics.

- Variables can be either continuous or discrete.

- Frequency-distribution tables and diagrams are useful methods of summarizing large amounts of data.

- When the number of measured values is large, data are usually grouped into classes or intervals. This technique is particularly helpful with continuous variables.

- A frequency diagram shows the frequencies of values in each individual interval, while a cumulative-frequency diagram shows the running total of frequencies from the lowest interval up.

- A relative-frequency diagram shows the frequency of each interval as a proportion of the whole data set.

- Categorical data can be presented in various forms, including bar graphs, circle graphs (or pie charts), and pictographs.

Communicate Your Understanding

1. a) What information does a histogram present?

   b) Explain why you cannot use categorical data in a histogram.

2. a) What is the difference between a frequency diagram and a cumulative-frequency diagram?

   b) What are the advantages of each of these diagrams?

3. a) What is the difference between a frequency diagram and a relative-frequency diagram?

   b) What information can be easily read from a frequency diagram?

   c) What information can be easily read from a relative-frequency diagram?

4. Describe the strengths and weaknesses of circle graphs and pictographs.

Practise

1. Explain the problem with the intervals in each of the following tables.

a)

Age (years)	Frequency
28–32	6
33–38	8
38–42	11
42–48	9
48–52	4

b)

Score (%)	Frequency
61–65	5
66–70	11
71–75	7
76–80	4
91–95	1

2. Would you choose a histogram or a bar graph with separated bars for the data listed below? Explain your choices.

   a) the numbers from 100 rolls of a standard die

   b) the distances 40 athletes throw a shot-put

   c) the ages of all players in a junior lacrosse league

   d) the heights of all players in a junior lacrosse league

3. A catering service conducted a survey asking respondents to choose from six different hot meals.

Meal Chosen	Number
Chicken cordon bleu	16
New York steak	20
Pasta primavera (vegetarian)	9
Lamb chop	12
Grilled salmon	10
Mushroom stir-fry with almonds (vegetarian)	5

   a) Create a circle graph to illustrate these data.

   b) Use the circle graph to determine what percent of the people surveyed chose vegetarian dishes.

   c) Sketch a pictograph for the data.

   d) Use the pictograph to determine whether more than half of the respondents chose red-meat dishes.

4. a) Estimate the number of hours you spent each weekday on each of the following activities: eating, sleeping, attending class, homework, a job, household chores, recreation, other.

   b) Present this information using a circle graph.

   c) Present the information using a pictograph.

Apply, Solve, Communicate

5. The examination scores for a biology class are shown below.

68	77	91	66	52	58	79	94	81
60	73	57	44	58	71	78	80	54
87	43	61	90	41	76	55	75	49

   a) Determine the range for these data.

   b) Determine a reasonable interval size and number of intervals.

   c) Produce a frequency table for the grouped data.

   d) Produce a histogram and frequency polygon for the grouped data.

   e) Produce a relative-frequency polygon for the data.

   f) Produce a cumulative-frequency polygon for the data.

   g) What do the frequency polygon, the relative-frequency polygon, and the cumulative-frequency polygon each illustrate best?

B

6. **a)** Sketch a bar graph to show the results you would expect if you were to roll a standard die 30 times.

   b) Perform the experiment or simulate it with software or the random-number generator of a graphing calculator. Record the results in a table.

   c) Produce a bar graph for the data you collected.

   d) Compare the bar graphs from a) and c). Account for any discrepancies you observe.

7. **Application** In order to set a reasonable price for a "bottomless" cup of coffee, a restaurant owner recorded the number of cups each customer ordered on a typical afternoon.

2	1	2	3	0	1	1	1	2	2
1	3	1	4	2	0	1	2	3	1

   a) Would you present these data in a grouped or ungrouped format? Explain your choice.

   b) Create a frequency table and diagram.

   c) Create a cumulative-frequency diagram.

   d) How can the restaurant owner use this information to set a price for a cup of coffee? What additional information would be helpful?

8. **Application** The list below shows the value of purchases, in dollars, by 30 customers at a clothing store.

55.40	48.26	28.31	14.12	88.90	34.45
51.02	71.87	105.12	10.19	74.44	29.05
43.56	90.66	23.00	60.52	43.17	28.49
67.03	16.18	76.05	45.68	22.76	36.73
39.92	112.48	81.21	56.73	47.19	34.45

   a) Would you present these data in a grouped or ungrouped format? Explain your choice.

   b) Create a frequency table and diagram.

   c) Create a cumulative-frequency diagram.

   d) How might the store owner use this information in planning sales promotions?

9. The speeds of 24 motorists ticketed for exceeding a 60-km/h limit are listed below.

75	72	66	80	75	70	71	82
69	70	72	78	90	75	76	80
75	96	91	77	76	84	74	79

   a) Construct a frequency-distribution table for these data.

   b) Construct a histogram and frequency polygon.

   c) Construct a cumulative-frequency diagram.

   d) How many of the motorists exceeded the speed limit by 15 km/h or less?

   e) How many exceeded the speed limit by over 20 km/h?

10. **Communication** This table summarizes the salaries for François' hockey team.

Salary ($)	Number of Players
300 000	2
500 000	3
750 000	8
900 000	6
1 000 000	2
1 500 000	1
3 000 000	1
4 000 000	1

    a) Reorganize these data into appropriate intervals and present them in a frequency table.

    b) Create a histogram for these data.

    c) Identify and explain any unusual features about this distribution.

11. Communication

a) What is the sum of all the relative frequencies for any set of data?

b) Explain why this sum occurs.

12.
The following relative-frequency polygon was constructed for the examination scores for a class of 25 students. Construct the frequency-distribution table for the students' scores.

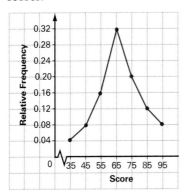

13. Inquiry/Problem Solving
The manager of a rock band suspects that MP3 web sites have reduced sales of the band's CDs. A survey of fans last year showed that at least 50% had purchased two or more of the band's CDs. A recent survey of 40 fans found they had purchased the following numbers of the band's CDs.

2	1	2	1	3	1	4	1	0	1
0	2	4	1	0	5	2	3	4	1
2	1	1	1	3	1	0	5	4	2
3	1	1	0	2	2	0	0	1	3

Does the new data support the manager's theory? Show the calculations you made to reach your conclusion, and illustrate the results with a diagram.

14. Inquiry/Problem Solving

a) What are the possible outcomes for a roll of two "funny dice" that have faces with the numbers 1, 1, 3, 5, 6, and 7?

b) Sketch a relative-frequency polygon to show the results you would expect if these dice were rolled 100 times.

c) Explain why your graph has the shape it does.

d) Use software or a graphing calculator to simulate rolling the funny dice 100 times, and draw a relative-frequency polygon for the results.

e) Account for any differences between the diagrams in parts b) and d).

15.
This cumulative-frequency diagram shows the distribution of the examination scores for a statistics class.

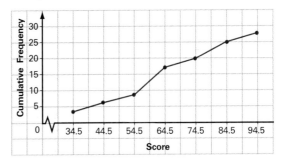

a) What interval contains the greatest number of scores? Explain how you can tell.

b) How many scores fall within this interval?

16.
Predict the shape of the relative-frequency diagram for the examination scores of a first-year university calculus class. Explain why you chose the shape you did. Assume that students enrolled in a wide range of programs take this course. State any other assumptions that you need to make.

Indices

In the previous section, you used tables and graphs of frequencies to summarize data. Indices are another way to summarize data and recognize trends. An **index** relates the value of a variable (or group of variables) to a base level, which is often the value on a particular date. The base level is set so that the index produces numbers that are easy to understand and compare. Indices are used to report on a wide variety of variables, including prices and wages, ultraviolet levels in sunlight, and even the readability of textbooks.

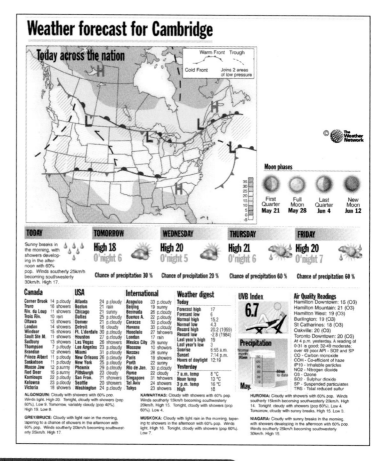

INVESTIGATE & INQUIRE: Consumer Price Index

The graph below shows Statistics Canada's consumer price index (CPI), which tracks the cost of over 600 items that would be purchased by a typical family in Canada. For this chart, the base is the cost of the same items in 1992.

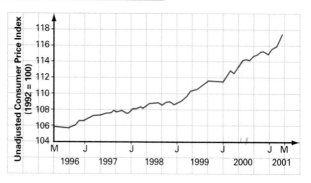

1. What trend do you see in this graph? Estimate the annual rate of increase.

2. Estimate the annual rate of increase for the period from 1992 to 1996. Do you think the difference between this rate and the one from 1996 to 2001 is significant? Why or why not?

3. What was the index value in February of 1998? What does this value tell you about consumer prices at that time?

4. What would be the best way to estimate what the consumer price index will be in May of 2003? Explain your reasoning.

5. Explain how the choice of the vertical scale in the graph emphasizes changes in the index. Do you think this emphasis could be misleading? Why or why not?

The best-known Canadian business index is the S&P/TSX Composite Index, managed for the Toronto Stock Exchange by Standard & Poor's Corporation. Introduced in May, 2002, this index is a continuation of the TSE 300 Composite Index®, which goes back to 1977. The S&P/TSX Composite Index is a measure of the total market value of the shares of over 200 of the largest companies traded on the Toronto Stock Exchange. The index is the current value of these stocks divided by their total value in a base year and then multiplied by a scaling factor. When there are significant changes (such as takeovers or bankruptcies) in any of the companies in the index, the scaling factor is adjusted so that the values of the index remain directly comparable to earlier values. Note that the composite index weights each company by the total value of its shares (its market capitalization) rather than by the price of the individual shares. The S&P/TSX Composite Index usually indicates trends for major Canadian corporations reasonably well, but it does not always accurately reflect the overall Canadian stock market.

Time-series graphs are often used to show how indices change over time. Such graphs plot variable values versus time and join the adjacent data points with straight lines.

Example 1 Stock Market Index

The following table shows the TSE 300 Composite Index® from 1971 to 2001.

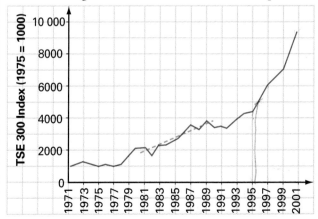

a) What does the notation "1975 = 1000" mean?

b) By what factor did the index grow over the period shown?

c) Estimate the rate of growth of the index during the 1980s.

Solution

a) The notation indicates that the index shows the stock prices *relative* to what they were in 1975. This 1975 base has been set at 1000. An index value of 2000 would mean that overall the stocks of the 300 companies in the index are selling for twice what they did in 1975.

b) From the graph, you can see that the index increased from about 1000 in 1971 to about 10 000 in 2001. Thus, the index increased by a factor of approximately 10 over this period.

c) To estimate the rate of growth of the index during the 1980s, approximate the time-series graph with a straight line during that 10-year interval. Then, calculate the slope of the line.

$$m = \frac{\text{rise}}{\text{run}}$$
$$\doteq \frac{3700 - 1700}{10}$$
$$= 200$$

WEB CONNECTION

www.mcgrawhill.ca/links/MDM12

For more information on stock indices, visit the above web site and follow the links. Write a brief description of the rules for inclusion in the various market indices.

The TSE 300 Composite Index® rose about 200 points a year during the 1980s.

Statistics Canada calculates a variety of carefully researched economic indices. For example, there are price indices for new housing, raw materials, machinery and equipment, industrial products, and farm products. Most of these indices are available with breakdowns by province or region and by specific categories, such as agriculture, forestry, or manufacturing. Statisticians, economists, and the media make extensive use of these indices. (See section 1.3 for information on how to access Statistics Canada data.)

The **consumer price index (CPI)** is the most widely reported of these economic indices because it is an important measure of inflation. **Inflation** is a general increase in prices, which corresponds to a decrease in the value of money. To measure the average change in retail prices across Canada, Statistics Canada monitors the retail prices of a set of over 600 goods and services including food, shelter, clothing, transportation, household items, health and personal care, recreation and education, and alcohol and tobacco products. These items are representative of purchases by typical Canadians and are weighted according to estimates of the total amount Canadians spend on each item. For example, milk has a weighting of 0.69% while tea has a weighting of only 0.06%.

Data in Action

Statistics Canada usually publishes the consumer price index for each month in the third week of the following month. Over 60 000 price quotations are collected for each update.

Example 2 Consumer Price Index

The following graph shows the amount by which the consumer price index changed since the same month of the previous year.

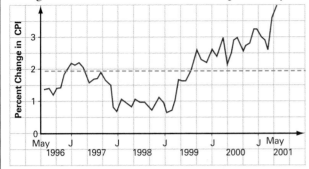

a) What does this graph tell you about changes in the CPI from 1996 to 2001?

b) Estimate the mean annual change in the CPI for this period.

Solution

a) Note that the graph above shows the annual changes in the CPI, unlike the graph on page 104, which illustrates the value of the CPI for any given month. From the above graph, you can see that the annual change in the CPI varied between 0.5% and 4% from 1996 to 2001. Overall, there is an upward trend in the annual change during this period.

b) You can estimate the mean annual change by drawing a horizontal line such that the total area between the line and the parts of the curve above it is approximately equal to the total area between the line and the parts of the curve below it. As shown above, this line meets the y-axis near 2%.

Thus, the mean annual increase in the CPI was roughly 2% from 1996 to 2001.

Project Prep

If your statistics project examines how a variable changes over time, a time-series graph may be an effective way to illustrate your findings.

The consumer price index and the cost of living index are not quite the same. The cost of living index measures the cost of maintaining a constant standard of living. If consumers like two similar products equally well, their standard of living does not change when they switch from one to the other. For example, if you like both apples and pears, you might start buying more apples and fewer pears if the price of pears went up while the price of apples was unchanged. Thus, your cost of living index increases less than the consumer price index does.

WEB CONNECTION

www.mcgrawhill.ca/links/MDM12

For more information about Statistics Canada indices, visit the above web site and follow the links to Statistics Canada.

Indices are also used in many other fields, including science, sociology, medicine, and engineering. There are even indices of the clarity of writing.

Example 3 Readability Index

The Gunning fog index is a measure of the readability of prose. This index estimates the years of schooling required to read the material easily.

Gunning fog index = 0.4(average words per sentence + percent "hard" words)

where "hard" words are all words over two syllables long except proper nouns, compounds of easy words, and verbs whose third syllable is *ed* or *es*.

a) Calculate the Gunning fog index for a book with an average sentence length of 8 words and a 20% proportion of hard words.

b) What are the advantages and limitations of this index?

Solution

a) Gunning fog index = 0.4(8 + 20)
$$= 11.2$$

The Gunning fog index shows that the book is written at a level appropriate for readers who have completed grade 11.

b) The Gunning fog index is easy to use and understand. It generates a grade-level rating, which is often more useful than a readability rating on an arbitrary scale, such as 1 to 10 or 1 to 100. However, the index assumes that bigger words and longer sentences always make prose harder to read. A talented writer could use longer words and sentences and still be more readable than writers who cannot clearly express their ideas. The Gunning fog index cannot, of course, evaluate literary merit.

Project Prep

You may want to use an index to summarize and compare sets of data in your statistics project.

WEB CONNECTION

www.mcgrawhill.ca/links/MDM12

Visit the above web site to find a link to a readability-index calculator. Determine the reading level of a novel of your choice.

Practise

1. Refer to the consumer price index graph on page 104.

a) By how many index points did the CPI increase from January, 1992 to January, 2000?

b) Express this increase as a percent.

c) Estimate what an item that cost

   i) $7.50 in 1992 cost in April, 1998

   ii) $55 in August, 1997 cost in May, 2000

Apply, Solve, Communicate

2. a) Explain why there is a wide variety of items in the CPI basket.

b) Is the percent increase for the price of each item in the CPI basket the same? Explain.

3. Refer to the graph of the TSE 300 Composite Index® on page 105.

a) When did this index first reach five times its base value?

b) Estimate the growth rate of the index from 1971 to 1977. What does this growth rate suggest about the Canadian economy during this period?

c) During what two-year period did the index grow most rapidly? Explain your answer.

d) Could a straight line be a useful mathematical model for the TSE 300 Composite Index®? Explain why or why not.

4. Communication

a) Define inflation.

b) In what way do the consumer price index and the new housing price index provide a measure of inflation?

c) How would you expect these two indices to be related?

d) Why do you think that they would be related in this way?

5. Application Consider the following time-series graph for the consumer price index.

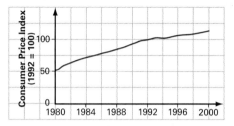

a) Identify at least three features of this graph that are different from the CPI graph on page 104.

b) Explain two advantages that the graph shown here has over the one on page 104.

c) Explain two disadvantages of the graph shown here compared to the one on page 104.

d) Estimate the year in which the CPI was at 50.

e) Explain the significance of the result in part d) in terms of prices in 1992.

6. Application The following graph illustrates the CPI both with and without energy price changes.

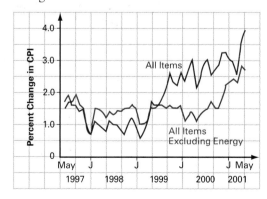

a) How is this graph different from the one on page 107?

b) Describe how the overall trend in energy costs compares to that of the CPI for the period shown.

c) What insight is gained by removing the energy component of the CPI?

d) Estimate the overall increase in the energy-adjusted CPI for the period shown.

e) Discuss how your result in part d) compares to the value found in part b) of Example 2.

7. François' agent wants to bargain for a better salary based on François' statistics for his first five seasons with the team.

a) Produce a time-series graph for François' goals, assists, and points over the past five years.

b) Calculate the mean number of goals, assists, and points per game played during each of François' five seasons.

c) Generate a new time-series graph based on the data from part b).

d) Which time-series graph will the agent likely use, and which will the team's manager likely use during the contract negotiations? Explain.

e) Explain the method or technology that you used to answer parts a) to d).

8. Aerial surveys of wolves in Algonquin Park produced the following estimates of their population density.

Year	Wolves/100 km²
1988–89	4.91
1989–90	2.47
1990–91	2.80
1991–92	3.62
1992–93	2.53
1993–94	2.23
1994–95	2.82
1995–96	2.75
1996–97	2.33
1997–98	3.04
1998–99	1.59

a) Using 1988–89 as a base, construct an index for these data.

b) Comment on any trends that you observe.

9. Use Statistics Canada web sites or other sources to find statistics for the following and describe any trends you notice.

a) the population of Canada

b) the national unemployment rate

c) the gross domestic product

10. **Inquiry/Problem Solving**

a) Use data from E-STAT or other sources to generate a time-series graph that shows the annual number of crimes in Canada for the period 1989–1999. If using E-STAT, look in the **Nation** section under **Justice/Crimes and Offences**.

b) Explain any patterns that you notice.

c) In what year did the number of crimes peak?

d) Suggest possible reasons why the number of crimes peaked in that year. What other statistics would you need to confirm whether these reasons are related to the peak in the number of crimes?

11. **a)** Use data from E-STAT or other sources to generate a time-series graph that shows the number of police officers in Canada for the period 1989–1999. If using E-STAT, look in the **Nation** section under **Justice/Police services**.

b) In what ways are the patterns in these data similar to the patterns in the data in question 10? In what ways are the patterns different?

c) In what year did the number of police officers peak?

d) Explain how this information could affect your answer to part d) of question 10.

12. **Communication** Use the Internet, a library, or other resources to research two indices not discussed in this section. Briefly describe what each index measures, recent trends in the index, and any explanation or rationale for these trends.

13. **Inquiry/Problem Solving** The pictograph below shows total greenhouse-gas emissions for each province and territory in 1996.

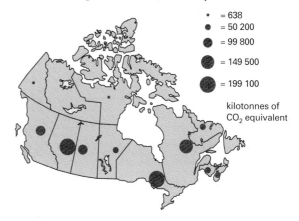

•	= 638
●	= 50 200
●	= 99 800
●	= 149 500
●	= 199 100

kilotonnes of CO_2 equivalent

a) Which two provinces have the highest levels of greenhouse-gas emissions?

b) Are the diameters or areas of the circles proportional to the numbers they represent? Justify your answer.

c) What are the advantages and disadvantages of presenting these data as a pictograph?

d) Which provinces have the highest levels of greenhouse-gas emissions per geographic area?

e) Is your answer to part d) what you would have expected? How can you account for such relatively high levels in these areas?

f) Research information from E-STAT or other sources to determine the greenhouse-gas emissions per person for each province.

14. The graph below shows the national unemployment rate from January, 1997, to June, 2001.

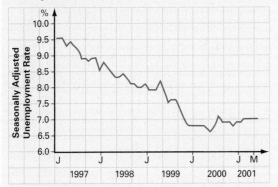

a) Describe the overall trend for the period shown.

b) When did the unemployment rate reach its lowest level?

c) Estimate the overall unemployment rate for the period shown.

d) Explain what the term *seasonally adjusted* means.

e) Who is more likely to use this graph in an election campaign, the governing party or an opposing party? Explain.

f) How might an opposing party produce a graph showing rising unemployment without changing the data? Why would they produce such a graph?

C

15. A *Pareto chart* is a type of frequency diagram in which the frequencies for categorical data are shown by connected bars arranged in descending order of frequency. In a random survey, commuters listed their most common method of travelling to the downtown of a large city.

a) Construct a Pareto chart for these data.

b) Describe the similarities and differences between a Pareto chart and other frequency diagrams.

Method	Number of Respondents
Automobile: alone	26
Automobile: car pool	35
Bus/Streetcar	52
Train	40
Bicycle/Walking	13

WEB CONNECTION

www.mcgrawhill.ca/links/MDM12

For more information about Pareto charts, visit the above web site and follow the links. Give two examples of situations where you would use a Pareto chart. Explain your reasoning.

16. Pick five careers of interest to you.

a) Use resources such as CANSIM II, E-STAT, newspapers, or the Internet to obtain information about entry-level income levels for these professions.

b) Choose an effective method to present your data.

c) Describe any significant information you discovered.

17. a) Research unemployment data for Ontario over the past 20 years.

b) Present the data in an appropriate form.

c) Conduct additional research to account for any trends or unusual features of the data.

d) Predict unemployment trends for both the short term and the long term. Explain your predictions.

Sampling Techniques

Who will win the next federal election? Are Canadians concerned about global warming? Should a Canadian city bid to host the next Olympic Games? Governments, political parties, advocacy groups, and news agencies often want to know the public's opinions on such questions. Since it is not feasible to ask every citizen directly, researchers often survey a much smaller group and use the results to estimate the opinions of the entire population.

INVESTIGATE & INQUIRE: Extrapolating From a Sample

1. Work in groups or as a class to design a survey to determine the opinions of students in your school on a subject such as favourite movies, extra-curricular activities, or types of music.

2. Have everyone in your class answer the survey.

3. Decide how to categorize and record the results. Could you refine the survey questions to get results that are easier to work with? Explain the changes you would make.

4. How could you organize and present the data to make it easier to recognize any patterns? Can you draw any conclusions from the data?

5. **a)** Extrapolate your data to estimate the opinions of the entire school population. Explain your method.

   b) Describe any reasons why you think the estimates in part a) may be inaccurate.

   c) How could you improve your survey methods to get more valid results?

In statistics, the term **population** refers to all individuals who belong to a group being studied. In the investigation above, the population is all the students in your school, and your class is a **sample** of that population. The population for a statistical study depends on the kind of data being collected.

Example 1 Identifying a Population

Identify the population for each of the following questions.

a) Whom do you plan to vote for in the next Ontario election?

b) What is your favourite type of baseball glove?

c) Do women prefer to wear ordinary glasses or contact lenses?

Solution

a) The population consists of those people in Ontario who will be eligible to vote on election day.

b) The population would be just those people who play baseball. However, you might want to narrow the population further. For example, you might be interested only in answers from local or professional baseball players.

c) The population is all women who use corrective lenses.

Once you have identified the population, you need to decide how you will obtain your data. If the population is small, it may be possible to survey the entire group. For larger populations, you need to use an appropriate sampling technique. If selected carefully, a relatively small sample can give quite accurate results.

The group of individuals who actually have a chance of being selected is called the **sampling frame.** The sampling frame varies depending on the sampling technique used. Here are some of the most commonly used sampling techniques.

Simple Random Sample

In a **simple random sample**, every member of the population has an equal chance of being selected and the selection of any particular individual does not affect the chances of any other individual being chosen. Choosing the sample randomly reduces the risk that selected members will not be representative of the whole population. You could select the sample by drawing names randomly or by assigning each member of the population a unique number and then using a random-number generator to determine which members to include.

Systematic Sample

For a **systematic sample**, you go through the population sequentially and select members at regular intervals. The sample size and the population size determine the sampling interval.

$$\text{interval} = \frac{\text{population size}}{\text{sample size}}$$

For example, if you wanted the sample to be a tenth of the population, you would select every tenth member of the population, starting with one chosen randomly from among the first ten in sequence.

Example 2 Designing a Systematic Sample

A telephone company is planning a marketing survey of its 760 000 customers. For budget reasons, the company wants a sample size of about 250.

a) Suggest a method for selecting a systematic sample.

b) What expense is most likely to limit the sample size?

Solution

a) First, determine the sampling interval.

$$\text{interval} = \frac{\text{population size}}{\text{sample size}}$$
$$= \frac{760\ 000}{250}$$
$$= 3040$$

The company could randomly select one of the first 3040 names on its list of customers and then choose every 3040th customer from that point on. For simplicity, the company might choose to select every 3000th customer instead.

b) The major cost is likely to be salaries for the staff to call and interview the customers.

Stratified Sample

Sometimes a population includes groups of members who share common characteristics, such as gender, age, or education level. Such groups are called **strata**. A **stratified sample** has the same proportion of members from each stratum as the population does.

Example 3 Designing a Stratified Sample

Before booking bands for the school dances, the students' council at Statsville High School wants to survey the music preferences of the student body. The following table shows the enrolment at the school.

Grade	Number of Students
9	255
10	232
11	209
12	184
Total	880

a) Design a stratified sample for a survey of 25% of the student body.

b) Suggest other ways to stratify this sample.

Solution

a) To obtain a stratified sample with the correct proportions, simply select 25% of the students in each grade level as shown on the right.

Grade	Number of Students	Relative Frequency	Number Surveyed
9	255	0.29	64
10	232	0.26	58
11	209	0.24	52
12	184	0.21	46
Total	880	1.00	220

b) The sample could be stratified according to gender or age instead of grade level.

Other Sampling Techniques

Cluster Sample: If certain groups are likely to be representative of the entire population, you can use a random selection of such groups as a **cluster sample**. For example, a fast-food chain could save time and money by surveying all its employees at randomly selected locations instead of surveying randomly selected employees throughout the chain.

Multi-Stage Sample: A **multi-stage sample** uses several levels of random sampling. If, for example, your population consisted of all Ontario households, you could first randomly sample from all cities and townships in Ontario, then randomly sample from all subdivisions or blocks within the selected cities and townships, and finally randomly sample from all houses within the selected subdivisions or blocks.

Voluntary-Response Sample: In a **voluntary-response sample,** the researcher simply invites any member of the population to participate in the survey. The results from the responses of such surveys can be skewed because the people who choose to respond are often not representative of the population. Call-in shows and mail-in surveys rely on voluntary-response samples.

Convenience Sample: Often, a sample is selected simply because it is easily accessible. While obviously not as random as some of the other techniques, such convenience samples can sometimes yield helpful information. The investigation at the beginning of this section used your class as a convenience sample.

Key Concepts

- A carefully selected sample can provide accurate information about a population.

- Selecting an appropriate sampling technique is important to ensure that the sample reflects the characteristics of the population. Randomly selected samples have a good chance of being representative of the population.

- The choice of sampling technique will depend on a number of factors, such as the nature of the population, cost, convenience, and reliability.

1. What are the advantages and disadvantages of using a sample to estimate the characteristics of a population?

2. Discuss whether a systematic sample is a random sample.

3. **a)** Explain the difference between stratified sampling and cluster sampling.

   b) Suggest a situation in which it would be appropriate to use each of these two sampling techniques.

Practise

1. Identify the population for each of the following questions.

   a) Who should be the next president of the students' council?

   b) Who should be next year's grade-10 representative on the student council?

   c) What is the your favourite soft drink?

   d) Which Beatles song was the best?

   e) How effective is a new headache remedy?

2. Classify the sampling method used in each of the following scenarios.

   a) A radio-show host invites listeners to call in with their views on banning smoking in restaurants.

   b) The Heritage Ministry selects a sample of recent immigrants such that the proportions from each country of origin are the same as for all immigrants last year.

   c) A reporter stops people on a downtown street to ask what they think of the city's lakefront.

   d) A school guidance counsellor arranges interviews with every fifth student on the alphabetized attendance roster.

   e) A statistician conducting a survey randomly selects 20 cities from across Canada, then 5 neighbourhoods from each of the cities, and then 3 households from each of the neighbourhoods.

   f) The province randomly chooses 25 public schools to participate in a new fundraising initiative.

3. What type(s) of sample would be appropriate for

   a) a survey of engineers, technicians, and managers employed by a company?

   b) determining the most popular pizza topping?

   c) measuring customer satisfaction for a department store?

Apply, Solve, Communicate

4. Natasha is organizing the annual family picnic and wants to arrange a menu that will appeal to children, teens, and adults. She estimates that she has enough time to survey about a dozen people. How should Natasha design a stratified sample if she expects 13 children, 8 teens, and 16 adults to attend the picnic?

5. **Communication** Find out, or estimate, how many students attend your school. Describe how you would design a systematic sample of these students. Assume that you can survey about 20 students.

6. The newly elected Chancellor of the Galactic Federation is interested in the opinions of all citizens regarding economic conditions in the galaxy. Unfortunately, she does not have the resources to visit every populated planet or to send delegates to them. Describe how the Chancellor might organize a multi-stage sample to carry out her survey.

7. **Communication** A community centre chooses 15 of its members at random and asks them to have each member of their families complete a short questionnaire.

   a) What type of sample is the community centre using?

   b) Are the 15 community-centre members a random sample of the community? Explain.

   c) To what extent are the family members randomly chosen?

8. **Application** A students' council is conducting a poll of students as they enter the cafeteria.

   a) What sampling method is the student council using?

   b) Discuss whether this method is appropriate for surveying students' opinions on
      i) the new mural in the cafeteria
      ii) the location for the graduation prom

   c) Would another sampling technique be better for either of the surveys in part b)?

9. **Application** The host of a call-in program invites listeners to comment on a recent trade by the Toronto Maple Leafs. One caller criticizes the host, stating that the sampling technique is not random. The host replies: "So what? It doesn't matter!"

   a) What sampling technique is the call-in show using?

   b) Is the caller's statement correct? Explain.

   c) Is the host's response mathematically correct? Why or why not?

10. Look in newspapers and periodicals or on the Internet for an article about a study involving a systematic, stratified, cluster, or multi-stage sample. Comment on the suitability of the sampling technique and the validity of the study. Present your answer in the form of a brief report. Include any suggestions you have for improving the study.

11. **Inquiry/Problem Solving** Design a data-gathering method that uses a combination of convenience and systematic sampling techniques.

12. **Inquiry/Problem Solving** Pick a professional sport that has championship playoffs each year.

   a) Design a multi-stage sample to gather your schoolmates' opinions on which team is likely to win the next championship.

   b) Describe how you would carry out your study and illustrate your findings.

   c) Research the media to find what the professional commentators are predicting. Do you think these opinions would be more valid than the results of your survey? Why or why not?

2.4 Bias in Surveys

The results of a survey can be accurate only if the sample is representative of the population and the measurements are objective. The methods used for choosing the sample and collecting the data must be free from **bias**. Statistical bias is any factor that favours certain outcomes or responses and hence systematically skews the survey results. Such bias is often unintentional. A researcher may inadvertently use an unsuitable method or simply fail to recognize a factor that prevents a sample from being fully random. Regrettably, some people deliberately bias surveys in order to get the results they want. For this reason, it is important to understand not only how to use statistics, but also how to recognize the misuse of statistics.

INVESTIGATE & INQUIRE: Bias in a Survey

1. What sampling technique is the pollster in this cartoon likely to be using?
2. What is wrong with his survey methods? How could he improve them?
3. Do you think the bias in this survey is intentional? Why or why not?
4. Will this bias seriously distort the results of the survey? Explain your reasoning.
5. What point is the cartoonist making about survey methods?
6. Sketch your own cartoon or short comic strip about data management.

Sampling bias occurs when the sampling frame does not reflect the characteristics of the population. Biased samples can result from problems with either the sampling technique or the data-collection method.

Example 1 Sampling Bias

Identify the bias in each of the following surveys and suggest how it could be avoided.

a) A survey asked students at a high-school football game whether a fund for extra-curricular activities should be used to buy new equipment for the football team or instruments for the school band.

b) An aid agency in a developing country wants to know what proportion of households have at least one personal computer. One of the agency's staff members conducts a survey by calling households randomly selected from the telephone directory.

Solution

a) Since the sample includes only football fans, it is not representative of the whole student body. A poor choice of sampling technique makes the results of the survey invalid. A random sample selected from the entire student body would give unbiased results.

b) There could be a significant number of households without telephones. Such households are unlikely to have computers. Since the telephone survey excludes these households, it will overestimate the proportion of households that have computers. By using a telephone survey as the data-collection method, the researcher has inadvertently biased the sample. Visiting randomly selected households would give a more accurate estimate of the proportion that have computers. However, this method of data collection would be more time-consuming and more costly than a telephone survey.

Non-response bias occurs when particular groups are under-represented in a survey because they choose not to participate. Thus, non-response bias is a form of sampling bias.

Example 2 Non-Response Bias

A science class asks every fifth student entering the cafeteria to answer a survey on environmental issues. Less than half agree to complete the questionnaire. The completed questionnaires show that a high proportion of the respondents are concerned about the environment and well-informed about environmental issues. What bias could affect these results?

Solution

The students who chose not to participate in the survey are likely to be those least interested in environmental issues. As a result, the sample may not be representative of all the students at the school.

To avoid non-response bias, researchers must ensure that the sampling process is truly random. For example, they could include questions that identify members of particular groups to verify that they are properly represented in the sample.

Measurement bias occurs when the data-collection method consistently either under- or overestimates a characteristic of the population. While random errors tend to cancel out, a consistent measurement error will skew the results of a survey. Often, measurement bias results from a data-collection process that affects the variable it is measuring.

Example 3 Measurement Bias

Identify the bias in each of the following surveys and suggest how it could be avoided.

a) A highway engineer suggests that an economical way to survey traffic speeds on an expressway would be to have the police officers who patrol the highway record the speed of the traffic around them every half hour.

b) As part of a survey of the "Greatest Hits of All Time," a radio station asks its listeners: Which was the best song by the Beatles?

   i) Help!   ii) Nowhere Man

   iii) In My Life   iv) Other:

c) A poll by a tabloid newspaper includes the question: "Do you favour the proposed bylaw in which the government will dictate whether you have the right to smoke in a restaurant?"

Solution

a) Most drivers who are speeding will slow down when they see a police cruiser. A survey by police cruisers would underestimate the average traffic speed. Here, the data-collection method would systematically decrease the variable it is measuring. A survey by unmarked cars or hidden speed sensors would give more accurate results.

b) The question was intended to remind listeners of some of the Beatles' early recordings that might have been overshadowed by their later hits. However, some people will choose one of the suggested songs as their answer even though they would not have thought of these songs without prompting. Such **leading questions** usually produce biased results. The survey would more accurately determine listeners' opinions if the question did not include any suggested answers.

c) This question distracts attention from the real issue, namely smoking in restaurants, by suggesting that the government will infringe on the respondents' rights. Such **loaded questions** contain wording or information intended to influence the respondents' answers. A question with straightforward neutral language will produce more accurate data. For example, the question could read simply: "Should smoking in restaurants be banned?"

Response bias occurs when participants in a survey deliberately give false or misleading answers. The respondents might want to influence the results unduly, or they may simply be afraid or embarrassed to answer sensitive questions honestly.

Project Prep

When gathering data for your statistics project, you will need to ensure that the sampling process is free from bias.

Example 4 Response Bias

A teacher has just explained a particularly difficult concept to her class and wants to check that all the students have grasped this concept. She realizes that if she asks those who did not understand to put up their hands, these students may be too embarrassed to admit that they could not follow the lesson. How could the teacher eliminate this response bias?

Solution

The teacher could say: "This material is very difficult. Does anyone want me to go over it again?" This question is much less embarrassing for students to answer honestly, since it suggests that it is normal to have difficulty with the material. Better still, she could conduct a survey that lets the students answer anonymously. The teacher could ask the students to rate their understanding on a scale of 1 to 5 and mark the ratings on slips of paper, which they would deposit in a box. The teacher can then use these ballots to decide whether to review the challenging material at the next class.

As the last two examples illustrate, careful wording of survey questions is essential for avoiding bias. Researchers can also use techniques such as follow-up questions and guarantees of anonymity to eliminate response bias. For a study to be valid, all aspects of the sampling process must be free from bias.

Key Concepts

- Sampling, measurement, response, and non-response bias can all invalidate the results of a survey.

- Intentional bias can be used to manipulate statistics in favour of a certain point of view.

- Unintentional bias can be introduced if the sampling and data-collection methods are not chosen carefully.

- Leading and loaded questions contain language that can influence the respondents' answers.

1. Explain the difference between a measurement bias and a sampling bias.

2. Explain how a researcher could inadvertently bias a study.

3. Describe how each of the following might use intentional bias
   a) the media
   b) a marketing department
   c) a lobby group

Practise

1. Classify the bias in each of the following scenarios.

   a) Members of a golf and country club are polled regarding the construction of a highway interchange on part of their golf course.

   b) A group of city councillors are asked whether they have ever taken part in an illegal protest.

   c) A random poll asks the following question: "The proposed casino will produce a number of jobs and economic activity in and around your city, and it will also generate revenue for the provincial government. Are you in favour of this forward-thinking initiative?"

   d) A survey uses a cluster sample of Toronto residents to determine public opinion on whether the provincial government should increase funding for the public transit.

Apply, Solve, Communicate

2. For each scenario in question 1, suggest how the survey process could be changed to eliminate bias.

3. **Communication** Reword each of the following questions to eliminate the measurement bias.

   a) In light of the current government's weak policies, do you think that it is time for a refreshing change at the next federal election?

   b) Do you plan to support the current government at the next federal election, in order that they can continue to implement their effective policies?

   c) Is first-year calculus as brutal as they say?

   d) Which of the following is your favourite male movie star?
      i) Al Pacino   ii) Keanu Reeves
      iii) Robert DeNiro   iv) Jack Nicholson
      v) Antonio Banderas vi) Other:

   e) Do you think that fighting should be eliminated from professional hockey so that skilled players can restore the high standards of the game?

4. **Communication**

   a) Write your own example of a leading question and a loaded question.

   b) Write an unbiased version for each of these two questions.

ACHIEVEMENT CHECK

Knowledge/Understanding	Thinking/Inquiry/Problem Solving	Communication	Application

5. A school principal wants to survey data-management students to determine whether having computer Internet access at home improves their success in this course.

a) What type of sample would you suggest? Why? Describe a technique for choosing the sample.

b) The following questions were drafted for the survey questionnaire. Identify any bias in the questions and suggest a rewording to eliminate the bias.

i) Can your family afford high-speed Internet access?

ii) Answer the question that follows your mark in data management.
Over 80%: How many hours per week do you spend on the Internet at home?
60–80%: Would home Internet access improve your mark in data management ?
Below 60%: Would increased Internet access at school improve your mark in data management?

c) Suppose the goal is to convince the school board that every data-management student needs daily access to computers and the Internet in the classroom. How might you alter your sampling technique to help achieve the desired results in this survey? Would these results still be statistically valid?

6. Application A talk-show host conducts an on-air survey about re-instituting capital punishment in Canada. Six out of ten callers voice their support for capital punishment. The next day, the host claims that 60% of Canadians are in favour of capital punishment. Is this claim statistically valid? Explain your reasoning.

C

7. a) Locate an article from a newspaper, periodical, or Internet site that involves a study that contains bias.

b) Briefly describe the study and its findings.

c) Describe the nature of the bias inherent in the study.

d) How has this bias affected the results of the study?

e) Suggest how the study could have eliminated the bias.

8. Inquiry/Problem Solving Do you think that the members of Parliament are a representative sample of the population? Why or why not?

2.5 Measures of Central Tendency

It is often convenient to use a central value to summarize a set of data. People frequently use a simple arithmetic average for this purpose. However, there are several different ways to find values around which a set of data tends to cluster. Such values are known as **measures of central tendency**.

INVESTIGATE & INQUIRE: Not Your Average Average

François is a NHL hockey player whose first major-league contract is up for renewal. His agent is bargaining with the team's general manager.

Agent: Based on François' strong performance, we can accept no less than the team's average salary.

Manager: Agreed, François deserves a substantial increase. The team is willing to pay François the team's average salary, which is $750 000 a season.

Agent: I'm certain that we calculated the average salary to be $1 000 000 per season. You had better check your arithmetic.

Manager: There is no error, my friend. Half of the players earn $750 000 or more, while half of the players receive $750 000 or less. $750 000 is a fair offer.

This table lists the current salaries for the team.

Salary ($)	Number of Players
300 000	2
500 000	3
750 000	8
900 000	6
1 000 000	2
1 500 000	1
3 000 000	1
4 000 000	1

1. From looking at the table, do you think the agent or the manager is correct? Explain why.

2. Find the mean salary for the team. Describe how you calculated this amount.

3. Find the median salary. What method did you use to find it?

4. Were the statements by François' agent and the team manager correct?

5. Explain the problem with the use of the term *average* in these negotiations.

In statistics, the three most commonly used measures of central tendency are the mean, median, and mode. Each of these measures has its particular advantages and disadvantages for a given set of data.

A **mean** is defined as the sum of the values of a variable divided by the number of values. In statistics, it is important to distinguish between the mean of a population and the mean of a sample of that population. The sample mean will approximate the actual mean of the population, but the two means could have different values. Different symbols are used to distinguish the two kinds of means: The Greek letter mu, μ, represents a population mean, while $\bar{x}$, read as "x-bar," represents a sample mean. Thus,

$$\mu = \frac{x_1 + x_2 + \ldots + x_N}{N} \quad \text{and} \quad \bar{x} = \frac{x_1 + x_2 + \ldots + x_n}{n}$$

$$= \frac{\Sigma x}{N} \qquad\qquad\qquad = \frac{\Sigma x}{n}$$

where Σx is the sum of all values of X in the population or sample, N is the number of values in the entire population, and n is the number of values in a sample. Note that Σ, the capital Greek letter sigma, is used in mathematics as a symbol for "the sum of." If no limits are shown above or below the sigma, the sum includes all of the data.

Usually, the mean is what people are referring to when they use the term *average* in everyday conversation.

The **median** is the middle value of the data when they are ranked from highest to lowest. When there is an even number of values, the median is the midpoint between the two middle values.

The **mode** is the value that occurs most frequently in a distribution. Some distributions do not have a mode, while others have several.

Some distributions have **outliers,** which are values distant from the majority of the data. Outliers have a greater effect on means than on medians. For example, the mean and median for the salaries of the hockey team in the investigation have substantially different values because of the two very high salaries for the team's star players.

Example 1 Determining Mean, Median, and Mode

Two classes that wrote the same physics examination had the following results.

Class A	71	82	55	76	66	71	90	84	95	64	71	70	83	45	73	51	68	
Class B	54	80	12	61	73	69	92	81	80	61	75	74	15	44	91	63	50	84

a) Determine the mean, median, and mode for each class.

b) Use the measures of central tendency to compare the performance of the two classes.

c) What is the effect of any outliers on the mean and median?

Solution

a) For class A, the mean is

$$\bar{x} = \frac{\Sigma x}{n}$$

$$= \frac{71 + 82 + \ldots + 68}{17}$$

$$= \frac{1215}{17}$$

$$= 71.5$$

WEB CONNECTION

www.mcgrawhill.ca/links/MDM12

For more information about means, medians, and modes, visit the above web site and follow the links. For each measure, give an example of a situation where that measure is the best indicator of the centre of the data.

When the marks are ranked from highest to lowest, the middle value is 71. Therefore, the median mark for class A is 71. The mode for class A is also 71 since this mark is the only one that occurs three times.

Similarly, the mean mark for class B is $\frac{54 + 80 + \ldots + 84}{18} = 64.4$. When the marks are ranked from highest to lowest, the two middle values are 69 and 73, so the median mark for class B is $\frac{69 + 73}{2} = 71$. There are two modes since the values 61 and 80 both occur twice. However, the sample is so small that all the values occur only once or twice, so these modes may not be a reliable measure.

b) Although the mean score for class A is significantly higher than that for class B, the median marks for the two classes are the same. Notice that the measures of central tendency for class A agree closely, but those for class B do not.

c) A closer examination of the raw data shows that, aside from the two extremely low scores of 15 and 12 in class B, the distributions are not all that different. Without these two outlying marks, the mean for class B would be 70.1, almost the same as the mean for class A. Because of the relatively small size of class B, the effect of the outliers on its mean is significant. However, the values of these outliers have no effect on the median for class B. Even if the two outlying marks were changed to numbers in the 60s, the median mark would not change because it would still be greater than the two marks.

The median is often a better measure of central tendency than the mean for small data sets that contain outliers. For larger data sets, the effect of outliers on the mean is less significant.

Example 2 Comparing Samples to a Population

Compare the measures of central tendency for each class in Example 1 to those for all the students who wrote the physics examination.

Solution 1   Using a Graphing Calculator

Use the STAT EDIT menu to check that lists L1 and L2 are clear. Then, enter the data for class A in L1 and the data for class B in L2. Next, use the augment(function from the LIST OPS menu to combine L1 and L2, and store the result in L3. You can use the mean(and median(functions from the LIST MATH menu to find the mean and median for each of the three lists. You can also find these measures by using the 1-Var Stats command from the STAT CALC menu. To find the modes, sort the lists with the SortA(function from the LIST OPS menu, and then scroll down through the lists to find the most frequent values. Alternatively, you can use STAT PLOT to display a histogram for each list and read the x-values for the tallest bars with the TRACE instruction.

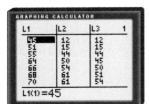 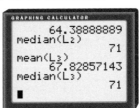 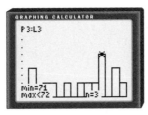

Note that the mean for class A overestimates the population mean, while the mean for class B underestimates it. The measures of central tendency for class A are reasonably close to those for the whole population of students who wrote the physics examination, but the two sets of measures are not identical. Because both of the low-score outliers happen to be in class B, it is a less representative sample of the population.

Solution 2   Using a Spreadsheet

Enter the data for class A and class B in separate columns. The AVG and MEAN functions in Corel® Quattro® Pro will calculate the mean for any range of cells you specify, as will the AVERAGE function in Microsoft® Excel.
In both spreadsheets, you can use the MEDIAN, and MODE functions to find the median and mode for each class and for the combined data for both classes. Note that all these functions ignore any blank cells in a specified range. The MODE function reports only one mode even if the data have two or more modes.

A:G5	▼ @ { }			@MODE(A3..B20)					
	A	B	C	D	E	F	G	H
1	MARKS			MEASURES OF CENTRAL TENDENCY				
2	Class A	Class B			Mean	Median	Mode	
3	71	54		Class A	71.47059	71	71	
4	82	80		Class B	64.38889	71	61	
5	55	12		Population	67.82857	71	71	
6	76	61						
7	66	73						
	71	69						

Solution 3   Using Fathom™

Drag the **case table** icon to the workspace and name the attribute for the first column Marks. Enter the data for class A and change the name of the **collection** from Collection1 to ClassA. Use the same method to enter the marks for class B into a collection called ClassB. To create a collection with the combined data, first open another **case table** and name the **collection** Both. Then, go back to the class A **case table** and use the Edit menu to select all cases and then copy them. Return to the Both **case table** and select Paste Cases from the Edit menu. Copy the cases from the class B table in the same way.

Now, right-click on the class A **collection** to open the **inspector**. Click the Measures tab, and create Mean, Median, and Mode measures. Use the Edit Formula menu to enter the formulas for these measures. Use the same procedure to find the **mean, median,** and **mode** for the other two collections. Note from the screen below that Fathom™ uses a complicated formula to find modes. See the Help menu or the Fathom™ section of Appendix B for details.

Project Prep

In your statistics project, you may find measures of central tendency useful for describing your data.

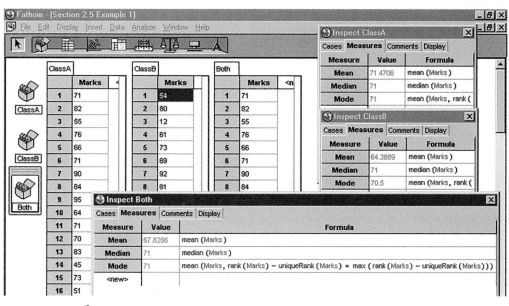

Chapter 8 discusses a method for calculating how representative of a population a sample is likely to be.

Sometimes, certain data within a set are more significant than others. For example, the mark on a final examination is often considered to be more important than the mark on a term test for determining an overall grade for a course. A **weighted mean** gives a measure of central tendency that reflects the relative importance of the data:

$$\bar{x}_w = \frac{w_1 x_1 + w_2 x_2 + \ldots + w_n x_n}{w_1 + w_2 + \ldots + w_n}$$

$$= \frac{\sum_i w_i x_i}{\sum_i w_i}$$

where $\sum_i w_i x_i$ is the sum of the weighted values and $\sum_i w_i$ is the sum of the various weighting factors.

Weighted means are often used in calculations of indices.

Example 3 Calculating a Weighted Mean

The personnel manager for Statsville Marketing Limited considers five criteria when interviewing a job applicant. The manager gives each applicant a score between 1 and 5 in each category, with 5 as the highest score. Each category has a weighting between 1 and 3. The following table lists a recent applicant's scores and the company's weighting factors.

Criterion	Score, x_i	Weighting Factor, w_i
Education	4	2
Job experience	2	2
Interpersonal skills	5	3
Communication skills	5	3
References	4	1

a) Determine the weighted mean score for this job applicant.

b) How does this weighted mean differ from the unweighted mean?

c) What do the weighting factors indicate about the company's hiring priorities?

Solution

a) To compute the weighted mean, find the sum of the products of each score and its weighting factor.

$$\bar{x}_w = \frac{\sum_i w_i x_i}{\sum_i w_i}$$

$$= \frac{2(4) + 2(2) + 3(5) + 3(5) + (1)4}{2 + 2 + 3 + 3 + 1}$$

$$= \frac{46}{11}$$

$$= 4.2$$

Therefore, this applicant had a weighted-mean score of approximately 4.2.

b) The unweighted mean is simply the sum of unweighted scores divided by 5.

$$\bar{x} = \frac{\sum x}{n}$$

$$= \frac{4 + 2 + 5 + 5 + 4}{5}$$

$$= 4$$

Without the weighting factors, this applicant would have a mean score of 4 out of 5.

c) Judging by these weighting factors, the company places a high importance on an applicant's interpersonal and communication skills, moderate importance on education and job experience, and some, but low, importance on references.

When a set of data has been grouped into intervals, you can approximate the mean using the formula

$$\mu \doteq \frac{\sum_i f_i m_i}{\sum_i f_i} \qquad \bar{x} \doteq \frac{\sum_i f_i m_i}{\sum_i f_i}$$

where m_i is the midpoint value of an interval and f_i the frequency for that interval.

You can estimate the median for grouped data by taking the midpoint of the interval within which the median is found. This interval can be found by analysing the cumulative frequencies.

Example 4   Calculating the Mean and Median for Grouped Data

A group of children were asked how many hours a day they spend watching television. The table at the right summarizes their responses.

Number of Hours	Number of Children, f_i
0–1	1
1–2	4
2–3	7
3–4	3
4–5	2
5–6	1

a) Determine the mean and median number of hours for this distribution.

b) Why are these values simply approximations?

Solution

a) First, find the midpoints and cumulative frequencies for the intervals. Then, use the midpoints and the frequencies for the intervals to calculate an estimate for the mean.

Number of Hours	Midpoint, x_i	Number of Children, f_i	Cumulative Frequency	$f_i x_i$
0–1	0.5	1	1	0.5
1–2	1.5	4	5	6
2–3	2.5	7	12	17.5
3–4	3.5	3	15	10.5
4–5	4.5	2	17	9
5–6	5.5	1	18	5.5
		$\sum_i f_i = 18$		$\sum_i f_i x_i = 49$

$$\bar{x} \doteq \frac{\sum_i f_i x_i}{\sum_i f_i}$$
$$= \frac{49}{18}$$
$$\doteq 2.7$$

Therefore, the mean time the children spent watching television is approximately 2.7 h a day.

To determine the median, you must identify the interval in which the middle value occurs. There are 18 data values, so the median is the mean of the ninth and tenth values. According to the cumulative-frequency column, both of these occur within the interval of 2–3 h. Therefore, an approximate value for the median is 2.5 h.

b) These values for the mean and median are approximate because you do not know where the data lie within each interval. For example, the child whose viewing time is listed in the first interval could have watched anywhere from 0 to 60 min of television a day. If the median value is close to one of the boundaries of the interval, then taking the midpoint of the interval as the median could give an error of almost 30 min.

Practise

1. For each set of data, calculate the mean, median, and mode.

   a) 2.4 3.5 1.9 3.0 3.5 2.4 1.6 3.8 1.2 2.4 3.1 2.7 1.7 2.2 3.3

   b) 10 15 14 19 18 17 12 10 14 15 18 20 9 14 11 18

2. **a)** List a set of eight values that has no mode.

   b) List a set of eight values that has a median that is not one of the data values.

   c) List a set of eight values that has two modes.

   d) List a set of eight values that has a median that is one of the data values.

Apply, Solve, Communicate

3. Stacey got 87% on her term work in chemistry and 71% on the final examination. What will her final grade be if the term mark counts for 70% and the final examination counts for 30%?

4. Communication Determine which measure of central tendency is most appropriate for each of the following sets of data. Justify your choice in each case.

  a) baseball cap sizes

  b) standardized test scores for 2000 students

  c) final grades for a class of 18 students

  d) lifetimes of mass-produced items, such as batteries or light bulbs

5. An interviewer rates candidates out of 5 for each of three criteria: experience, education, and interview performance. If the first two criteria are each weighted twice as much as the interview, determine which of the following candidates should get the job.

Criterion	Nadia	Enzo	Stephan
Experience	4	5	5
Education	4	4	3
Interview	4	3	4

6. Determine the effect the two outliers have on the mean mark for all the students in Example 2. Explain why this effect is different from the effect the outliers had on the mean mark for class B.

7. Application The following table shows the grading system for Xabbu's calculus course.

Term Mark	Overall Mark
Knowledge and understanding (K/U) 35%	Term mark 70%
Thinking, inquiry, problem solving (TIPS) 25%	Final examination 30%
Communication (C) 15%	
Application (A) 25%	

  a) Determine Xabbu's term mark if he scored 82% in K/U, 71% in TIPS, 85% in C, and 75% in A.

  b) Determine Xabbu's overall mark if he scored 65% on the final examination.

8. Application An academic award is to be granted to the student with the highest overall score in four weighted categories. Here are the scores for the three finalists.

Criterion	Weighting	Paulo	Janet	Jamie
Academic achievement	3	4	3	5
Extra-curricular activities	2	4	4	4
Community service	2	2	5	3
Interview	1	5	5	4

  a) Calculate each student's mean score without considering the weighting factors.

  b) Calculate the weighted-mean score for each student.

  c) Who should win the award? Explain.

9. Al, a shoe salesman, needs to restock his best-selling sandal. Here is a list of the sizes of the pairs he sold last week. This sandal does not come in half-sizes.

10	7	6	8	7	10	5	10	7	9
11	4	6	7	10	10	7	8	10	7
9	7	10	4	7	7	10	11		

  a) Determine the three measures of central tendency for these sandals.

  b) Which measure has the greatest significance for Al? Explain.

  c) What other value is also significant?

  d) Construct a histogram for the data. What might account for the shape of this histogram?

10. Communication Last year, the mean number of goals scored by a player on Statsville's soccer team was 6.

  a) How many goals did the team score last year if there were 15 players on the team?

  b) Explain how you arrived at the answer for part a) and show why your method works.

11. Inquiry/Problem Solving The following table shows the salary structure of Statsville Plush Toys, Inc. Assume that salaries exactly on an interval boundary have been placed in the higher interval.

Salary Range ($000)	Number of Employees
20–30	12
30–40	24
40–50	32
50–60	19
60–70	9
70–80	3
80–90	0
90–100	1

a) Determine the approximate mean salary for an employee of this firm.

b) Determine the approximate median salary.

c) How much does the outlier influence the mean and median salaries? Use calculations to justify your answer.

12. Inquiry/Problem Solving A group of friends and relatives get together every Sunday for a little pick-up hockey. The ages of the 30 regulars are shown below.

22	28	32	45	48	19	20	52	50	21
30	46	21	38	45	49	18	25	23	46
51	24	39	48	28	20	50	33	17	48

a) Determine the mean, median, and mode for this distribution.

b) Which measure best describes these data? Explain your choice.

c) Group these data into six intervals and produce a frequency table.

d) Illustrate the grouped data with a frequency diagram. Explain why the shape of this frequency diagram could be typical for such groups of hockey players.

e) Produce a cumulative-frequency diagram.

f) Determine a mean, median, and mode for the grouped data. Explain any differences between these measures and the ones you calculated in part a).

13. The **modal interval** for grouped data is the interval that contains more data than any other interval.

a) Determine the modal interval(s) for your data in part d) of question 12.

b) Is the modal interval a useful measure of central tendency for this particular distribution? Why or why not?

14. a) Explain the effect outliers have on the median of a distribution. Use examples to support your explanation.

b) Explain the effect outliers have on the mode of a distribution. Consider different cases and give examples of each.

C

15. The harmonic mean is defined as $\left(\sum_i \frac{1}{nx_i} \right)^{-1}$, where n is the number of values in the set of data.

a) Use a harmonic mean to find the average price of gasoline for a driver who bought $20 worth at 65¢/L last week and another $20 worth at 70¢/L this week.

b) Describe the types of calculations for which the harmonic mean is useful.

16. The geometric mean is defined as $\sqrt[n]{x_1 \times x_2 \times \ldots \times x_n}$, where n is the number of values in the set of data.

a) Use the geometric mean to find the average annual increase in a labour contract that gives a 4% raise the first year and a 2% raise for the next three years.

b) Describe the types of calculations for which the geometric mean is useful.

Measures of Spread

The measures of central tendency indicate the central values of a set of data. Often, you will also want to know how closely the data cluster around these centres.

INVESTIGATE & INQUIRE: Spread in a Set of Data

For a game of basketball, a group of friends split into two randomly chosen teams. The heights of the players are shown in the table below.

Falcons		Ravens	
Player	Height (cm)	Player	Height (cm)
Laura	183	Sam	166
Jamie	165	Shannon	163
Deepa	148	Tracy	168
Colleen	146	Claudette	161
Ingrid	181	Maria	165
Justiss	178	Amy	166
Sheila	154	Selena	166

1. Judging by the raw data in this table, which team do you think has a height advantage? Explain why.

2. Do the measures of central tendency confirm that the teams are mismatched? Why or why not?

3. Explain how the distributions of heights on the two teams might give one of them an advantage. How could you use a diagram to illustrate the key difference between the two teams?

The **measures of spread** or **dispersion** of a data set are quantities that indicate how closely a set of data clusters around its centre. Just as there are several measures of central tendency, there are also different measures of spread.

Standard Deviation and Variance

A **deviation** is the difference between an individual value in a set of data and the mean for the data.

For a population,
deviation $= x - \mu$

For a sample,
deviation $= x - \bar{x}$

The larger the size of the deviations, the greater the spread in the data. Values less than the mean have negative deviations. If you simply add up all the deviations for a data set, they will cancel out. You could use the sum of the absolute values of the deviations as a measure of spread. However, statisticians have shown that a root-mean-square quantity is a more useful measure of spread. The **standard deviation** is the square root of the mean of the squares of the deviations.

The lowercase Greek letter sigma, σ, is the symbol for the standard deviation of a population, while the letter s stands for the standard deviation of a sample.

Population standard deviation

$$\sigma = \sqrt{\frac{\sum(x - \mu)^2}{N}}$$

Sample standard deviation

$$s = \sqrt{\frac{\sum(x - \bar{x})^2}{n - 1}}$$

where N is the number of data in the population and n is the number in the sample.

Note that the formula for s has $n - 1$ in the denominator instead of n. This denominator compensates for the fact that a sample taken from a population tends to underestimate the deviations in the population. Remember that the sample mean, $\bar{x}$, is not necessarily equal to the population mean, μ. Since $\bar{x}$ is the central value of the sample, the sample data cluster closer to $\bar{x}$ than to μ. When n is large, the formula for s approaches that for σ.

Also note that the standard deviation gives greater weight to the larger deviations since it is based on the *squares* of the deviations.

The mean of the squares of the deviations is another useful measure. This quantity is called the **variance** and is equal to the square of the standard deviation.

Population variance

$$\sigma^2 = \frac{\sum(x - \mu)^2}{N}$$

Sample variance

$$s^2 = \frac{\sum(x - \bar{x})^2}{n - 1}$$

Example 1 Using a Formula to Calculate Standard Deviations

Use means and standard deviations to compare the distribution of heights for the two basketball teams listed in the table on page 136.

Solution

Since you are considering the teams as two separate populations, use the mean and standard deviation formulas for populations. First, calculate the mean height for the Falcons.

$$\mu = \frac{\Sigma x}{N}$$
$$= \frac{1155}{7}$$
$$= 165$$

Next, calculate all the deviations and their squares.

Falcons	Height (cm)	Deviation, $x - \mu$	$(x - \mu)^2$
Laura	183	18	324
Jamie	165	0	0
Deepa	148	−17	289
Colleen	146	−19	361
Ingrid	181	16	256
Justiss	178	13	169
Sheila	154	−11	121
Sum	**1155**	**0**	**1520**

Now, you can determine the standard deviation.

$$\sigma = \sqrt{\frac{\Sigma(x - \mu)^2}{N}}$$
$$= \sqrt{\frac{1520}{7}}$$
$$= 14.7$$

Therefore, the Falcons have a mean height of 165 cm with a standard deviation of 14.7 cm.

Similarly, you can determine that the Ravens also have a mean height of 165 cm, but their standard deviation is only 2.1 cm. Clearly, the Falcons have a much greater spread in height than the Ravens. Since the two teams have the same mean height, the difference in the standard deviations indicates that the Falcons have some players who are taller than any of the Ravens, but also some players who are shorter.

If you were to consider either of the basketball teams in the example above as a sample of the whole group of players, you would use the formula for s to calculate the team's standard deviation. In this case, you would be using the sample to estimate the characteristics of a larger population. However, the teams are very small samples, so they could have significant random variations, as the difference in their standard deviations demonstrates.

For large samples the calculation of standard deviation can be quite tedious. However, most business and scientific calculators have built-in functions for such calculations, as do spreadsheets and statistical software.

See Appendix B for more detailed information about technology functions and keystrokes.

Example 2 Using Technology to Calculate Standard Deviations

A veterinarian has collected data on the life spans of a rare breed of cats.

Life Spans (in years)													
16	18	19	12	11	15	20	21	18	15	16	13	16	22
18	19	17	14	9	14	15	19	20	15	15			

Determine the mean, standard deviation, and the variance for these data.

Solution 1 Using a Graphing Calculator

Use the ClrList command to make sure list L1 is clear, then enter the data into it. Use the 1-Var Stats command from the STAT CALC menu to calculate a set of statistics including the mean and the standard deviation. Note that the calculator displays both a sample standard deviation, Sx, and a population standard deviation, σx. Use Sx since you are dealing with a sample in this case. Find the variance by calculating the square of Sx.

The mean life span for this breed of cats is about 16.3 years with a standard deviation of 3.2 years and a variance of 10.1. Note that variances are usually stated without units. The units for this variance are years squared.

Solution 2 Using a Spreadsheet

Enter the data into your spreadsheet program. With Corel® Quattro® Pro, you can use the AVG, STDS, and VARS functions to calculate the mean, sample standard deviation, and sample variance. In Microsoft® Excel, the equivalent functions are AVERAGE, STDEV, and VAR.

Microsoft Excel - Cats

File Edit View Insert Format Tools Data Window Help

D3 = =VAR(A3:A27)

	A	B	C	D	E	F
1	Life Span		Mean	16.28		
2	(in years)		Standard deviation	3.182242		
3	16		Variance	10.12667		
4	18					
5	19					
6	12					
7	11					
8	15					
9	20					
10	21					
11	18					
12	15					
13	16					
14	13					
15	16					
16	22					
17	18					
18	19					
19	17					
20	14					
21	9					

Solution 3   Using Fathom™

Drag a new **case table** onto the workspace, name the attribute for the first column Lifespan, and enter the data. Right-click to open the **inspector**, and click the Measures tab. Create Mean, StdDev, and Variance measures and select the formulas for the **mean**, **standard deviation**, and **variance** from the Edit Formula/ Functions/Statistical/One Attribute menu.

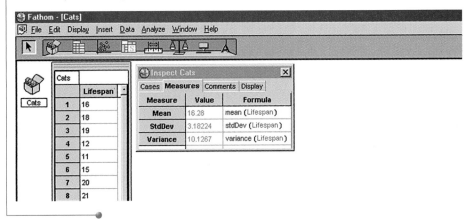

If you are working with grouped data, you can estimate the standard deviation using the following formulas.

For a population,

$$\sigma \doteq \sqrt{\frac{\Sigma f_i (m_i - \mu)^2}{N}}$$

For a sample,

$$s \doteq \sqrt{\frac{\Sigma f_i (m_i - \bar{x})^2}{n - 1}}$$

where f_i is the frequency for a given interval and m_i is the midpoint of the interval. However, calculating standard deviations from raw, ungrouped data will give more accurate results.

> **Project Prep**
>
> In your statistics project, you may wish to use an appropriate measure of spread to describe the distribution of your data.

Quartiles and Interquartile Ranges

Quartiles divide a set of ordered data into four groups with equal numbers of values, just as the median divides data into two equally sized groups. The three "dividing points" are the first quartile (Q_1), the median (sometimes called the second quartile or Q_2), and the third quartile (Q_3). Q_1 and Q_3 are the medians of the lower and upper halves of the data.

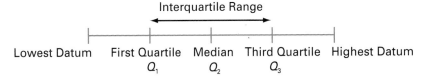

Recall that when there are an even number of data, you take the midpoint between the two middle values as the median. If the number of data below the median is even, Q_1 is the midpoint between the two middle values in this half of the data. Q_3 is determined in a similar way.

The **interquartile range** is $Q_3 - Q_1$, which is the range of the middle half of the data. The larger the interquartile range, the larger the spread of the central half of the data. Thus, the interquartile range provides a measure of spread. The **semi-interquartile range** is one half of the interquartile range. Both these ranges indicate how closely the data are clustered around the median.

A **box-and-whisker plot** of the data illustrates these measures. The box shows the first quartile, the median, and the third quartile. The ends of the "whiskers" represent the lowest and highest values in the set of data. Thus, the length of the box shows the interquartile range, while the left whisker shows the range of the data below the first quartile, and the right whisker shows the range above the third quartile.

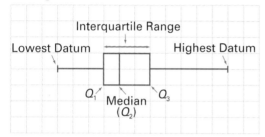

A **modified box-and-whisker plot** is often used when the data contain outliers. By convention, any point that is at least 1.5 times the box length away from the box is classified as an outlier. A modified box-and-whisker plot shows such outliers as separate points instead of including them in the whiskers. This method usually gives a clearer illustration of the distribution.

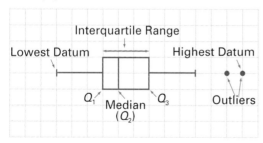

Example 3 Determining Quartiles and Interquartile Ranges

A random survey of people at a science-fiction convention asked them how many times they had seen *Star Wars*. The results are shown below.

3 4 2 8 10 5 1 15 5 16 6 3 4 9 12 3 30 2 10 7

a) Determine the median, the first and third quartiles, and the interquartile and semi-interquartile ranges. What information do these measures provide?

b) Prepare a suitable box plot of the data.

c) Compare the results in part a) to those from last year's survey, which found a median of 5.1 with an interquartile range of 8.0.

Solution 1   Using Pencil and Paper

a) First, put the data into numerical order.

1 2 2 3 3 3 4 4 5 5 6 7 8 9 10 10 12 15 16 30

The median is either the middle datum or, as in this case, the mean of the two middle data:

$$\text{median} = \frac{5+6}{2}$$
$$= 5.5$$

The median value of 5.5 indicates that half of the people surveyed had seen *Star Wars* less than 5.5 times and the other half had seen it more than 5.5 times.

To determine Q_1, find the median of the lower half of the data. Again, there are two middle values, both of which are 3. Therefore, $Q_1 = 3$.

Similarly, the two middle values of the upper half of the data are both 10, so $Q_3 = 10$.

Since Q_1 and Q_3 are the boundaries for the central half of the data, they show that half of the people surveyed have seen *Star Wars* between 3 and 10 times.

$$Q_3 - Q_1 = 10 - 3$$
$$= 7$$

Therefore, the interquartile range is 7. The semi-interquartile range is half this value, or 3.5. These ranges indicate the spread of the central half of the data.

b) The value of 30 at the end of the ordered data is clearly an outlier. Therefore, a modified box-and-whisker plot will best illustrate this set of data.

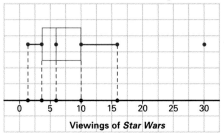

Viewings of *Star Wars*

c) Comparing the two surveys shows that the median number of viewings is higher this year and the data are somewhat less spread out.

Solution 2 Using a Graphing Calculator

a) Use the STAT EDIT menu to enter the data into a list. Use the 1-Var Stats command from the CALC EDIT menu to calculate the statistics for your list. Scroll down to see the values for the median, Q_1, and Q_3. Use the values for Q_1 and Q_3 to calculate the interquartile and semi-interquartile ranges.

b) Use STAT PLOT to select a modified box plot of your list. Press GRAPH to display the box-and-whisker plot and adjust the window settings, if necessary.

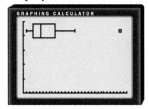

Solution 3 Using Fathom™

a) Drag a new **case table** onto the workspace, create an attribute called StarWars, and enter your data. Open the **inspector** and create Median, Q1, Q3, and IQR measures. Use the Edit Formula/Functions/Statistical/One Attribute menu to enter the formulas for the **median**, **quartiles**, and **interquartile range**.

b) Drag the **graph icon** onto the workspace, then drop the **StarWars** attribute on the *x*-axis of the graph. Select **Box Plot** from the drop-down menu in the upper right corner of the graph.

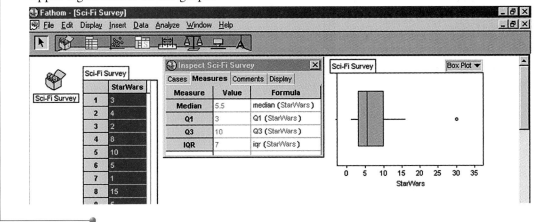

Although a quartile is, strictly speaking, a single value, people sometimes speak of a datum being *within* a quartile. What they really mean is that the datum is in the quarter whose upper boundary is the quartile. For example, if a value x_1 is "within the first quartile," then $x_1 \leq Q_1$. Similarly, if x_2 is "within the third quartile," then the median $\leq x_2 \leq Q_3$.

Example 4 Classifying Data by Quartiles

In a survey of low-risk mutual funds, the median annual yield was 7.2%, while Q_1 was 5.9% and Q_3 was 8.3%. Describe the following funds in terms of quartiles.

Mutual Fund	Annual Yield (%)
XXY Value	7.5
YYZ Dividend	9.0
ZZZ Bond	7.2

Solution

The yield for the XXY Value fund was between the median and Q_3. You might see this fund described as being in the third quartile or having a third-quartile yield.

YYZ Dividend's yield was above Q_3. This fund might be termed a fourth- or top-quartile fund.

ZZZ Bond's yield was equal to the median. This fund could be described as a median fund or as having median performance.

Percentiles

Percentiles are similar to quartiles, except that percentiles divide the data into 100 intervals that have equal numbers of values. Thus, k percent of the data are less than or equal to kth percentile, P_k, and $(100 - k)$ percent are greater than or equal to P_k. Standardized tests often use percentiles to convert raw scores to scores on a scale from 1 to 100. As with quartiles, people sometimes use the term *percentile* to refer to the intervals rather than their boundaries.

Example 5 Percentiles

An audio magazine tested 60 different models of speakers and gave each one an overall rating based on sound quality, reliability, efficiency, and appearance. The raw scores for the speakers are listed in ascending order below.

35	47	57	62	64	67	72	76	83	90
38	50	58	62	65	68	72	78	84	91
41	51	58	62	65	68	73	79	86	92
44	53	59	63	66	69	74	81	86	94
45	53	60	63	67	69	75	82	87	96
45	56	62	64	67	70	75	82	88	98

a) If the Audio Maximizer Ultra 3000 scored at the 50th percentile, what was its raw score?

b) What is the 90th percentile for these data?

c) Does the SchmederVox's score of 75 place it at the 75th percentile?

Solution

a) Half of the raw scores are less than or equal to the 50th percentile and half are greater than or equal to it. From the table, you can see that 67 divides the data in this way. Therefore, the Audio Maximizer Ultra 3000 had a raw score of 67.

b) The 90th percentile is the boundary between the lower 90% of the scores and the top 10%. In the table, you can see that the top 10% of the scores are in the 10th column. Therefore, the 90th percentile is the midpoint between values of 88 and 90, which is 89.

c) First, determine 75% of the number of raw scores.

$$60 \times 75\% = 45$$

There are 45 scores less than or equal to the 75th percentile. Therefore, the 75th percentile is the midpoint between the 45th and 46th scores. These two scores are 79 and 81, so the 75th percentile is 80. The SchmederVox's score of 75 is below the 75th percentile.

Z-Scores

A **z-score** is the number of standard deviations that a datum is from the mean. You calculate the z-score by dividing the deviation of a datum by the standard deviation.

For a population,

$$z = \frac{x - \mu}{\sigma}$$

For a sample,

$$z = \frac{x - \bar{x}}{s}$$

Variable values below the mean have negative z-scores, values above the mean have positive z-scores, and values equal to the mean have a zero z-score. Chapter 8 describes z-scores in more detail.

Example 6 Determining Z-Scores

Determine the z-scores for the Audio Maximizer Ultra 3000 and SchmederVox speakers.

Solution

You can use a calculator, spreadsheet, or statistical software to determine that the mean is 68.1 and the standard deviation is 15.2 for the speaker scores in Example 4.

Now, use the mean and standard deviation to calculate the z-scores for the two speakers.

For the Audio Maximizer Ultra 3000,

$$z = \frac{x - \bar{x}}{s}$$

$$= \frac{67 - 68.1}{15.2}$$

$$= -0.072$$

For the SchmederVox,

$$z = \frac{x - \bar{x}}{s}$$

$$= \frac{75 - 68.1}{15.2}$$

$$= 0.46$$

The Audio Maximizer Ultra 3000 has a z-score of -0.072, indicating that it is approximately 7% of a standard deviation below the mean. The SchmederVox speaker has a z-score of 0.46, indicating that it is approximately half a standard deviation above the mean.

Key Concepts

- The variance and the standard deviation are measures of how closely a set of data clusters around its mean. The variance and standard deviation of a sample may differ from those of the population the sample is drawn from.

- Quartiles are values that divide a set of ordered data into four intervals with equal numbers of data, while percentiles divide the data into 100 intervals.

- The interquartile range and semi-interquartile range are measures of how closely a set of data clusters around its median.

- The z-score of a datum is a measure of how many standard deviations the value is from the mean.

Communicate Your Understanding

1. Explain how the term *root-mean-square* applies to the calculation of the standard deviation.

2. Why does the semi-interquartile range give only an approximate measure of how far the first and third quartiles are from the median?

3. Describe the similarities and differences between the standard deviation and the semi-interquartile range.

4. Are the median, the second quartile, and the 50th percentile always equal? Explain why or why not.

Practise

1. Determine the mean, standard deviation, and variance for the following samples.

a) Scores on a data management quiz (out of 10 with a bonus question):

5	7	9	6	5	10	8	2
11	8	7	7	6	9	5	8

b) Costs for books purchased including taxes (in dollars):

12.55	15.31	21.98	45.35	19.81
33.89	29.53	30.19	38.20	

2. Determine the median, Q_1, Q_3, the interquartile range, and semi-interquartile range for the following sets of data.

a) Number of home runs hit by players on the Statsville little league team:

6	4	3	8	9	11	6	5	15

b) Final grades in a geography class:

88	56	72	67	59	48	81	62
90	75	75	43	71	64	78	84

3. For a recent standardized test, the median was 88, Q_1 was 67, and Q_3 was 105. Describe the following scores in terms of quartiles.

a) 8

b) 81

c) 103

4. What percentile corresponds to

a) the first quartile?

b) the median?

c) the third quartile?

5. Convert these raw scores to z-scores.

18	15	26	20	21

Apply, Solve, Communicate

B

6. The board members of a provincial organization receive a car allowance for travel to meetings. Here are the distances the board logged last year (in kilometres).

44	18	125	80	63	42	35	68	52
75	260	96	110	72	51			

a) Determine the mean, standard deviation, and variance for these data.

b) Determine the median, interquartile range, and semi-interquartile range.

c) Illustrate these data using a box-and-whisker plot.

d) Identify any outliers.

7. The nurses' union collects data on the hours worked by operating-room nurses at the Statsville General Hospital.

Hours Per Week	Number of Employees
12	1
32	5
35	7
38	8
42	5

a) Determine the mean, variance, and standard deviation for the nurses' hours.

b) Determine the median, interquartile range, and semi-interquartile range.

c) Illustrate these data using a box-and-whisker plot.

8. Application

a) Predict the changes in the standard deviation and the box-and-whisker plot if the outlier were removed from the data in question 7.

b) Remove the outlier and compare the new results to your original results.

c) Account for any differences between your prediction and your results in part b).

9. Application Here are the current salaries for François' team.

Salary ($)	Number of Players
300 000	2
500 000	3
750 000	8
900 000	6
1 000 000	2
1 500 000	1
3 000 000	1
4 000 000	1

a) Determine the standard deviation, variance, interquartile range, and semi-interquartile range for these data.

b) Illustrate the data with a modified box-and-whisker plot.

c) Determine the z-score of François' current salary of $300 000.

d) What will the new z-score be if François' agent does get him a million-dollar contract?

10. Communication Carol's golf drives have a mean of 185 m with a standard deviation of 25 m, while her friend Chi-Yan shoots a mean distance of 170 m with a standard deviation of 10 m. Explain which of the two friends is likely to have a better score in a round of golf. What assumptions do you have to make for your answer?

11. Under what conditions will Q_1 equal one of the data points in a distribution?

12. a) Construct a set of data in which $Q_1 = Q_3$ and describe a situation in which this equality might occur.

b) Will such data sets always have a median equal to Q_1 and Q_3? Explain your reasoning.

13. Is it possible for a set of data to have a standard deviation much smaller than its semi-interquartile range? Give an example or explain why one is not possible.

14. Inquiry/Problem Solving A business-travellers' association rates hotels on a variety of factors including price, cleanliness, services, and amenities to produce an overall score out of 100 for each hotel. Here are the ratings for 50 hotels in a major city.

39	50	56	60	65	68	73	77	81	87
41	50	56	60	65	68	74	78	81	89
42	51	57	60	66	70	74	78	84	91
44	53	58	62	67	71	75	79	85	94
48	55	59	63	68	73	76	80	86	96

a) What score represents

   i) the 50th percentile?

   ii) the 95th percentile?

b) What percentile corresponds to a rating of 50?

c) The travellers' association lists hotels above the 90th percentile as "highly recommended" and hotels between the 75th and 90th percentiles as "recommended." What are the minimum scores for the two levels of recommended hotels?

✓ ACHIEVEMENT CHECK			
Knowledge/ Understanding	Thinking/Inquiry/ Problem Solving	Communication	Application

15. a) A data-management teacher has two classes whose midterm marks have identical means. However, the standard deviations for each class are significantly different. Describe what these measures tell you about the two classes.

b) If two sets of data have the same mean, can one of them have a larger standard deviation and a smaller interquartile range than the other? Give an example or explain why one is not possible.

16. Show that $\sum(x - \bar{x}) = 0$ for any distribution.

17. a) Show that $s = \sqrt{\dfrac{n(\sum x^2) - (\sum x)^2}{n(n-1)}}$.

(Hint: Use the fact that $\sum x = n\bar{x}$.)

b) What are two advantages of using the formula in part a) for calculating standard deviations?

18. Communication The **midrange** of a set of data is defined as half of the sum of the highest value and the lowest value. The incomes for the employees of Statsville Lawn Ornaments Limited are listed below (in thousands of dollars).

| 28 | 34 | 49 | 22 | 50 | 31 | 55 | 32 | 73 | 21 |
| 63 | 112 | 35 | 19 | 44 | 28 | 59 | 85 | 47 | 39 |

a) Determine the midrange and interquartile range for these data.

b) What are the similarities and differences between these two measures of spread?

19. The **mean absolute deviation** of a set of data is defined as $\dfrac{\sum|x - \bar{x}|}{n}$, where $|x - \bar{x}|$ is the absolute value of the difference between each data point and the mean.

a) Calculate the mean absolute deviation and the standard deviation for the data in question 18.

b) What are the similarities and differences between these two measures of spread?

Career Connection

Statistician

Use of statistics today is so widespread that there are numerous career opportunities for statisticians in a broad range of fields. Governments, medical-research laboratories, sports agencies, financial groups, and universities are just a few of the many organizations that employ statisticians. Current trends suggest an ongoing need for statisticians in many areas.

A statistician is engaged in the collection, analysis, presentation, and interpretation of data in a variety of forms. Statisticians provide insight into which data are likely to be reliable and whether valid conclusions or predictions can be drawn from them. A research statistician might develop new statistical techniques or applications.

Because computers are essential for analysing large amounts of data, a statistician should possess a strong background in computers as well as mathematics. Many positions call for a minimum of a bachelor's or master's degree. Research at a university or work for a consulting firm usually requires a doctorate.

WEB CONNECTION

www.mcgrawhill.ca/links/MDM12

For more information about a career as a statistician and other careers related to mathematics, visit the above web site and follow the links.

Review of Key Concepts

2.1 Data Analysis With Graphs
Refer to the Key Concepts on page 100.

1. The following data show monthly sales of houses by a real-estate agency.

6	5	7	6	8	3	5	4	6
7	5	9	5	6	6	7		

   a) Construct an ungrouped frequency table for this distribution.

   b) Create a frequency diagram.

   c) Create a cumulative-frequency diagram.

2. A veterinary study recorded the masses in grams of 25 kittens at birth.

240	300	275	350	280	260	320
295	340	305	280	265	300	275
315	285	320	325	275	270	290
245	235	305	265			

   a) Organize these data into groups.

   b) Create a frequency table and histogram.

   c) Create a frequency polygon.

   d) Create a relative-frequency diagram.

3. A class of data-management students listed their favourite board games.

Game	Frequency
Pictionary®	10
Chess	5
Trivial Pursuit®	8
MONOPOLY®	3
Balderdash®	6
Other	4

   a) What type of data does this table show? Explain your reasoning.

   b) Graph these data using an appropriate format.

   c) Explain why you chose the type of graph you did.

2.2 Indices
Refer to the Key Concepts on page 109.

The following graph shows four categories from the basket of goods and services used to calculate the consumer price index.

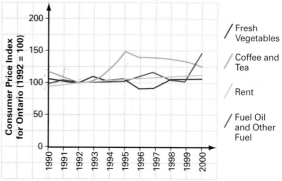

4. a) What is this type of graph called?

   b) Which of the four categories had the greatest increase during the period shown?

   c) Why do all four graphs intersect at 1992?

   d) Which category was

   i) the most volatile?

   ii) the least volatile?

   e) Suggest reasons for this difference in volatility.

5. a) If a tin of coffee cost $5.99 in 1992, what would you expect it to cost in

   i) 1995?

   ii) 1990?

   b) What rent would a typical tenant pay in 2000 for an apartment that had a rent of $550 per month in 1990?

   c) What might you expect to pay for broccoli in 2000, if the average price you paid in 1996 was $1.49 a bunch?

2.3 Sampling Techniques

Refer to the Key Concepts on page 116.

6. **a)** Explain the difference between a stratified sample and a systematic sample.

   b) Describe a situation where a convenience sample would be an appropriate technique.

   c) What are the advantages and disadvantages of a voluntary-response sample?

7. Suppose you are conducting a survey that you would like to be as representative as possible of the entire student body at your school. However, you have time to visit only six classes and to process data from a total of 30 students.

   a) What sampling technique would you use?

   b) Describe how you would select the students for your sample.

8. Drawing names from a hat and using a random-number generator are two ways to obtain a simple random sample. Describe two other ways of selecting a random sample.

2.4 Bias in Surveys

Refer to the Key Concepts on page 122.

9. Identify the type of bias in each of the following situations and state whether the bias is due to the sampling technique or the method of data collection.

   a) A survey asks a group of children whether or not they should be allowed unlimited amounts of junk food.

   b) A teachers asks students to raise their hands if they have ever told a harmless lie.

   c) A budding musician plays a new song for family members and friends to see if it is good enough to record professionally.

   d) Every fourth person entering a public library is asked: "Do you think Carol Shields should receive the Giller prize for her brilliant and critically acclaimed new novel?"

10. For each situation in question 9, suggest how the statistical process could be changed to remove the bias.

2.5 Measures of Central Tendency

Refer to the Key Concepts on page 133.

11. **a)** Determine the mean, median, and mode for the data in question 1.

    b) Which measure of central tendency best describes these data? Explain your reasoning.

12. **a)** Use your grouped data from question 2 to estimate the mean and median masses for the kittens.

    b) Determine the actual mean and median masses from the raw data.

    c) Explain any differences between your answers to parts a) and b).

13. **a)** For what type of "average" will the following statement always be true?

    "There are as many people with below-average ages as there are with above-average ages. "

    b) Is this statement likely to be true for either of the other measures of central tendency discussed in this chapter? Why or why not?

14. Angela is applying to a university engineering program that weights an applicant's eight best grade-12 marks as shown in the following table.

Subjects	Weighting
Calculus, chemistry, geometry and discrete mathematics, physics	3
Computer science, data management, English	2
Other	1

Angela's grade-12 final marks are listed below.

Subject	Mark	Subject	Mark
Calculus	95	Computer science	84
English	89	Chemistry	90
Geometry and discrete mathematics	94	Mathematics of data management	87
Physical education	80	Physics	92

a) Calculate Angela's weighted average.

b) Calculate Angela's unweighted average.

c) Explain why the engineering program would use this weighting system.

15. Describe three situations where the mode would be the most appropriate measure of central tendency.

2.6 Measures of Spread

Refer to the Key Concepts on page 147.

16. a) Determine the standard deviation, the interquartile range, and the semi-interquartile range for the data in question 1.

b) Create a box-and-whisker plot for these data.

c) Are there any outliers in the data? Justify your answer.

17. a) Explain why you cannot calculate the semi-interquartile range if you know only the difference between either Q_3 and the median or median and Q_1.

b) Explain how you could determine the semi-interquartile range if you did know both of the differences in part a).

18. a) For the data in question 2, determine

  i) the first and third quartiles

  ii) the 10th, 25th, 75th, and 90th percentiles

b) Would you expect any of the values in part a) to be equal? Why or why not?

19. The scores on a precision-driving test for prospective drivers at a transit company have a mean of 100 and a standard deviation of 15.

a) Determine the z-score for each of the following raw scores.

  i) 85   ii) 135   iii) 100   iv) 62

b) Determine the raw score corresponding to each of the following z-scores.

  i) 1   ii) −2   iii) 1.5   iv) −1.2

20. Dr. Simba's fourth-year class in animal biology has only 12 students. Their scores on the midterm examination are shown below.

50	71	65	54	84	69	82
67	52	52	86	85		

a) Calculate the mean and median for these data. Compare these two statistics.

b) Calculate the standard deviation and the semi-interquartile range. Compare these statistics and comment on what you notice.

c) Which measure of spread is most suitable for describing this data set? Explain why.

Chapter Test

Use the following set of data-management final examination scores to answer questions 1 through 5.

92	48	59	62	66	98	70	70	55	63
70	97	61	53	56	64	46	69	58	64

1. a) Group these data into intervals and create a frequency table.

   b) Produce a frequency diagram and a frequency polygon.

   c) Produce a cumulative-frequency diagram.

2. Determine the

   a) three measures of central tendency

   b) standard deviation and variance

   c) interquartile and semi-interquartile ranges

3. a) Produce a modified box-and-whisker plot for this distribution.

   b) Identify any outliers.

   c) Identify and explain any other unusual features of this graph.

4. Explain which of the three measures of central tendency is most appropriate to describe this distribution of marks and why the other two measures are not appropriate.

5. Students with scores above the 90th percentile receive a book prize.

   a) How many students will receive prizes?

   b) What are these students' scores?

6. An interview committee graded three short-listed candidates for a management position as shown below. The scores are on a scale of 1 to 5, with 5 as the top score.

Criterion	Weight	Clarise	Pina	Steven
Education	2	3	3	4
Experience	2	4	5	3
Interpersonal skills	3	3	3	5
First interview	1	5	4	3

Who should the committee hire based on these data? Justify your choice.

7. Describe the type of sample used in each of the following scenarios.

   a) A proportionate number of boys and girls are randomly selected from a class.

   b) A software company randomly chooses a group of schools in a particular school district to test a new timetable program.

   c) A newspaper prints a questionnaire and invites its readers to mail in their responses.

   d) A telephone-survey company uses a random-number generator to select which households to call.

   e) An interviewer polls people passing by on the street.

8. A group of 8 children in a day-care centre are to be interviewed about their favourite games. Describe how you would select a systematic sample if there are 52 children at the centre.

9. **a)** Identify the bias in the following surveys and explain the effect it could have on their results.

   i) Parents of high-school students were asked: "Do you think that students should be released from school a half hour early on Friday, free to run around and get into trouble?"

   ii) Audience members at an investment workshop were asked to raise their hands if they had been late with a bill payment within the last six months.

   iii) A random survey of corporate executives asked: "Do you favour granting a cable-television licence for a new economics and business channel?"

   b) Suggest how to eliminate the bias in each of the surveys in part a).

10. A mutual-fund company proudly advertises that all of its funds have "first-quartile performance." What mathematical errors has the company made in this advertisement?

ACHIEVEMENT CHECK

Knowledge/Understanding	Thinking/Inquiry/Problem Solving	Communication	Application

11. The graph below shows the stock price for an Ontario technology company over a one-month period in 2001.

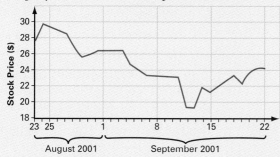

a) When did the stock reach its lowest value during the period shown? Suggest a possible reason for this low point.

b) Compare the percent drop in stock price from September 1 to September 8 to the drop during the following week.

c) Sketch a new graph and provide a commentary that the company could use to encourage investors to buy the company's stock.

CHAPTER 3

Statistics of Two Variables

Specific Expectations	Section
Define the correlation coefficient as a measure of the fit of a scatter graph to a linear model.	3.1, 3.2, 3.3, 3.5
Calculate the correlation coefficient for a set of data, using graphing calculators or statistical software.	3.1, 3.2, 3.3, 3.5
Demonstrate an understanding of the distinction between cause-effect relationships and the mathematical correlation between variables.	3.1, 3.2, 3.3, 3.4, 3.5
Describe possible misuses of regression.	3.2, 3.3, 3.5
Explain examples of the use and misuse of statistics in the media.	3.5
Assess the validity of conclusions made on the basis of statistical studies, by analysing possible sources of bias in the studies and by calculating and interpreting additional statistics, where possible.	3.2, 3.3, 3.4, 3.5
Demonstrate an understanding of the purpose and the use of a variety of sampling techniques.	3.4, 3.5
Organize and summarize data from secondary sources, using technology.	3.1, 3.2, 3.3, 3.4, 3.5
Locate data to answer questions of significance or personal interest, by searching well-organized databases.	3.1, 3.2, 3.4, 3.5
Use the Internet effectively as a source for databases.	3.1, 3.2, 3.4, 3.5

Chapter Problem

Job Prospects

Gina is in her second year of business studies at university and she is starting to think about a job upon graduation. She has two primary concerns—the job market and expected income. Gina does some research at the university's placement centre and finds employment statistics for graduates of her program and industry surveys of entry-level salaries.

Year	Number of Graduates	Number Hired Upon Graduation	Mean Starting Salary ($000)
1992	172	151	26
1993	180	160	27
1994	192	140	28
1995	170	147	27.5
1996	168	142	27
1997	176	155	26.5
1998	180	160	27
1999	192	162	29
2000	200	172	31
2001	220	180	34

1. How could Gina graph this data to estimate

   a) her chances of finding a job in her field when she graduates in two years?

   b) her starting salary?

2. What assumptions does Gina have to make for her predictions? What other factors could affect the accuracy of Gina's estimates?

This chapter introduces statistical techniques for measuring relationships between two variables. As you will see, these techniques will enable Gina to make more precise estimates of her job prospects.

Two-variable statistics have an enormous range of applications including industrial processes, medical studies, and environmental issues—in fact, almost any field where you need to determine if a change in one variable affects another.

Review of Prerequisite Skills

If you need help with any of the skills listed in purple below, refer to Appendix A.

1. **Scatter plots** For each of the following sets of data, create a scatter plot and describe any patterns you see.

a)
x	y
3	18
5	15
8	12
9	10
12	8
15	4
17	1

b)
x	y
4	6
7	2
13	17
14	5
23	19
24	11
25	30
33	21
36	29
40	39
42	26
46	32

2. **Scatter plots** For each plot in question 1,

   i) graph the line of best fit and calculate its equation

   ii) estimate the x- and y-intercepts

   iii) estimate the value of y when x = 7

3. **Graphing linear equations** Determine the slope and y-intercept for the lines defined by the following equations, and then graph the lines.

   a) $y = 3x - 4$ b) $y = -2x + 6$

   c) $12x - 6y = 7$

4. **Graphing quadratic functions** Graph the following functions and estimate any x- and y-intercepts.

   a) $y = 2x^2$

   b) $y = x^2 + 5x - 6$

   c) $y = -3x^2 + x + 2$

5. **Graphing exponential functions**

   a) Identify the base and the numerical coefficient for each of the following functions.

      i) $y = 0.5(3)^x$ ii) $y = 2^x$ iii) $y = 100(0.5)^x$

   b) Graph each of the functions in part a).

   c) Explain what happens to the value of x as the curves in part b) approach the x-axis.

6. **Sigma notation** Calculate each sum without the use of technology.

   a) $\sum_{i=1}^{8} i$ b) $\sum_{i=1}^{5} i^2$

7. **Sigma notation** Given $\bar{x} = 2.5$, calculate each sum without the use of technology.

   a) $\sum_{i=1}^{6} (i - \bar{x})$ b) $\sum_{i=1}^{4} (i - \bar{x})^2$

8. **Sigma notation**

   a) Repeat questions 6 and 7 using appropriate technology such as a graphing calculator or a spreadsheet.

   b) Explain the method that you chose.

9. **Sampling (Chapter 2)** Briefly explain each of the following terms.

   a) simple random sample

   b) systematic sample

   c) outlier

10. **Bias (Chapter 2)**

   a) Explain the term *measurement bias*.

   b) Give an example of a survey method containing unintentional measurement bias.

   c) Give an example of a survey method containing intentional measurement bias.

   d) Give an example of sampling bias.

Scatter Plots and Linear Correlation

Does smoking cause lung cancer? Is job performance related to marks in high school? Do pollution levels affect the ozone layer in the atmosphere? Often the answers to such questions are not clear-cut, and inferences have to be made from large sets of data. Two-variable statistics provide methods for detecting relationships between variables and for developing mathematical models of these relationships.

The visual pattern in a graph or plot can often reveal the nature of the relationship between two variables.

INVESTIGATE & INQUIRE: Visualizing Relationships Between Variables

A study examines two new obedience-training methods for dogs. The dogs were randomly selected to receive from 5 to 16 h of training in one of the two training programs. The dogs were assessed using a performance test graded out of 20.

Rogers Method		Laing System	
Hours	Score	Hours	Score
10	12	8	10
15	16	6	9
7	10	15	12
12	15	16	7
8	9	9	11
5	8	11	7
8	11	10	9
16	19	10	6
10	14	8	15

1. Could you determine which of the two training systems is more effective by comparing the mean scores? Could you calculate another statistic that would give a better comparison? Explain your reasoning.

2. Consider how you could plot the data for the Rogers Method. What do you think would be the best method? Explain why.

3. Use this method to plot the data for the Rogers Method. Describe any patterns you see in the plotted data.

4. Use the same method to plot the data for the Laing System and describe any patterns you see.

5. Based on your data plots, which training method do you think is more effective? Explain your answer.

6. Did your plotting method make it easy to compare the two sets of data? Are there ways you could improve your method?

7. a) Suggest factors that could influence the test scores but have not been taken into account.

   b) How could these factors affect the validity of conclusions drawn from the data provided?

In data analysis, you are often trying to discern whether one variable, the **dependent** (or **response**) **variable**, is affected by another variable, the **independent** (or **explanatory**) **variable**. Variables have a **linear correlation** if changes in one variable tend to be proportional to changes in the other. Variables X and Y have a **perfect positive** (or **direct**) **linear correlation** if Y increases at a constant rate as X increases. Similarly, X and Y have a **perfect negative** (or **inverse**) **linear correlation** if Y decreases at a constant rate as X increases.

A **scatter plot** shows such relationships graphically, usually with the independent variable as the horizontal axis and the dependent variable as the vertical axis. The **line of best fit** is the straight line that passes as close as possible to all of the points on a scatter plot. The stronger the correlation, the more closely the data points cluster around the line of best fit.

Example 1 Classifying Linear Correlations

Classify the relationship between the variables X and Y for the data shown in the following diagrams.

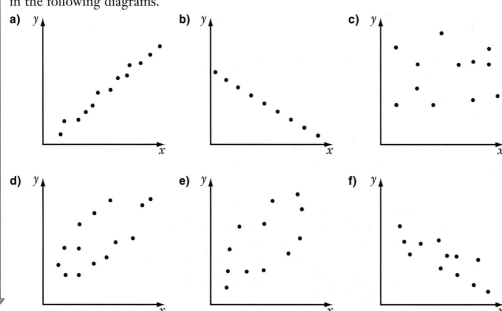

a)   b)   c)

d)   e)   f)

Solution

a) The data points are clustered around a line that rises to the right (positive slope), indicating definitely that Y increases as X increases. Although the points are not perfectly lined up, there is a *strong positive linear correlation* between X and Y.

b) The data points are all exactly on a line that slopes down to the right, so Y decreases as X increases. In fact, the changes in Y are *exactly* proportional to the changes in X. There is a *perfect negative linear correlation* between X and Y.

c) No discernible linear pattern exists. As X increases, Y appears to change randomly. Therefore, there is *zero linear correlation* between X and Y.

d) A definite positive trend exists, but it is not as clear as the one in part a). Here, X and Y have a *moderate positive linear correlation*.

e) A slight positive trend exists. X and Y have a *weak positive linear correlation*.

f) A definite negative trend exists, but it is hard to classify at a glance. Here, X and Y have a *moderate or strong negative linear correlation*.

As Example 1 shows, a scatter plot often can give only a rough indication of the correlation between two variables. Obviously, it would be useful to have a more precise way to measure correlation. Karl Pearson (1857–1936) developed a formula for estimating such a measure. Pearson, who also invented the term *standard deviation*, was a key figure in the development of modern statistics.

The Correlation Coefficient

To develop a measure of correlation, mathematicians first defined the **covariance** of two variables in a sample:

$$s_{XY} = \frac{1}{n-1} \sum (x - \bar{x})(y - \bar{y})$$

where n is the size of the sample, x represents individual values of the variable X, y represents individual values of the variable Y, $\bar{x}$ is the mean of X, and $\bar{y}$ is the mean of Y.

Recall from Chapter 2 that the symbol $\sum$ means "the sum of." Thus, the covariance is the sum of the *products* of the deviations of x and y for all the data points divided by $n - 1$. The covariance depends on how the deviations of the two variables are related. For example, the covariance will have a large positive value if both $x - \bar{x}$ and $y - \bar{y}$ tend to be large at the same time, and a negative value if one tends to be positive when the other is negative.

The **correlation coefficient**, **r**, is the covariance divided by the product of the standard deviations for X and Y:

$$r = \frac{s_{XY}}{s_X \times s_Y}$$

where s_X is the standard deviation of X and s_Y is the standard deviation of Y.

This coefficient gives a quantitative measure of the strength of a linear correlation. In other words, the correlation coefficient indicates how closely the data points cluster around the line of best fit. The correlation coefficient is also called the **Pearson product-moment coefficient of correlation (PPMC)** or **Pearson's r**.

The correlation coefficient always has values in the range from −1 to 1. Consider a perfect positive linear correlation first. For such correlations, changes in the dependent variable Y are directly proportional to changes in the independent variable X, so $Y = aX + b$, where a is a positive constant. It follows that

$$s_{XY} = \frac{1}{n-1}\sum (x - \overline{x})(y - \overline{y})$$
$$= \frac{1}{n-1}\sum (x - \overline{x})[(ax + b) - (a\overline{x} + b)]$$
$$= \frac{1}{n-1}\sum (x - \overline{x})(ax - a\overline{x})$$
$$= \frac{1}{n-1}\sum a(x - \overline{x})^2$$
$$= a\frac{\sum (x - \overline{x})^2}{n-1}$$
$$= a s_X^2$$

$$s_Y = \sqrt{\frac{\sum (y - \overline{y})^2}{n-1}}$$
$$= \sqrt{\frac{\sum [(ax + b) - (a\overline{x} + b)]^2}{n-1}}$$
$$= \sqrt{\frac{\sum (ax - a\overline{x})^2}{n-1}}$$
$$= \sqrt{\frac{a^2 \sum (x - \overline{x})^2}{n-1}}$$
$$= a\sqrt{\frac{\sum (x - \overline{x})^2}{n-1}}$$
$$= a s_X$$

Substituting into the equation for the correlation coefficient gives

$$r = \frac{s_{XY}}{s_X s_Y}$$
$$= \frac{a s_X^2}{s_X (a s_X)}$$
$$= 1$$

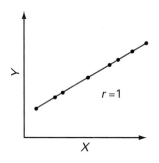

Similarly, $r = -1$ for a perfect negative linear correlation.

For two variables with no correlation, Y is equally likely to increase or decrease as X increases. The terms in $\sum(x - \bar{x})(y - \bar{y})$ are randomly positive or negative and tend to cancel each other. Therefore, the correlation coefficient is close to zero if there is little or no correlation between the variables. For moderate linear correlations, the summation terms partially cancel out.

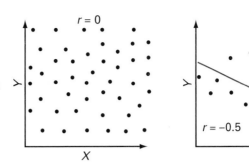

The following diagram illustrates how the correlation coefficient corresponds to the strength of a linear correlation.

Using algebraic manipulation and the fact that $\sum x = n\bar{x}$, Pearson showed that

$$r = \frac{n\sum xy - (\sum x)(\sum y)}{\sqrt{[n\sum x^2 - (\sum x)^2][n\sum y^2 - (\sum y)^2]}}$$

where n is the number of data points in the sample, x represents individual values of the variable X, and y represents individual values of the variable Y. (Note that $\sum x^2$ is the sum of the squares of all the individual values of X, while $(\sum x)^2$ is the square of the sum of all the individual values.)

Like the alternative formula for standard deviations (page 150), this formula for r avoids having to calculate all the deviations individually. Many scientific and statistical calculators have built-in functions for calculating the correlation coefficient.

It is important to be aware that increasing the number of data points used in determining a correlation improves the accuracy of the mathematical model. Some of the examples and exercise questions have a fairly small set of data in order to simplify the computations. Larger data sets can be found in the e-book that accompanies this text.

Example 2 Applying the Correlation Coefficient Formula

A farmer wants to determine whether there is a relationship between the mean temperature during the growing season and the size of his wheat crop. He assembles the following data for the last six crops.

Mean Temperature (°C)	Yield (tonnes/hectare)
4	1.6
8	2.4
10	2.0
9	2.6
11	2.1
6	2.2

a) Does a scatter plot of these data indicate any linear correlation between the two variables?

b) Compute the correlation coefficient.

c) What can the farmer conclude about the relationship between the mean temperatures during the growing season and the wheat yields on his farm?

Solution

a) The farmer wants to know whether the crop yield depends on temperature. Here, temperature is the independent variable, X, and crop yield is the dependent variable, Y. The scatter plot has a somewhat positive trend, so there appears to be a moderate positive linear correlation.

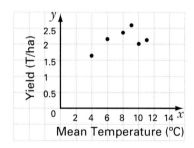

b) To compute r, set up a table to calculate the quantities required by the formula.

Temperature, x	Yield, y	x^2	y^2	xy
4	1.6	16	2.56	6.4
8	2.4	64	5.76	19.2
10	2.0	100	4.00	20.0
9	2.6	81	6.76	23.4
11	2.1	121	4.41	23.1
6	2.2	36	4.84	13.2
$\sum x = 48$	$\sum y = 12.9$	$\sum x^2 = 418$	$\sum y^2 = 28.33$	$\sum xy = 105.3$

Now compute r, using the formula:

$$r = \frac{n\Sigma(xy) - (\Sigma x)(\Sigma y)}{\sqrt{[n\Sigma x^2 - (\Sigma x)^2][n\Sigma y^2 - (\Sigma y)^2]}}$$
$$= \frac{6(105.3) - (48)(12.9)}{\sqrt{[6(418) - (48)^2][6(28.33) - (12.9)^2]}}$$
$$= \frac{631.8 - 619.2}{\sqrt{(2508 - 2304)(169.98 - 166.41)}}$$
$$= \frac{12.6}{26.99}$$
$$= 0.467$$

Data in Action

From 1992 to 2001, Canada produced an average of 27 million tonnes of wheat a year. About 70% of this crop was exported.

The correlation coefficient for crop yield versus mean temperature is approximately 0.47, which confirms a moderate positive linear correlation.

c) It appears that the crop yield tends to increase somewhat as the mean temperature for the growing season increases. However, the farmer cannot conclude that higher temperatures *cause* greater crop yields. Other variables could account for the correlation. For example, the lower temperatures could be associated with heavy rains, which could lower yields by flooding fields or leaching nutrients from the soil.

The important principle that a correlation does not prove the existence of a cause-and-effect relationship between two variables is discussed further in section 3.4.

Example 3 Using Technology to Determine Correlation Coefficients

Determine whether there is a linear correlation between horsepower and fuel consumption for these five vehicles by creating a scatter plot and calculating the correlation coefficient.

Vehicle	Horsepower, x	Fuel Consumption (L/100 km), y
Midsize sedan	105	6.7
Minivan	170	23.5
Small sports utility vehicle	124	5.9
Midsize motorcycle	17	3.4
Luxury sports car	296	8.4

Solution 1 Using a Graphing Calculator

Use the ClrList command to make sure lists L1 and L2 are clear, then enter the horsepower data in L1 and the fuel consumption figures in L2.

See Appendix B for more details on the graphing calculator and software functions used in this section.

To display a scatter plot, first make sure that all functions in the Y= editor are either clear or turned off. Then, use STAT PLOT to select PLOT1.

Turn the plot on, select the scatter-plot icon, and enter L1 for XLIST and L2 for YLIST. (Some of these settings may already be in place.) From the ZOOM menu, select 9:ZoomStat. The calculator will automatically optimize the window settings and display the scatter plot.

To calculate the correlation coefficient, from the CATALOG menu, select DiagnosticOn, then select the LinReg(ax+b) instruction from the STAT CALC menu. The calculator will perform a series of statistical calculations using the data in lists L1 and L2. The last line on the screen shows that the correlation coefficient is approximately 0.353.

Therefore, there is a moderate linear correlation between horsepower and fuel consumption for the five vehicles.

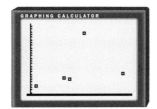

Solution 2   Using a Spreadsheet

Set up three columns and enter the data from the table above. Highlight the numerical data and use your spreadsheet's Chart feature to display a scatter plot. Both Corel® Quattro® Pro and Microsoft® Excel have a CORREL function that allows you to calculate the correlation coefficient easily. The scatter plot and correlation coefficient indicate a moderate correlation between horsepower and fuel consumption.

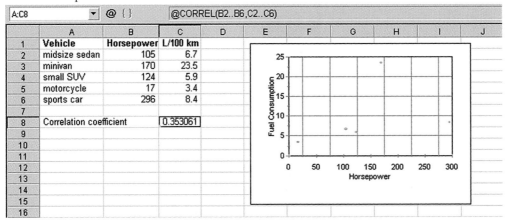

Solution 3   Using Fathom™

Create a new **collection** by setting up a **case table** with three attributes: Vehicle, Hp, and FuelUse. Enter the data for the five cases. To create a **scatter plot**, drag the **graph icon** onto the work area and drop the Hp attribute on the x-axis and the FuelUse attribute on the y-axis.

To calculate the **correlation coefficient**, right-click on the **collection** and select Inspect Collection. Select the Measures tab and name a new measure PPMC. Right-click this measure and select Edit Formula, then Functions/Statistical/Two Attributes/correlation. When you enter the Hp and FuelUse attributes in the correlation function, Fathom™ will calculate the correlation coefficient for these data.

Again, the scatter plot and correlation coefficient show a moderate linear correlation.

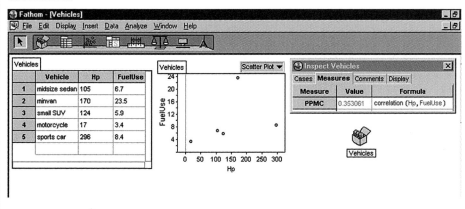

Project Prep

For your statistics project, you may be investigating the linear correlation between two variables. A graphing calculator or computer software may be a valuable aid for this analysis.

Notice that the scatter plots in Example 3 have an outlier at (170, 23.5). Without this data point, you would have a strong positive linear correlation. Section 3.2 examines the effect of outliers in more detail.

Key Concepts

- Statistical studies often find linear correlations between two variables.

- A scatter plot can often reveal the relationship between two variables. The independent variable is usually plotted on the horizontal axis and the dependent variable on the vertical axis.

- Two variables have a linear correlation if changes in one variable tend to be proportional to changes in the other. Linear correlations can be positive or negative and vary in strength from zero to perfect.

- The correlation coefficient, r, is a quantitative measure of the correlation between two variables. Negative values indicate negative correlations while positive values indicate positive correlations. The greater the absolute value of r, the stronger the linear correlation, with zero indicating no correlation at all and 1 indicating a perfect correlation.

- Manual calculations of correlation coefficients can be quite tedious, but a variety of powerful technology tools are available for such calculations.

Practise

1. Classify the type of linear correlation that you would expect with the following pairs of variables.

   a) hours of study, examination score

   b) speed in excess of the speed limit, amount charged on a traffic fine

   c) hours of television watched per week, final mark in calculus

   d) a person's height, sum of the digits in the person's telephone number

   e) a person's height, the person's strength

2. Identify the independent variable and the dependent variable in a correlational study of

   a) heart disease and cholesterol level

   b) hours of basketball practice and free-throw success rate

   c) amount of fertilizer used and height of plant

   d) income and level of education

   e) running speed and pulse rate

Apply, Solve, Communicate

3. For a week prior to their final physics examination, a group of friends collect data to see whether time spent studying or time spent watching TV had a stronger correlation with their marks on the examination.

Hours Studied	Hours Watching TV	Examination Score
10	8	72
11	7	67
15	4	81
14	3	93
8	9	54
5	10	66

   a) Create a scatter plot of hours studied versus examination score. Classify the linear correlation.

   b) Create a similar scatter plot for the hours spent watching TV.

   c) Which independent variable has a stronger correlation with the examination scores? Explain.

d) Calculate the correlation coefficient for hours studied versus examination score and for hours watching TV versus examination score. Do these answers support your answer to c)? Explain.

4. Application Refer to the tables in the investigation on page 159.

   a) Determine the correlation coefficient and classify the linear correlation for the data for each training method.

   b) Suppose that you interchanged the dependent and independent variables, so that the test scores appear on the horizontal axis of a scatter plot and the hours of training appear on the vertical axis. Predict the effect this change will have on the scatter plot and the correlation coefficient for each set of data.

   c) Test your predictions by plotting the data and calculating the correlation coefficients with the variables reversed. Explain any differences between your results and your predictions in part b).

5. A company studied whether there was a relationship between its employees' years of service and number of days absent. The data for eight randomly selected employees are shown below.

Employee	Years of Service	Days Absent Last Year
Jim	5	2
Leah	2	6
Efraim	7	3
Dawn	6	3
Chris	4	4
Cheyenne	8	0
Karrie	1	2
Luke	10	1

   a) Create a scatter plot for these data and classify the linear correlation.

   b) Calculate the correlation coefficient.

c) Does the computed r-value agree with the classification you made in part a)? Explain why or why not.

d) Identify any outliers in the data.

e) Suggest possible reasons for any outliers identified in part d).

6. Application Six classmates compared their arm spans and their scores on a recent mathematics test as shown in the following

Arm Span (m)	Score
1.5	82
1.4	71
1.7	75
1.6	66
1.6	90
1.8	73

   a) Illustrate these data with a scatter plot.

   b) Determine the correlation coefficient and classify the linear correlation.

   c) What can the students conclude from their data?

7. a) Use data in the table on page 157 to create a scatter plot that compares the size of graduating classes in Gina's program to the number of graduates who found jobs.

   b) Classify the linear correlation.

   c) Determine the linear correlation coefficient.

8. a) Search sources such as E-STAT, CANSIM II, the Internet, newspapers, and magazines for pairs of variables that exhibit

   i) a strong positive linear correlation

   ii) a strong negative linear correlation

   iii) a weak or zero linear correlation

   b) For each pair of variables in part a), identify the independent variable and the dependent variable.

9. Find a set of data for two variables known to have a perfect positive linear correlation. Use these data to demonstrate that the correlation coefficient for such variables is 1. Alternatively, find a set of data with a perfect negative correlation and show that the correlation coefficient is −1.

10. **Communication**

   a) Would you expect to see a correlation between the temperature at an outdoor track and the number of people using the track? Why or why not?

   b) Sketch a typical scatter plot of this type of data.

   c) Explain the key features of your scatter plot.

11. **Inquiry/Problem Solving** Refer to data tables in the investigation on page 159.

   a) How could the Rogers Training Company graph the data so that their training method looks particularly good?

   b) How could Laing Limited present the same data in a way that favours their training system?

   c) How could a mathematically knowledgeable consumer detect the distortions in how the two companies present the data?

12. **Inquiry/Problem Solving**

   a) Prove that interchanging the independent and dependent variables does not change the correlation coefficient for any set of data.

   b) Illustrate your proof with calculations using a set of data selected from one of the examples or exercise questions in this section.

13. a) Search sources such as newspapers, magazines, and the Internet for a set of two-variable data with

   i) a moderate positive linear correlation

   ii) a moderate negative correlation

   iii) a correlation in which $|r| > 0.9$

   b) Outline any conclusions that you can make from each set of data. Are there any assumptions inherent in these conclusions? Explain.

   c) Pose at least two questions that could form the basis for further research.

14. a) Sketch scatter plots of three different patterns of data that you think would have zero linear correlation.

   b) Explain why r would equal zero for each of these patterns.

   c) Use Fathom™ or a spreadsheet to create a scatter plot that looks like one of your patterns and calculate the correlation coefficient. Adjust the data points to get r as close to zero as you can.

3.2 Linear Regression

Regression is an analytic technique for determining the relationship between a dependent variable and an independent variable. When the two variables have a linear correlation, you can develop a simple mathematical model of the relationship between the two variables by finding a line of best fit. You can then use the equation for this line to make predictions by **interpolation** (estimating between data points) and **extrapolation** (estimating beyond the range of the data).

INVESTIGATE & INQUIRE: Modelling a Linear Relationship

A university would like to construct a mathematical model to predict first-year marks for incoming students based on their achievement in grade 12. A comparison of these marks for a random sample of first-year students is shown below.

Grade 12 Average	85	90	76	78	88	84	76	96	86	85
First-Year Average	74	83	68	70	75	72	64	91	78	86

1. **a)** Construct a scatter plot for these data. Which variable should be placed on the vertical axis? Explain.

   b) Classify the linear correlation for this data, based on the scatter plot.

2. **a)** Estimate and draw a line of best fit for the data.

   b) Measure the slope and y-intercept for this line, and write an equation for it in the form $y = mx + b$.

3. Use this linear model to predict

   a) the first-year average for a student who had an 82 average in grade 12

   b) the grade-12 average for a student with a first-year average of 60

4. **a)** Use software or the linear regression instruction of a graphing calculator to find the slope and y-intercept for the line of best fit. (Note that most graphing calculators use a instead of m to represent slope.)

   b) Are this slope and y-intercept close to the ones you measured in question 2? Why or why not?

c) Estimate how much the new values for slope and *y*-intercept will change your predictions in question 3. Check your estimate by recalculating your predictions using the new values and explain any discrepancies.

5. List the factors that could affect the accuracy of these mathematical models. Which factor do you think is most critical? How could you test how much effect this factor could have?

It is fairly easy to "eyeball" a good estimate of the line of best fit on a scatter plot when the linear correlation is strong. However, an analytic method using a **least-squares fit** gives more accurate results, especially for weak correlations.

Consider the line of best fit in the following scatter plot. A dashed blue line shows the **residual** or vertical deviation of each data point from the line of best fit. The residual is the difference between the values of *y* at the data point and at the point that lies on the line of best fit and has the same *x*-coordinate as the data point. Notice that the residuals are positive for points above the line and negative for points below the line. The boxes show the squares of the residuals.

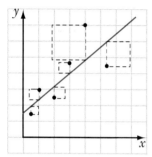

For the line of best fit in the least-squares method,
- the sum of the residuals is zero (the positive and negative residuals cancel out)
- the sum of the squares of the residuals has the least possible value

Although the algebra is daunting, it can be shown that this line has the equation

$$y = ax + b, \text{ where } a = \frac{n(\Sigma xy) - (\Sigma x)(\Sigma y)}{n(\Sigma x^2) - (\Sigma x)^2} \text{ and } b = \overline{y} - a\overline{x}$$

Recall from Chapter 2 that $\overline{x}$ is the mean of *x* and $\overline{y}$ is the mean of *y*. Many statistics texts use an equation with the form $y = a + bx$, so you may sometimes see the equations for *a* and *b* reversed.

Example 1 Applying the Least-Squares Formula

This table shows data for the full-time employees of a small company.

a) Use a scatter plot to classify the correlation between age and income.

b) Find the equation of the line of best fit analytically.

c) Predict the income for a new employee who is 21 and an employee retiring at age 65.

Age (years)	Annual Income ($000)
33	33
25	31
19	18
44	52
50	56
54	60
38	44
29	35

Solution

a) The scatter plot suggests a strong positive linear correlation between age and income level.

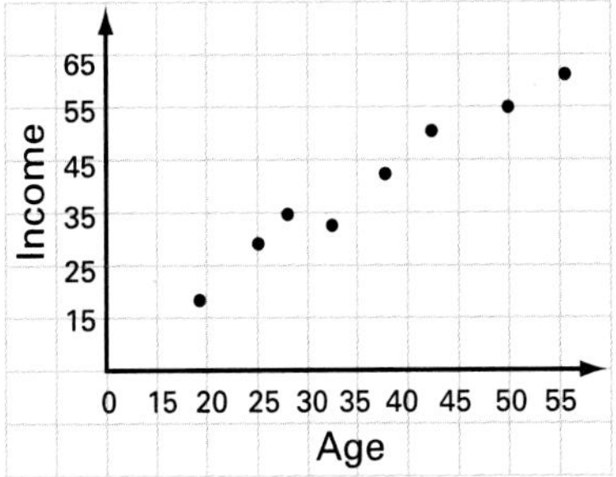

b) To determine the equation of the line of best fit, organize the data into a table and compute the sums required for the formula.

Age, x	Income, y	x^2	xy
33	33	1089	1089
25	31	625	775
19	18	361	342
44	52	1936	2288
50	56	2500	2800
54	60	2916	3240
38	44	1444	1672
29	35	841	1015
$\sum x = 292$	$\sum y = 329$	$\sum x^2 = 11\ 712$	$\sum xy = 13\ 221$

Substitute these totals into the formula for a.

$$a = \frac{n(\sum xy) - (\sum x)(\sum y)}{n(\sum x^2) - (\sum x)^2}$$

$$= \frac{8(13\ 221) - (292)(329)}{8(11\ 712) - (292)^2}$$

$$= \frac{9700}{8432}$$

$$\doteq 1.15$$

To determine b, you also need the means of x and y.

$$\bar{x} = \frac{\Sigma x}{n} \qquad \bar{y} = \frac{\Sigma y}{n} \qquad b = \bar{y} - a\bar{x}$$
$$= \frac{292}{8} \qquad = \frac{329}{8} \qquad = 41.125 - 1.15(36.5)$$
$$\qquad\qquad\qquad\qquad\qquad = -0.85$$
$$= 36.5 \qquad = 41.125$$

Now, substitute the values of a and b into the equation for the line of best fit.

$$y = ax + b$$
$$= 1.15x - 0.85$$

Therefore, the equation of the line of best fit is $y = 1.15x - 0.85$.

c) Use the equation of the line of best fit as a model.

For a 21-year-old employee,

$$y = ax + b$$
$$= 1.15(21) - 0.85$$
$$= 23.3$$

For a 65-year-old employee,

$$y = ax + b$$
$$= 1.15(65) - 0.85$$
$$= 73.9$$

Therefore, you would expect the new employee to have an income of about $23 300 and the retiring employee to have an income of about $73 900. Note that the second estimate is an extrapolation beyond the range of the data, so it could be less accurate than the first estimate, which is interpolated between two data points.

Note that the slope a indicates only how y varies with x on the line of best fit. The slope does not tell you anything about the strength of the correlation between the two variables. It is quite possible to have a weak correlation with a large slope or a strong correlation with a small slope.

Example 2 Linear Regression Using Technology

Researchers monitoring the numbers of wolves and rabbits in a wildlife reserve think that the wolf population depends on the rabbit population since wolves prey on rabbits. Over the years, the researchers collected the following data.

Year	1994	1995	1996	1997	1998	1999	2000	2001
Rabbit Population	61	72	78	76	65	54	39	43
Wolf Population	26	33	42	49	37	30	24	19

a) Determine the line of best fit and the correlation coefficient for these data.

b) Graph the data and the line of best fit. Do these data support the researchers' theory?

Solution 1   Using a Graphing Calculator

a) You can use the calculator's linear regression instruction to find both the line of best fit and the correlation coefficient. Since the theory is that the wolf population depends on the rabbit population, the rabbit population is the independent variable and the wolf population is the dependent variable.

Use the **STAT EDIT** menu to enter the rabbit data into list L1 and the wolf data into L2. Set DiagnosticOn, and then use the **STAT CALC** menu to select LinReg(ax+b).

The equation of the line of best fit is $y = 0.58x - 3.1$ and the correlation coefficient is 0.87.

b) Store the equation for the line of best fit as a function, Y1. Then, use the **STAT PLOT** menu to set up the scatter plot. By displaying both Y1 and the scatter plot, you can see how closely the data plots are distributed around the line of best fit.

The correlation coefficient and the scatter plot show a strong positive linear correlation between the variables. This correlation supports the researchers' theory, but does not prove that changes in the rabbit population are the cause of the changes in the wolf population.

Solution 2   Using a Spreadsheet

Set up a table with the data for the rabbit and wolf populations. You can calculate the correlation coefficient with the **CORREL** function. Use the Chart feature to create a scatter plot.

In Corel® Quattro® Pro, you can find the equation of the line of best fit by selecting **Tools/Numeric Tools/Regression**. Enter the cell ranges for the data, and the program will display regression calculations including the constant (*b*), the *x*-coefficient (or slope, *a*), and r^2.

In Microsoft® Excel, you can find the equation of the line of best fit by selecting Chart/Add Trendline. Check that the default setting is Linear. Select the straight line that appears on your chart, then click Format/Selected Trendline/Options. Check the Display equation on chart box. You can also display r^2.

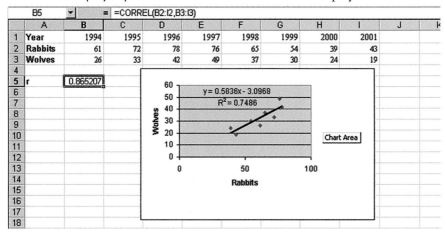

Solution 3 Using Fathom™

Drag a new **case table** to the workspace, create attributes for **Year**, **Rabbits**, and **Wolves**, and enter the data. Drag a new **graph** to the workspace, then drag the **Rabbits** attribute to the x-axis and the **Wolves** attribute to the y-axis. From the Graph menu, select **Least Squares Line**. Fathom™ will display r^2 and the equation for the **line of best fit**. To calculate the correlation coefficient directly, select **Inspect Collection**, click the **Measures** tab, then create a new measure by selecting **Functions/Statistical/Two Attributes/correlation** and entering **Rabbits** and **Wolves** as the attributes.

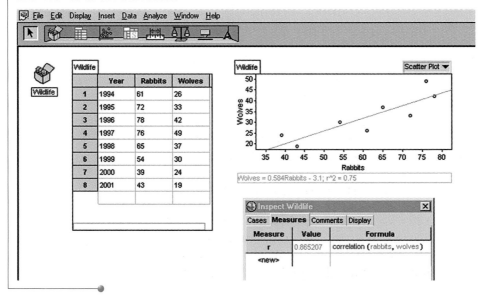

Project Prep

When analysing two-variable data for your statistics project, you may wish to develop a linear model, particularly if a strong linear correlation is evident.

In Example 2, the sample size is small, so you should be cautious about making generalizations from it. Small samples have a greater chance of not being representative of the whole population. Also, outliers can seriously affect the results of a regression on a small sample.

Example 3 The Effect of Outliers

To evaluate the performance of one of its instructors, a driving school tabulates the number of hours of instruction and the driving-test scores for the instructor's students.

Instructional Hours	10	15	21	6	18	20	12
Student's Score	78	85	96	75	84	45	82

a) What assumption is the management of the driving school making? Is this assumption reasonable?

b) Analyse these data to determine whether they suggest that the instructor is an effective teacher.

c) Comment on any data that seem unusual.

d) Determine the effect of any outliers on your analysis.

Solution

a) The management of the driving school is assuming that the correlation between instructional hours and test scores is an indication of the instructor's teaching skills. Such a relationship could be difficult to prove definitively. However, the assumption would be reasonable if the driving school has found that some instructors have consistently strong correlations between the time spent with their students and the students' test scores while other instructors have consistently weaker correlations.

b) The number of hours of instruction is the independent variable. You could analyse the data using any of the methods in the two previous examples. For simplicity, a spreadsheet solution is shown here.

Except for an obvious outlier at (20, 45), the scatter plot below indicates a strong positive linear correlation. At first glance, it appears that the number of instructional hours is positively correlated to the students' test scores. However, the linear regression analysis yields a line of best fit with the equation $y = -0.13x + 80$ and a correlation coefficient of -0.05.

These results indicate that there is virtually a zero linear correlation, and the line of best fit even has a negative slope! The outlier has a dramatic impact on the regression results because it is distant from the other data points and the sample size is quite small. Although the scatter plot looked

favourable, the regression analysis suggests that the instructor's lessons had no positive effect on the students' test results.

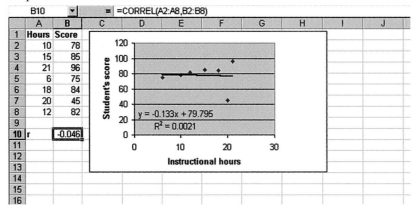

c) The fact that the outlier is substantially below all the other data points suggests that some special circumstance may have caused an abnormal result. For instance, there might have been an illness or emotional upset that affected this one student's performance on the driving test. In that case, it would be reasonable to exclude this data point when evaluating the driving instructor.

d) Remove the outlier from your data table and repeat your analysis.

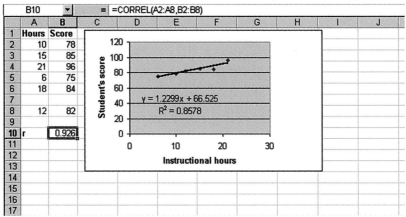

Notice that the line of best fit is now much closer to the data points and has a positive slope. The correlation coefficient, r, is 0.93, indicating a strong positive linear correlation between the number of instructional hours and the driver's test scores. This result suggests that the instructor may be an effective teacher after all. It is quite possible that the original analysis was not a fair evaluation. However, to do a proper evaluation, you would need a larger set of data, more information about the outlier, or, ideally, both.

As Example 3 demonstrates, outliers can skew a regression analysis, but they could also simply indicate that the data really do have large variations. A comprehensive analysis of a set of data should look for outliers, examine their possible causes and their effect on the analysis, and discuss whether they should be excluded from the calculations. As you observed in Chapter 2, outliers have less effect on larger samples.

Project Prep

If your statistics project involves a linear relationship that contains outliers, you will need to consider carefully their impact on your results, and how you will deal with them.

WEB CONNECTION

www.mcgrawhill.ca/links/MDM12

Visit the above web site and follow the links to learn more about linear regression. Describe an application of linear regression that interests you.

Key Concepts

- Linear regression provides a means for analytically determining a line of best fit. In the least-squares method, the line of best fit is the line which minimizes the sum of the squares of the residuals while having the sum of the residuals equal zero.

- You can use the equation of the line of best fit to predict the value of one of the two variables given the value of the other variable.

- The correlation coefficient is a measure of how well a regression line fits a set of data.

- Outliers and small sample sizes can reduce the accuracy of a linear model.

Communicate Your Understanding

1. What does the correlation coefficient reveal about the line of best fit generated by a linear regression?

2. Will the correlation coefficient always be negative when the slope of the line of best fit is negative? Explain your reasoning.

3. Describe the problem that outliers present for a regression analysis and outline what you could do to resolve this problem.

Practise

1. Identify any outliers in the following sets of data and explain your choices.

a)

X	25	34	43	55	92	105	16
Y	30	41	52	66	18	120	21

b)

X	5	7	6	6	4	8
Y	304	99	198	205	106	9

2. a) Perform a linear regression analysis to generate the line of best fit for each set of data in question 1.

b) Repeat the linear regressions in part a), leaving out any outliers.

c) Compare the lines of best fit in parts a) and b).

Apply, Solve, Communicate

3. Use the formula for the method of least squares to verify the slope and intercept values you found for the data in the investigation on page 171. Account for any discrepancies.

4. Use software or a graphing calculator to verify the regression results in Example 1.

5. Application The following table lists the heights and masses for a group of fire-department trainees.

Height (cm)	Mass (kg)
177	91
185	88
173	82
169	79
188	87
182	85
175	79

a) Create a scatter plot and classify the linear correlation.

b) Apply the method of least squares to generate the equation of the line of best fit.

c) Predict the mass of a trainee whose height is 165 cm.

d) Predict the height of a 79-kg trainee.

e) Explain any discrepancy between your answer to part d) and the actual height of the 79-kg trainee in the sample group.

6. A random survey of a small group of high-school students collected information on the students' ages and the number of books they had read in the past year.

Age (years)	Books Read
16	5
15	3
18	8
17	6
16	4
15	4
14	5
17	15

a) Create a scatter plot for this data. Classify the linear correlation.

b) Determine the correlation coefficient and the equation of the line of best fit.

c) Identify the outlier.

d) Repeat part b) with the outlier excluded.

e) Does removing the outlier improve the linear model? Explain.

f) Suggest other ways to improve the model.

g) Do your results suggest that the number of books a student reads depends on the student's age? Explain.

7. Application Market research has provided the following data on the monthly sales of a licensed T-shirt for a popular rock band.

Price ($)	Monthly Sales
10	2500
12	2200
15	1600
18	1200
20	800
24	250

a) Create a scatter plot for these data.

b) Use linear regression to model these data.

c) Predict the sales if the shirts are priced at $19.

d) The vendor has 1500 shirts in stock and the band is going to finish its concert tour in a month. What is the maximum price the vendor can charge and still avoid having shirts left over when the band stops touring?

8. Communication MDM Entertainment has produced a series of TV specials on the lives of great mathematicians. The executive producer wants to know if there is a linear correlation between production costs and revenue from the sales of broadcast rights. The costs and gross sales revenue for productions in 2001 and 2002 were as follows (amounts in millions of dollars).

2001		2002	
Cost ($M)	Sales ($M)	Cost ($M)	Sales ($M)
5.5	15.4	2.7	5.2
4.1	12.1	1.9	1.0
1.8	6.9	3.4	3.4
3.2	9.4	2.1	1.9
4.2	1.5	1.4	1.5

a) Create a scatter plot using the data for the productions in 2001. Do there appear to be any outliers? Explain.

b) Determine the correlation coefficient and the equation of the line of best fit.

c) Repeat the linear regression analysis with any outliers removed.

d) Repeat parts a) and b) using the data for the productions in 2002.

e) Repeat parts a) and b) using the combined data for productions in both 2001 and 2002. Do there still appear to be any outliers?

f) Which of the four linear equations do you think is the best model for the relationship between production costs and revenue? Explain your choice.

g) Explain why the executive producer might choose to use the equation from part d) to predict the income from MDM's 2003 productions.

9. At Gina's university, there are 250 business students who expect to graduate in 2006.

a) Model the relationship between the total number of graduates and the number hired by performing a linear regression on the data in the table on page 157. Determine the equation of the line of best fit and the correlation coefficient.

b) Use this linear model to predict how many graduates will be hired in 2006.

c) Identify any outliers in this scatter plot and suggest possible reasons for an outlier. Would any of these reasons justify excluding the outlier from the regression calculations?

d) Repeat part a) with the outlier removed.

e) Compare the results in parts a) and d). What assumptions do you have to make?

10. Communication Refer to Example 2, which describes population data for wolves and rabbits in a wildlife reserve. An alternate theory has it that the rabbit population depends on the wolf population since the wolves prey on the rabbits.

a) Create a scatter plot of rabbit population versus wolf population and classify the linear correlation. How are your data points related to those in Example 2?

b) Determine the correlation coefficient and the equation of the line of best fit. Graph this line on your scatter plot.

c) Is the equation of the line of best fit the inverse of that found in Example 2? Explain.

d) Plot both populations as a time series. Can you recognize a pattern or relationship between the two series? Explain.

e) Does the time series suggest which population is the dependent variable? Explain.

11. The following table lists the mathematics of data management marks and grade 12 averages for a small group of students.

Mathematics of Data Management Mark	Grade 12 Average
74	77
81	87
66	68
53	67
92	85
45	55
80	76

a) Using Fathom™ or *The Geometer's Sketchpad®*,

i) create a scatter plot for these data

ii) add a moveable line to the scatter plot and construct the geometric square for the deviation of each data point from the moveable line

iii) generate a dynamic sum of the areas of these squares

iv) manoeuvre the moveable line to the position that minimizes the sum of the areas of the squares.

v) record the equation of this line

b) Determine the equation of the line of best fit for this set of data.

c) Compare the equations you found in parts a) and b). Explain any differences or similarities.

12. Application Use E-STAT or other sources to obtain the annual consumer price index figures from 1914 to 2000.

a) Download this information into a spreadsheet or statistical software, or enter it into a graphing calculator. (If you use a graphing calculator, enter the data from every third year.) Find the line of best fit and comment on whether a straight line appears to be a good model for the data.

b) What does the slope of the line of best fit tell you about the rate of inflation?

c) Find the slope of the line of best fit for the data for just the last 20 years, and then repeat the calculation using only the data for the last 5 years.

d) What conclusions can you make by comparing the three slopes? Explain your reasoning.

Knowledge/ Understanding	Thinking/Inquiry/ Problem Solving	Communication	Application

13. The Worldwatch Institute has collected the following data on concentrations of carbon dioxide (CO_2) in the atmosphere.

Year	CO_2 Level (ppm)
1975	331
1976	332
1977	333.7
1978	335.3
1979	336.7
1980	338.5
1981	339.8
1982	341
1983	342.6
1984	344.3
1985	345.7
1986	347
1987	348.8
1988	351.4
1989	352.7
1990	354
1991	355.5
1992	356.2
1993	357
1994	358.8
1995	360.7

a) Use technology to produce a scatter plot of these data and describe any correlation that exists.

b) Use a linear regression to find the line of best fit for the data. Discuss the reliability of this model.

c) Use the regression equation to predict the level of atmospheric CO_2 that you would expect today.

d) Research current CO_2 levels. Are the results close to the predicted level? What factors could have affected the trend?

14. Suppose that a set of data has a perfect linear correlation except for two outliers, one above the line of best fit and the other an equal distance below it. The residuals of these two outliers are equal in magnitude, but one is positive and the other negative. Would you agree that a perfect linear correlation exists because the effects of the two residuals cancel out? Support your opinion with mathematical reasoning and a diagram.

15. Inquiry/Problem Solving Recall the formulas for the line of best fit using the method of least squares that minimizes the squares of vertical deviations.

a) Modify these formulas to produce a line of best fit that minimizes the squares of *horizontal* deviations.

b) Do you think your modified formulas will produce the same equation as the regular least-squares formula?

c) Use your modified formula to calculate a line of best fit for one of the examples in this section. Does your line have the same equation as the line of best fit in the example? Is your equation the inverse of the equation in the example? Explain why or why not.

16. a) Calculate the residuals for all of the data points in Example 3 on page 177. Make a plot of these residuals versus the independent variable, X, and comment on any pattern you see.

b) Explain how you could use such residual plots to detect outliers.

Non-Linear Regression

Many relationships between two variables follow patterns that are not linear. For example, square-law, exponential, and logarithmic relationships often appear in the natural sciences. **Non-linear regression** is an analytical technique for finding a curve of best fit for data from such relationships. The equation for this curve can then be used to model the relationship between the two variables.

As you might expect, the calculations for curves are more complicated than those for straight lines. Graphing calculators have built-in regression functions for a variety of curves, as do some spreadsheets and statistical programs. Once you enter the data and specify the type of curve, these technologies can automatically find the best-fit curve of that type. They can also calculate the coefficient of determination, r^2, which is a useful measure of how closely a curve fits the data.

INVESTIGATE & INQUIRE: Bacterial Growth

A laboratory technician monitors the growth of a bacterial culture by scanning it every hour and estimating the number of bacteria. The initial population is unknown.

Time (h)	0	1	2	3	4	5	6	7
Population	?	10	21	43	82	168	320	475

1. **a)** Create a scatter plot and classify the linear correlation.

   b) Determine the correlation coefficient and the line of best fit.

   c) Add the line of best fit to your scatter plot. Do you think this line is a satisfactory model? Explain why or why not.

2. **a)** Use software or a graphing calculator to find a curve of best fit with a

   i) quadratic regression of the form $y = ax^2 + bx + c$

   ii) cubic regression of the form $y = ax^3 + bx^2 + cx + d$

   b) Graph these curves onto a scatter plot of the data.

   c) Record the equation and the coefficient of determination, r^2, for the curves.

   d) Use the equations to estimate the initial population of the bacterial culture. Do these estimates seem reasonable? Why or why not?

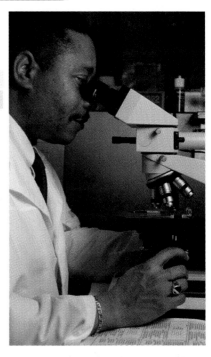

See Appendix B for details on using technology for non-linear regressions.

3. **a)** Perform an exponential regression on the data. Graph the curve of best fit and record its equation and coefficient of determination.

   b) Use this model to estimate the initial population.

   c) Do you think the exponential equation is a better model for the growth of the bacterial culture than the quadratic or cubic equations? Explain your reasoning.

Recall that Pearson's correlation coefficient, r, is a measure of the linearity of the data, so it can indicate only how closely a straight line fits the data. However, the **coefficient of determination, r^2**, is defined such that it applies to any type of regression curve.

$$r^2 = \frac{\text{variation in } y \text{ explained by variation in } x}{\text{total variation in } y}$$

$$= \frac{\Sigma(y_{est} - \overline{y})^2}{\Sigma(y - \overline{y})^2}$$

where $\overline{y}$ is the mean y value, y_{est} is the value estimated by the best-fit curve for a given value of x, and y is the actual observed value for a given value of x.

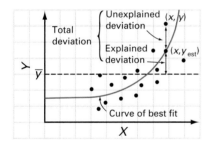

The total variation is the sum of the squares of the deviations for all of the individual data points.

The coefficient of determination can have values from 0 to 1. If the curve is a perfect fit, then y_{est} and y will be identical for each value of x. In this case, the variation in x accounts for all of the variation in y, so $r^2 = 1$. Conversely, if the curve is a poor fit, the total of $(y_{est} - \overline{y})^2$ will be much smaller than the total of $(y - \overline{y})^2$, since the variation in x will account for only a small part of the total variation in y. Therefore, r^2 will be close to 0. For any given type of regression, the curve of best fit will be the one that has the highest value for r^2.

For graphing calculators and Microsoft® Excel, the procedures for non-linear regression are almost identical to those for linear regression. At present, Corel® Quattro® Pro and Fathom™ do not have built-in functions for non-linear regression.

Exponential Regression

Exponential regressions produce equations with the form $y = ab^x$ or $y = ae^{kx}$, where $e = 2.718\ 28\ldots$, an irrational number commonly used as the base for exponents and logarithms. These two forms are equivalent, and it is straightforward to convert from one to the other.

Example 1 Exponential Regression

Generate an exponential regression for the bacterial culture in the investigation on page 184. Graph the curve of best fit and determine its equation and the coefficient of determination.

Solution 1 Using a Graphing Calculator

Use the ClrList command from the STAT EDIT menu to clear lists L1 and L2, and then enter the data. Set DiagnosticOn so that regression calculations will display the coefficient of determination. From the STAT CALC menu, select the non-linear regression function ExpReg. If you do not enter any list names, the calculator will use L1 and L2 by default.

The equation for the curve of best fit is $y = 5.70(1.93)^x$, and the coefficient of determination is $r^2 = 0.995$. Store the equation as Y1. Use STAT PLOT to display a scatter plot of the data along with Y1. From the ZOOM menu, select 9:ZoomStat to adjust the window settings automatically.

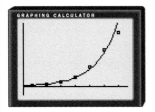

Solution 2 Using a Spreadsheet

Enter the data into two columns. Next, highlight these columns and use the Chart feature to create an x-y scatter plot.

Select Chart/Add Trendline and then choose Expontenial regression. Then, select the curve that appears on your chart, and click Format/Selected Trendline/Options. Check the option boxes to display the equation and r^2.

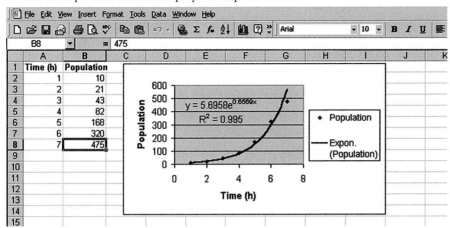

The equation of the best-fit curve is $y = 5.7e^{0.66x}$ and the coefficient of determination is $r^2 = 0.995$. This equation appears different from the one found with the graphing calculator. In fact, the two forms are equivalent, since $e^{0.66} \doteq 1.93$.

Power and Polynomial Regression

In **power regressions**, the curve of best fit has an equation with the form $y = ax^b$.

Example 2 Power Regression

For a physics project, a group of students videotape a ball dropped from the top of a 4-m high ladder, which they have marked every 10 cm. During playback, they stop the videotape every tenth of a second and compile the following table for the distance the ball travelled.

Time (s)	0.1	0.2	0.3	0.4	0.5	0.6	0.7	0.8	0.9	1.0
Distance (m)	0.05	0.2	0.4	0.8	1.2	1.7	2.4	3.1	3.9	4.9

a) Does a linear model fit the data well?

b) Use a power regression to find a curve of best fit for the data. Does the power-regression curve fit the data more closely than the linear model does?

c) Use the equation for the regression curve to predict

   i) how long the ball would take to fall 10 m

   ii) how far the ball would fall in 5 s

Solution 1 Using a Graphing Calculator

a) Although the linear correlation coefficient is 0.97, a scatter plot of the data shows a definite curved pattern. Since $b = -1.09$, the linear model predicts an initial position of about -1.1 m and clearly does not fit the first part of the data well. Also, the pattern in the scatter plot suggests the linear model could give inaccurate predictions for times beyond 1 s.

 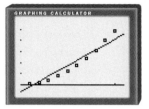

b) From the STAT CALC menu, select the non-linear regression function PwrReg and then follow the same steps as in Example 1.

 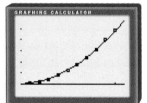

The equation for the curve of best fit is $y = 4.83x^2$. The coefficient of determination and a graph on the scatter plot show that the quadratic curve is almost a perfect fit to the data.

c) Substitute the known values into the equation for the quadratic curve of best fit:

  i) $10 = 4.83x^2$ **ii)** $y = 4.83(5)^2$

      $x^2 = \dfrac{10}{4.83}$ $= 4.83(25)$

        $x = \sqrt{\dfrac{10}{4.83}}$ $= 121$

         $= 1.4$

The quadratic model predicts that
i) the ball would take approximately 1.4 s to fall 10 m
ii) the ball would fall 121 m in 5 s

Solution 2 Using a Spreadsheet

a) As in Solution 1, the scatter plot shows that a curve might be a better model.

b) Use the Chart feature as in Example 1, but select **Power** when adding the trend line.

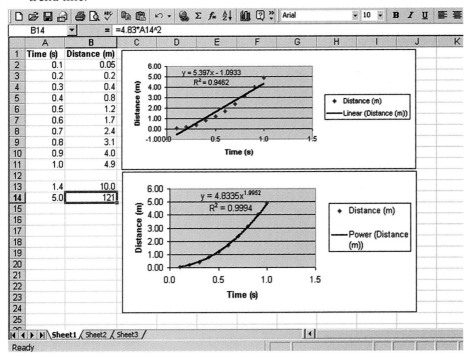

The equation for the curve of best fit is $y = 4.83x^2$. The graph and the value for r^2 show that the quadratic curve is almost a perfect fit to the data.

c) Use the equation for the curve of best fit to enter formulas for the two values you want to predict, as shown in cells **A13** and **B14** in the screen above.

Example 3 Polynomial Regression

Suppose that the laboratory technician takes further measurements of the bacterial culture in Example 1.

Time (h)	8	9	10	11	12	13	14
Population	630	775	830	980	1105	1215	1410

a) Discuss the effectiveness of the exponential model from Example 1 for the new data.

b) Find a new exponential curve of best fit.

c) Find a better curve of best fit. Comment on the effectiveness of the new model.

Solution

a) If you add the new data to the scatter plot, you will see that the exponential curve determined earlier, $y = 5.7(1.9)^x$, is no longer a good fit.

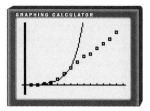

b) If you perform a new exponential regression on all 14 data points, you obtain the equation $y = 18(1.4)^x$ with a coefficient of determination of $r^2 = 0.88$. From the graph, you can see that this curve is not a particularly good fit either.

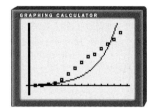

Because of the wide range of non-linear regression options, you can insist on a fairly high value of r^2 when searching for a curve of best fit to model the data.

c) If you perform a quadratic regression, you get a much better fit with the equation $y = 4.0x^2 + 55x - 122$ and a coefficient of determination of $r^2 = 0.986$.

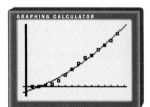

This quadratic model will probably serve well for interpolating between most of the data shown, but may not be accurate for times before 3 h and after 14 h. At some point between 2 h and 3 h, the curve intersects the x-axis, indicating a negative population prior to this time. Clearly the quadratic model is not accurate in this range.

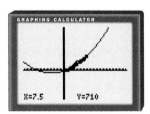

Similarly, if you zoom out, you will notice a problem beyond 14 h. The rate of change of the quadratic curve continues to increase after 14 h, but the trend of the data does not suggest such an increase. In fact, from 7 h to 14 h the trend appears quite linear.

It is important to recognize the limitations of regression curves. One interesting property of polynomial regressions is that for a set of n data points, a polynomial function of degree $n - 1$ can be produced which perfectly fits the data, that is, with $r^2 = 1$.

For example, you can determine the equation for a line (a first-degree polynomial) with two points and the equation for a quadratic (a second-degree polynomial) with three points. However, these polynomials are not always the best models for the data. Often, these curves can give inaccurate predictions when extrapolated.

Sometimes, you can find that several different types of curves fit closely to a set of data. Extrapolating to an initial or final state may help determine which model is the most suitable. Also, the mathematical model should show a logical relationship between the variables.

> **Project Prep**
>
> Non-linear models may be useful when you are analysing two-variable data in your statistics project.

Key Concepts

- Some relationships between two variables can be modelled using non-linear regressions such as quadratic, cubic, power, polynomial, and exponential curves.

- The coefficient of determination, r^2, is a measure of how well a regression curve fits a set of data.

- Sometimes more than one type of regression curve can provide a good fit for data. To be an effective model, however, the curve must be useful for extrapolating beyond the data.

Communicate Your Understanding

1. A data set for two variables has a linear correlation coefficient of 0.23. Does this value preclude a strong correlation between the variables? Explain why or why not.

2. A best-fit curve for a set of data has a coefficient of determination of $r^2 = 0.76$. Describe some techniques you can use to improve the model.

Practise

1. Match each of the following coefficients of determination with one of the diagrams below.

a) 0   b) 0.5   c) 0.9   d) 1

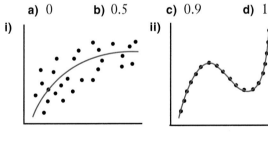

2. For each set of data use software or a graphing calculator to find the equation and coefficient of determination for a curve of best fit.

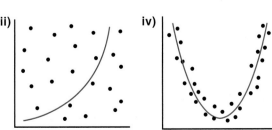

a) x	y	b) x	y	c) x	y
−2.8	0.6	−2.7	1.6	1.1	2.5
−3.5	−5.8	−3.5	−3	3.5	11
−2	3	−2.2	3	2.8	8.6
−1	6	−0.5	−0.5	2.3	7
0.2	4	0	1.3	0	1
1	1	0.6	4.7	3.8	14
−1.5	5	−1.8	1.7	1.4	4.2
1.4	−3.1	−3.8	−7	−4	0.2
0.7	3	−1.3	0.6	−1.3	0.6
−0.3	6.1	0.8	7	3	12
−3.3	−3.1	0.5	2.7	4.1	17
−4	−7	−1	1.5	2.2	5
2	−5.7	−3	−1.1	−2.7	0.4

3.3 Non-Linear Regression • MHR **191**

Apply, Solve, Communicate

 B

3. The heights of a stand of pine trees were measured along with the area under the cone formed by their branches.

Height (m)	Area (m²)
2.0	5.9
1.5	3.4
1.8	4.8
2.4	8.6
2.2	7.3
1.2	2.1
1.8	4.9
3.1	14.4

a) Create a scatter plot of these data.

b) Determine the correlation coefficient and the equation of the line of best fit.

c) Use a power regression to calculate a coefficient of determination and an equation for a curve of best fit.

d) Which model do you think is more accurate? Explain why.

e) Use the more accurate model to predict

  i) the area under a tree whose height is 2.7 m

  ii) the height of a tree whose area is 30 m²

f) Suggest a reason why the height and circumference of a tree might be related in the way that the model in part d) suggests.

4. **Application** The biologist Max Kleiber (1893–1976) pioneered research on the metabolisms of animals. In 1932, he determined the relationship between an animal's mass and its energy requirements or basal metabolic rate (BMR). Here are data for eight animals.

Animal	Mass (kg)	BMR (kJ/day)
Frog	0.018	0.050
Squirrel	0.90	1.0
Cat	3.0	2.6
Monkey	7.0	4.0
Baboon	30	14
Human	60	25
Dolphin	160	44
Camel	530	116

a) Create a scatter plot and explain why Kleiber thought a power-regression curve would fit the data.

b) Use a power regression to find the equation of the curve of best fit. Can you rewrite the equation so that it has exponents that are whole numbers? Do so, if possible, or explain why not.

c) Is this power equation a good mathematical model for the relationship between an animal's mass and its basal metabolic rate? Explain why or why not.

d) Use the equation of the curve of best fit to predict the basal metabolic rate of

  i) a 15-kg dog

  ii) a 2-tonne whale

5. **Application** As a sample of a radioactive element decays into more stable elements, the amount of radiation it gives off decreases. The level of radiation can be used to estimate how much of the original element remains. Here are measurements for a sample of radium-227.

Time (h)	Radiation Level (%)
0	100
1	37
2	14
3	5.0
4	1.8
5	0.7
6	0.3

a) Create a scatter plot for these data.

b) Use an exponential regression to find the equation for the curve of best fit.

c) Is this equation a good model for the radioactive decay of this element? Explain why or why not.

d) A half-life is the time it takes for half of the sample to decay. Use the regression equation to estimate the half-life of radium-227.

6. a) Create a time-series graph for the mean starting salary of the graduates who find jobs. Describe the pattern that you see.

b) Use non-linear regression to construct a curve of best fit for the data. Record the equation of the curve and the coefficient of determination.

c) Comment on whether this equation is a good model for the graduates' starting salaries.

7. An engineer testing the transmitter for a new radio station measures the radiated power at various distances from the transmitter. The engineer's readings are in microwatts per square metre.

Distance (km)	Power Level ($\mu W/m^2$)
2.0	510
5.0	78
8.0	32
10.0	19
12.0	14
15.0	9
20.0	5

a) Find an equation for a curve of best fit for these data that has a coefficient of determination of at least 0.98.

b) Use the equation for this curve of best fit to estimate the power level at a distance of

   i) 1.0 km from the transmitter

   ii) 4.0 km from the transmitter

   iii) 50.0 km from the transmitter

8. Communication *Logistic* curves are often a good model for population growth. These curves have equations with the form

$$y = \frac{c}{1 + ae^{-bx}}, \text{ where } a, b, \text{ and } c \text{ are constants.}$$

Consider the following data for the bacterial culture in Example 1:

Time (h)	0	1	2	3	4	5
Population	?	10	21	43	82	168

Time (h)	6	7	8	9	10	11
Population	320	475	630	775	830	980

Time (h)	12	13	14	15	16	17
Population	1105	1215	1410	1490	1550	1575

Time (h)	18	19	20
Population	1590	1600	1600

a) Use software or a graphing calculator to find the equation and coefficient of determination for the logistic curve that best fits the data for the bacteria population from 1 to 20 h.

b) Graph this curve on a scatter plot of the data.

c) How well does this curve appear to fit the entire data set? Describe the shape of the curve.

d) Write a brief paragraph to explain why you think a bacterial population may exhibit this type of growth pattern.

9. Inquiry/Problem Solving The following table shows the estimated population of a crop-destroying insect.

Year	Population (billions)
1995	100
1996	130
1997	170
1998	220
1999	285
2000	375
2001	490

a) Determine an exponential curve of best fit for the population data.

b) Suppose that 100 million of an arachnid that preys on the insect are imported from overseas in 1995. Assuming the arachnid population doubles every year, estimate when it would equal 10% of the insect population.

c) What further information would you need in order to estimate the population of the crop-destroying insect once the arachnids have been introduced?

d) Write an expression for the size of this population.

10. Use technology to calculate the coefficient of determination for two of the linear regression examples in section 3.2. Is there any relationship between these coefficients of determination and the linear correlation coefficients for these examples?

11. Inquiry/Problem Solving Use a software program, such as Microsoft® Excel, to analyse these two sets of data:

Data Set A		Data Set B	
x	y	x	y
2	5	2	6
4	7	4	5
6	2	7	-4
8	5	9	1
		12	2

a) For each set of data,

   i) determine the degree of polynomial regression that will generate a perfectly fit regression curve

   ii) perform the polynomial regression and record the value of r^2 and the equation of the regression curve

b) Assess the effectiveness of the best-fit polynomial curve as a model for the trend of the set of data.

c) For data set B,

   i) explain why the best-fit polynomial curve is an unsatisfactory model

   ii) generate a better model and record the value of r^2 and the equation of your new best-fit curve

   iii) explain why this curve is a better model than the polynomial curve found in part a)

3.4 Cause and Effect

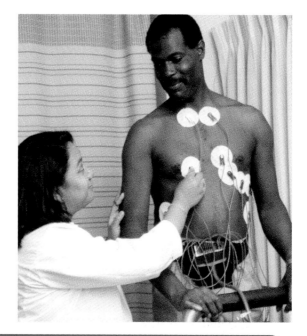

Usually, the main reason for a correlational study is to find evidence of a cause-and-effect relationship. A health researcher may wish to prove that even mild exercise reduces the risk of heart disease. A chemical company developing an oil additive would like to demonstrate that it improves engine performance. A school board may want to know whether calculators help students learn mathematics. In each of these cases, establishing a strong correlation between the variables is just the first step in determining whether one affects the other.

INVESTIGATE & INQUIRE: Correlation Versus Cause and Effect

1. List the type of correlation that you would expect to observe between the following pairs of variables. Also list whether you think the correlation is due to a cause-and-effect relationship or some other factor.

   a) hours spent practising at a golf driving range, golf drive distance

   b) hours spent practising at a golf driving range, golf score

   c) size of corn harvest, size of apple harvest

   d) score on a geometry test, score on an algebra test

   e) income, number of CDs purchased

2. Compare your list with those of your classmates and discuss any differences. Would you change your list because of factors suggested by your classmates?

3. Suggest how you could verify whether there is a cause-and-effect relationship between each pair of variables.

A strong correlation does not prove that the changes in one variable cause changes in the other. There are various types and degrees of causal relationships between variables.

Cause-and-Effect Relationship: A change in X produces a change in Y. Such relationships are sometimes clearly evident, especially in physical processes. For example, increasing the height from which you drop an object increases its impact velocity. Similarly, increasing the speed of a production line increases the number of items produced each day (and, perhaps, the rate of defects).

Common-Cause Factor: An external variable causes two variables to change in the same way. For example, suppose that a town finds that its revenue from parking fees at the public beach each summer correlates with the local tomato harvest. It is extremely unlikely that cars parked at the beach have any effect on the tomato crop. Instead good weather is a common-cause factor that increases both the tomato crop and the number of people who park at the beach.

Reverse Cause-and-Effect Relationship: The dependent and independent variables are reversed in the process of establishing causality. For example, suppose that a researcher observes a positive linear correlation between the amount of coffee consumed by a group of medical students and their levels of anxiety. The researcher theorizes that drinking coffee causes nervousness, but instead finds that nervous people are more likely to drink coffee.

Accidental Relationship: A correlation exists without any causal relationship between variables. For example, the number of females enrolled in undergraduate engineering programs and the number of "reality" shows on television both increased for several years. These two variables have a positive linear correlation, but it is likely entirely coincidental.

Presumed Relationship: A correlation does not seem to be accidental even though no cause-and-effect relationship or common-cause factor is apparent. For example, suppose you found a correlation between people's level of fitness and the number of adventure movies they watched. It seems logical that a physically fit person might prefer adventure movies, but it would be difficult to find a common cause or to prove that the one variable affects the other.

Example 1 Causal Relationships

Classify the relationships in the following situations.
a) The rate of a chemical reaction increases with temperature.
b) Leadership ability has a positive correlation with academic achievement.
c) The prices of butter and motorcycles have a strong positive correlation over many years.
d) Sales of cellular telephones had a strong negative correlation with ozone levels in the atmosphere over the last decade.
e) Traffic congestion has a strong correlation with the number of urban expressways.

Solution

a) Cause-and-effect relationship: Higher temperatures cause faster reaction rates.

b) Presumed relationship: A positive correlation between leadership ability and academic achievement seems logical, yet there is no apparent common-cause factor or cause-and-effect relationship.

c) Common-cause factor: Inflation has caused parallel increases in the prices of butter and motorcycles over the years.

d) Accidental relationship: The correlation between sales of cellular telephones and ozone levels is largely coincidental. However, it is possible that the chemicals used to manufacture cellular telephones cause a small portion of the depletion of the ozone layer.

e) Cause-and-effect relationship and reverse cause-and-effect relationship: Originally expressways were built to relieve traffic congestion, so traffic congestion did lead to the construction of expressways in major cites throughout North America. However, numerous studies over the last 20 years have shown that urban expressways cause traffic congestion by encouraging more people to use cars.

As Example 1 demonstrates, several types of causal relationships can be involved in the same situation. Determining the nature of causal relationships can be further complicated by the presence of **extraneous variables** that affect either the dependent or the independent variable. Here, *extraneous* means external rather than irrelevant.

For example, you might expect to see a strong positive correlation between term marks and final examination results for students in your class since both these variables are affected by each student's aptitude and study habits. However, there are extraneous factors that could affect the examination results, including the time each student had for studying before the examination, the individual examination schedules, and varying abilities to work well under pressure.

In order to reduce the effect of extraneous variables, researchers often compare an **experimental group** to a **control group**. These two groups should be as similar as possible, so that extraneous variables will have about the same effect on both groups. The researchers vary the independent variable for the experimental group but not for the control group. Any *difference* in the dependent variables for the two groups can then be attributed to the changes in the independent variable.

Example 2 Using a Control Group

A medical researcher wants to test a new drug believed to help smokers overcome the addictive effects of nicotine. Fifty people who want to quit smoking volunteer for the study. The researcher carefully divides the volunteers into two groups, each with an equal number of moderate and heavy smokers. One group is given nicotine patches with the new drug, while the second group uses ordinary nicotine patches. Fourteen people in the first group quit smoking completely, as do nine people in the second group.

a) Identify the experimental group, the control group, the independent variable, and the dependent variable.

b) Can the researcher conclude that the new drug is effective?

c) What further study should the researcher do?

Solution

a) The experimental group consists of the volunteers being given nicotine patches with the new drug, while the control group consists of the volunteers being given the ordinary patches. The independent variable is the presence of the new drug, and the dependent variable is the number of volunteers who quit smoking.

b) The results of the study are promising, but the researcher has not proven that the new drug is effective. The sample size is relatively small, which is prudent for an early trial of a new drug that could have unknown side-effects. However, the sample is small enough that the results could be affected by random statistical fluctuations or extraneous variables, such as the volunteers' work environments, previous attempts to quit, and the influence of their families and friends.

c) Assuming that the new drug does not have any serious side-effects, the researcher should conduct further studies with larger groups and try to select the experimental and control groups to minimize the effect of all extraneous variables. The researcher might also conduct a study with several experimental groups that receive different dosages of the new drug.

When designing a study or interpreting a correlation, you often need background knowledge and insight to recognize the causal relationships present. Here are some techniques that can help determine whether a correlation is the result of a cause-and-effect relationship.

- Use sampling methods that hold the extraneous variables constant.

- Conduct similar investigations with different samples and check for consistency in the results.

- Remove, or account for, possible common-cause factors.

The later chapters in this book introduce probability theory and some statistical methods for a more quantitative approach to determining cause-and-effect relationships.

Project Prep

In your statistics project, you may wish to consider cause-and-effect relationships and extraneous variables that could affect your study.

Key Concepts

- Correlation does not necessarily imply a cause-and-effect relationship. Correlations can also result from common-cause factors, reverse cause-and-effect relationships, accidental relationships, and presumed relationships.

- Extraneous variables can invalidate conclusions based on correlational evidence.

- Comparison with a control group can help remove the effect of extraneous variables in a study.

Communicate Your Understanding

1. Why does a strong linear correlation not imply cause and effect?

2. What is the key characteristic of a reverse cause-and-effect relationship?

3. Explain the difference between a common-cause factor and an extraneous variable.

4. Why are control groups used in statistical studies?

Practise

1. Identify the most likely type of causal relationship between each of the following pairs of variables. Assume that a strong positive correlation has been observed with the first variable as the independent variable.

  a) alcohol consumption, incidence of automobile accidents

  b) score on physics examination, score on calculus examination

  c) increase in pay, job performance

  d) population of rabbits, consumer price index

  e) number of scholarships received, number of job offers upon graduation

  f) coffee consumption, insomnia

  e) funding for athletic programs, number of medals won at Olympic games

2. For each of the following common-cause relationships, identify the common-cause factor. Assume a positive correlation between each pair of variables.

a) number of push-ups performed in one minute, number of sit-ups performed in one minute

b) number of speeding tickets, number of accidents

c) amount of money invested, amount of money spent

Apply, Solve, Communicate

3. A civil engineer examining traffic flow problems in a large city observes that the number of traffic accidents is positively correlated with traffic density and concludes that traffic density is likely to be a major cause of accidents. What alternative conclusion should the engineer consider?

B

4. **Communication** An elementary school is testing a new method for teaching grammar. Two similar classes are taught the same material, one with the established method and the other with the new method. When both classes take the same test, the class taught with the established method has somewhat higher marks.

a) What extraneous variables could influence the results of this study?

b) Explain whether the study gives the school enough evidence to reject the new method.

c) What further studies would you recommend for comparing the two teaching methods?

5. **Communication** An investor observes a positive correlation between the stock price of two competing computer companies. Explain what type of causal relationship is likely to account for this correlation.

6. **Application** A random survey of students at Statsville High School found that their interest in computer games is positively correlated with their marks in mathematics.

a) How would you classify this causal relationship?

b) Suppose that a follow-up study found that students who had increased the time they spent playing computer games tended to improve their mathematics marks. Assuming that this study held all extraneous variables constant, would you change your assessment of the nature of the causal relationship? Explain why or why not.

7. a) The net assets of Custom Industrial Renovations Inc., an industrial construction contractor, has a strong negative linear correlation with those of MuchMega-Fun, a toy distributor. How would you classify the causal relationship between these two variables?

b) Suppose that the two companies are both subsidiaries of Diversified Holdings Ltd., which often shifts investment capital between them. Explain how this additional information could change your interpretation of the correlation in part a).

8. **Communication** Aunt Gisele simply cannot sleep unless she has her evening herbal tea. However, the package for the tea does not list any ingredients known to induce sleep. Outline how you would conduct a study to determine whether the tea really does help people sleep.

9. Find out what a *double-blind* study is and briefly explain the advantages of using this technique in studies with a control group.

10. a) The data on page 157 show a positive correlation between the size of the graduating class and the number of

graduates hired. Does this correlation mean that increasing the number of graduates causes a higher demand for them? Explain your answer.

b) A recession during the first half of the 1990s reduced the demand for business graduates. Review the data on page 157 and describe any trends that may be caused by this recession.

Review the data on page 157

![checkmark] **ACHIEVEMENT CHECK**

Knowledge/ Understanding	Thinking/Inquiry/ Problem Solving	Communication	Application

11. The table below lists numbers of divorces and personal bankruptcies in Canada for the years 1976 through 1985.

Year	Divorces	Bankruptcies
1976	54 207	10 049
1977	55 370	12 772
1978	57 155	15 938
1979	59 474	17 876
1980	62 019	21 025
1981	67 671	23 036
1982	70 436	30 643
1983	68 567	26 822
1984	65 172	22 022
1985	61 976	19 752

a) Create a scatter plot and classify the linear correlation between the number of divorces and the number of bankruptcies.

b) Perform a regression analysis. Record the equation of the line of best fit and the correlation coefficient.

c) Identify an external variable that could be a common-cause factor.

d) Describe what further investigation you could do to analyse the possible relationship between divorces and bankruptcies.

12. Search the E-STAT, CANSIM II, or other databases for a set of data on two variables with a positive linear correlation that you believe to be accidental. Explain your findings and reasoning.

13. Use a library, the Internet, or other resources to find information on the Hawthorne effect and the placebo effect. Briefly explain what these effects are, how they can affect a study, and how researchers can avoid having their results skewed by these effects.

14. **Inquiry/Problem Solving** In a behavioural study of responses to violence, an experimental group was shown violent images, while a control group was shown neutral images. From the initial results, the researchers suspect that the gender of the people in the groups may be an extraneous variable. Suggest how the study could be redesigned to

a) remove the extraneous variable

b) determine whether gender is part of the cause-and-effect relationship

15. Look for material in the media or on the Internet that incorrectly uses correlational evidence to claim that a cause-and-effect relationship exists between the two variables. Briefly describe

a) the nature of the correlational study

b) the cause and effect claimed or inferred

c) the reasons why cause and effect was not properly proven, including any extraneous variables that were not accounted for

d) how the study could be improved

3.5 Critical Analysis

Newspapers and radio and television news programs often run stories involving statistics. Indeed, the news media often commission election polls or surveys on major issues. Although the networks and major newspapers are reasonably careful about how they present statistics, their reporters and editors often face tight deadlines and lack the time and mathematical knowledge to thoroughly critique statistical material. You should be particularly careful about accepting statistical evidence from sources that could be biased. Lobby groups and advertisers like to use statistics because they appear scientific and objective. Unfortunately, statistics from such sources are sometimes flawed by unintentional or, occasionally, entirely deliberate bias. To judge the conclusions of a study properly, you need information about its sampling and analytical methods.

THE DUPLEX — BY GLENN MCCOY

INVESTIGATE & INQUIRE: Statistics in the Media

1. Find as many instances as you can of statistical claims made in the media or on the Internet, including news stories, features, and advertisements. Collect newspaper and magazine clippings, point-form notes of radio and television stories, and printouts of web pages.

2. Compare the items you have collected with those found by your classmates. What proportion of the items provide enough information to show that they used valid statistical methods?

3. Select several of the items. For each one, discuss
   a) the motivation for the statistical study
   b) whether the statistical evidence justifies the claim being made

WEB CONNECTION

www.mcgrawhill.ca/links/MDM12

Visit the above web site and follow the links to learn more about how statistics can be misused. Describe two examples of the misuse of statistics.

The examples in this section illustrate how you can apply analytical tools to assess the results of statistical studies.

Example 1 Sample Size and Technique

A manager wants to know if a new aptitude test accurately predicts employee productivity. The manager has all 30 current employees write the test and then compares their scores to their productivities as measured in the most recent performance reviews. The data is ordered alphabetically by employee surname. In order to simplify the calculations, the manager selects a systematic sample using every seventh employee. Based on this sample, the manager concludes that the company should hire only applicants who do well on the aptitude test. Determine whether the manager's analysis is valid.

Solution

A linear regression of the systematic sample produces a line of best fit with the equation $y = 0.55x + 33$ and a correlation coefficient of $r = 0.98$, showing a strong linear correlation between productivity and scores on the aptitude test. Thus, these calculations seem to support the manager's conclusion. However, the manager has made the questionable assumption that a systematic sample will be representative of the population. The sample is so small that statistical fluctuations could seriously affect the results.

Test Score	Productivity
98	78
57	81
82	83
76	44
65	62
72	89
91	85
87	71
81	76
39	71
50	66
75	90
71	48
89	80
82	83
95	72
56	72
71	90
68	74
77	51
59	65
83	47
75	91
66	77
48	63
61	58
78	55
70	73
68	75
64	69

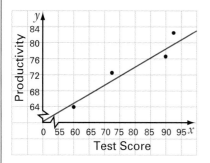

```
GRAPHING CALCULATOR
LinReg
y=ax+b
a=.5517362329
b=33.12837601
r²=.9568700048
r=.9781973241
```

Examine the raw data. A scatter plot with all 30 data points does not show any clear correlation at all. A linear regression yields a line of best fit with the equation $y = 0.15x + 60$ and a correlation coefficient of only 0.15.

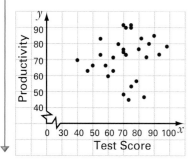

```
GRAPHING CALCULATOR
LinReg
y=ax+b
a=.146371507
b=60.7905258
r²=.0237875505
r=.1542321317
```

Thus, the new aptitude test will probably be useless for predicting employee productivity. Clearly, the sample was far from representative. The manager's choice of an inappropriate sampling technique has resulted in a sample size too small to make any valid conclusions.

In Example 1, the manager should have done an analysis using all of the data available. Even then the data set is still somewhat small to use as a basis for a major decision such as changing the company's hiring procedures. Remember that small samples are also particularly vulnerable to the effects of outliers.

Example 2 Extraneous Variables and Sample Bias

An advertising blitz by SuperFast Computer Training Inc. features profiles of some of its young graduates. The number of months of training that these graduates took, their job titles, and their incomes appear prominently in the advertisements.

Graduate	Months of Training	Income ($000)
Sarah, software developer	9	85
Zack, programmer	6	63
Eli, systems analyst	8	72
Yvette, computer technician	5	52
Kulwinder, web-site designer	6	66
Lynn, network administrator	4	60

a) Analyse the company's data to determine the strength of the linear correlation between the amount of training the graduates took and their incomes. Classify the linear correlation and find the equation of the linear model for the data.

b) Use this model to predict the income of a student who graduates from the company's two-year diploma program after 20 months of training. Does this prediction seem reasonable?

c) Does the linear correlation show that SuperFast's training accounts for the graduates' high incomes? Identify possible extraneous variables.

d) Discuss any problems with the sampling technique and the data.

Solution

a) The scatter plot for income versus months of training shows a definite positive linear correlation. The regression line is $y = 5.44x + 31.9$, and the correlation coefficient is 0.90. There appears to be a strong positive correlation between the amount of training and income.

Arial 10 **B** *I* <u>U</u> Normal

A:C9 +C12+C13*B9

	A	B	C
1	Graduate	Training	Income
2	Sarah	9	85
3	Zack	6	63
4	Eli	8	72
5	Yvette	5	52
6	Kulwinder	6	66
7	Lynn	4	60
8			
9		20	140.71
10			
11		Regression Output:	
12	Constant		31.865
13	X Coefficient(s)		5.4423
14	R Squared		0.8055
15	R		0.8975
16			
17			

Scatter plot: Income ($000) (vertical axis, 50 to 85) versus Training (months) (horizontal axis, 4 to 9).

b) As shown in cell C9 in the screen above, substituting 20 months into the linear regression equation predicts an income of approximately

$$y = 5.44(20) + 31.9$$
$$= 141$$

Therefore, the linear model predicts that a graduate who has taken 20 months of training will make about $141 000 a year. This amount is extremely high for a person with a two-year diploma and little or no job experience. The prediction suggests that the linear model may not be accurate, especially when applied to the company's longer programs.

c) Although the correlation between SuperFast's training and the graduates' incomes appears to be quite strong, the correlation by itself does not prove that the training causes the graduates' high incomes. A number of extraneous variables could contribute to the graduates' success, including experience prior to taking the training, aptitude for working with computers, access to a high-end computer at home, family or social connections in the industry, and the physical stamina to work very long hours.

d) The sample is small and could have intentional bias. There is no indication that the individuals in the advertisements were randomly chosen from the population of SuperFast's students. Quite likely, the company carefully selected the best success stories in order to give potential customers inflated expectations of future earnings. Also, the company shows youthful graduates, but does not actually state that the graduates earned their high incomes immediately after graduation. It may well have taken the graduates years of hard work to reach the income levels listed in the advertisements. Further, the amounts given are incomes, not salaries. The income of a graduate working for a small start-up company might include stock options that could turn out to be worthless. In short, the advertisements do not give you enough information to properly evaluate the data.

Example 2 had several fairly obvious extraneous variables. However, extraneous variables are sometimes difficult to recognize. Such **hidden** or **lurking variables** can also invalidate conclusions drawn from statistical results.

Example 3 Detecting a Hidden Variable

An arts council is considering whether to fund the start-up of a local youth orchestra. The council has a limited budget and knows that the number of youth orchestras in the province has been increasing. The council needs to know whether starting another youth orchestra will help the development of young musicians. One measure of the success of such programs is the number of youth-orchestra players who go on to professional orchestras. The council has collected the following data.

Year	Number of Youth Orchestras	Number of Players Becoming Professionals
1991	10	16
1992	11	18
1993	12	20
1994	12	23
1995	14	26
1996	14	32
1997	16	13
1998	16	16
1999	18	20
2000	20	26

a) Does a linear regression allow you to determine whether the council should fund a new youth orchestra? Can you draw any conclusions from other analysis?

b) Suppose you discover that one of the country's professional orchestras went bankrupt in 1997. How does this information affect your analysis?

Solution

a) A scatter plot of the number of youth-orchestra members who go on to play professionally versus the number of youth orchestras shows that there may be a weak positive linear correlation. The correlation coefficient is 0.16, indicating that the linear correlation is very weak. Therefore, you might conclude that starting another youth orchestra will not help the development of young musicians. However, notice that the data points seem to form two clusters in the scatter plot, one on the left side and the other on the right. This unusual pattern suggests the presence of a hidden variable, which could affect your analysis. You will need more information to determine the nature and effect of the possible hidden variable.

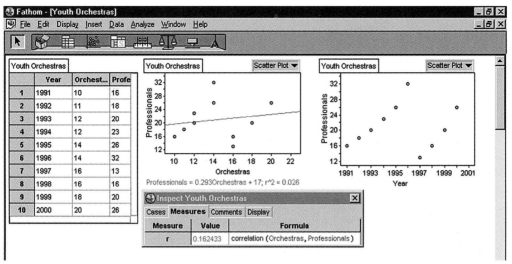

You have enough data to produce a time-series graph of the numbers of young musicians who go on to professional orchestras. This graph also has two clusters of data points. The numbers rise from 1991 to 1996, drop substantially in 1997, and then rise again. This pattern suggests that something unusual happened in 1997.

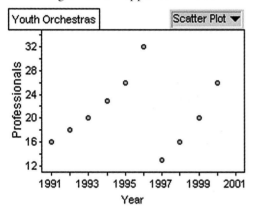

b) The collapse of a major orchestra means both that there is one less orchestra hiring young musicians and that about a hundred experienced players are suddenly available for work with the remaining professional orchestras. The resulting drop in the number of young musicians hired by professional orchestras could account for the clustering of data points you observed in part a). Because of the change in the number of jobs available for young musicians, it makes sense to analyse the clusters separately.

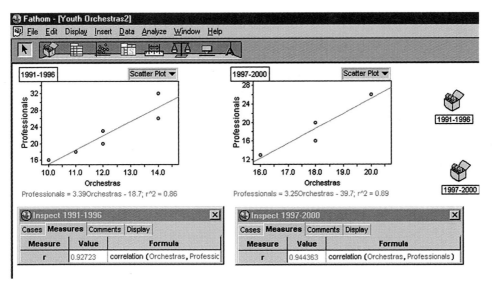

Observe that the two sets of data both exhibit a strong linear correlation. The correlation coefficients are 0.93 for the data prior to 1997 and 0.94 for the data from 1997 on. The number of players who go on to professional orchestras is strongly correlated to the number of youth orchestras. So, funding the new orchestra may be a worthwhile project for the arts council.

The presence of a hidden variable, the collapse of a major orchestra, distorted the data and masked the underlying pattern. However, splitting the data into two sets results in smaller sample sizes, so you still have to be cautious about drawing conclusions.

When evaluating claims based on statistical studies, you must assess the methods used for collecting and analysing the data. Some critical questions are:

- Is the sampling process free from intentional and unintentional bias?

- Could any outliers or extraneous variables influence the results?

- Are there any unusual patterns that suggest the presence of a hidden variable?

- Has causality been inferred with only correlational evidence?

Project Prep

When collecting and analysing data for your statistics project, you can apply the concepts in this section to ensure that your conclusions are valid.

Key Concepts

- Although the major media are usually responsible in how they present statistics, you should be cautious about accepting any claim that does not include information about the sampling technique and analytical methods used.

- Intentional or unintentional bias can invalidate statistical claims.

- Small sample sizes and inappropriate sampling techniques can distort the data and lead to erroneous conclusions.

- Extraneous variables must be eliminated or accounted for.

- A hidden variable can skew statistical results and yet still be hard to detect.

Communicate Your Understanding

1. Explain how a small sample size can lead to invalid conclusions.

2. A city councillor states that there are problems with the management of the police department because the number of reported crimes in the city has risen despite increased spending on law enforcement. Comment on the validity of this argument.

3. Give an example of a hidden variable not mentioned in this section, and explain why this variable would be hard to detect.

Apply, Solve, Communicate

1. An educational researcher discovers that levels of mathematics anxiety are negatively correlated with attendance in mathematics class. The researcher theorizes that poor attendance causes mathematics anxiety. Suggest an alternate interpretation of the evidence.

2. A survey finds a correlation between the proportion of high school students who own a car and the students' ages. What hidden variable could affect this study?

3. A student compares height and grade average with four friends and collects the following data.

Height (cm)	Grade Average (%)
171	73
145	91
162	70
159	81
178	68

From this table, the student concludes that taller students tend to get lower marks.

a) Does a regression analysis support the student's conclusion?

b) Why are the results of this analysis invalid?

c) How can the student get more accurate results?

4. Inquiry/Problem Solving A restaurant chain randomly surveys its customers several times a year. Since the surveys show that the level of customer satisfaction is rising over time, the company concludes that its customer service is improving. Discuss the validity of the surveys and the conclusion based on these surveys.

5. Application A teacher offers the following data to show that good attendance is important.

Days Absent	Final Grade
8	72
2	75
0	82
11	68
15	66
20	30

A student with a graphing calculator points out that the data indicate that anyone who misses 17 days or more is in danger of failing the course.

a) Show how the student arrived at this conclusion.

b) Identify and explain the problems that make this conclusion invalid.

c) Outline statistical methods to avoid these problems.

6. Using a graphing calculator, Gina found the cubic curve of best fit for the salary data in the table on page 157. This curve has a coefficient of determination of 0.98, indicating an almost perfect fit to the data. The equation of the cubic curve is

starting salary
$= 0.0518y^3 - 310y^2 + 618\ 412y - 411\ 344\ 091$

where the salary is given in thousands of dollars and y is the year of graduation.

a) What mean starting salary does this model predict for Gina's class when they graduate in 2005?

b) Is this prediction realistic? Explain.

c) Explain why this model generated such an inaccurate prediction despite having a high value for the coefficient of determination.

d) Suggest methods Gina could use to make a more accurate prediction.

7. Communication Find a newspaper or magazine article, television commercial, or web page that misuses statistics of two variables. Perform a critical analysis using the techniques in this chapter. Present your findings in a brief report.

8. Application A manufacturing company keeps records of its overall annual production and its number of employees. Data for a ten-year period are shown below.

Year	Number of Employees	Production (000)
1992	158	75
1993	165	81
1994	172	84
1995	148	68
1996	130	58
1997	120	51
1998	98	50
1999	105	57
2000	110	62
2001	120	70

a) Create a scatter plot to see if there is a linear correlation between annual production and number of employees. Classify the correlation.

b) At some point, the company began to lay off workers. When did these layoffs begin?

c) Does the scatter plot suggest the presence of a hidden variable? Could the layoffs account for the pattern you see? Explain why or why not.

d) The company's productivity is its annual production divided by the number of

employees. Create a time-series graph for the company's productivity.

e) Find the line of best fit for the graph in part d).

f) The company has adopted a better management system. When do you think the new system was implemented? Explain your reasoning.

9. Search E-STAT, CANSIM II, or other sources for time-series data for the price of a commodity such as gasoline, coffee, or computer memory. Analyse the data and comment on any evidence of a hidden variable. Conduct further research to determine if there are any hidden variables. Write a brief report outlining your analysis and conclusions.

10. **Inquiry/Problem Solving** A study conducted by Stanford University found that behavioural counselling for people who had suffered a heart attack reduced the risk of a further heart attack by 45%. Outline how you would design such a study. List the independent and dependent variables you would use and describe how you would account for any extraneous variables.

Review of Key Concepts

3.1 Scatter Plots and Linear Correlation

Refer to the Key Concepts on page 167.

1. a) Classify the linear correlation in each scatter plot shown below.

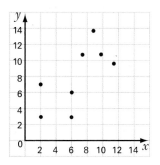

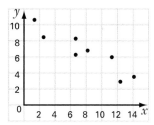

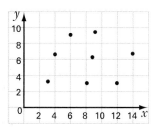

b) Determine the correlation coefficient for data points in the scatter plots in part a).

c) Do these correlation coefficients agree with your answers in part a)?

2. A survey of a group of randomly selected students compared the number of hours of television they watched per week with their grade averages.

Hours Per Week	12	10	5	3	15	16	8
Grade Average (%)	70	85	82	88	65	75	68

a) Create a scatter plot for these data. Classify the linear correlation.

b) Determine the correlation coefficient.

c) Can you make any conclusions about the effect that watching television has on academic achievement? Explain.

3.2 Linear Regression

Refer to the Key Concepts on page 179.

3. Use the method of least squares to find the equation for the line of best fit for the data in question 2.

4. The scores for players' first and second games at a bowling tournament are shown below.

First Game	169	150	202	230	187	177	164
Second Game	175	162	195	241	185	235	171

a) Create a scatter plot for these data.

b) Determine the correlation coefficient and the line of best fit.

c) Identify any outliers.

d) Repeat part b) with the outliers removed.

e) A player scores 250 in the first game. Use both linear models to predict this player's score for the second game. How far apart are the two predictions?

3.3 Non-Linear Regression

Refer to the Key Concepts on page 191.

5. An object is thrown straight up into the air. The table below shows the height of the object as it ascends.

Time (s)	0	0.1	0.2	0.3	0.4	0.5	0.6
Height (m)	0	1	1.8	2.6	3.2	3.8	4.2

a) Create a scatter plot for these data.

b) Perform a non-linear regression for these data. Record the equation of the curve of best fit and the coefficient of determination.

c) Use your model to predict the maximum height of the object.

d) Use your model to predict how long the object will be in the air.

e) Do you think that your model is accurate? Explain.

6. The table shows the distance travelled by a car as a function of time.

Time (s)	Distance (m)
0	0
2	6
4	22
6	50
8	90
10	140
12	190
14	240
16	290
18	340
20	380
22	410
24	430
26	440
28	440

a) Determine a curve of best fit to model the data.

b) Do you think the equation for this curve of best fit is a good model for the situation? Explain your reasoning.

c) Describe what the driver did between 0 and 28 s.

3.4 Cause and Effect
Refer to the Key Concepts on page 199.

7. Define or explain the following terms and provide an example of each one.

a) common-cause factor

b) reverse cause-and-effect relationship

c) extraneous variable

8. a) Explain the relationship between experimental and control groups.

b) Why is a control group needed in some statistical studies?

9. a) Explain the difference between an accidental relationship and a presumed relationship.

b) Provide an example of each.

10. The price of eggs is positively correlated with wages. Explain why you cannot conclude that raising the price of eggs should produce a raise in pay.

11. An educational researcher compiles data on Internet use and scholastic achievement for a random selection of students, and observes a strong positive linear correlation. She concludes that Internet use improves student grades. Comment on the validity of this conclusion.

3.5 Critical Analysis
Refer to the Key Cconcepts on page 209.

12. A teacher is trying to determine whether a new spelling game enhances learning. In his gifted class, he finds a strong positive correlation between use of the game and spelling-test scores. Should the teacher recommend the use of the game in all English classes at his school? Explain your answer.

13. a) Explain what is meant by the term *hidden variable*.

b) Explain how you might detect the presence of a hidden variable in a set of data.

Chapter Test

Category	Knowledge/ Understanding	Thinking/Inquiry/ Problem Solving	Communication	Application
Questions	All	5, 7, 10	1, 5, 6, 8, 10	3, 4, 7, 10

1. Explain or define each of the following terms.

   a) perfect negative linear correlation

   b) experimental research

   c) outlier

   d) extraneous variable

   e) hidden variable

2. Match the following.

Correlation Type	Coefficient, r
a) strong negative linear	1
b) direct	0.6
c) weak positive linear	0.3
d) moderate positive linear	−0.8
e) perfect negative linear	−1

3. The following set of data relates mean word length and recommended age level for a set of children's books.

Recommended Age	Mean Word Length
4	3.5
6	5.5
5	4.6
6	5.0
7	5.2
9	6.5
8	6.1
5	4.9

   a) Create a scatter plot and classify the linear correlation.

   b) Determine the correlation coefficient.

   c) Determine the line of best fit.

   d) Use this model to predict the average word length in a book recommended for 12-year olds.

Use the following information in order to answer questions 4–6.

Jerome has kept track of the hours he spent studying and his marks on examinations.

Subject	Hours Studied	Mark
Mathematics, grade 9	5	70
English, grade 9	3	65
Science, grade 9	4	68
Geography, grade 9	4	72
French, grade 9	2	38
Mathematics, grade 10	7	74
English, grade 10	5	69
Science, grade 10	6	71
History, grade 10	5	75
Mathematics, grade 11	12	76
English, grade 11	9	74
Physics, grade 11	14	78

4. **a)** Create a scatter plot for Jerome's data and classify the linear correlation.

   b) Perform a regression analysis. Identify the equation of the line of best fit as y_1, and record the correlation coefficient.

   c) Identify any outliers.

   d) Repeat part b) with the outlier removed. Identify this line as y_2.

5. Which of the two linear models found in question 4 gives a more optimistic prediction for Jerome's upcoming biology examination? Explain.

6. a) Identify at least three extraneous variables in Jerome's study.

b) Suggest some ways that Jerome might improve the validity of his study.

7. A phosphorescent material can glow in the dark by absorbing energy from light and then gradually re-emitting it. The following table shows the light levels for a phosphorescent plastic.

Time (h)	Light Level (lumens)
0	0.860
1	0.695
2	0.562
3	0.455
4	0.367
5	0.305
6	0.247

a) Create a scatter plot for the data.

b) Perform a quadratic regression. Record the equation of the curve of best fit and the coefficient of determination.

c) Repeat part b) for an exponential regression.

d) Compare how well these two models fit the data.

e) According to each model, what will be the light level after 10 h?

f) Which of these two models is superior for extrapolating beyond 6 h? Explain.

8. Explain how you could minimize the effects of extraneous variables in a correlation study.

9. Provide an example of a reverse cause-and-effect relationship.

✔ **ACHIEVEMENT CHECK**

Knowledge/Understanding	Thinking/Inquiry/Problem Solving	Communication	Application

10. The table shown on the right contains data from the Ontario Road Safety Annual Report for 1999.

a) Organize the data so that the age intervals are consistent. Create a scatter plot of the proportion of drivers involved in collisions versus age.

b) Perform a regression analysis. Record the equations of the curves of best fit for each regression you try as well as the coefficient of determination.

c) In Ontario, drivers over 80 must take vision and knowledge tests every two years to renew their licences. However, these drivers no longer have to take road tests as part of the review. Advocacy groups for seniors had lobbied the Ontario government for this change. How could such groups have used your data analysis to support their position?

Age	Licensed Drivers	Number of Collisions	% of Drivers in Age Group in Collisions
16	85 050	1 725	2.0
17	105 076	7 641	7.3
18	114 056	9 359	8.2
19	122 461	9 524	7.8
20	123 677	9 320	7.5
21–24	519 131	36 024	6.9
25–34	1 576 673	90 101	5.7
35–44	1 895 323	90 813	4.8
45–54	1 475 588	60 576	4.1
55–64	907 235	31 660	3.5
65–74	639 463	17 598	2.8
75 and older	354 581	9 732	2.7
Total	7 918 314	374 073	4.7

Wrap-Up

Implementing Your Action Plan

1. Look up the most recent census data from Statistics Canada. Pick a geographical region and study the data on age of all respondents by gender. Conjecture a relationship between age and the relative numbers of males and females. Use a table and a graph to organize and present the data. Does the set of data support your conjecture?

2. You may want to compare the data you analysed in step 1 to the corresponding data for other regions of Canada or for other countries. Identify any significant similarities or differences between the data sets. Suggest reasons for any differences you notice.

3. Access data on life expectancies in Canada for males and females from the 1920s to the present. Do life expectancies appear to be changing over time? Is there a correlation between these two variables? If so, use regression analysis to predict future life expectancies for males and females in Canada.

4. Access census data on life expectancies in the various regions of Canada. Select another attribute from the census data and conjecture whether there is a correlation between this variable and life expectancies. Analyse data from different regions to see if the data support your conjecture.

Suggested Resources

- Statistics Canada web sites and publications
- Embassies and consulates
- United Nations web sites and publications such as UNICEF's CyberSchoolbus and World Health Organization reports
- Statistical software (the Fathom™ sample documents include census data for Beverly Hills, California)
- Spreadsheets
- Graphing calculators

WEB CONNECTION

www.mcgrawhill.ca/links/MDM12

Visit the web site above to find links to various census databases.

Evaluating Your Project

To help assess your own project, consider the following questions.

1. Are the data you selected appropriate?

2. Are your representations of the data effective?

3. Are the mathematical models that you used reliable?

4. Who would be interested in your findings? Is there a potential market for this information?

5. Are there questions that arose from your research that warrant further investigation? How would you go about addressing these issues in a future project?

6. If you were to do this project again, what would you do differently? Why?

Section 9.4 describes methods for evaluating your own work.

Presentation

Present the findings of your investigation in one or more of the following forms:

- written report
- oral presentation
- computer presentation (using software such as Corel® Presentations™ or Microsoft® PowerPoint®)
- web page
- display board

Remember to include a bibliography. See section 9.5 and Appendix D for information on how to prepare a presentation.

Preparing for the Culminating Project

Applying Project Skills

Throughout this statistics project, you have developed skills in statistical research and analysis that may be helpful in preparing your culminating project:

- making a conjecture or hypothesis
- using technology to access, organize, and analyse data
- applying a variety of statistical tools
- comparing two sets of data
- presenting your findings

Keeping on Track

At this point, you should have a good idea of the basic nature of your culminating project. You should have identified the issue that will be the focus of your project and begun to gather relevant data. Section 9.2 provides suggestions to help you clearly define your task. Your next steps are to develop and implement an action plan.

Make sure there are enough data to support your work. Decide on the best way to organize and present the data. Then, determine what analysis you need to do. As you begin to work with the data, you may find that they are not suitable or that further research is necessary. Your analysis may lead to a new approach or topic that you would like to pursue. You may find it necessary to refine or alter the focus of your project. Such changes are a normal part of the development and implementation process.

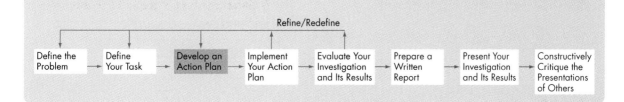

Refine/Redefine

Define the Problem → Define Your Task → Develop an Action Plan → Implement Your Action Plan → Evaluate Your Investigation and Its Results → Prepare a Written Report → Present Your Investigation and Its Results → Constructively Critique the Presentations of Others

1. Let $A = \begin{bmatrix} 7 & 3 \\ 0 & -2 \\ -5 & 4 \end{bmatrix}$, $B = \begin{bmatrix} 8 & 1 \\ -5 & 4 \end{bmatrix}$, and

$C = \begin{bmatrix} -8 & 0 \\ 5 & 6 \\ 9 & -3 \end{bmatrix}$. Calculate, if possible,

a) $-2(A + C)$ **b)** AC

c) $(BA)^t$ **d)** B^2

e) C^2 **f)** B^{-1}

2. a) Describe the iterative process used to generate the table below.

b) Continue the process until all the cells are filled.

17	16	15	14	13	
18	5	4	3	12	
	6	1	2	11	
	7	8	9	10	

3. Which of the following would you consider to be databases? Explain your reasoning.

a) a novel

b) school attendance records

c) the home page of a web site

d) an advertising flyer from a department store

4. What sampling techniques are most likely to be used for the following surveys? Explain each of your choices.

a) a radio call-in show

b) a political poll

c) a scientific study

5. Classify the type of linear correlation that you would expect for each pair of variables.

a) air temperature, altitude

b) income, athletic ability

c) people's ages from 1 to 20 years, their masses

d) people's ages from 21 to 40 years, their masses

6. Identify the most likely causal relationship between each of the following pairs of variables.

a) grade point average, starting salary upon graduation

b) grade in chemistry, grade in physics

c) sales of symphony tickets, carrot harvest

d) monthly rainfall, monthly umbrella sales

7. a) Sketch a map that can be coloured using only three colours.

b) Reconfigure your map as a network.

8. State whether each of the following networks is

   i) connected   **ii)** traceable   **iii)** planar

Provide evidence for your decisions.

a)
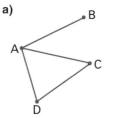

b)
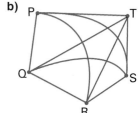

9. Use a tree diagram to represent the administrative structure of a school that has a principal, vice-principals, department heads, assistant heads, and teachers.

10. A renowned jazz pianist living in Toronto often goes on tours in the United States. For the tour shown below, which city has the most routes

    a) with exactly one stopover?

    b) with no more than two stopovers?

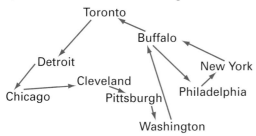

11. The following are responses to a survey that asked: "On average, how many hours per week do you read for pleasure?"

1 3 0 0 7 2 0 1 10 5 2 2 2 0 1 4 0 8 3 1 3
0 0 2 15 4 9 1 6 7 0 3 3 14 5 7 0 1 1 0 10 0

    Use a spreadsheet to

    a) sort the data from smallest to largest value

    b) determine the mean hours of pleasure reading

    c) organize the data into a frequency table with appropriate intervals

    d) make a histogram of the information in part c)

12. The annual incomes of 40 families surveyed at random are shown in the table.

Income ($000)						
28.5	38	61	109	42	56	19
27	44.5	81	36	39	51	40.5
67	28	60	87	58	120	111
73	65	34	54	16.5	135	70.5
59	47	92	38	55	84.5	107
71	59	26.5	76	50		

    a) Group these data into 8 to 12 intervals and create a frequency table.

b) Create a histogram and a cumulative-frequency diagram for the data.

c) What proportion of the families surveyed earn an annual income of $60 000 or less?

13. Classify the bias in each of the following situations. Explain your reasoning in each case.

    a) At a financial planning seminar, the audience were asked to raise their hands if they had ever considered declaring bankruptcy.

    b) A supervisor asked an employee if he would mind working late for a couple of hours on Friday evening.

    c) A survey asked neighbourhood dog-owners if dogs should be allowed to run free in the local park.

    d) An irascible talk show host listed the mayor's blunders over the last year and invited listeners to call in and express their opinions on whether the mayor should resign.

14. The scores in a recent bowling tournament are shown in the following table.

150	260	213	192	176	204	138	214	298	188
168	195	225	170	260	254	195	177	149	224
260	222	167	182	207	221	185	163	112	189

    a) Calculate the mean, median, and mode for this distribution. Which measure would be the most useful? Which would be the least useful? Explain your choices.

    b) Determine the standard deviation, first quartile, third quartile, and interquartile range.

    c) Explain what each of the quantities in part b) tells you about the distribution of scores.

    d) What score is the 50th percentile for this distribution?

e) Is the player who scored 222 above the 80th percentile? Explain why or why not.

15. The players on a school baseball team compared their batting averages and the hours they spent at the batting practice.

Batting Average	Practice Hours
0.220	20
0.215	18
0.185	15
0.170	14
0.200	18
0.245	22
0.230	19
0.165	15
0.205	17

a) Identify the independent variable and dependent variable. Explain your choices.

b) Produce a scatter plot for the data and classify the linear correlation.

c) Determine the correlation coefficient and the equation of the line of best fit.

d) Use this linear model to predict the batting average for players who had batting practice for

i) 16 h   **ii)** 13 h   **iii)** 35 h

e) Discuss how accurate you think each of these predictions will be.

16. Describe a method you could use to detect outliers in a sample.

17. A bright, young car salesperson has made the following gross sales with her first employer.

Year	Gross Sales ($ millions)
1997	0.8
1998	1.1
1999	1.6
2000	2.3
2001	3.5
2002	4.7

a) Create a time-series graph for these data.

b) Based on this graph, what level of sales would you predict for 2003?

c) List three factors that could affect the accuracy of your prediction.

d) Compute an index value for the sales each year using the 1997 sales as a base. What information do the index values provide?

e) Suppose that this salesperson is thinking of changing jobs. Outline how she could use the sales index to convince other employers to hire her.

18. The following time-series graph shows the Consumer Price Index (CPI) for the period 1971 to 2001.

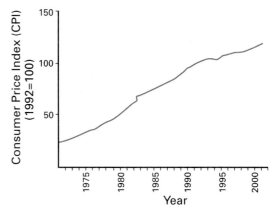

a) What is the base for this index? When did the CPI equal half of this base value?

b) Approximately how many times did the average price of goods double from 1971 to 1992?

c) Which decade on this graph had the highest rate of inflation? Explain your answer.

d) Estimate the overall rate of inflation for the period from 1971 to 2001.

Designing a Game

Background
Many games introduce elements of chance with random processes. For example, card games use shuffled cards, board games often use dice, and bingo uses randomly selected numbers.

Your Task
Design and then analyse a game for two or more players, involving some form of random process. One of the players may assume the role of dealer or game master.

Developing an Action Plan
You will need to decide on one or more instruments of chance, such as dice, cards, coins, coloured balls, a random-number generator, a spinner, or a nail maze. Recommend a method of tracking progress or keeping score, such as a game board or tally sheet. Create the rules of the game. Submit a proposal to your teacher outlining the concept and purpose of your game.

CHAPTER 4

Permutations and Organized Counting

Specific Expectations	Section
Represent complex tasks or issues, using diagrams.	4.1
Solve introductory counting problems involving the additive and multiplicative counting principles.	4.1, 4.2, 4.3
Express the answers to permutation and combination problems, using standard combinatorial symbols.	4.2, 4.3
Evaluate expressions involving factorial notation, using appropriate methods.	4.2, 4.3
Solve problems, using techniques for counting permutations where some objects may be alike.	4.3
Identify patterns in Pascal's triangle and relate the terms of Pascal's triangle to values of $\binom{n}{r}$, to the expansion of a binomial, and to the solution of related problems.	4.4, 4.5
Communicate clearly, coherently, and precisely the solutions to counting problems.	4.1, 4.2, 4.3, 4.4, 4.5

Chapter Problem

Students' Council Elections

Most high schools in Ontario have a students' council comprised of students from each grade. These students are elected representatives, and a part of their function is to act as a liaison between the staff and the students. Often, these students are instrumental in fundraising and in coordinating events, such as school dances and sports.

A students' council executive could consist of a president, vice-president, secretary, treasurer, social convenor, fundraising chair, and four grade representatives. Suppose ten students have been nominated to fill these positions. Five of the nominees are from grade 12, three are from grade 11, and the other two are a grade 9 and a grade 10 student.

1. In how many ways could the positions of president and vice-president be filled by these ten students if all ten are eligible for these positions? How many ways are there if only the grades 11 and 12 students are eligible?

2. The grade representatives must represent their current grade level. In how many ways could the grade representative positions be filled?

You could answer both of these questions by systematically listing all the possibilities and then counting them. In this chapter, you will learn easier and more powerful techniques that can also be applied to much more complex situations.

If you need help with any of the skills listed in **purple** below, refer to Appendix A.

1. **Tree diagrams** Draw a tree diagram to illustrate the number of ways a quarter, a dime, and a nickel can come up heads or tails if you toss one after the other.

2. **Tree diagrams**
   a) Draw a tree diagram to illustrate the possible outcomes of tossing a coin and rolling a six-sided die.
   b) How many possible outcomes are there?

3. **Number patterns** The manager of a grocery store asks a stock clerk to arrange a display of canned vegetables in a triangular pyramid like the one shown. Assume all cans are the same size and shape.

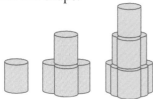

   a) How many cans is the tallest complete pyramid that the clerk can make with 100 cans of vegetables?
   b) How many cans make up the base level of the pyramid in part a)?
   c) How many cans are in the full pyramid in part a)?
   d) What is the sequence of the numbers of cans in the levels of the pyramid?

4. **Number patterns** What is the greatest possible number of rectangles that can be drawn on a
   a) 1 by 5 grid?   b) 2 by 5 grid?

   c) 3 by 5 grid?   d) 4 by 5 grid?

5. **Evaluating expressions** Evaluate each expression given $x = 5$, $y = 4$, and $z = 3$.
   a) $\dfrac{8y(x + 2)(y + 2)(z + 2)}{(x - 3)(y + 3)(z + 2)}$
   b) $\dfrac{(x - 2)^3(y + 2)^2(z + 1)^2}{y(x + 1)(y - 1)^2}$
   c) $\dfrac{(x + 4)(y - 2)(z + 3)}{(y - 1)(x - 3)z} + \dfrac{(x - 1)^2(z + 1)y}{(x - 3)^4(y + 4)}$

6. **Order of operations** Evaluate.
   a) $5(4) + (-1)^3(3)^2$
   b) $\dfrac{(10 - 2)^2(10 - 3)^2}{(10 - 2)^2 - (10 - 3)^2}$
   c) $\dfrac{6(6 - 1)(6 - 2)(6 - 3)(6 - 4)(6 - 5)}{3(3 - 1)(3 - 2)}$
   d) $\dfrac{50(50 - 1)(50 - 2)\ldots(50 - 49)}{48(48 - 1)(48 - 2)\ldots(48 - 47)}$
   e) $\dfrac{12 \times 11 \times 10 \times 9}{6^2} + \dfrac{10 \times 9 \times 8 \times 7}{2^4}$
   $- \dfrac{8 \times 7 \times 6 \times 5}{42}$

7. **Simplifying expressions** Simplify.
   a) $\dfrac{x^2 - xy + 2x}{2x}$ b) $\dfrac{(4x + 8)^2}{16}$
   c) $\dfrac{14(3x^2 + 6)}{7 \times 6}$
   d) $\dfrac{x(x - 1)(x - 2)(x - 3)}{x^2 - 2x}$
   e) $\dfrac{2y + 1}{x} + \dfrac{16y + 4}{4x}$

Organized Counting

The techniques and mathematical logic for counting possible arrangements or outcomes are useful for a wide variety of applications. A computer programmer writing software for a game or industrial process would use such techniques, as would a coach planning a starting line-up, a conference manager arranging a schedule of seminars, or a school board trying to make the most efficient use of its buses.

Combinatorics is the branch of mathematics dealing with ideas and methods for counting, especially in complex situations. These techniques are also valuable for probability calculations, as you will learn in Chapter 6.

INVESTIGATE & INQUIRE: Licence Plates

Until 1997, most licence plates for passenger cars in Ontario had three numbers followed by three letters. Suppose the provincial government had wanted all the vehicles registered in Ontario to have plates with the letters O, N, and T.

1. Draw a diagram to illustrate all the possibilities for arranging these three letters assuming that the letters can be repeated. How many possibilities are there?

2. How could you calculate the number of possible three-letter groups without listing them all?

3. Predict how many three-letter groups the letters O, N, T, and G can form.

4. How many three-letter groups do you think there would be if you had a choice of five letters?

5. Suggest a general strategy for counting all the different possibilities in situations like those above.

When you have to make a series of choices, you can usually determine the total number of possibilities without actually counting each one individually.

Example 1 Travel Itineraries

Martin lives in Kingston and is planning a trip to Vienna, Austria. He checks a web site offering inexpensive airfares and finds that if he travels through London, England, the fare is much lower. There are three flights available from Toronto to London and two flights from London to Vienna. If Martin can take a bus, plane, or train from Kingston to Toronto, how many ways can he travel from Kingston to Vienna?

Solution

You can use a tree diagram to illustrate and count Martin's choices. This diagram suggests another way to determine the number of options Martin has for his trip.

Martin's Choices

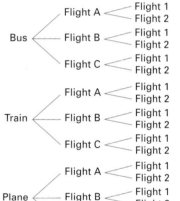

Choices for the first portion of trip: 3
Choices for the second portion of trip: 3
Choices for the third portion of trip: 2
Total number of choices: $3 \times 3 \times 2 = 18$

In all, Martin has 18 ways to travel from Kingston to Vienna.

Example 2 Stereo Systems

Javon is looking at stereos in an electronics store. The store has five types of receivers, four types of CD players, and five types of speakers. How many different choices of stereo systems does this store offer?

Solution

For each choice of receiver, Javon could choose any one of the CD players. Thus, there are $5 \times 4 = 20$ possible combinations of receivers and CD players. For each of these combinations, Javon could then choose one of the five kinds of speakers.

The store offers a total of $5 \times 4 \times 5 = 100$ different stereo systems.

These types of counting problems illustrate the **fundamental** or **multiplicative counting principle**:

> If a task or process is made up of stages with separate choices, the total number of choices is $m \times n \times p \times \ldots$, where m is the number of choices for the first stage, n is the number of choices for the second stage, p is the number of choices for the third stage, and so on.

Example 3 Applying the Fundamental Counting Principle

A school band often performs at benefits and other functions outside the school, so its members are looking into buying band uniforms. The band committee is considering four different white shirts, dress pants in grey, navy, or black, and black or grey vests with the school crest. How many different designs for the band uniform is the committee considering?

Project Prep

You can use the fundamental or multiplicative counting principle to help design the game for your probability project.

Solution

First stage: choices for the white shirts, $m = 4$
Second stage: choices for the dress pants, $n = 3$
Third stage: choices for the vests, $p = 2$
The total number of possibilities is
$m \times n \times p = 4 \times 3 \times 2$
$= 24$
The band committee is considering 24 different possible uniforms.

In some situations, an **indirect method** makes a calculation easier.

Example 4 Indirect Method

Leora, a triathlete, has four pairs of running shoes loose in her gym bag. In how many ways can she pull out two unmatched shoes one after the other?

Solution

You can find the number of ways of picking unmatched shoes by subtracting the number of ways of picking matching ones from the total number of ways of picking any two shoes.

There are eight possibilities when Leora pulls out the first shoe, but only seven when she pulls out the second shoe. By the fundamental counting principle, the number of ways Leora can pick any two shoes out of the bag is $8 \times 7 = 56$. She could pick each of the matched pairs in two ways: left shoe then right shoe or right shoe then left shoe. Thus, there are $4 \times 2 = 8$ ways of picking a matched pair.

Leora can pull out two unmatched shoes in $56 - 8 = 48$ ways.

Sometimes you will have to count several subsets of possibilities separately.

Example 5 Signal Flags

Sailing ships used to send messages with signal flags flown from their masts. How many different signals are possible with a set of four distinct flags if a minimum of two flags is used for each signal?

Solution

A ship could fly two, three, or four signal flags.

Signals with two flags: $4 \times 3 = 12$
Signals with three flags: $4 \times 3 \times 2 = 24$
Signals with four flags: $4 \times 3 \times 2 \times 1 = 24$
Total number of signals: $12 + 24 + 24 = 60$

Thus, the total number of signals possible with these flags is 60.

In Example 5, you were counting actions that could not occur at the same time. When counting such **mutually exclusive** actions, you can apply the **additive counting principle** or **rule of sum**:

If one mutually exclusive action can occur in m ways, a second in n ways, a third in p ways, and so on, then there are $m + n + p$ ... ways in which one of these actions can occur.

Key Concepts

- Tree diagrams are a useful tool for organized counting.

- If you can choose from m items of one type and n items of another, there are $m \times n$ ways to choose one item of each type (fundamental or multiplicative counting principle).

- If you can choose from either m items of one type or n items of another type, then the total number of ways you can choose an item is $m + n$ (additive counting principle).

- Both the multiplicative and the additive counting principles also apply to choices of three or more types of items.

- Sometimes an indirect method provides an easier way to solve a problem.

1. Explain the fundamental counting principle in your own words and give an example of how you could apply it.

2. Are there situations where the fundamental counting principle does not apply? If so, give one example.

3. Can you always use a tree diagram for organized counting? Explain your reasoning.

Practise

1. Construct a tree diagram to illustrate the possible contents of a sandwich made from white or brown bread, ham, chicken, or beef, and mustard or mayonnaise. How many different sandwiches are possible?

2. In how many ways can you roll either a sum of 4 or a sum of 11 with a pair of dice?

3. In how many ways can you draw a 6 or a face card from a deck of 52 playing cards?

4. How many ways are there to draw a 10 or a queen from the 24 cards in a euchre deck, which has four 10s and four queens?

5. Use tree diagrams to answer the following:

   a) How many different soccer uniforms are possible if there is a choice of two types of shirts, three types of shorts, and two types of socks?

   b) How many different three-scoop cones can be made from vanilla, chocolate, and strawberry ice cream?

   c) Suppose that a college program has six elective courses, three on English literature and three on the other arts. If the college requires students to take one of the English courses and one of the other arts courses, how many pairs of courses will satisfy these requirements?

Apply, Solve, Communicate

6. Ten different books and four different pens are sitting on a table. One of each is selected. Should you use the rule of sum or the product rule to count the number of possible selections? Explain your reasoning.

7. **Application** A grade 9 student may build a timetable by selecting one course for each period, with no duplication of courses. Period 1 must be science, geography, or physical education. Period 2 must be art, music, French, or business. Periods 3 and 4 must each be mathematics or English.

   a) Construct a tree diagram to illustrate the choices for a student's timetable.

   b) How many different timetables could a student choose?

8. A standard die is rolled five times. How many different outcomes are possible?

9. A car manufacturer offers three kinds of upholstery material in five different colours for this year's model. How many upholstery options would a buyer have? Explain your reasoning.

10. **Communication** In how many ways can a student answer a true-false test that has six questions. Explain your reasoning.

11. The final score of a soccer game is 6 to 3. How many different scores were possible at half-time?

12. A large room has a bank of five windows. Each window is either open or closed. How many different arrangements of open and closed windows are there?

13. Application A Canadian postal code uses six characters. The first, third, and fifth are letters, while the second, fourth, and sixth are digits. A U.S.A. zip code contains five characters, all digits.

a) How many codes are possible for each country?

b) How many more possible codes does the one country have than the other?

14. When three-digit area codes were introduced in 1947, the first digit had to be a number from 2 to 9 and the middle digit had to be either 1 or 0. How many area codes were possible under this system?

15. Asha builds new homes and offers her customers a choice of brick, aluminium siding, or wood for the exterior, cedar or asphalt shingles for the roof, and radiators or forced-air for the heating system. How many different configurations is Asha offering?

16. a) In how many ways could you choose two fives, one after the other, from a deck of cards?

b) In how many ways could you choose a red five and a spade, one after the other?

c) In how many ways could you choose a red five or a spade?

d) In how many ways could you choose a red five or a heart?

e) Explain which counting principles you could apply in parts a) to d).

17. Ten students have been nominated for a students' council executive. Five of the nominees are from grade 12, three are from grade 11, and the other two are from grades 9 and 10.

a) In how many ways could the nominees fill the positions of president and vice-president if all ten are eligible for these senior positions?

b) How many ways are there to fill these positions if only grade 11 and grade 12 students are eligible?

18. Communication

a) How many different licence plates could be made using three numbers followed by three letters?

b) In 1997, Ontario began issuing licence plates with four letters followed by three numbers. How many different plates are possible with this new system?

c) Research the licence plate formats used in the other provinces. Compare and contrast these formats briefly and suggest reasons for any differences between the formats.

19. In how many ways can you arrange the letters of the word *think* so that the *t* and the *h* are separated by at least one other letter?

20. Application Before the invention of the telephone, Samuel Morse (1791–1872) developed an efficient system for sending messages as a series of dots and dashes (short or long pulses). International code, a modified version of Morse code, is still widely used.

a) How many different characters can the international code represent with one to four pulses?

b) How many pulses would be necessary to represent the 72 letters of the Cambodian alphabet using a system like Morse code?

21. Ten finalists are competing in a race at the Canada Games.

a) In how many different orders can the competitors finish the race?

b) How many ways could the gold, silver, and bronze medals be awarded?

c) One of the finalists is a friend from your home town. How many of the possible finishes would include your friend winning a medal?

d) How many possible finishes would leave your friend out of the medal standings?

e) Suppose one of the competitors is injured and cannot finish the race. How does that affect your previous answers?

f) How would the competitor's injury affect your friend's chances of winning a medal? Explain your reasoning. What assumptions have you made?

C

22. A locksmith has ten types of blanks for keys. Each blank has five different cutting positions and three different cutting depths at each position, except the first position, which only has two depths. How many different keys are possible with these blanks?

23. Communication How many 5-digit numbers are there that include the digit 5 and exclude the digit 8? Explain your solution.

24. Inquiry/Problem Solving Your school is purchasing a new type of combination lock for the student lockers. These locks have 40 positions on their dials and use a three-number combination.

a) How many combinations are possible if consecutive numbers cannot be the same?

b) Are there any assumptions that you have made? Explain.

c) Assuming that the first number must be dialled clockwise from 0, how many different combinations are possible?

d) Suppose the first number can also be dialled counterclockwise from 0. Explain the effect this change has on the number of possible combinations.

e) If you need four numbers to open the lock, how many different combinations are possible?

25. Inquiry/Problem Solving In chess, a knight can move either two squares horizontally plus one vertically or two squares vertically plus one horizontally.

a) If a knight starts from one corner of a standard 8×8 chessboard, how many different squares could it reach after

i) one move?

ii) two moves?

iii) three moves?

b) Could you use the fundamental counting principle to calculate the answers for part a)? Why or why not?

Factorials and Permutations

In many situations, you need to determine the number of different orders in which you can choose or place a set of items.

Consider how many different ways a president and a vice-president could be chosen from eight members of a students' council.

1. **a)** Have one person in your class make two signs, writing *President* on one and *Vice-President* on the other. Now, choose two people to stand at the front of the class. Using the signs to indicate which person holds each position, decide in how many ways you can choose a president and a vice-president from the two people at the front of the class.

   b) Choose three students to be at the front of the class. Again using the signs to indicate who holds each position, determine how many ways you can choose a president and a vice-president from the three people at the front of the class.

   c) Repeat the process with four students. Do you see a pattern in the number of ways a president and a vice-president can be chosen from the different sizes of groups? If so, what is the pattern? If not, continue the process with five students and then with six students.

   d) When you see a pattern, predict how many ways a president and a vice-president can be chosen from the eight members of the students' council.

   e) Suggest other ways of simulating the selection of a president and a vice-president for the students' council.

2. Suppose that each of the eight members of the students' council has to give a brief speech at an assembly. Consider how you could determine the number of different orders in which they could speak.

   a) Choose two students from your class and list all the possible orders in which they could speak.

   b) Choose three students and list all the possible orders in which they could speak.

   c) Repeat this process with four students.

   d) Is there an easy method to organize the list so that you could include all the possibilities?

   e) Is this method related to your results in question 1? Explain.

   f) Can you use your method to predict the number of different orders in which eight students could give speeches?

Many counting and probability calculations involve the product of a series of consecutive integers. You can use **factorial** notation to write such expressions more easily. For any natural number n,

$$n! = n \times (n - 1) \times (n - 2) \times (n - 3) \times \ldots \times 3 \times 2 \times 1$$

This expression is read as *n factorial*.

Example 1   Evaluating Factorials

Calculate each factorial.

a) 2!   **b)** 4!   **c)** 8!

Solution

a) $2! = 2 \times 1$
       $= 2$

b) $4! = 4 \times 3 \times 2 \times 1$
       $= 24$

c) $8! = 8 \times 7 \times 6 \times 5 \times 4 \times 3 \times 2 \times 1$
       $= 40\ 320$

As you can see from Example 1, $n!$ increases dramatically as n becomes larger. However, calculators and computer software provide an easy means of calculating the larger factorials. Most scientific and graphing calculators have a factorial key or function.

Example 2 Using Technology to Evaluate Factorials

Calculate.

a) 21!   **b)** 53!   **c)** 70!

Solution 1   Using a Graphing Calculator

Enter the number on the home screen and then use the ! function on the
MATH PRB menu to calculate the factorial.

a) $21! = 21 \times 20 \times 19 \times 18 \times \ldots \times 2 \times 1$
    $= 5.1091 \times 10^{19}$

b) $53! = 53 \times 52 \times 51 \times \ldots \times 3 \times 2 \times 1$
    $= 4.2749 \times 10^{69}$

c) Entering 70! on a graphing calculator gives an ERR:OVERFLOW message since
$70! > 10^{100}$ which is the largest number the calculator can handle. In fact, 69! is
the largest factorial you can calculate directly on TI-83 series calculators.

Solution 2   Using a Spreadsheet

Both Corel® Quattro® Pro and Microsoft® Excel have a built-in factorial
function with the syntax FACT(*n*).

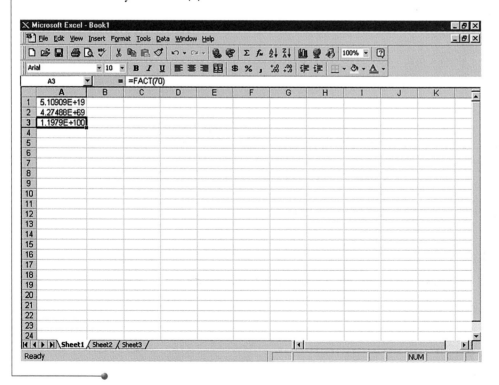

Example 3 Evaluating Factorial Expressions

Evaluate.

a) $\dfrac{10!}{5!}$ **b)** $\dfrac{83!}{79!}$

Solution

In both these expressions, you can divide out the common terms in the numerator and denominator.

a)
$$\frac{10!}{5!} = \frac{10 \times 9 \times 8 \times 7 \times 6 \times 5 \times 4 \times 3 \times 2 \times 1}{5 \times 4 \times 3 \times 2 \times 1}$$
$$= 10 \times 9 \times 8 \times 7 \times 6$$
$$= 30\ 240$$

b)
$$\frac{83!}{79!} = \frac{83 \times 82 \times 81 \times 80 \times 79 \times 78 \times \ldots \times 2 \times 1}{79 \times 78 \times \ldots \times 2 \times 1}$$
$$= 83 \times 82 \times 81 \times 80$$
$$= 44\ 102\ 880$$

Note that by dividing out the common terms, you can use a calculator to evaluate this expression even though the factorials are too large for the calculator.

Example 4 Counting Possibilities

The senior choir has rehearsed five songs for an upcoming assembly. In how many different orders can the choir perform the songs?

Solution

There are five ways to choose the first song, four ways to choose the second, three ways to choose the third, two ways to choose the fourth, and only one way to choose the final song. Using the fundamental counting principle, the total number of different ways is

$$5 \times 4 \times 3 \times 2 \times 1 = 5!$$
$$= 120$$

The choir can sing the five songs in 120 different orders.

Example 5 Indirect Method

In how many ways could ten questions on a test be arranged, if the easiest question and the most difficult question

a) are side-by-side?

b) are not side-by-side?

Solution

a) Treat the easiest question and the most difficult question as a unit making nine items that are to be arranged. The two questions can be arranged in 2! ways within their unit.

$9! \times 2! = 725\ 760$

The questions can be arranged in 725 760 ways if the easiest question and the most difficult question are side-by-side.

b) Use the indirect method. The number of arrangements with the easiest and most difficult questions separated is equal to the total number of possible arrangements less the number with the two questions side-by-side:

$$10! - 9! \times 2! = 3\ 628\ 800 - 725\ 760$$
$$= 2\ 903\ 040$$

The questions can be arranged in 2 903 040 ways if the easiest question and the most difficult question are not side-by-side.

A **permutation** of n distinct items is an arrangement of all the items in a definite order. The total number of such permutations is denoted by $_nP_n$ or $P(n, n)$.

There are n possible ways of choosing the first item, $n - 1$ ways of choosing the second, $n - 2$ ways of choosing the third, and so on. Applying the fundamental counting principle as in Example 5 gives

$$_nP_n = n \times (n - 1) \times (n - 2) \times (n - 3) \times \dots \times 3 \times 2 \times 1$$
$$= n!$$

Example 6 Applying the Permutation Formula

In how many different orders can eight nominees for the students' council give their speeches at an assembly?

Solution

$$_8P_8 = 8!$$
$$= 8 \times 7 \times 6 \times 5 \times 4 \times 3 \times 2 \times 1$$
$$= 40\ 320$$

There are 40 320 different orders in which the eight nominees can give their speeches.

Example 7 Student Government

In how many ways could a president and a vice-president be chosen from a group of eight nominees?

Solution

Using the fundamental counting principle, there are 8×7, or 56, ways to choose a president and a vice-president.

A permutation of n distinct items taken r at a time is an arrangement of r of the n items in a definite order. Such permutations are sometimes called r-arrangements of n items. The total number of possible arrangements of r items out of a set of n is denoted by $_nP_r$ or $P(n, r)$.

There are n ways of choosing the first item, $n - 1$ ways of choosing the second item, and so on down to $n - r + 1$ ways of choosing the rth item. Using the fundamental counting principle,

$$_nP_r = n(n - 1)(n - 2)\ldots(n - r + 1)$$

It is often more convenient to rewrite this expression in terms of factorials.

$$_nP_r = \frac{n!}{(n - r)!}$$

The denominator divides out completely, as in Example 3, so these two ways of writing $_nP_r$ are equivalent.

Project Prep

The permutations formula could be a useful tool for your probability project.

Example 8 Applying the Permutation Formula

In a card game, each player is dealt a face down "reserve" of 13 cards that can be turned up and used one by one during the game. How many different sequences of reserve cards could a player have?

Solution 1   Using Pencil and Paper

Here, you are taking 13 cards from a deck of 52.

$$_{52}P_{13} = \frac{52!}{(52 - 13)!}$$

$$= \frac{52!}{39!}$$

$$= 52 \times 51 \times 50 \times \ldots \times 41 \times 40$$

$$= 3.9542 \times 10^{21}$$

There are approximately 3.95×10^{21} different sequences of reserve cards a player could have.

Solution 2   Using a Graphing Calculator

Use the nPr function on the MATH PRB menu.

There are approximately 3.95×10^{21} different sequences of reserve cards a player could turn up during one game.

Solution 3   Using a Spreadsheet

Both Corel® Quattro® Pro and Microsoft® Excel have a permutations function
with the syntax PERMUT(n,r).

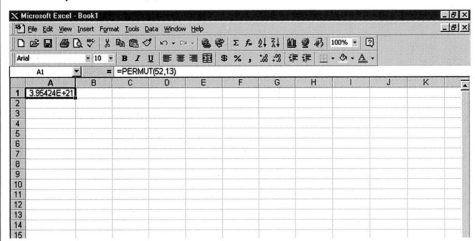

There are approximately 3.95×10^{21} different sequences of reserve cards a
player could turn up during one game.

Key Concepts

- A factorial indicates the multiplication of consecutive natural numbers.
  $n! = n(n - 1)(n - 2) \times \ldots \times 1$.

- The number of permutations of n distinct items chosen n at a time in a
  definite order is $_nP_n = n!$

- The number of permutations of r items taken from n distinct items is

  $$_nP_r = \frac{n!}{(n - r)!}.$$

Communicate Your Understanding

1. Explain why it is convenient to write the expression for the number of
   possible permutations in terms of factorials.

2. a) Is $(-3)!$ possible? Explain your answer.

   b) In how many ways can you order an empty list, or zero items? What does
      this tell you about the value of $0!$? Check your answer using a calculator.

Practise

1. Express in factorial notation.

 a) $6 \times 5 \times 4 \times 3 \times 2 \times 1$

 b) $8 \times 7 \times 6 \times 5 \times 4 \times 3 \times 2 \times 1$

 c) $3 \times 2 \times 1$

 d) $9 \times 8 \times 7 \times 6 \times 5 \times 4 \times 3 \times 2 \times 1$

2. Evaluate.

 a) $\dfrac{7!}{4!}$ **b)** $\dfrac{11!}{9!}$

 c) $\dfrac{8!}{5!\,2!}$ **d)** $\dfrac{15!}{3!\,8!}$

 e) $\dfrac{85!}{82!}$ **f)** $\dfrac{14!}{4!\,5!}$

3. Express in the form $_nP_r$.

 a) $6 \times 5 \times 4$

 b) $9 \times 8 \times 7 \times 6$

 c) $20 \times 19 \times 18 \times 17$

 d) $101 \times 100 \times 99 \times 98 \times 97$

 e) $76 \times 75 \times 74 \times 73 \times 72 \times 71 \times 70$

4. Evaluate without using technology.

 a) $P(10, 4)$ **b)** $P(16, 4)$ **c)** $_5P_2$

 d) $_9P_4$ **e)** $7!$

5. Use either a spreadsheet or a graphing or scientific calculator to verify your answers to question 4.

Apply, Solve, Communicate

6. a) How many ways can you arrange the letters in the word *factor*?

 b) How many ways can Ismail arrange four different textbooks on the shelf in his locker?

 c) How many ways can Laura colour 4 adjacent regions on a map if she has a set of 12 coloured pencils?

B

7. Simplify each of the following in factorial form. Do not evaluate.

 a) $12 \times 11 \times 10 \times 9!$

 b) $72 \times 7!$

 c) $(n + 4)(n + 5)(n + 3)!$

8. Communication Explain how a factorial is an iterative process.

9. Seven children are to line up for a photograph.

 a) How many different arrangements are possible?

 b) How many arrangements are possible if Brenda is in the middle?

 c) How many arrangements are possible if Ahmed is on the far left and Yen is on the far right?

 d) How many arrangements are possible if Hanh and Brian must be together?

10. A 12-volume encyclopedia is to be placed on a shelf. How many incorrect arrangements are there?

11. In how many ways can the 12 members of a volleyball team line up, if the captain and assistant captain must remain together?

12. Ten people are to be seated at a rectangular table for dinner. Tanya will sit at the head of the table. Henry must not sit beside either Wilson or Nancy. In how many ways can the people be seated for dinner?

13. Application Joanne prefers classical and pop music. If her friend Charlene has five classical CDs, four country and western CDs, and seven pop CDs, in how many orders can Joanne and Charlene play the CDs Joanne likes?

14. In how many ways can the valedictorian, class poet, and presenter of the class gift be chosen from a class of 20 students?

15. **Application** If you have a standard deck of 52 cards, in how many different ways can you deal out

   a) 5 cards? **b)** 10 cards?

   c) 5 red cards? **d)** 4 queens?

16. **Inquiry/Problem Solving** Suppose you are designing a coding system for data relayed by a satellite. To make transmissions errors easier to detect, each code must have no repeated digits.

   a) If you need 60 000 different codes, how many digits long should each code be?

   b) How many ten-digit codes can you create if the first three digits must be 1, 3, or 6?

17. Arnold Schoenberg (1874–1951) pioneered serialism, a technique for composing music based on a tone row, a sequence in which each of the 12 tones in an octave is played only once. How many tone rows are possible?

18. Consider the students' council described on page 223 at the beginning of this chapter.

   a) In how many ways can the secretary, treasurer, social convenor, and fundraising chair be elected if all ten nominees are eligible for any of these positions?

   b) In how many ways can the council be chosen if the president and vice-president must be grade 12 students and the grade representatives must represent their current grade level?

19. **Inquiry/Problem Solving** A student has volunteered to photograph the school's championship basketball team for the yearbook. In order to get the perfect picture, the student plans to photograph the ten players and their coach lined up in every possible order. Determine whether this plan is practical.

ACHIEVEMENT CHECK			
Knowledge/ Understanding	Thinking/Inquiry/ Problem Solving	Communication	Application

20. Wayne has a briefcase with a three-digit combination lock. He can set the combination himself, and his favourite digits are 3, 4, 5, 6, and 7. Each digit can be used at most once.

   a) How many permutations of three of these five digits are there?

   b) If you think of each permutation as a three-digit number, how many of these numbers would be odd numbers?

   c) How many of the three-digit numbers are even numbers and begin with a 4?

   d) How many of the three-digit numbers are even numbers and do *not* begin with a 4?

   e) Is there a connection among the four answers above? If so, state what it is and why it occurs.

C

21. TI-83 series calculators use the definition $\left(-\dfrac{1}{2}\right)! = \sqrt{\pi}$. Research the origin of this definition and explain why it is useful for mathematical calculations.

22. **Communication** How many different ways can six people be seated at a round table? Explain your reasoning.

23. What is the highest power of 2 that divides evenly into 100! ?

24. A committee of three teachers are to select the winner from among ten students nominated for special award. The teachers each make a list of their top three choices in order. The lists have only one name in common, and that name has a different rank on each list. In how many ways could the teachers have made their lists?

4.3 Permutations With Some Identical Items

Often, you will deal with permutations in which some items are identical.

1. In their mathematics class, John and Jenn calculate the number of permutations of all the letters of their first names.

   a) How many permutations do you think John finds?

   b) List all the permutations of John's name.

   c) How many permutations do you think Jenn finds?

   d) List all the permutations of Jenn's name.

   e) Why do you think there are different numbers of permutations for the two names?

2. a) List all the permutations of the letters in your first name. Is the number of permutations different from what you would calculate using the $_nP_n = n!$ formula? If so, explain why.

   b) List and count all the permutations of a word that has two identical pairs of letters. Compare your results with those your classmates found with other words. What effect do the identical letters have on the number of different permutations?

   c) Predict how many permutations you could make with the letters in the word *googol*. Work with several classmates to verify your prediction by writing out and counting all of the possible permutations.

3. Suggest a general formula for the number of permutations of a word that has two or more identical letters.

As the investigation above suggests, you can develop a general formula for permutations in which some items are identical.

Example 1 Permutations With Some Identical Elements

Compare the different permutations for the words *DOLE*, *DOLL*, and *LOLL*.

Solution

The following are all the permutations of *DOLE*:

DOLE	DOEL	DLOE	DLEO	DEOL	DELO
ODLE	ODEL	OLDE	OLED	OEDL	OELD
LODE	LOED	LDOE	LDEO	LEOD	LEDO
EOLD	EODL	ELOD	ELDO	EDOL	EDLO

There are 24 permutations of the four letters in *DOLE*. This number matches what you would calculate using $_4P_4 = 4!$

To keep track of the permutations of the letters in the word *DOLL*, use a subscript to distinguish the one *L* from the other.

$DOLL_1$	DOL_1L	$DLOL_1$	DLL_1O	DL_1OL	DL_1LO
$ODLL_1$	ODL_1L	$OLDL_1$	OLL_1D	OL_1DL	OL_1LD
$LODL_1$	LOL_1D	$LDOL_1$	LDL_1O	LL_1OD	LL_1DO
L_1OLD	L_1ODL	L_1LOD	L_1LDO	L_1DOL	L_1DLO

Of the 24 arrangements listed here, only 12 are actually different from each other. Since the two *L*s are in fact identical, each of the permutations shown in black is duplicated by one of the permutations shown in red. If the two *L*s in a permutation trade places, the resulting permutation is the same as the original one. The two *L*s can trade places in $_2P_2 = 2!$ ways.

Thus, the number of different arrangements is

$$\frac{4!}{2!} = \frac{24}{2}$$
$$= 12$$

In other words, to find the number of permutations, you divide the total number of arrangements by the number of ways in which you can arrange the identical letters. For the letters in *DOLL*, there are four ways to choose the first letter, three ways to choose the second, two ways to choose the third, and one way to choose the fourth. You then divide by the 2! or 2 ways that you can arrange the two *L*s.

Similarly, you can use subscripts to distinguish the three *L*s in *LOLL*, and then highlight the duplicate arrangements.

L_2OLL_1	L_2OL_1L	L_2LOL_1	L_2LL_1O	L_2L_1OL	L_2L_1LO
OL_2LL_1	OL_2L_1L	OLL_2L_1	OLL_1L_2	OL_1L_2L	OL_1LL_2
LOL_2L_1	LOL_1L_2	LL_2OL_1	LL_2L_1O	LL_1OL_2	LL_1L_2O
L_1OLL_2	L_1OL_2L	L_1LOL_2	L_1LL_2O	L_1L_2OL	L_1L_2LO

The arrangements shown in black are the only different ones. As with the other two words, there are 24 possible arrangements if you distinguish between the identical *L*s. Here, the three identical *L*s can trade places in $_3P_3 = 3!$ ways.

Thus, the number of permutations is $\dfrac{4!}{3!} = 4$.

You can generalize the argument in Example 1 to show that the number of permutations of a set of *n* items of which *a* are identical is $\dfrac{n!}{a!}$.

Example 2 Tile Patterns

Tanisha is laying out tiles for the edge of a mosaic. How many patterns can she make if she uses four yellow tiles and one each of blue, green, red, and grey tiles?

Solution

Here, $n = 8$ and $a = 4$.

$$\dfrac{8!}{4!} = 8 \times 7 \times 6 \times 5$$

$$= 1680$$

Tanisha can make 1680 different patterns with the eight tiles.

Example 3 Permutation With Several Sets of Identical Elements

The word *bookkeeper* is unusual in that it has three consecutive double letters. How many permutations are there of the letters in *bookkeeper*?

Solution

If each letter were different, there would be 10! permutations, but there are two *o*s, two *k*s, and three *e*s. You must divide by 2! twice to allow for the duplication of the *o*s and *k*s, and then divide by 3! to allow for the three *e*s:

$$\dfrac{10!}{2!2!3!} = \dfrac{10 \times 9 \times 8 \times 7 \times 6 \times 5 \times 4}{2 \times 2}$$

$$= 151\,200$$

There are 151 200 permutations of the letters in *bookkeeper*.

> The number of permutations of a set of *n* objects containing *a* identical objects of one kind, *b* identical objects of a second kind, *c* identical objects of a third kind, and so on is $\dfrac{n!}{a!b!c!\ldots}$.

Example 4 Applying the Formula for Several Sets of Identical Elements

Barbara is hanging a display of clothing imprinted with the school's crest on a line on a wall in the cafeteria. She has five sweatshirts, three T-shirts, and four pairs of sweatpants. In how many ways can Barbara arrange the display?

Solution

Here, $a = 5$, $b = 3$, $c = 4$, and the total number of items is 12.

So,

$$\frac{n!}{a!b!c!} = \frac{12!}{5!3!4!}$$

$$= 27\ 720$$

Barbara can arrange the display in 27 720 different ways.

Project Prep

The game you design for your probability project could involve permutations of identical objects.

Key Concepts

- When dealing with permutations of n items that include a identical items of one type, b identical items of another type, and so on, you can use the formula $\dfrac{n!}{a!b!c!\ldots}$.

Communicate Your Understanding

1. Explain why there are fewer permutations of a given number of items if some of the items are identical.

2. **a)** Explain why the formula for the numbers of permutations when some items are identical has the denominator $a!b!c!\ldots$ instead of $a \times b \times c\ldots$.

   b) Will there ever be cases where this denominator is larger than the numerator? Explain.

   c) Will there ever be a case where the formula does not give a whole number answer? What can you conclude about the denominator and the numerator? Explain your reasoning.

Practise

1. Identify the indistinguishable items in each situation.

    a) The letters of the word *mathematics* are arranged.

    b) Dina has six notebooks, two green and four white.

    c) The cafeteria prepares 50 chicken sandwiches, 100 hamburgers, and 70 plates of French fries.

    d) Thomas and Richard, identical twins, are sitting with Marianna and Megan.

2. How many permutations are there of all the letters in each name?

    a) Inverary   **b)** Beamsville

    c) Mattawa   **d)** Penetanguishene

3. How many different five-digit numbers can be formed using three 2s and two 5s?

4. How many different six-digit numbers are possible using the following numbers?

    a) 1, 2, 3, 4, 5, 6   **b)** 1, 1, 1, 2, 3, 4

    c) 1, 3, 3, 4, 4, 5   **d)** 6, 6, 6, 6, 7, 8

Apply, Solve, Communicate

5. **Communication** A coin is tossed eight times. In how many different orders could five heads and three tails occur? Explain your reasoning.

6. **Inquiry/Problem Solving** How many 7-digit even numbers less than 3 000 000 can be formed using all the digits 1, 2, 2, 3, 5, 5, 6?

7. Kathryn's soccer team played a good season, finishing with 16 wins, 3 losses, and 1 tie. In how many orders could these results have happened? Explain your reasoning.

8. **a)** Calculate the number of permutations for each of the jumbled words in this puzzle.

    b) Estimate how long it would take to solve this puzzle by systematically writing out the permutations.

Unscramble these four Jumbles, one letter to each square, to form four ordinary words.

SUYFS

YATTS

SPEEXO

HAREMM

A real pro

WHAT A GOOD HISTORY TEACHER SHOULD BE.

Now arrange the circled letters to form the surprise answer, as suggested by the above cartoon.

A " ⬡⬡⬡⬡ " ⬡⬡⬡⬡⬡⬡

WEB CONNECTION

www.mcgrawhill.ca/links/MDM12

For more word jumbles and other puzzles, visit the above web site and follow the links. Find or generate two puzzles for a classmate to solve.

9. **Application** Roberta is a pilot for a small airline. If she flies to Sudbury three times, Timmins twice, and Thunder Bay five times before returning home, how many different itineraries could she follow? Explain your reasoning.

10. After their training run, six members of a track team split a bag of assorted doughnuts. How many ways can the team share the doughnuts if the bag contains

    a) six different doughnuts?

    b) three each of two varieties?

    c) two each of three varieties?

11. As a project for the photography class, Haseeb wants to create a linear collage of photos of his friends. He creates a template with 20 spaces in a row. If Haseeb has 5 identical photos of each of 4 friends, in how many ways can he make his collage?

12. **Communication** A used car lot has four green flags, three red flags, and two blue flags in a bin. In how many ways can the owner arrange these flags on a wire stretched across the lot? Explain your reasoning.

13. **Application** Malik wants to skateboard over to visit his friend Gord who lives six blocks away. Gord's house is two blocks west and four blocks north of Malik's house. Each time Malik goes over, he likes to take a different route. How many different routes are there for Malik if he only travels west or north?

14. Fran is working on a word puzzle and is looking for four-letter "scrambles" from the clue word *calculate*.

   a) How many of the possible four-letter scrambles contain four different letters?

   b) How many contain two *a*s and one other pair of identical letters?

   c) How many scrambles consist of any two pairs of identical letters?

   d) What possibilities have you not yet taken into account? Find the number of scrambles for each of these cases.

   e) What is the total number of four-letter scrambles taking all cases into account?

15. Ten students have been nominated for the positions of secretary, treasurer, social convenor, and fundraising chair. In how many ways can these positions be filled if the Norman twins are running and plan to switch positions on occasion for fun since no one can tell them apart?

16. **Inquiry/Problem Solving** In how many ways can all the letters of the word *CANADA* be arranged if the consonants must always be in the order in which they occur in the word itself?

C

17. Glen works part time stocking shelves in a grocery store. The manager asks him to make a pyramid display using 72 cans of corn, 36 cans of peas, and 57 cans of carrots. Assume all the cans are the same size and shape. On his break, Glen tries to work out how many different ways he could arrange the cans into a pyramid shape with a triangular base.

   a) Write a formula for the number of different ways Glen could stack the cans in the pyramid.

   b) Estimate how long it will take Glen to calculate this number of permutations by hand.

   c) Use computer software or a calculator to complete the calculation.

18. How many different ways are there of arranging seven green and eight brown bottles in a row, so that exactly one pair of green bottles is side-by-side?

19. In how many ways could a class of 18 students divide into groups of 3 students each?

4.4 Pascal's Triangle

The array of numbers shown below is called Pascal's triangle in honour of French mathematician, Blaise Pascal (1623–1662). Although it is believed that the 14th century Chinese mathematician Chu Shi-kie knew of this array and some of its applications, Pascal discovered it independently at age 13. Pascal found many mathematical uses for the array, especially in probability theory.

Pascal's method for building his triangle is a simple iterative process similar to those described in, section 1.1. In Pascal's triangle, each term is equal to the sum of the two terms immediately above it. The first and last terms in each row are both equal to 1 since the only term immediately above them is also always a 1.

If $t_{n,r}$ represents the term in row n, position r, then $t_{n,r} = t_{n-1,r-1} + t_{n-1,r}$.

For example, $t_{6,2} = t_{5,1} + t_{5,2}$. Note that both the row and position labelling begin with 0.

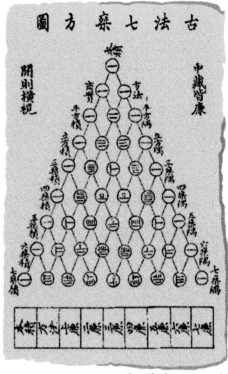

Chu Shi-kie's triangle

			1				Row 0
		1		1			Row 1
	1		2		1		Row 2
1		3		3		1	Row 3

```
              1                      Row 0              t_{0,0}
           1     1                   Row 1           t_{1,0}   t_{1,1}
        1     2     1                Row 2        t_{2,0}  t_{2,1}  t_{2,2}
     1     3     3     1             Row 3     t_{3,0}  t_{3,1}  t_{3,2}  t_{3,3}
  1     4     6     4     1          Row 4  t_{4,0} t_{4,1} t_{4,2} t_{4,3} t_{4,4}
1   5    10    10    5    1          Row 5 t_{5,0} t_{5,1} t_{5,2} t_{5,3} t_{5,4} t_{5,5}
1  6  15  20  15  6  1               Row 6 t_{6,0} t_{6,1} t_{6,2} t_{6,3} t_{6,4} t_{6,5} t_{6,6}
```

WEB CONNECTION

www.mcgrawhill.ca/links/MDM12

Visit the above web site and follow the links to learn more about Pascal's triangle. Write a brief report about an application or an aspect of Pascal's triangle that interests you.

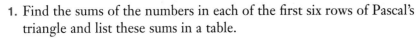

1. Find the sums of the numbers in each of the first six rows of Pascal's triangle and list these sums in a table.

2. Predict the sum of the entries in

   a) row 7   **b)** row 8   **c)** row 9

3. Verify your predictions by calculating the sums of the numbers in rows 7, 8, and 9.

4. Predict the sum of the entries in row n of Pascal's triangle.

5. List any other patterns you find in Pascal's triangle. Compare your list with those of your classmates. Do their lists suggest further patterns you could look for?

In his book *Mathematical Carnival*, Martin Gardner describes Pascal's triangle as "so simple that a 10-year old can write it down, yet it contains such inexhaustible riches and links with so many seemingly unrelated aspects of mathematics, that it is surely one of the most elegant of number arrays."

Example 1 Pascal's Method

a) The first six terms in row 25 of Pascal's triangle are 1, 25, 300, 2300, 12 650, and 53 130. Determine the first six terms in row 26.

b) Use Pascal's method to write a formula for each of the following terms:

   i) $t_{12,5}$

   ii) $t_{40,32}$

   iii) $t_{n+1,r+1}$

Solution

a)
$$t_{26,1} = 1 \qquad\qquad t_{26,2} = 1 + 25 \qquad\qquad t_{26,3} = 25 + 300$$
$$= 26 \qquad\qquad\qquad = 325$$

$$t_{26,4} = 300 + 2300 \qquad t_{26,5} = 2300 + 12\ 650 \qquad t_{26,6} = 12\ 650 + 53\ 130$$
$$= 2600 \qquad\qquad\qquad = 14\ 950 \qquad\qquad\qquad = 65\ 780$$

b) i) $t_{12,5} = t_{11,4} + t_{11,5}$

   ii) $t_{40,32} = t_{39,31} + t_{39,32}$

   iii) $t_{n+1,r+1} = t_{n,r} + t_{n,r+1}$

Example 2 Row Sums

Which row in Pascal's triangle has the sum of its terms equal to 32 768?

Solution

From the investigation on page 248, you know that the sum of the
terms in any row n is 2^n. Dividing 32 768 by 2 repeatedly, you find that
$32\ 768 = 2^{15}$. Thus, it is row 15 of Pascal's triangle that has terms totalling 32 768.

Example 3 Divisibility

Determine whether $t_{n,2}$ is divisible by $t_{n,1}$ in each row of Pascal's triangle.

Solution

Row	$\dfrac{t_{n,2}}{t_{n,1}}$	Divisible?
0 and 1	n/a	n/a
2	0.5	no
3	1	yes
4	1.5	no
5	2	yes
6	2.5	no
7	3	yes

It appears that $t_{n,2}$ is divisible by $t_{n,1}$ only in odd-numbered rows.
However, $2t_{n,2}$ is divisible by $t_{n,1}$ in all rows that have three or more terms.

Example 4 Triangular Numbers

Coins can be arranged in the shape of an equilateral triangle as shown.

a) Continue the pattern to determine the numbers of coins in triangles
with four, five, and six rows.

b) Locate these numbers in Pascal's triangle.

c) Relate Pascal's triangle to the number of coins in a triangle with n rows.

d) How many coins are in a triangle with 12 rows?

Solution

a) The numbers of coins in the triangles follow the pattern $1 + 2 + 3 + \dots$ as shown in the table below.

b) The numbers of coins in the triangles match the entries on the third diagonal of Pascal's triangle.

Number of Rows	Number of Coins	Term in Pascal's Triangle
1	1	$t_{2,2}$
2	3	$t_{3,2}$
3	6	$t_{4,2}$
4	10	$t_{5,2}$
5	15	$t_{6,2}$
6	21	$t_{7,2}$

```
                    1
                  1   1
                1   2   1
              1   3   3   1
            1   4   6   4   1
          1   5  10  10   5   1
        1   6  15  20  15   6   1
      1   7  21  35  35  21   7   1
```

c) Compare the entries in the first and third columns of the table. The row number of the term from Pascal's triangle is always one greater than the number of rows in the equilateral triangle. The position of the term in the row, r, is always 2. Thus, the number of coins in a triangle with n rows is equal to the term $t_{n+1,2}$ in Pascal's triangle.

d) $t_{12+1,2} = t_{13,2}$
$$= 78$$

A triangle with 12 rows contains 78 coins.

Numbers that correspond to the number of items stacked in a triangular array are known as **triangular numbers**. Notice that the nth triangular number is also the sum of the first n positive integers.

Example 5 Perfect Squares

Can you find a relationship between perfect squares and the sums of pairs of entries in Pascal's triangle?

Solution

Again, look at the third diagonal in Pascal's triangle.

n	n^2	Entries in Pascal's Triangle	Terms in Pascal's Triangle
1	1	1	$t_{2,2}$
2	4	$1 + 3$	$t_{2,2} + t_{3,2}$
3	9	$3 + 6$	$t_{3,2} + t_{4,2}$
4	16	$6 + 10$	$t_{4,2} + t_{5,2}$

Each perfect square greater than 1 is equal to the sum of a pair of adjacent terms on the third diagonal of Pascal's triangle: $n^2 = t_{n,2} + t_{n+1,2}$ for $n > 1$.

- Each term in Pascal's triangle is equal to the sum of the two adjacent terms in the row immediately above: $t_{n,r} = t_{n-1,r-1} + t_{n-1,r}$ where $t_{n,r}$ represents the rth term in row n.

- The sum of the terms in row n of Pascal's triangle is 2^n.

- The terms in the third diagonal of Pascal's triangle are triangular numbers. Many other number patterns occur in Pascal's triangle.

Communicate Your Understanding

1. Describe the symmetry in Pascal's triangle.

2. Explain why the triangular numbers in Example 4 occur in Pascal's triangle.

Practise

1. For future use, make a diagram of the first 12 rows of Pascal's triangle.

2. Express as a single term from Pascal's triangle.

a) $t_{7,2} + t_{7,3}$

b) $t_{51,40} + t_{51,41}$

c) $t_{18,12} - t_{17,12}$

d) $t_{n,r} - t_{n-1,r}$

3. Determine the sum of the terms in each of these rows in Pascal's triangle.

a) row 12

b) row 20

c) row 25

d) row $(n - 1)$

4. Determine the row number for each of the following row sums from Pascal's triangle.

a) 256 **b)** 2048

c) 16 384 **d)** 65 536

Apply, Solve, Communicate

5. Inquiry/Problem Solving

a) Alternately add and subtract the terms in each of the first seven rows of Pascal's triangle and list the results in a table similar to the one below.

Row	Sum/Difference	Result
0	1	1
1	$1 - 1$	0
2	$1 - 2 + 1$	0
3	$1 - 3 + 3 - 1$	0
⋮		

b) Predict the result of alternately adding and subtracting the entries in the eighth row. Verify your prediction.

c) Predict the result for the nth row.

6. a) Predict the sum of the squares of the terms in the nth row of Pascal's triangle.

b) Predict the result of alternately adding and subtracting the squares of the terms in the nth row of Pascal's triangle.

7. Communication

a) Compare the first four powers of 11 with entries in Pascal's triangle. Describe any pattern you notice.

b) Explain how you could express row 5 as a power of 11 by regrouping the entries.

c) Demonstrate how to express rows 6 and 7 as powers of 11 using the regrouping method from part b). Describe your method clearly.

8. a) How many diagonals are there in

i) a quadrilateral?

ii) a pentagon?

iii) a hexagon?

b) Find a relationship between entries in Pascal's triangle and the maximum number of diagonals in an n-sided polygon.

c) Use part b) to predict how many diagonals are in a heptagon and an octagon. Verify your prediction by drawing these polygons and counting the number of possible diagonals in each.

9. Make a conjecture about the divisibility of the terms in prime-numbered rows of Pascal's triangle. Confirm that your conjecture is valid up to row 11.

10. a) Which rows of Pascal's triangle contain only odd numbers? Is there a pattern to these rows?

b) Are there any rows that have only even numbers?

c) Are there more even or odd entries in Pascal's triangle? Explain how you arrived at your answer.

11. Application Oranges can be piled in a tetrahedral shape as shown. The first pile contains one orange, the second contains four oranges, the third contains ten oranges, and so on. The numbers of items in such stacks are known as **tetrahedral numbers.**

a) Relate the number of oranges in the nth pile to entries in Pascal's triangle.

b) What is the 12th tetrahedral number?

12. a) Relate the sum of the squares of the first n positive integers to entries in Pascal's triangle.

b) Use part a) to predict the sum of the squares of the first ten positive integers. Verify your prediction by adding the numbers.

13. Inquiry/Problem Solving A straight line drawn through a circle divides it into two regions.

a) Determine the maximum number of regions formed by n straight lines drawn through a circle. Use Pascal's triangle to help develop a formula.

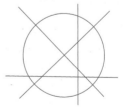

b) What is the maximum number of regions inside a circle cut by 15 lines?

14. Describe how you would set up a spreadsheet to calculate the entries in Pascal's triangle.

15. The Fibonacci sequence is 1, 1, 2, 3, 5, 8, 13, 21, ... Each term is the sum of the previous two terms. Find a relationship between the Fibonacci sequence and the following version of Pascal's triangle.

```
1
1  1
1  2  1
1  3  3  1
1  4  6  4  1
1  5  10 10 5  1
1  6  15 20 15 6  1
1  7  21 35 35 21 7  1
...
```

16. Application Toothpicks are laid out to form triangles as shown below. The first triangle contains 3 toothpicks, the second contains 9 toothpicks, the third contains 18 toothpicks, and so on.

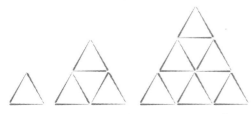

   a) Relate the number of toothpicks in the *n*th triangle to entries in Pascal's triangle.

   b) How many toothpicks would the 10th triangle contain?

17. Design a 3-dimensional version of Pascal's triangle. Use your own criteria for the layers. The base may be any regular geometric shape, but each successive layer must have larger dimensions than the one above it.

18. a) Write the first 20 rows of Pascal's triangle on a sheet of graph paper, placing each entry in a separate square.

   b) Shade in all the squares containing numbers divisible by 2.

   c) Describe, in detail, the patterns produced.

   d) Repeat this process for entries divisible by other whole numbers. Observe the resulting patterns and make a conjecture about the divisibility of the terms in Pascal's triangle by various whole numbers.

19. Communication

   a) Describe the iterative process used to generate the terms in the triangle below.

$$\frac{1}{1}$$
$$\frac{1}{2} \quad \frac{1}{2}$$
$$\frac{1}{3} \quad \frac{1}{6} \quad \frac{1}{3}$$
$$\frac{1}{4} \quad \frac{1}{12} \quad \frac{1}{12} \quad \frac{1}{4}$$
$$\frac{1}{5} \quad \frac{1}{20} \quad \frac{1}{30} \quad \frac{1}{20} \quad \frac{1}{5}$$
$$\frac{1}{6} \quad \frac{1}{30} \quad \frac{1}{60} \quad \frac{1}{60} \quad \frac{1}{30} \quad \frac{1}{6}$$

   b) Write the entries for the next two rows.

   c) Describe three patterns in this triangle.

   d) Research why this triangle is called the harmonic triangle. Briefly explain the origin of the name, listing your source(s).

Applying Pascal's Method

The iterative process that generates the terms in Pascal's triangle can also be applied to counting paths or routes between two points. Consider water being poured into the top bucket in the diagram. You can use Pascal's method to count the different paths that water overflowing from the top bucket could take to each of the buckets in the bottom row.

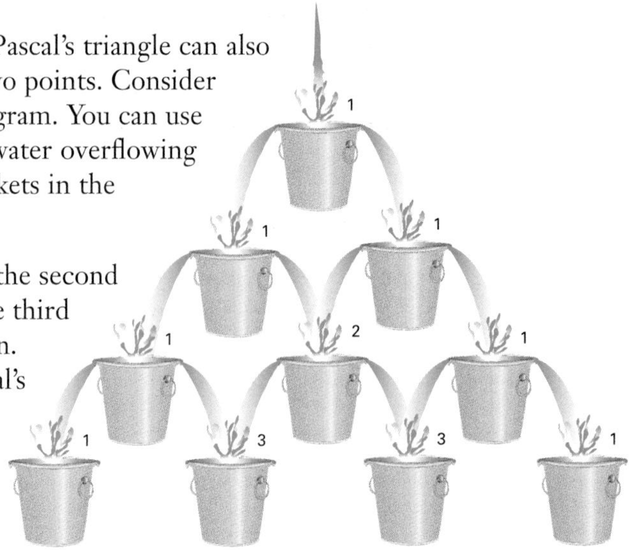

The water has one path to each of the buckets in the second row. There is one path to each outer bucket of the third row, but two paths to the middle bucket, and so on. The numbers in the diagram match those in Pascal's triangle because they were derived using the same method—Pascal's method.

INVESTIGATE & INQUIRE: Counting Routes

Suppose you are standing at the corner of Pythagoras Street and Kovalevsky Avenue, and want to reach the corner of Fibonacci Terrace and Euler Boulevard. To avoid going out of your way, you would travel only east and south. Notice that you could start out by going to the corner of either Euclid Street and Kovalevsky Avenue or Pythagoras Street and de Fermat Drive.

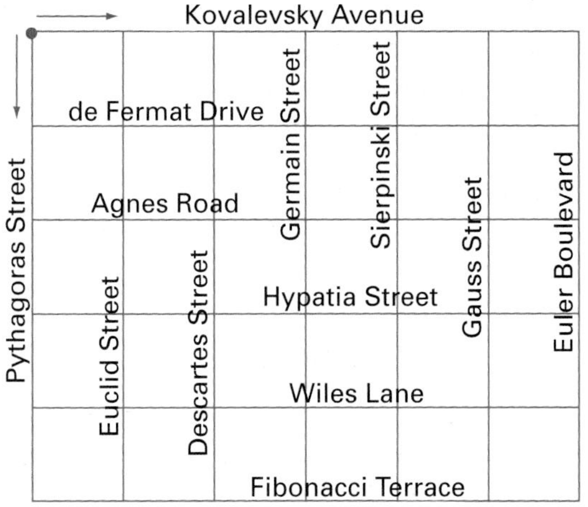

1. How many routes are possible to the corner of Euclid Street and de Fermat Drive from your starting point? Sketch the street grid and mark the number of routes onto it.

2. **a)** Continue to travel only east or south. How many routes are possible from the start to the corner of

   i) Descartes Street and Kovalevsky Avenue?

   ii) Pythagoras Street and Agnes Road?

   iii) Euclid Street and Agnes Road?

   iv) Descartes Street and de Fermat Drive?

   v) Descartes Street and Agnes Road?

   b) List the routes you counted in part a).

3. Consider your method and the resulting numbers. How do they relate to Pascal's triangle?

4. Continue to mark the number of routes possible on your sketch until you have reached the corner of Fibonacci Terrace and Euler Boulevard. How many different routes are possible?

5. Describe the process you used to find the number of routes from Pythagoras Street and Kovalevsky Avenue to Fibonacci Terrace and Euler Boulevard.

Example 1 Counting Paths in an Array

Determine how many different paths will spell *PASCAL* if you start at the top and proceed to the next row by moving diagonally left or right.

```
        P
      A   A
    S   S   S
  C   C   C   C
    A   A   A
      L   L
```

Solution

Starting at the top, record the number of possible paths moving diagonally to the left and right as you proceed to each different letter. For instance, there is one path from *P* to the left *A* and one path from *P* to the right *A*. There is one path from an *A* to the left *S*, two paths from an *A* to the middle *S*, and one path from an *A* to the right *S*.

Continuing with this counting reveals that there are 10 different paths leading to each *L*. Therefore, a total of 20 paths spell *PASCAL*.

```
              P
          1 A     A 1
      1 S     2 S     S 1
    1 C     3 C     3 C     C 1
      4 A     6 A     A 4
        10 L     L 10
```

Example 2 Counting Paths on a Checkerboard

On the checkerboard shown, the checker can travel only diagonally upward. It cannot move through a square containing an X. Determine the number of paths from the checker's current position to the top of the board.

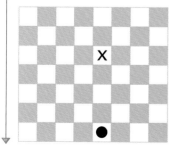

Solution

Use Pascal's method to find the number of paths to each successive position. There is one path possible into each of the squares diagonally adjacent to the checker's starting position. From the second row there are four paths to the third row: one path to the third square from the left, two to the fifth square, and one to the seventh square. Continue this process for the remaining four rows. The square containing an X gets a zero or no number since there are no paths through this blocked square.

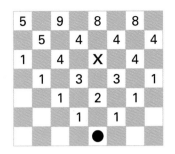

From left to right, there are 5, 9, 8, and 8 paths to the white squares at the top of the board, making a total of 30 paths.

Key Concepts

- Pascal's method involves adding two neighbouring terms in order to find the term below.

- Pascal's method can be applied to counting paths in a variety of arrays and grids.

Communicate Your Understanding

1. Suggest a context in which you could apply Pascal's method, other than those in the examples above.

2. Which of the numbers along the perimeter of a map tallying possible routes are always 1? Explain.

Practise

1. Fill in the missing numbers using Pascal's method.

```
              495
           825
  3003  2112
```

2. In the following arrangements of letters, start from the top and proceed to the next row by moving diagonally left or right. How many different paths will spell each word?

a)
```
            P
          A   A
        T   T   T
      T   T   T   T
      E   E   E   E
    R   R   R   R   R
  N   N   N   N   N   N
S   S   S   S   S   S   S
```

b)
```
            M
          A   A
        T   T   T
      H   H   H   H
    E   E   E   E   E
  M   M   M   M   M   M
A   A   A   A   A   A   A
  T   T   T   T   T   T
    I   I   I   I   I
      C   C   C   C
        S   S   S
```

c)
```
          T
        R   R
      I   I   I
    A   A   A   A
      N   N   N
        G   G
      L   L   L
    E   E   E   E
```

3. The first nine terms of a row of Pascal's triangle are shown below. Determine the first nine terms of the previous and next rows.

1 16 120 560 1820 4368 8008 11 440 12 870

Apply, Solve, Communicate

B

4. Determine the number of possible routes from A to B if you travel only south or east.

a)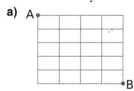

b)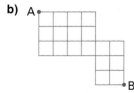

c)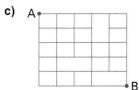

5. Sung is three blocks east and five blocks south of her friend's home. How many different routes are possible if she walks only west or north?

6. Ryan lives four blocks north and five blocks west of his school. Is it possible for him to take a different route to school each day, walking only south and east? Assume that there are 194 school days in a year.

7. A checker is placed on a checkerboard as shown. The checker may move diagonally upward. Although it cannot move into a square with an X, the checker may jump over the X into the diagonally opposite square.

Answer in back off by one.

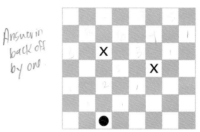

a) How many paths are there to the top of the board?

b) How many paths would there be if the checker could move both diagonally and straight upward?

8. **Inquiry/Problem Solving**

a) If a checker is placed as shown below, how many possible paths are there for that checker to reach the top of the game board? Recall that checkers can travel only diagonally on the white squares, one square at a time, moving upward.

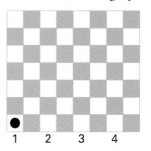

b) When a checker reaches the opposite side, it becomes a "king." If the starting squares are labelled 1 to 4, from left to right, from which starting square does a checker have the most routes to become a king? Verify your statement.

9. Application The following diagrams represent communication networks between a company's computer centres in various cities.

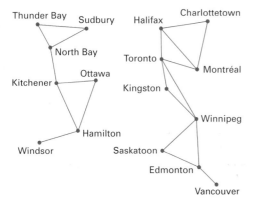

a) How many routes are there from Windsor to Thunder Bay?

b) How many routes are there from Ottawa to Sudbury?

c) How many routes are there from Montréal to Saskatoon?

d) How many routes are there from Vancouver to Charlottetown?

e) If the direction were reversed, would the number of routes be the same for parts a) to d)? Explain.

10. To outfox the Big Bad Wolf, Little Red Riding Hood mapped all the paths through the woods to Grandma's house. How many different routes could she take, assuming she always travels from left to right?

11. Communication A popular game show uses a more elaborate version of the Plinko board shown below. Contestants drop a peg into one of the slots at the top of the upright board. The peg is equally likely to go left or right at each post it encounters.

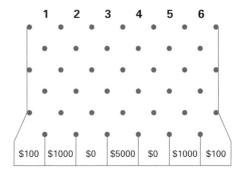

a) Into which slot should contestants drop their pegs to maximize their chances of winning the $5000 prize? Which slot gives contestants the least chance of winning this prize? Justify your answers.

b) Suppose you dropped 100 pegs into the slots randomly, one at a time. Sketch a graph of the number of pegs likely to wind up in each compartment at the bottom of the board. How is this graph related to those described in earlier chapters?

12. Inquiry/Problem Solving

a) Build a new version of Pascal's triangle, using the formula for $t_{n,r}$ on page 247, but start with $t_{0,0} = 2$.

b) Investigate this triangle and state a conjecture about its terms.

c) State a conjecture about the sum of the terms in each row.

13. Inquiry/Problem Solving Develop a formula relating $t_{n,r}$ of Pascal's triangle to the terms in row $n - 3$.

14. The grid below shows the streets in Anya's neighbourhood.

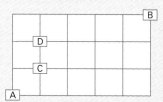

a) If she only travels east and north, how many different routes can Anya take from her house at intersection A to her friend's house at intersection B?

b) How many of the routes in part a) have only one change of direction?

c) Suppose another friend lives at intersection C. How many ways can Anya travel from A to B, meeting her friend at C along the way?

d) How many ways can she travel to B without passing through C? Explain your reasoning.

e) If Anya takes any route from A to B, is she more likely to pass through intersection C or D? Explain your reasoning.

15. Develop a general formula to determine the number of possible routes to travel *n* blocks north and *m* blocks west.

16. Inquiry/Problem Solving In chess, a knight moves in L-shaped jumps consisting of two squares along a row or column plus one square at a right angle. On a standard 8×8 chessboard, the starting position for a knight is the second square of the bottom row. If the knight travels upward on every move, how many routes can it take to the top of the board?

17. Inquiry/Problem Solving Water is poured into the top bucket of a triangular stack of 2-L buckets. When each bucket is full, the water overflows equally on both sides into the buckets immediately below. How much water will have been poured into the top bucket when at least one of the buckets in the bottom row is full?

18. Application Is it possible to arrange a pyramid of buckets such that the bottom layer will fill evenly when water overflows from the bucket at the top of the pyramid?

19. Application Enya is standing in the centre square of a 9 by 9 grid. She travels outward one square at a time, moving diagonally or along a row or column. How many different paths can Enya follow to the perimeter?

20. Communication Describe how a chessboard path activity involving Pascal's method is related to network diagrams like those in section 1.5. Would network diagrams for such activities be planar? Explain.

Review of Key Concepts

4.1 Organized Counting
Refer to the Key Concepts on page 228.

1. A restaurant has a daily special with soup or salad for an appetizer; fish, chicken, or a vegetarian dish for the entrée; and cake, ice cream, or fruit salad for dessert. Use a tree diagram to illustrate all the different meals possible with this special.

2. A theatre company has a half-price offer for students who buy tickets for at least three of the eight plays presented this season. How many choices of three plays would a student have?

3. In how many different orders can a photographer pose a row of six people without having the tallest person beside the shortest one?

4. A transporter truck has three compact cars, a station wagon, and a minivan on its trailer. In how many ways can the driver load the shipment so that one of the heavier vehicles is directly over the rear axle of the trailer?

4.2 Factorials and Permutations
Refer to the Key Concepts on page 238.

5. For what values of n is $n!$ less than 2^n? Justify your answer.

6. A band has recorded five hit singles. In how many different orders could the band play three of these five songs at a concert?

7. In how many ways could a chairperson, treasurer, and secretary be chosen from a 12-member board of directors?

4.3 Permutations With Some Identical Items
Refer to the Key Concepts on page 244.

8. How many different ten-digit telephone numbers contain four 2s, three 3s, and three 7s?

9. a) How many permutations are there of the letters in the word *baseball*?
   b) How many begin with the letter *a*?
   c) How many end with the letter *e*?

10. Find the number of 4×4 patterns you can make using eight white, four grey, and four blue floor tiles.

4.4 Pascal's Triangle
Refer to the Key Concepts on page 251.

11. Write out the first five rows of Pascal's triangle.

12. What is the sum of the entries in the seventh row of Pascal's triangle?

13. Describe three patterns in Pascal's triangle.

4.5 Applying Pascal's Method
Refer to the Key Concepts on page 256.

14. Explain why Pascal's method can be considered an iterative process.

15. How many paths through the array shown will spell *SIERPINSKI*?

```
          S
       I     I
     E    E    E
       R     R
     P    P    P
   I    I    I    I
     N    N    N
   S    S    S    S
     K    K    K
       I     I
```

Chapter Test

1. Natasha tosses four coins one after the other.
   a) In how many different orders could heads or tails occur.
   b) Draw a tree diagram to illustrate all the possible results.
   c) Explain how your tree diagram corresponds to your calculation in part a).

2. Evaluate the following by first expressing each in terms of factorials.
   a) $_{15}P_6$ b) $P(6, 2)$ c) $_7P_3$
   d) $_9P_9$ e) $P(7, 0)$

3. Suppose you are designing a remote control that uses short, medium, or long pulses of infrared light to send control signals to a device.
   a) How many different control codes can you define using
      i) three pulses?
      ii) one, two, or three pulses?
   b) Explain how the multiplicative and additive counting principles apply in your calculations for part a).

4. a) How many four-digit numbers can you form with the digits 1, 2, 3, 4, 5, 6, and 7 if no digit is repeated?
   b) How many of these four-digit numbers are odd numbers?
   c) How many of them are even numbers?

5. How many ways are there to roll either a 6 or a 12 with two dice?

6. How many permutations are there of the letters of each of the following words?
   a) data   b) management   c) microwave

7. A number of long, thin sticks are lying in a pile at odd angles such that the sticks cross each other.
   a) Relate the maximum number of intersection points of n sticks to entries in Pascal's triangle.
   b) What is the maximum number of intersection points with six overlapping sticks?

8. At a banquet, four couples are sitting along one side of a table with men and women alternating.
   a) How many seating arrangements are possible for these eight people?
   b) How many arrangements are possible if each couple sits together? Explain your reasoning.
   c) How many arrangements are possible if no one is sitting beside his or her partner?
   d) Explain why the answers from parts b) and c) do not add up to the answer from part a).

CHAPTER 5

Combinations and the Binomial Theorem

Specific Expectations	Section
Use Venn diagrams as a tool for organizing information in counting problems.	5.1
Solve introductory counting problems involving the additive and multiplicative counting principles.	5.1, 5.2, 5.3
Express answers to permutation and combination problems, using standard combinatorial symbols.	5.1, 5.2, 5.3
Evaluate expressions involving factorial notation, using appropriate methods.	5.2, 5.3
Solve problems, using techniques for counting combinations.	5.2, 5.3
Identify patterns in Pascal's triangle and relate the terms of Pascal's triangle to values of $\binom{n}{r}$, to the expansion of a binomial, and to the solution of related problems.	5.4
Communicate clearly, coherently, and precisely the solutions to counting problems.	5.1, 5.2, 5.3, 5.4

Chapter Problem

Radio Programming

Jeffrey works as a DJ at a local radio station. He does the drive shift from 16 00 to 20 00, Monday to Friday. Before going on the air, he must choose the music he will play during these four hours.

The station has a few rules that Jeffrey must follow, but he is allowed quite a bit of leeway. Jeffrey must choose all his music from the top 100 songs for the week and he must play at least 12 songs an hour. In his first hour, all his choices must be from the top-20 list.

1. In how many ways can Jeffrey choose the music for his first hour?

2. In how many ways can he program the second hour if he chooses at least 10 songs that are in positions 15 to 40 on the charts?

3. Over his 4-h shift, he will play at least 48 songs from the top 100. In how many ways can he choose these songs?

In these questions, Jeffrey can play the songs in any order. Such questions can be answered with the help of combinatorics, the branch of mathematics introduced in Chapter 4. However, the permutations in Chapter 4 dealt with situations where the order of items was important. Now, you will learn techniques you can apply in situations where order is not important.

Review of Prerequisite Skills

If you need help with any of the skills listed in **purple** below, refer to Appendix A.

1. **Factorials (section 4.2)** Evaluate.

   a) $8!$

   b) $\dfrac{8!}{5!}$

   c) $\dfrac{24!}{22!}$

   d) $3! \times 4!$

2. **Permutations (section 4.2)** Evaluate mentally.

   a) $_5P_5$

   b) $_{10}P_2$

   c) $_{12}P_1$

   d) $_7P_3$

3. **Permutations (section 4.2)** Evaluate manually.

   a) $_{10}P_5$

   b) $P(16, 2)$

   c) $_{10}P_{10}$

   d) $P(8, 5)$

4. **Permutations (section 4.2)** Evaluate using software or a calculator.

   a) $_{50}P_{25}$

   b) $P(37, 16)$

   c) $_{29}P_{29}$

   d) $_{46}P_{23}$

5. **Organized counting (section 4.1)** Every Canadian aircraft has five letters in its registration. The first letter must be C, the second letter must be F or G, and the last three letters have no restrictions. If repeated letters are allowed, how many aircraft can be registered with this system?

6. **Applying permutations (Chapter 4)**

   a) How many arrangements are there of three different letters from the word *kings*?

   b) How many arrangements are there of all the letters of the word *management*?

   c) How many ways could first, second, and third prizes be awarded to 12 entrants in a mathematics contest?

7. **Exponent laws** Use the exponent laws to simplify each of the following.

   a) $(-3y)^0$

   b) $(-4x)^3$

   c) $15(7x)^4(4y)^2$

   d) $21(x^3)^2\left(\dfrac{1}{x^2}\right)^5$

   e) $(4x^0y)^2(3x^2y)^3$

   f) $\left(\dfrac{1}{2}\right)^4(3x^2)(2y)^3$

   g) $(-3xy)(-5x^2y)^2$

   h) $\left(\dfrac{1}{3}\right)^0(-2xy)^3$

8. **Simplifying expressions** Expand and simplify.

   a) $(x - 5)^2$

   b) $(5x - y)^2$

   c) $(x^2 + 5)^2$

   d) $(x + 3)(x - 5)^2$

   e) $(x^2 - y)^2$

   f) $(2x + 3)^2$

   g) $(x - 4)^2(x - 2)$

   h) $(2x^2 + 3y)^2$

   i) $(2x + 1)^2(x - 2)$

   j) $(x + y)(x - 2y)^2$

9. **Sigma notation** Rewrite the following using sigma notation.

   a) $1 + 2 + 4 + 8 + 16$

   b) $x + 2x^2 + 3x^3 + 4x^4 + 5x^5$

   c) $\dfrac{1}{2} + \dfrac{1}{3} + \dfrac{1}{4} + \dfrac{1}{5} + \dots$

10. **Sigma notation** Expand.

   a) $\displaystyle\sum_{n=2}^{5} 2n$

   b) $\displaystyle\sum_{n=1}^{4} \dfrac{x^n}{n!}$

   c) $\displaystyle\sum_{n=1}^{5} (2^n + n^2)$

Organized Counting With Venn Diagrams

In Chapter 4, you used tree diagrams as a tool for counting items when the order of the items was important. This section introduces a type of diagram that helps you organize data about groups of items when the order of the items is not important.

A group of students meet regularly to plan the dances at Vennville High School. Amar, Belinda, Charles, and Danica are on the dance committee, and Belinda, Charles, Edith, Franco, and Geoff are on the students' council. Hans and Irena are not members of either group, but they attend meetings as reporters for the school newspaper.

1. Draw two circles to represent the dance committee and the students' council. Where on the diagram would you put initials representing the students who are

   a) on the dance committee?

   b) on the students' council?

   c) on the dance committee and the students' council?

   d) not on either the dance committee or the students' council?

2. Redraw your diagram marking on it the number of initials in each region. What relationships can you see among these numbers?

Your sketch representing the dance committee and the students' council is a simple example of a **Venn diagram**. The English logician John Venn (1834–1923) introduced such diagrams as a tool for analysing situations where there is some overlap among groups of items, or **sets**. Circles represent different sets and a rectangular box around the circles represents the **universal set**, *S*, from which all the items are drawn. This box is usually labelled with an *S* in the top left corner.

The items in a set are often called the **elements** or **members** of the set. The size of a circle in a Venn diagram does not have to be proportional to the number of elements in the set the circle represents. When some items in a set are also elements of another set, these items are **common elements** and the sets are shown as overlapping circles. If *all* elements of a set C are also elements of set A, then C is a **subset** of A. A Venn diagram would show this set C as a region contained within the circle for set A.

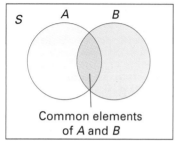

The common elements are a subset of both A and B.

WEB CONNECTION

www.mcgrawhill.ca/links/MDM12

To learn more about Venn diagrams, visit the above web site and follow the links. Describe an example of how Venn diagrams can be used to organize information.

You can use Venn diagrams to organize information for situations in which the number of items in a group are important but the order of the items is not.

Example 1 Common Elements

There are 10 students on the volleyball team and 15 on the basketball team. When planning a field trip with both teams, the coach has to arrange transportation for a total of only 19 students.

a) Use a Venn diagram to illustrate this situation.

b) Explain why you cannot use the additive counting principle to find the total number of students on the teams.

c) Determine how many students are on both teams.

d) Determine the number of students in the remaining regions of your diagram and explain what these regions represent.

Solution

a) Some students must be on both the volleyball and the basketball team. Draw a box with an S in the top left-hand corner. Draw and label two overlapping circles to represent the volleyball and basketball teams.

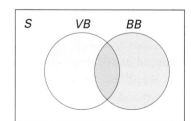

b) The additive counting principle (or rule of sum) applies only to mutually exclusive events or items. However, it is possible for students to be on both teams. If you simply add the 10 students on the volleyball team to 15 students on the basketball team, you get a total of 25 students because the students who play on both teams have been counted twice.

c) The difference between the total in part b) and the total number of students actually on the two teams is equal to the number of students who are members of both teams. Thus, $25 - 19 = 6$ students play on both teams. In the Venn diagram, these 6 students are represented by the area where the two circles overlap.

d) There are $10 - 6 = 4$ students in the section of the *VB* circle that does not overlap with the *BB* circle. These are the students who play only on the volleyball team. Similarly, the non-overlapping portion of the *BB* circle represents the $15 - 6 = 9$ students who play only on the basketball team.

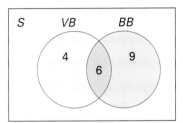

Example 1 illustrates the **principle of inclusion and exclusion**. If you are counting the total number of elements in two groups or sets that have common elements, you must subtract the common elements so that they are not included twice.

Principle of Inclusion and Exclusion for Two Sets

For sets *A* and *B*, the total number of elements in either *A* or *B* is the number in *A* plus the number in *B* minus the number in both *A* and *B*.

$n(A \text{ or } B) = n(A) + n(B) - n(A \text{ and } B),$
where $n(X)$ represents the numbers of elements in a set *X*.

The set of all elements in either set *A* or set *B* is the **union** of *A* and *B*, which is often written as $A \cup B$. Similarly, the set of all elements in both *A* and *B* is the **intersection** of *A* and B, written as $A \cap B$. Thus the principle of inclusion and exclusion for two sets can also be stated as

$n(A \cup B) = n(A) + n(B) - n(A \cap B)$

Note that the additive counting principle (or rule of sum) could be considered a special case of the principle of inclusion and exclusion that applies only when sets *A* and *B* have no elements in common, so that $n(A \text{ and } B) = 0$. The principle of inclusion and exclusion can also be applied to three or more sets.

Example 2 Applying the Principle of Inclusion and Exclusion

A drama club is putting on two one-act plays. There are 11 actors in the Feydeau farce and 7 in the Molière piece.

a) If 3 actors are in both plays, how many actors are there in all?

b) Use a Venn diagram to calculate how many students are in only one of the two plays.

Solution

a) Calculate the number of students in both plays using the principle of inclusion and exclusion.

n(total) = n(Feydeau) + n(Molière) − n(Feydeau and Molière)
$$= 11 + 7 - 3$$
$$= 15$$

There are 15 students involved in the two one-act plays.

b) There are 3 students in the overlap between the two circles. So, there must be 11 − 3 = 8 students in the region for Feydeau only and 7 − 3 = 4 students in the region for Molière only.

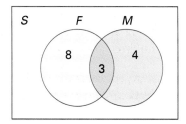

Thus, a total of 8 + 4 = 12 students are in only one of the two plays.

As in the first example, using a Venn diagram can clarify the relationships between several sets and subsets.

Example 3 Working With Three Sets

Of the 140 grade 12 students at Vennville High School, 52 have signed up for biology, 71 for chemistry, and 40 for physics. The science students include 15 who are taking both biology and chemistry, 8 who are taking chemistry and physics, 11 who are taking biology and physics, and 2 who are taking all three science courses.

a) How many students are not taking any of these three science courses?

b) Illustrate the enrolments with a Venn diagram.

Solution

a) Extend the principle of inclusion and exclusion to three sets. Total the numbers of students in each course, subtract the numbers of students taking two courses, then add the number taking all three. This procedure subtracts out the students who have been counted twice because they are in two

courses, and then adds back those who were subtracted twice because they were in all three courses.

For simplicity, let *B* stand for biology, *C* stand for chemistry, and *P* stand for physics. Then, the total number of students taking at least one of these three courses is

$$n(\text{total}) = n(B) + n(C) + n(P) - n(B \text{ and } C) - n(C \text{ and } P) - n(B \text{ and } P) + n(B \text{ and } C \text{ and } P)$$
$$= 52 + 71 + 40 - 15 - 8 - 11 + 2$$
$$= 131$$

There are 131 students taking one or more of the three science courses. To find the number of grade 12 students who are not taking any of these science courses, subtract 131 from the total number of grade 12 students.

Thus, $140 - 131 = 9$ students are not taking any of these three science courses in grade 12.

b) For this example, it is easiest to start with the overlap among the three courses and then work outward. Since there are 2 students taking all three courses, mark 2 in the centre of the diagram where the three circles overlap.

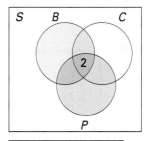

Next, consider the adjacent regions representing the students who are taking exactly two of the three courses.

Biology and chemistry: Of the 15 students taking these two courses, 2 are also taking physics, so 13 students are taking only biology and chemistry.
Chemistry and physics: 8 students less the 2 in the centre region leaves 6.
Biology and physics: $11 - 2 = 9$.

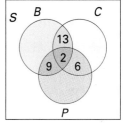

Now, consider the regions representing students taking only one of the science courses.

Biology: Of the 52 students taking this course, $13 + 2 + 9 = 24$ are in the regions overlapping with the other two courses, leaving 28 students who are taking biology only.
Chemistry: 71 students less the $13 + 2 + 6$ leaves 50.
Physics: $40 - (9 + 2 + 6) = 23$.

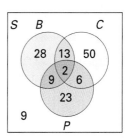

Adding all the numbers within the circles gives a total of 131. Thus, there must be $140 - 131 = 9$ grade 12 students who are not taking any of the three science courses, which agrees with the answer found in part a).

Key Concepts

- Venn diagrams can help you visualize the relationships between sets of items, especially when the sets have some items in common.

- The principle of inclusion and exclusion gives a formula for finding the number of items in the union of two or more sets. For two sets, the formula is $n(A \text{ or } B) = n(A) + n(B) - n(A \text{ and } B)$.

Communicate Your Understanding

1. Describe the principal use of Venn diagrams.

2. Is the universal set the same for all Venn diagrams? Explain why or why not.

3. Explain why the additive counting principle can be used in place of the principle of inclusion and exclusion for mutually exclusive sets.

Practise

1. Let set A consist of an apple, an orange, and a pear and set B consist of the apple and a banana.

   a) List the elements of
      i) A and B
      ii) A or B
      iii) S
      iv) $S \cap B$
      v) $A \cup B \cup S$

   b) List the value of
      i) $n(A) + n(B)$
      ii) $n(A \text{ or } B)$
      iii) $n(S)$
      iv) $n(A \cup B)$
      v) $n(S \cap A)$

   c) List all subsets containing exactly two elements for
      i) A
      ii) B
      iii) $A \cup B$

2. A recent survey of a group of students found that many participate in baseball, football, and soccer. The Venn diagram below shows the results of the survey.

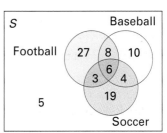

a) How many students participated in the survey?

b) How many of these students play both soccer and baseball?

c) How many play only one sport?

d) How many play football and soccer?

e) How many play all three sports?

f) How many do not play soccer?

Apply, Solve, Communicate

3. Of the 220 graduating students in a school, 110 attended the semi-formal dance and 150 attended the formal dance. If 58 students attended both events, how many graduating students did not attend either of the two dances? Illustrate your answer with a Venn diagram.

4. Application A survey of 1000 television viewers conducted by a local television station produced the following data:
 • 40% watch the news at 12 00
 • 60% watch the news at 18 00
 • 50% watch the news at 23 00
 • 25% watch the news at 12 00 and at 18 00
 • 20% watch the news at 12 00 and at 23 00
 • 20% watch the news at 18 00 and at 23 00
 • 10% watch all three news broadcasts

a) What percent of those surveyed watch at least one of these programs?

b) What percent watch none of these news broadcasts?

c) What percent view the news at 12 00 and at 18 00, but not at 23 00?

d) What percent view only one of these shows?

e) What percent view exactly two of these shows?

5. Suppose the Canadian Embassy in the Netherlands has 32 employees, all of whom speak both French and English. In addition, 22 of the employees speak German and 15 speak Dutch. If there are 10 who speak both German and Dutch, how many of the employees speak neither German nor Dutch? Illustrate your answer with a Venn diagram.

6. Application There are 900 employees at CantoCrafts Inc. Of these, 615 are female, 345 are under 35 years old, 482 are single, 295 are single females, 187 are singles under 35 years old, 190 are females under 35 years old, and 120 are single females under 35 years old. Use a Venn diagram to determine how many employees are married males who are at least 35 years old.

7. Communication A survey of 100 people who volunteered information about their reading habits showed that
 • 75 read newspapers daily
 • 35 read books at least once a week
 • 45 read magazines regularly
 • 25 read both newspapers and books
 • 15 read both books and magazines
 • 10 read newspapers, books, and magazines

a) Construct a Venn diagram to determine the maximum number of people in the survey who read both newspapers and magazines.

b) Explain why you cannot determine exactly how many of the people surveyed read both newspapers and magazines.

8. Jeffrey works as a DJ at a local radio station. On occasion, he chooses some of the songs he will play based on the phone-in requests received by the switchboard the previous day. Jeffrey's list of 200 possible selections includes

- all the songs in the top 100
- 134 hard-rock songs
- 50 phone-in requests
- 45 hard-rock songs in the top 100
- 20 phone-in requests in the top 100
- 24 phone-in requests for hard-rock songs

Use a Venn diagram to determine

a) how many phone-in requests were for hard-rock songs in the top 100

b) how many of the songs in the top 100 were neither phone-in requests nor hard-rock selections

9. Inquiry/Problem Solving The Vennville junior hockey team has 12 members who can play forward, 8 who can play defence, and 2 who can be goalies. What is the smallest possible size of the team if

a) no one plays more than one position?

b) no one plays both defence and forward?

c) three of the players are able to play defence or forward?

d) both the goalies can play forward but not defence?

10. Inquiry/Problem Solving Use the principle of inclusion and exclusion to develop a formula for the number of elements in

a) three sets   **b)** four sets   **c)** n sets

Combinations

In Chapter 4, you learned about permutations—arrangements in which the order of the items is specified. However, in many situations, order does not matter. For example, in many card games, what is in your hand is important, but the order in which it was dealt is not.

Suppose the students at a secondary school elect a council of eight members, two from each grade. This council then chooses two of its members as co-chairpersons. How could you calculate the number of different pairs of members who could be chosen as the co-chairs?

Choose someone in the class to record your answers to the following questions on a blackboard or an overhead projector.

a) Start with the simplest case. Choose two students to stand at the front of the class. In how many ways can you choose two co-chairs from this pair of students?

b) Choose three students to be at the front of the class. In how many ways can you choose two co-chairs from this trio?

c) In how many ways can you choose two co-chairs from a group of four students?

d) In how many ways can you choose two co-chairs from a group of five students? Do you see a pattern developing? If so, what is it? If not, try choosing from a group of six students and then from a group of seven students while continuing to look for a pattern.

e) When you see a pattern, predict the number of ways two co-chairs can be chosen from a group of eight students.

f) Can you suggest how you could find the answers for this investigation from the numbers of permutations you found in the investigation in section 4.2?

In the investigation on the previous page, you were dealing with a situation in which you were selecting two people from a group, but the order in which you chose the two did not matter. In a permutation, there is a difference between selecting, say, Bob as president and Margot as vice-president as opposed to selecting Margot as president and Bob as vice-president. If you select Bob and Margot as co-chairs, the order in which you select them does not matter since they will both have the same job.

A selection from a group of items without regard to order is called a **combination**.

Example 1 Comparing Permutations and Combinations

a) In how many ways could Alana, Barbara, Carl, Domenic, and Edward fill the positions of president, vice-president, and secretary?

b) In how many ways could these same five people form a committee with three members? List the ways.

c) How are the numbers of ways in parts a) and b) related?

Solution

a) Since the positions are different, order is important. Use a permutation, $_nP_r$. There are five people to choose from, so $n = 5$. There are three people being chosen, so $r = 3$. The number of permutations is $_5P_3 = 60$.

There are 60 ways Alana, Barbara, Carl, Domenic, and Edward could fill the positions of president, vice-president, and secretary.

b) The easiest way to find all committee combinations is to write them in an ordered fashion. Let A represent Alana, B represent Barbara, C represent Carl, D represent Domenic, and E represent Edward.

The possible combinations are:
ABC   ABD   ABE   ACD   ACE
ADE   BCD   BCE   BDE   CDE

All other possible arrangements include the same three people as one of the combinations listed above. For example, ABC is the same as ACB, BAC, BCA, CAB, and CBA since order is not important.

So, there are only ten ways Alana, Barbara, Carl, Domenic and Edward can form a three-person committee.

c) In part a), there were 60 possible permutations, while in part b), there were 10 possible combinations. The difference is a factor of 6. This factor is $_3P_3 = 3!$, the number of possible arrangements of the three people in each combination. Thus,

$$\text{number of combinations} = \frac{\text{number of permutations}}{\text{number of permutations of the objects selected}}$$

$$= \frac{_5P_3}{3!}$$

$$= \frac{60}{6}$$

$$= 10$$

Combinations of n distinct objects taken r at a time

The number of combinations of r objects chosen from a set of n distinct objects is

$$_nC_r = \frac{_nP_r}{r!}$$

$$= \frac{\frac{n!}{(n-r)!}}{r!}$$

$$= \frac{n!}{(n-r)!r!}$$

The notations $_nC_r$, $C(n, r)$, and $\binom{n}{r}$ are all equivalent. Many people prefer the form $\binom{n}{r}$ when a number of combinations are multiplied together. The symbol $_nC_r$ is used most often in this text since it is what appears on most scientific and graphing calculators.

Example 2 Applying the Combinations Formula

How many different sampler dishes with 3 different flavours could you get at an ice-cream shop with 31 different flavours?

Solution

There are 31 flavours, so $n = 31$. The sampler dish has 3 flavours, so $r = 3$.

$$_{31}C_3 = \frac{31!}{(31-3)!\,3}$$

$$= \frac{31!}{28!3!}$$

$$= \frac{31 \times 30 \times 29}{3 \times 2}$$

$$= 4495$$

There are 4495 possible sampler combinations.

Note that the number of combinations in Example 2 was easy to calculate because the number of items chosen, r, was quite small.

Example 3 Calculating Numbers of Combinations Manually

A ballet choreographer wants 18 dancers for a scene.

a) In how many ways can the choreographer choose the dancers if the company has 20 dancers? 24 dancers?

b) How would you advise the choreographer to choose the dancers?

Solution

a) When n and r are close in value, $_nC_r$ can be calculated mentally. With $n = 20$ and $r = 18$,

$$_{20}C_{18} = \frac{20!}{(20 - 18)! \, 18!}$$

$$= \frac{20 \times 19}{2!} \qquad 20 \div 2 = 10$$

$$= 190 \qquad \text{Then, } 10 \times 19 = 190$$

The choreographer could choose from 190 different combinations of the 20 dancers.

With $n = 24$ and $r = 18$, $_nC_r$ can be calculated manually or with a basic calculator once you have divided out the common terms in the factorials.

$$_{24}C_{18} = \frac{24!}{(24 - 18)! \, 18!}$$

$$= \frac{24 \times 23 \times 22 \times 21 \times 20 \times 19}{6!}$$

$$= \frac{24 \times 23 \times 22 \times 21 \times 20 \times 19}{6 \times 5 \times 4 \times 3 \times 2 \times 1}$$

$$= 23 \times 11 \times 7 \times 4 \times 19$$

$$= 134\ 596$$

With the 4 additional dancers, the choreographer now has a choice of 134 596 combinations.

b) From part a), you can see that it would be impractical for the choreographer to try every possible combination. Instead the choreographer could use an indirect method and try to decide which dancers are least likely to be suitable for the scene.

Even though there are fewer permutations of n objects than there are combinations, the numbers of combinations are often still too large to calculate manually.

Example 4 Using Technology to Calculate Numbers of Combinations

Each player in a bridge game receives a hand of 13 cards dealt from a standard deck. How many different bridge hands are possible?

For details of calculator and software functions, refer to Appendix B.

Solution 1 Using a Graphing Calculator

Here, the order in which the player receives the cards does not matter. What you want to determine is the number of different combinations of cards a player could have once the dealing is complete. So, the answer is simply $_{52}C_{13}$. You can evaluate $_{52}C_{13}$ by using the nCr function on the MATH PRB menu of a graphing calculator. This function is similar to the nPr function used for permutations.

```
GRAPHING CALCULATOR
52 nCr 13
    6.350135596E11
■
```

There are about 635 billion possible bridge hands.

Solution 2 Using a Spreadsheet

Most spreadsheet programs have a combinations function for calculating numbers of combinations. In Microsoft® Excel, this function is the COMBIN(n,r) function. In Corel® Quattro® Pro, this function is the COMB(r,n) function.

A1		=	=COMBIN(52,13)				
	A	B	C	D	E	F	G	H
1	6.35014E+11							
2								
3								
4								
5								
6								
7								
8								

You now have a variety of methods for finding numbers of combinations—paper-and-pencil calculations, factorials, scientific or graphing calculators, and software. When appropriate, you can also apply both of the counting principles described in Chapter 4.

Example 5 Using the Counting Principles With Combinations

Ursula runs a small landscaping business. She has on hand 12 kinds of rose bushes, 16 kinds of small shrubs, 11 kinds of evergreen seedlings, and 18 kinds of flowering lilies. In how many ways can Ursula fill an order if a customer wants

a) 15 different varieties consisting of 4 roses, 3 shrubs, 2 evergreens, and 6 lilies?

b) either 4 different roses or 6 different lilies?

Project Prep

Techniques for calculating numbers of combinations could be helpful for designing the game in your probability project, especially if your game uses cards.

5.2 Combinations • MHR   **277**

Solution

a) The order in which Ursula chooses the plants does not matter.

The number of ways of choosing the roses is $_{12}C_4$.
The number of ways of choosing the shrubs is $_{16}C_3$.
The number of ways of choosing the evergreens is $_{11}C_2$.
The number of ways of choosing the lilies is $_{18}C_6$.

Since varying the rose selection for each different selection of the shrubs, evergreens, and lilies produces a different choice of plants, you can apply the fundamental (multiplicative) counting principle. Multiply the series of combinations to find the total number of possibilities.

$$_{12}C_4 \times {}_{16}C_3 \times {}_{11}C_2 \times {}_{18}C_6 = 495 \times 560 \times 55 \times 18\ 564$$
$$= 2.830\ 267\ 44 \times 10^{11}$$

Ursula has over 283 billion ways of choosing the plants for her customer.

b) Ursula can choose the 4 rose bushes in $_{12}C_4$ ways.

She can choose the 6 lilies in $_{18}C_6$ ways.

Since the customer wants *either* the rose bushes *or* the lilies, you can apply the additive counting principle to find the total number of possibilities.

$$_{12}C_4 + {}_{18}C_6 = 495 + 18\ 564$$
$$= 19\ 059$$

Ursula can fill the order for either roses or lilies in 19 059 ways.

As you can see, even relatively simple situations can produce very large numbers of combinations.

Key Concepts

- A combination is a selection of objects in which order is not important.

- The number of combinations of n distinct objects taken r at a time is denoted as $_nC_r$, $C(n, r)$, or $\binom{n}{r}$ and is equal to $\dfrac{n!}{(n-r)!\ r!}$.

- The multiplicative and additive counting principles can be applied to problems involving combinations.

1. Explain why n objects have more possible permutations than combinations. Use a simple example to illustrate your explanation.

2. Explain whether you would use combinations, permutations, or another method to calculate the number of ways of choosing

   a) three items from a menu of ten items

   b) an appetizer, an entrée, and a dessert from a menu with three appetizers, four entrées, and five desserts

3. Give an example of a combination expression you could calculate

   a) by hand

   b) algebraically

   c) only with a calculator or computer

Practise

1. Evaluate using a variety of methods. Explain why you chose each method.

   a) $_{21}C_{19}$ b) $_{30}C_{28}$

   c) $_{18}C_5$ d) $_{16}C_3$

   e) $_{19}C_4$ f) $_{25}C_{20}$

2. Evaluate the following pairs of combinations and compare their values.

   a) $_{11}C_1, _{11}C_{10}$

   b) $_{11}C_2, _{11}C_9$

   c) $_{11}C_3, _{11}C_8$

Apply, Solve, Communicate

3. **Communication** In how many ways could you choose 2 red jellybeans from a package of 15 red jellybeans? Explain your reasoning.

4. How many ways can 4 cards be chosen from a deck of 52, if the order in which they are chosen does not matter?

5. How many groups of 3 toys can a child choose to take on a vacation from a toy box containing 11 toys?

6. How many sets of 6 questions for a test can be chosen from a list of 22 questions?

7. In how many ways can a teacher select 5 students from a class of 23 to make a bulletin-board display? Explain your reasoning.

8. As a promotion, a video store decides to give away posters for recently released movies.

   a) If posters are available for 27 recent releases, in how many ways could the video-store owner choose 8 different posters for the promotion?

   b) Are you able to calculate the number of ways mentally? Why or why not?

9. Communication A club has 11 members.

a) How many different 2-member committees could the club form?

b) How many different 3-member committees could the club form?

c) In how many ways can a club president, treasurer, and secretary be chosen?

d) By what factor do the answers in parts b) and c) differ? How do you account for this difference?

10. Fritz has a deck of 52 cards, and he may choose any number of these cards, from none to all. Use a spreadsheet or Fathom™ to calculate and graph the number of combinations for each of Fritz's choices.

11. Application A track club, a swim club, and a cycling club are forming a joint committee to organize a triathlon. The committee will have two members from each club. In how many ways can the committee be formed if ten runners, eight swimmers, and seven cyclists volunteer to serve on it?

12. In how many ways can a jury of 6 men and 6 women be chosen from a group of 10 men and 15 women?

13. Inquiry/Problem Solving There are 15 technicians and 11 chemists working in a research laboratory. In how many ways could they form a 5-member safety committee if the committee

a) may be chosen in any way?

b) must have exactly one technician?

c) must have exactly one chemist?

d) must have exactly two chemists?

e) may be all technicians or all chemists?

14. Jeffrey, a DJ at a local radio station, is choosing the music he will play on his shift. He must choose all his music from the top 100 songs for the week and he must play at least 12 songs an hour. In his first hour, all his choices must be from the top-20 list.

a) In how many ways can Jeffrey choose the songs for his first hour if he wants to choose exactly 12 songs?

b) In how many ways can Jeffrey choose the 12 songs if he wants to pick 8 of the top 10 and 4 from the songs listed from 11 to 20 on the chart?

c) In how many ways can Jeffrey choose either 12 or 13 songs to play in the first hour of his shift?

d) In how many ways can Jeffrey choose the songs if he wants to play up to 15 songs in the first hour?

15. The game of euchre uses only 24 of the cards from a standard deck. How many different five-card euchre hands are possible?

16. Application A taxi is shuttling 11 students to a concert. The taxi can hold only 4 students. In how many ways can 4 students be chosen for

a) the taxi's first trip?

b) the taxi's second trip?

17. Diane is making a quilt. She needs three pieces with a yellow undertone, two pieces with a blue undertone, and four pieces with a white undertone. If she has six squares with a yellow undertone, five with a blue undertone, and eight with a white undertone to choose from, in how many ways can she choose the squares for the quilt?

18. **Inquiry/Problem Solving** At a family reunion, everyone greets each other with a handshake. If there are 20 people at the reunion, how many handshakes take place?

19. A basketball team consists of five players—one centre, two forwards, and two guards. The senior squad at Vennville Central High School has two centres, six forwards, and four guards.

   a) How many ways can the coach pick the two starting guards for a game?

   b) How many different starting lineups are possible if all team members play their specified positions?

   c) How many of these starting lineups include Dana, the team's 185-cm centre?

   d) Some coaches designate the forwards as power forward and small forward. If all six forwards are adept in either position, how would this designation affect the number of possible starting lineups?

   e) As the league final approaches, the centre Dana, forward Ashlee, and guard Hollie are all down with a nasty flu. Fortunately, the five healthy forwards can also play the guard position. If the coach can assign these players as either forwards or guards, will the number of possible starting lineups be close to the number in part b)? Support your answer mathematically.

   f) Is the same result achieved if the forwards play their regular positions but the guards can play as either forwards or guards? Explain your answer.

20. In the game of bridge, each player is dealt a hand of 13 cards from a standard deck of 52 cards.

   a) By what factor does the number of possible bridge hands differ from the number of ways a bridge hand could be dealt to a player? Explain your reasoning.

   b) Use combinations to write an expression for the number of bridge hands that have exactly five clubs, two spades, three diamonds, and three hearts.

   c) Use combinations to write an expression for the number of bridge hands that have exactly five hearts.

   d) Use software or a calculator to evaluate the expressions in parts b) and c).

C

21. There are 18 students involved in the class production of *Arsenic and Old Lace*.

   a) In how many ways can the teacher cast the play if there are five male roles and seven female roles and the class has nine male and nine female students?

   b) In how many ways can the teacher cast the play if Jean will play the young female part only if Jovane plays the male lead?

   c) In how many ways can the teacher cast the play if all the roles could be played by either a male or a female student?

22. A large sack contains six basketballs and five volleyballs. Find the number of combinations of four balls that can be chosen from the sack if

   a) they may be any type of ball

   b) two must be volleyballs and two must be basketballs

   c) all four must be volleyballs

   d) none may be volleyballs

Problem Solving With Combinations

In the last section, you considered the number of ways of choosing r items from a set of n distinct items. This section will examine situations where you want to know the total number of possible combinations of any size that you could choose from a given number of items, some of which may be identical.

1. **a)** How many different sums of money can you create with a penny and a nickel? List these sums.

   b) How many different sums can you create with a penny, a nickel, and a dime? List them.

   c) Predict how many different sums you can create with a penny, a nickel, a dime, and a quarter. Test your conjecture by listing the possible sums.

2. **a)** How many different sums of money can you create with two pennies and a dime? List them.

   b) How many different sums can you create with three pennies and a dime?

   c) Predict how many sums you can create with four pennies and a dime. Test your conjecture. Can you see a pattern developing? If so, what is it?

Example 1 All Possible Combinations of Distinct Items

An artist has an apple, an orange, and a pear in his refrigerator. In how many ways can the artist choose one or more pieces of fruit for a still-life painting?

Solution

The artist has two choices for each piece of fruit: either include it in the painting or leave it out. Thus, the artist has a total of $2 \times 2 \times 2 = 8$ choices. Note that one of these choices is to leave out the apple, the orange, *and* the pear. However, the artist wants at least one piece of fruit in his painting. Thus, he has $2^3 - 1 = 7$ combinations to choose from.

You can apply the same logic to any group of distinct items.

> The total number of combinations containing at least one item chosen from a group of n distinct items is $2^n - 1$.

Remember that combinations are subsets of the group of n objects. A **null set** is a set that has no elements. Thus,

> A set with n distinct elements has 2^n subsets including the null set.

Example 2 Applying the Formula for Numbers of Subsets

In how many ways can a committee with at least one member be appointed from a board with six members?

Solution

The board could choose 1, 2, 3, 4, 5, or 6 people for the committee, so $n = 6$. Since the committee must have at least one member, use the formula that excludes the null set.

$$2^6 - 1 = 64 - 1$$
$$= 63$$

There are 63 ways to choose a committee of at least one person from a six-member board.

Example 3 All Possible Combinations With Some Identical Items

Kate is responsible for stocking the coffee room at her office. She can purchase up to three cases of cookies, four cases of soft drinks, and two cases of coffee packets without having to send the order through the accounting department. How many different direct purchases can Kate make?

Solution

Kate can order more than one of each kind of item, so this situation involves combinations in which some items are alike.
• Kate may choose to buy three or two or one or no cases of cookies, so she has four ways to choose cookies.
• Kate may choose to buy four or three or two or one or no cases of soft drinks, so she has five ways to choose soft drinks.
• Kate may choose to buy two or one or no cases of coffee packets, so she has three ways to choose coffee.

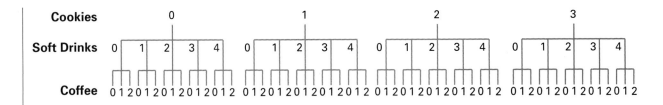

As shown on the first branch of the diagram above, one of these choices is purchasing *no* cookies, *no* soft drinks, and *no* coffee. Since this choice is not a purchase at all, subtract it from the total number of choices.

Thus, Kate can make $4 \times 5 \times 3 - 1 = 59$ different direct purchases.

In a situation where you can choose all, some, or none of the p items available, you have $p + 1$ choices. You can then apply the fundamental (multiplicative) counting principle if you have successive choices of different kinds of items. Always consider whether the choice of not picking any items makes sense. If it does not, subtract 1 from the total.

Combinations of Items in Which Some are Alike

If at least one item is chosen, the total number of selections that can be made from p items of one kind, q items of another kind, r items of another kind, and so on, is $(p + 1)(q + 1)(r + 1) \ldots - 1$

Having identical elements in a set reduces the number of possible combinations when you choose r items from that set. You cannot calculate this number by simply dividing by a factorial as you did with permutations in section 4.3. Often, you have to consider a large number of cases individually. However, some situations have restrictive conditions that make it much easier to count the number of possible combinations.

Example 4 Combinations With Some Identical Items

The director of a short documentary has found five rock songs, two blues tunes, and three jazz pieces that suit the theme of the film. In how many ways can the director choose three pieces for the soundtrack if she wants the film to include some jazz?

Solution 1   Counting Cases

The director can select exactly one, two, or three jazz pieces.

Case 1:   *One jazz piece*
> The director can choose the one jazz piece in $_3C_1$ ways and two of the seven non-jazz pieces in $_7C_2$ ways. Thus, there are $_3C_1 \times {_7C_2} = 63$ combinations of music with one jazz piece.

Case 2:   *Two jazz pieces*
> The director can choose the two jazz pieces in $_3C_2$ ways and one of the seven non-jazz pieces in $_7C_1$ ways. There are $_3C_2 \times {_7C_1} = 21$ combinations with two jazz pieces.

Case 3:   *Three jazz pieces*
> The director can choose the three jazz pieces and none of the seven non-jazz pieces in only one way: $_3C_3 \times {_7C_0} = 1$.

The total number of combinations with at least one jazz piece is $63 + 21 + 1 = 85$.

Solution 2   Indirect Method

You can find the total number of possible combinations of three pieces of music and subtract those that do not have any jazz.

The total number of ways of choosing any three pieces from the ten available is $_{10}C_3 = 120$. The number of ways of not picking any jazz, that is, choosing only from the non-jazz pieces is $_7C_3 = 35$.

Thus, the number of ways of choosing at least one jazz piece is $120 - 35 = 85$.

Here is a summary of ways to approach questions involving choosing or selecting objects.

Is order important?

Yes: Use permutations. Can the same objects be selected more than once (like digits for a telephone number)?
> **Yes:** Use the fundamental counting principle.
> **No:** Are some of the objects identical?
>> **Yes:** Use the formula $\dfrac{n!}{a!b!c!\dots}$.
>>
>> **No:** Use $_nP_r = \dfrac{n!}{(n-r)!}$.

No: Use combinations. Are you choosing exactly r objects?
> **Yes**: Could some of the objects be identical?
>> **Yes:** Count the individual cases.
>>
>> **No:** Use $_nC_r = \dfrac{n!}{(n-r)!r!}$
>
> **No:** Are some of the objects identical?
>> **Yes:** Use $(p+1)(q+1)(r+1) - 1$ to find the total number of combinations with at least one object.
>>
>> **No:** Use 2^n to find the total number of combinations; subtract 1 if you do not want to include the null set.

Key Concepts

- Use the formula $(p + 1)(q + 1)(r + 1) \ldots - 1$ to find the total number of selections of at least one item that can be made from p items of one kind, q of a second kind, r of a third kind, and so on.

- A set with n distinct elements has 2^n subsets including the null set.

- For combinations with some identical elements, you often have to consider all possible cases individually.

- In a situation where you must choose *at least* one particular item, either consider the total number of choices available minus the number without the desired item or add all the cases in which it is possible to have the desired item.

Communicate Your Understanding

1. Give an example of a situation where you would use the formula $(p + 1)(q + 1)(r + 1) \ldots - 1$. Explain why this formula applies.

2. Give an example of a situation in which you would use the expression $2^n - 1$. Explain your reasoning.

3. Using examples, describe two different ways to solve a problem where *at least* one particular item must be chosen. Explain why both methods give the same answer.

Practise

1. How many different sums of money can you make with a penny, a dime, a one-dollar coin, and a two-dollar coin?

2. How many different sums of money can be made with one $5 bill, two $10 bills, and one $50 bill?

3. How many subsets are there for a set with
   a) two distinct elements?
   b) four distinct elements?
   c) seven distinct elements?

4. In how many ways can a committee with eight members form a subcommittee with at least one person on it?

5. Determine whether the following questions involve permutations or combinations and list any formulas that would apply.
   a) How many committees of 3 students can be formed from 12 students?
   b) In how many ways can 12 runners finish first, second, and third in a race?
   c) How many outfits can you assemble from three pairs of pants, four shirts, and two pairs of shoes?
   d) How many two-letter arrangements can be formed from the word *star*?

Apply, Solve, Communicate

6. Seven managers and eight sales representatives volunteer to attend a trade show. Their company can afford to send five people. In how many ways could they be selected

a) without any restriction?

b) if there must be at least one manager and one sales representative chosen?

7. Application A cookie jar contains three chocolate-chip, two peanut-butter, one lemon, one almond, and five raisin cookies.

a) In how many ways can you reach into the jar and select some cookies?

b) In how many ways can you select some cookies, if you must include at least one chocolate-chip cookie?

8. A project team of 6 students is to be selected from a class of 30.

a) How many different teams can be selected?

b) Pierre, Gregory, and Miguel are students in this class. How many of the teams would include these 3 students?

c) How many teams would not include Pierre, Gregory, and Miguel?

9. The game of euchre uses only the 9s, 10s, jacks, queens, kings, and aces from a standard deck of cards. How many five-card hands have

a) all red cards?

b) at least two red cards?

c) at most two red cards?

10. If you are dealing from a standard deck of 52 cards,

a) how many different 4-card hands could have at least one card from each suit?

b) how many different 5-card hands could have at least one spade?

c) how many different 5-card hands could have at least two face cards (jacks, queens, or kings)?

11. The number 5880 can be factored into prime divisors as $2 \times 2 \times 2 \times 3 \times 5 \times 7 \times 7$.

a) Determine the total number of divisors of 5880.

b) How many of the divisors are even?

c) How many of the divisors are odd?

12. Application A theme park has a variety of rides. There are seven roller coasters, four water rides, and nine story rides. If Stephanie wants to try one of each type of ride, how many different combinations of rides could she choose?

13. Shuwei finds 11 shirts in his size at a clearance sale. How many different purchases could Shuwei make?

14. Communication Using the summary on page 285, draw a flow chart for solving counting problems.

15. a) How many different teams of 4 students could be chosen from the 15 students in the grade-12 Mathematics League?

b) How many of the possible teams would include the youngest student in the league?

c) How many of the possible teams would exclude the youngest student?

16. Inquiry/Problem Solving

a) Use combinations to determine how many diagonals there are in

    i) a pentagon   **ii)** a hexagon

b) Draw sketches to verify your answers in part a).

17. A school is trying to decide on new school colours. The students can choose three colours from gold, black, green, blue, red, and white, but they know that another school has already chosen black, gold, and red. How many different combinations of three colours can the students choose?

18. Application The social convenor has 12 volunteers to work at a school dance. Each dance requires 2 volunteers at the door, 4 volunteers on the floor, and 6 floaters. Joe and Jim have not volunteered before, so the social convenor does not want to assign them to work together. In how many ways can the volunteers be assigned?

19. Jeffrey is a DJ at a local radio station. For the second hour of his shift, he must choose all his music from the top 100 songs for the week. Jeffery will play exactly 12 songs during this hour.

Chapter Problem

a) How many different stacks of discs could Jeffrey pull from the station's collection if he chooses at least 10 songs that are in positions 15 to 40 on the charts?

b) Jeffrey wants to start his second hour with a hard-rock song and finish with a pop classic. How many different play lists can Jeffrey prepare if he has chosen 4 hard rock songs, 5 soul pieces, and 3 pop classics?

c) Jeffrey has 8 favourite songs currently on the top 100 list. How many different subsets of these songs could he choose to play during his shift?

✔ **ACHIEVEMENT CHECK**

Knowledge/ Understanding	Thinking/Inquiry/ Problem Solving	Communication	Application

20. There are 52 white keys on a piano. The lowest key is A. The keys are designated A, B, C, D, E, F, and G in succession, and then the sequence of letters repeats, ending with a C for the highest key.

a) If five notes are played simultaneously, in how many ways could the notes all be

   i) As?   **ii)** Gs?

   iii) the same letter? **iv)** different letters?

b) If the five keys are played in order, how would your answers in part a) change?

21. Communication

a) How many possible combinations are there for the letters in the three circles for each of the clue words in this puzzle?

b) Explain why you cannot answer part a) with a single $_nC_r$ calculation for each word.

Unscramble these four Jumbles, one letter to each square, to form four ordinary words.

ON THE AIR

DEVEL

VEENT

PAPNYS

SIFOSY

WHAT THE NERVOUS DISC JOCKEY LIVES ON.

Now arrange the circled letters to form the surprise answer, as suggested by the above cartoon.

22. Determine the number of ways of selecting four letters, without regard for order, from the word *parallelogram*.

 C

23. Inquiry/Problem Solving Suppose the artist in Example 1 of this section had two apples, two oranges, and two pears in his refrigerator. How many combinations does he have to choose from if he wants to paint a still-life with

a) two pieces of fruit?

b) three pieces of fruit?

c) four pieces of fruit?

24. How many different sums of money can be formed from one $2 bill, three $5 bills, two $10 bills, and one $20 bill?

Blaise Pascal

5.4 The Binomial Theorem

Recall that a binomial is a polynomial with just two terms, so it has the form $a + b$. Expanding $(a + b)^n$ becomes very laborious as n increases. This section introduces a method for expanding powers of binomials. This method is useful both as an algebraic tool and for probability calculations, as you will see in later chapters.

INVESTIGATE & INQUIRE: Patterns in the Binomial Expansion

1. Expand each of the following and simplify fully.

   a) $(a + b)^1$ **b)** $(a + b)^2$ **c)** $(a + b)^3$

   d) $(a + b)^4$ **e)** $(a + b)^5$

2. Study the terms in each of these expansions. Describe how the degree of each term relates to the power of the binomial.

3. Compare the terms in Pascal's triangle to the expansions in question 1. Describe any pattern you find.

4. Predict the terms in the expansion of $(a + b)^6$.

In section 4.4, you found a number of patterns in Pascal's triangle. Now that you are familiar with combinations, there is another important pattern that you can recognize. Each term in Pascal's triangle corresponds to a value of $_nC_r$.

$$1$$
$$1 \quad 1$$
$$1 \quad 2 \quad 1$$
$$1 \quad 3 \quad 3 \quad 1$$
$$1 \quad 4 \quad 6 \quad 4 \quad 1$$
$$1 \quad 5 \quad 10 \quad 10 \quad 5 \quad 1$$

$$_0C_0$$
$$_1C_0 \quad _1C_1$$
$$_2C_0 \quad _2C_1 \quad _2C_2$$
$$_3C_0 \quad _3C_1 \quad _3C_2 \quad _3C_3$$
$$_4C_0 \quad _4C_1 \quad _4C_2 \quad _4C_3 \quad _4C_4$$
$$_5C_0 \quad _5C_1 \quad _5C_2 \quad _5C_3 \quad _5C_4 \quad _5C_5$$

Comparing the two triangles shown on page 289, you can see that $t_{n,r} = {}_nC_r$. Recall that Pascal's method for creating his triangle uses the relationship

$$t_{n,r} = t_{n-1, \, r-1} + t_{n-1, \, r}.$$

So, this relationship must apply to combinations as well.

Pascal's Formula

$${}_nC_r = {}_{n-1}C_{r-1} + {}_{n-1}C_r$$

Proof:

$$
\begin{aligned}
{}_{n-1}C_{r-1} + {}_{n-1}C_r &= \frac{(n-1)!}{(r-1)!(n-r)!} + \frac{(n-1)!}{r!(n-r-1)!} \\
&= \frac{r(n-1)!}{r(r-1)!(n-r)!} + \frac{(n-1)!(n-r)}{r!(n-r)(n-r-1)!} \\
&= \frac{r(n-1)!}{r!(n-r)!} + \frac{(n-1)!(n-r)}{r!(n-r)!} \\
&= \frac{(n-1)!}{r!(n-r)!}[r + (n-r)] \\
&= \frac{(n-1)! \times n}{r!(n-r)!} \\
&= \frac{n!}{r!(n-r)!} \\
&= {}_nC_r
\end{aligned}
$$

This proof shows that the values of ${}_nC_r$ do indeed follow the pattern that creates Pascal's triangle. It follows that ${}_nC_r = t_{n,r}$ for all the terms in Pascal's triangle.

● **Example 1 Applying Pascal's Formula to Combinations**

Rewrite each of the following using Pascal's formula.
a) ${}_{12}C_8$ **b)** ${}_{19}C_5 + {}_{19}C_6$

Solution

a) ${}_{12}C_8 = {}_{11}C_7 + {}_{11}C_8$ **b)** ${}_{19}C_5 + {}_{19}C_6 = {}_{20}C_6$

As you might expect from the investigation at the beginning of this section, the coefficients of each term in the expansion of $(a + b)^n$ correspond to the terms in row n of Pascal's triangle. Thus, you can write these coefficients in combinatorial form.

The Binomial Theorem

$$(a + b)^n = {}_nC_0a^n + {}_nC_1a^{n-1}b + {}_nC_2a^{n-2}b^2 + \ldots + {}_nC_ra^{n-r}b^r + \ldots + {}_nC_nb^n$$

$$\text{or } (a + b)^n = \sum_{r=0}^{n} {}_nC_ra^{n-r}b^r$$

Example 2 Applying the Binomial Theorem

Use combinations to expand $(a + b)^6$.

Solution

$$(a + b)^6 = \sum_{r=0}^{6} {}_6C_ra^{6-r}b^r$$

$$= {}_6C_0a^6 + {}_6C_1a^5b + {}_6C_2a^4b^2 + {}_6C_3a^3b^3 + {}_6C_4a^2b^4 + {}_6C_5ab^5 + {}_6C_6b^6$$

$$= a^6 + 6a^5b + 15a^4b^2 + 20a^3b^3 + 15a^2b^4 + 6ab^5 + b^6$$

Example 3 Binomial Expansions Using Pascal's Triangle

Use Pascal's triangle to expand
a) $(2x - 1)^4$
b) $(3x - 2y)^5$

Solution

a) Substitute $2x$ for a and -1 for b. Since the exponent is 4, use the terms in row 4 of Pascal's triangle as the coefficients: 1, 4, 6, 4, and 1. Thus,

$$(2x - 1)^4 = 1(2x)^4 + 4(2x)^3(-1) + 6(2x)^2(-1)^2 + 4(2x)(-1)^3 + 1(-1)^4$$

$$= 16x^4 + 4(8x^3)(-1) + 6(4x^2)(1) + 4(2x)(-1) + 1$$

$$= 16x^4 - 32x^3 + 24x^2 - 8x + 1$$

b) Substitute $3x$ for a and $-2y$ for b, and use the terms from row 5 as coefficients.

$$(3x - 2y)^5 = 1(3x)^5 + 5(3x)^4(-2y) + 10(3x)^3(-2y)^2 + 10(3x)^2(-2y)^3 + 5(3x)(-2y)^4 + 1(-2y)^5$$

$$= 243x^5 - 810x^4y + 1080x^3y^2 - 720x^2y^3 + 240xy^4 - 32y^5$$

Example 4 Expanding Binomials Containing Negative Exponents

Use the binomial theorem to expand and simplify $\left(x + \dfrac{2}{x^2}\right)^4$.

Solution

Substitute x for a and $\dfrac{2}{x^2}$ for b.

$$\left(x + \frac{2}{x^2}\right)^4 = \sum_{r=0}^{4} {}_4C_r x^{4-r}\left(\frac{2}{x^2}\right)^r$$

$$= {}_4C_0 x^4 + {}_4C_1 x^3\left(\frac{2}{x^2}\right) + {}_4C_2 x^2\left(\frac{2}{x^2}\right)^2 + {}_4C_3 x\left(\frac{2}{x^2}\right)^3 + {}_4C_4\left(\frac{2}{x^2}\right)^4$$

$$= 1x^4 + 4x^3\left(\frac{2}{x^2}\right) + 6x^2\left(\frac{4}{x^4}\right) + 4x\left(\frac{8}{x^6}\right) + 1\left(\frac{16}{x^8}\right)$$

$$= x^4 + 8x + 24x^{-2} + 32x^{-5} + 16x^{-8}$$

Example 5 Patterns With Combinations

Using the patterns in Pascal's triangle from the investigation and Example 4 in section 4.4, write each of the following in combinatorial form.

a) the sum of the terms in row 5 and row 6

b) the sum of the terms in row n

c) the first 5 triangular numbers

d) the nth triangular number

Solution

a) *Row 5:*

$$1 + 5 + 10 + 10 + 5 + 1$$
$$= {}_5C_0 + {}_5C_1 + {}_5C_2 + {}_5C_3 + {}_5C_4 + {}_5C_5$$
$$= 32$$
$$= 2^5$$

Row 6:

$$1 + 6 + 15 + 20 + 15 + 6 + 1$$
$$= {}_6C_0 + {}_6C_1 + {}_6C_2 + {}_6C_3 + {}_6C_4 + {}_6C_5 + {}_6C_6$$
$$= 64$$
$$= 2^6$$

b) From part a) it appears that ${}_nC_0 + {}_nC_1 + \ldots + {}_nC_n = 2^n$.

Using the binomial theorem,
$$2^n = (1 + 1)^n$$
$$= {}_nC_0 \times 1^n + {}_nC_1 \times 1^{n-1} \times 1 + \ldots + {}_nC_n \times 1^n$$
$$= {}_nC_0 + {}_nC_1 + \ldots + {}_nC_n$$

c)

n	Triangular Numbers	Combinatorial Form
1	1	${}_2C_2$
2	3	${}_3C_2$
3	6	${}_4C_2$
4	10	${}_5C_2$
5	15	${}_6C_2$

d) The nth triangular number is ${}_{n+1}C_2$.

Example 6 Factoring Using the Binomial Theorem

Rewrite $1 + 10x^2 + 40x^4 + 80x^6 + 80x^8 + 32x^{10}$ in the form $(a + b)^n$.

Solution

There are six terms, so the exponent must be 5.
The first term of a binomial expansion is a^n, so a must be 1.
The final term is $32x^{10} = (2x^2)^5$, so $b = 2x^2$.
Therefore, $1 + 10x^2 + 40x^4 + 80x^6 + 80x^8 + 32x^{10} = (1 + 2x^2)^5$

Key Concepts

- The coefficients of the terms in the expansion of $(a + b)^n$ correspond to the terms in row n of Pascal's triangle.

- The binomial $(a + b)^n$ can also be expanded using combinatorial symbols:
$$(a + b)^n = {}_nC_0a^n + {}_nC_1a^{n-1}b + {}_nC_2a^{n-2}b^2 + \ldots + {}_nC_nb^n \text{ or } \sum_{r=0}^{n} {}_nC_ra^{n-r}b^r$$

- The degree of each term in the binomial expansion of $(a + b)^n$ is n.

- Patterns in Pascal's triangle can be summarized using combinatorial symbols.

Communicate Your Understanding

1. Describe how Pascal's triangle and the binomial theorem are related.

2. a) Describe how you would use Pascal's triangle to expand $(2x + 5y)^9$.
   b) Describe how you would use the binomial theorem to expand $(2x + 5y)^9$.

3. Relate the sum of the terms in the nth row of Pascal's triangle to the total number of subsets of a set of n elements. Explain the relationship.

Practise

A

1. Rewrite each of the following using Pascal's formula.

   a) ${}_{17}C_{11}$
   b) ${}_{43}C_{36}$
   c) ${}_{n+1}C_{r+1}$
   d) ${}_{32}C_4 + {}_{32}C_5$
   e) ${}_{15}C_{10} + {}_{15}C_9$
   f) ${}_nC_r + {}_nC_{r+1}$
   g) ${}_{18}C_9 - {}_{17}C_9$
   h) ${}_{24}C_8 - {}_{23}C_7$
   i) ${}_nC_r - {}_{n-1}C_{r-1}$

2. Determine the value of k in each of these terms from the binomial expansion of $(a + b)^{10}$.

   a) $210a^6b^k$
   b) $45a^kb^8$
   c) $252a^kb^k$

3. How many terms would be in the expansion of the following binomials?

   a) $(x + y)^{12}$
   b) $(2x - 3y)^5$
   c) $(5x - 2)^{20}$

4. For the following terms from the expansion of $(a + b)^{11}$, state the coefficient in both ${}_nC_r$ and numeric form.

   a) a^2b^9
   b) a^{11}
   c) a^6b^5

Apply, Solve, Communicate

5. Using the binomial theorem and patterns in Pascal's triangle, simplify each of the following.

a) $_9C_0 + {}_9C_1 + \ldots + {}_9C_9$

b) $_{12}C_0 - {}_{12}C_1 + {}_{12}C_2 - \ldots - {}_{12}C_{11} + {}_{12}C_{12}$

c) $\displaystyle\sum_{r=0}^{15} {}_{15}C_r$ d) $\displaystyle\sum_{r=0}^{n} {}_nC_r$

6. If $\displaystyle\sum_{r=0}^{n} {}_nC_r = 16\ 384$, determine the value of n.

7. a) Write formulas in combinatorial form for the following. (Refer to section 4.4, if necessary.)

   i) the sum of the squares of the terms in the nth row of Pascal's triangle

   ii) the result of alternately adding and subtracting the squares of the terms in the nth row of Pascal's triangle

   iii) the number of diagonals in an n-sided polygon

b) Use your formulas from part a) to determine

   i) the sum of the squares of the terms in row 15 of Pascal's triangle

   ii) the result of alternately adding and subtracting the squares of the terms in row 12 of Pascal's triangle

   iii) the number of diagonals in a 14-sided polygon

8. How many terms would be in the expansion of $(x^2 + x)^8$?

9. Use the binomial theorem to expand and simplify the following.

a) $(x + y)^7$ b) $(2x + 3y)^6$

c) $(2x - 5y)^5$ d) $(x^2 + 5)^4$

e) $(3a^2 + 4c)^7$ f) $5(2p - 6c^2)^5$

10. Communication

a) Find and simplify the first five terms of the expansion of $(3x + y)^{10}$.

b) Find and simplify the first five terms of the expansion of $(3x - y)^{10}$.

c) Describe any similarities and differences between the terms in parts a) and b).

11. Use the binomial theorem to expand and simplify the following.

a) $\left(x^2 - \dfrac{1}{x}\right)^5$ b) $\left(2y + \dfrac{3}{y^2}\right)^4$

c) $(\sqrt{x} + 2x^2)^6$ d) $\left(k + \dfrac{k}{m^2}\right)^5$

e) $\left(\sqrt{y} - \dfrac{2}{\sqrt{y}}\right)^7$ f) $2\left(3m^2 - \dfrac{2}{\sqrt{m}}\right)^4$

12. Application Rewrite the following expansions in the form $(a + b)^n$, where n is a positive integer.

a) $x^6 + 6x^5y + 15x^4y^2 + 20\ x^3y^3 + 15x^2y^4 + 6xy^5 + y^6$

b) $y^{12} + 8y^9 + 24y^6 + 32y^3 + 16$

c) $243a^5 - 405a^4b + 270a^3b^2 - 90a^2b^3 + 15ab^4 - b^5$

13. Communication Use the binomial theorem to simplify each of the following. Explain your results.

a) $\left(\dfrac{1}{2}\right)^5 + 5\left(\dfrac{1}{2}\right)^5 + 10\left(\dfrac{1}{2}\right)^5 + 10\left(\dfrac{1}{2}\right)^5$
$+ 5\left(\dfrac{1}{2}\right)^5 + \left(\dfrac{1}{2}\right)^5$

b) $(0.7)^7 + 7(0.7)^6(0.3) + 21(0.7)^5(0.3)^2 + \ldots + (0.3)^7$

c) $7^9 - 9 \times 7^8 + 36 \times 7^7 - \ldots - 7^0$

14. a) Expand $\left(x + \dfrac{2}{x}\right)^4$ and compare it with the expansion of $\dfrac{1}{x^4}(x^2 + 2)^4$.

b) Explain your results.

15. Use your knowledge of algebra and the binomial theorem to expand and simplify each of the following.

  a) $(25x^2 + 30xy + 9y^2)^3$

  b) $(3x - 2y)^5(3x + 2y)^5$

16. Application

  a) Calculate an approximation for $(1.2)^9$ by expanding $(1 + 0.2)^9$.

  b) How many terms do you have to evaluate to get an approximation accurate to two decimal places?

17. In a trivia contest, Adam has drawn a topic he knows nothing about, so he makes random guesses for the ten true/false questions. Use the binomial theorem to help find

  a) the number of ways that Adam can answer the test using exactly four *true*s

  b) the number of ways that Adam can answer the test using at least one *true*

ACHIEVEMENT CHECK

Knowledge/ Understanding	Thinking/Inquiry/ Problem Solving	Communication	Application

18. a) Expand $(h + t)^5$.

  b) Explain how this expansion can be used to determine the number of ways of getting exactly h heads when five coins are tossed.

  c) How would your answer in part b) change if six coins are being tossed? How would it change for n coins? Explain.

19. Find the first three terms, ranked by degree of the terms, in each expansion.

  a) $(x + 3)(2x + 5)^4$

  b) $(2x + 1)^2(4x - 3)^5$

  c) $(x^2 - 5)^9(x^3 + 2)^6$

20. Inquiry/Problem Solving

  a) Use the binomial theorem to expand $(x + y + z)^2$ by first rewriting it as $[x + (y + z)]^2$.

  b) Repeat part a) with $(x + y + z)^3$.

  c) Using parts a) and b), predict the expansion of $(x + y + z)^4$. Verify your prediction by using the binomial theorem to expand $(x + y + z)^4$.

  d) Write a formula for $(x + y + z)^n$.

  e) Use your formula to expand and simplify $(x + y + z)^5$.

21. a) In the expansion of $(x + y)^5$, replace x and y with B and G, respectively. Expand and simplify.

  b) Assume that a couple has an equal chance of having a boy or a girl. How would the expansion in part a) help find the number of ways of having k girls in a family with five children?

  c) In how many ways could a family with five children have exactly three girls?

  d) In how many ways could they have exactly four boys?

22. A simple code consists of a string of five symbols that represent different letters of the alphabet. Each symbol is either a dot (•) or a dash (–).

  a) How many different letters are possible using this code?

  b) How many coded letters will contain exactly two dots?

  c) How many different coded letters will contain at least one dash?

Review of Key Concepts

5.1 Organized Counting With Venn Diagrams

Refer to the Key Concepts on page 270.

1. Which regions in the diagram below correspond to
   a) the union of sets A and B?
   b) the intersection of sets B and C?
   c) $A \cap C$?
   d) either B or S?

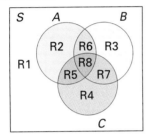

2. a) Write the equation for the number of elements contained in either of two sets.
   b) Explain why the principle of inclusion and exclusion subtracts the last term in this equation.
   c) Give a simple example to illustrate your explanation.

3. A survey of households in a major city found that
   - 96% had colour televisions
   - 65% had computers
   - 51% had dishwashers
   - 63% had colour televisions and computers
   - 49% had colour televisions and dishwashers
   - 31% had computers and dishwashers
   - 30% had all three
   a) List the categories of households not included in these survey results.

   b) Use a Venn diagram to find the proportion of households in each of these categories.

5.2 Combinations

Refer to the Key Concepts on page 278.

4. Evaluate the following and indicate any calculations that could be done manually.
   a) $_{41}C_8$
   b) $_{33}C_{15}$
   c) $_{25}C_{17}$
   d) $_{50}C_{10}$
   e) $_{10}C_8$
   f) $_{15}C_{13}$
   g) $_5C_4$
   h) $_{25}C_{24}$
   i) $_{15}C_{11}$
   j) $_{25}C_{20}$
   k) $_{16}C_8$
   l) $_{30}C_{26}$

5. A track and field club has 12 members who are runners and 10 members who specialize in field events. The club has been invited to send a team of 3 runners and 2 field athletes to an out-of-town meet. How many different teams could the club send?

6. A bridge hand consists of 13 cards. How many bridge hands include 5 cards of one suit, 6 cards of a second, and 2 cards of a third?

7. Explain why combination locks should really be called permutation locks.

5.3 Problem Solving With Combinations

Refer to the Key Concepts on page 286.

8. At Subs Galore, you have a choice of lettuce, onions, tomatoes, green peppers, mushrooms, cheese, olives, cucumbers, and hot peppers on your submarine sandwich. How many ways can you "dress" your sandwich?

9. Ballots for municipal elections usually list candidates for several different positions. If a resident can vote for a mayor, two councillors, a school trustee, and a hydro commissioner, how many combinations of positions could the resident choose to mark on the ballot?

10. There are 12 questions on an examination, and each student must answer 8 questions including at least 4 of the first 5 questions. How many different combinations of questions could a student choose to answer?

11. Naomi invites eight friends to a party on short notice, so they may not all be able to come. How many combinations of guests could attend the party?

12. In how many ways could 15 different books be divided equally among 3 people?

13. The camera club has five members, and the mathematics club has eight. There is only one member common to both clubs. In how many ways could a committee of four people be formed with at least one member from each club?

5.4 The Binomial Theorem
Refer to the Key Concepts on page 293.

14. Without expanding $(x + y)^5$, determine
   a) the number of terms in the expansion
   b) the value of k in the term $10x^k y^2$

15. Use Pascal's triangle to expand
   a) $(x + y)^8$
   b) $(4x - y)^6$
   c) $(2x + 5y)^4$
   d) $(7x - 3)^5$

16. Use the binomial theorem to expand
   a) $(x + y)^6$
   b) $(6x - 5y)^4$
   c) $(5x + 2y)^5$
   d) $(3x - 2)^6$

17. Write the first three terms of the expansion of
   a) $(2x + 5y)^7$
   b) $(4x - y)^6$

18. Describe the steps in the binomial expansion of $(2x - 3y)^6$.

19. Find the last term in the binomial expansion of $\left(\dfrac{1}{x^2} + 2x \right)^5$.

20. Find the middle term in the binomial expansion of $\left(\sqrt{x} + \dfrac{5}{\sqrt{x}} \right)^8$.

21. In the expansion of $(a + x)^6$, the first three terms are $1 + 3 + 3.75$. Find the values of a and x.

22. Use the binomial theorem to expand and simplify $(y^2 - 2)^6 (y^2 + 2)^6$.

23. Write $1024x^{10} - 3840x^8 + 5760x^6 - 4320x^4 + 1620x^2 - 243$ in the form $(a + b)^n$. Explain your steps.

Chapter Test

Category	Knowledge/ Understanding	Thinking/Inquiry/ Problem Solving	Communication	Application
Questions	All	12	6, 12	5, 6, 7, 8, 9

1. Evaluate each of the following. List any calculations that require a calculator.

   a) $_{25}C_{25}$

   b) $_{52}C_1$

   c) $_{12}C_3$

   d) $_{40}C_{15}$

2. Rewrite each of the following as a single combination.

   a) $_{10}C_7 + _{10}C_8$

   b) $_{23}C_{15} - _{22}C_{14}$

3. Use Pascal's triangle to expand

   a) $(3x - 4)^4$

   b) $(2x + 3y)^7$

4. Use the binomial theorem to expand

   a) $(8x - 3)^5$

   b) $(2x - 5y)^6$

5. A student fundraising committee has 14 members, including 7 from grade 12. In how many ways can a 4-member subcommittee for commencement awards be formed if

   a) there are no restrictions?

   b) the subcommittee must be all grade-12 students?

   c) the subcommittee must have 2 students from grade 12 and 2 from other grades?

   d) the subcommittee must have no more than 3 grade-12 students?

6. A track club has 20 members.

   a) In how many ways can the club choose 3 members to help officiate at a meet?

   b) In how many ways can the club choose a starter, a marshal, and a timer?

   c) Should your answers to parts a) and b) be the same? Explain why or why not.

7. Statistics on the grade-12 courses taken by students graduating from a secondary school showed that

   • 85 of the graduates had taken a science course
   • 75 of the graduates had taken a second language
   • 41 of the graduates had taken mathematics
   • 43 studied both science and a second language
   • 32 studied both science and mathematics
   • 27 had studied both a second language and mathematics
   • 19 had studied all three subjects

   a) Use a Venn diagram to determine the minimum number of students who could be in this graduating class.

   b) How many students studied mathematics, but neither science nor a second language?

8. A field-hockey team played seven games and won four of them. There were no ties.

a) How many arrangements of the four wins and three losses are possible?

b) In how many of these arrangements would the team have at least two wins in a row?

9. A restaurant offers an all-you-can-eat Chinese buffet with the following items:

- egg roll, wonton soup
- chicken wings, chicken balls, beef, pork
- steamed rice, fried rice, chow mein
- chop suey, mixed vegetables, salad
- fruit salad, custard tart, almond cookie

a) How many different combinations of items could you have?

b) The restaurant also has a lunch special with your choice of one item from each group. How many choices do you have with this special?

10. In the expansion of $(1 + x)^n$, the first three terms are $1 - 0.9 + 0.36$. Find the values of x and n.

11. Use the binomial theorem to expand and simplify $(4x^2 - 12x + 9)^3$.

12. A small transit bus has 8 window seats and 12 aisle seats. Ten passengers board the bus and select seats at random. How many seating arrangements have all the window seats occupied if which passenger is in a seat

a) does not matter?   **b)** matters?

ACHIEVEMENT CHECK

Knowledge/Understanding	Thinking/Inquiry/Problem Solving	Communication	Application

13. The students' council is having pizza at their next meeting. There are 20 council members, 6 of whom are vegetarian. A committee of 3 will order six pizzas from a pizza shop that has a special price for large pizzas with up to three toppings. The shop offers ten different toppings.

a) How many different pizza committees can the council choose if there must be at least one vegetarian and one non-vegetarian on the committee?

b) In how many ways could the committee choose *exactly* three toppings for a pizza?

c) In how many ways could the committee choose *up to* three toppings for a pizza?

d) The committee wants as much variety as possible in the toppings. They decide to order each topping exactly once and to have at least one topping on each pizza. Describe the different cases possible when distributing the toppings in this way.

e) For one of these cases, determine the number of ways of choosing and distributing the ten toppings.

CHAPTER 6

Introduction to Probability

Specific Expectations	Section
Use Venn diagrams as a tool for organizing information in counting problems.	6.5
Solve problems, using techniques for counting permutations where some objects may be alike.	6.3
Solve problems, using techniques for counting combinations.	6.3
Solve probability problems involving combinations of simple events, using counting techniques.	6.3, 6.4, 6.5, 6.6
Interpret probability statements, including statements about odds, from a variety of sources.	6.1, 6.2, 6.3, 6.4, 6.5, 6.6
Design and carry out simulations to estimate probabilities in situations for which the calculation of the theoretical probabilities is difficult or impossible.	6.3
Assess the validity of some simulation results by comparing them with the theoretical probabilities, using the probability concepts developed in the course.	6.3
Represent complex tasks or issues, using diagrams.	6.1, 6.5
Represent numerical data, using matrices, and demonstrate an understanding of terminology and notation related to matrices.	6.6
Demonstrate proficiency in matrix operations, including addition, scalar multiplication, matrix multiplication, the calculation of row sums, and the calculation of column sums, as necessary to solve problems, with and without the aid of technology.	6.6
Solve problems drawn from a variety of applications, using matrix methods.	6.6

Chapter Problem

Genetic Probabilities

Biologists are studying a deer population in a provincial conservation area. The biologists know that many of the bucks (male deer) in the area have an unusual "cross-hatched" antler structure, which seems to be genetic in origin. Of 48 randomly tagged deer, 26 were does (females), 22 were bucks, and 7 of the bucks had cross-hatched antlers.

Several of the does have small bald patches on their hides. This condition also seems to have some genetic element. Careful long-term study has found that female offspring of does with bald patches have a 65% likelihood of developing the condition

themselves, while offspring of healthy does have only a 20% likelihood of developing it. Currently, 30% of the does have bald patches.

1. Out of ten deer randomly captured, how many would you expect to have either cross-hatched antlers or bald patches?

2. Do you think that the proportion of does with the bald patches will increase, decrease, or remain relatively stable?

In this chapter, you will learn methods that the biologists could use to calculate probabilities from their samples and to make predictions about the deer population.

Review of Prerequisite Skills

If you need help with any of the skills listed in purple below, refer to Appendix A.

1. **Fractions, percents, decimals** Express each decimal as a percent.
   a) 0.35
   b) 0.04
   c) 0.95
   d) 0.008
   e) 0.085
   f) 0.375

2. **Fractions, percents, decimals** Express each percent as a decimal.
   a) 15%
   b) 3%
   c) 85%
   d) 6.5%
   e) 26.5%
   f) 75.2%

3. **Fractions, percents, decimals** Express each percent as a fraction in simplest form.
   a) 12%
   b) 35%
   c) 67%
   d) 4%
   e) 0.5%
   f) 98%

4. **Fractions, percents, decimals** Express each fraction as a percent. Round answers to the nearest tenth, if necessary.
   a) $\frac{1}{4}$
   b) $\frac{13}{15}$
   c) $\frac{11}{14}$
   d) $\frac{7}{10}$
   e) $\frac{4}{9}$
   f) $\frac{13}{20}$

5. **Tree diagrams** A coin is flipped three times. Draw a tree diagram to illustrate all possible outcomes.

6. **Tree diagrams** In the game of backgammon, you roll two dice to determine how you can move your counters. Suppose you roll first one die and then the other and you need to roll 9 or more to move a counter to safety. Use a tree diagram to list the different rolls in which
   a) you make at least 9
   b) you fail to move your counter to safety

7. **Fundamental counting principle (section 4.1)** Benoit is going skating on a cold wintry day. He has a toque, a watch cap, a beret, a heavy scarf, a light scarf, leather gloves, and wool gloves. In how many different ways can Benoit dress for the cold weather?

8. **Additive counting principle (section 4.1)** How many 13-card bridge hands include either seven hearts or eight diamonds?

9. **Venn diagrams (section 5.1)**
   a) List the elements for each of the following sets for whole numbers from 1 to 10 inclusive.
      i) E, the set of even numbers
      ii) O, the set of odd numbers
      iii) C, the set of composite numbers
      iv) P, the set of perfect squares
   b) Draw a diagram to illustrate how the following sets are related.
      i) E and O
      ii) E and C
      iii) O and P
      iv) E, C, and P

10. Principle of inclusion and exclusion (section 5.1)

   a) Explain the principle of inclusion and exclusion.

   b) A gift store stocks baseball hats in red or green colours. Of the 35 hats on display on a given day, 20 are green. As well, 18 of the hats have a grasshopper logo on the brim. Suppose 11 of the red hats have logos. How many hats are red, or have logos, or both?

11. Factorials (section 4.2) Evaluate.

   a) $6!$

   b) $0!$

   c) $\dfrac{16!}{14!}$

   d) $\dfrac{12!}{9! \, 3!}$

   e) $\dfrac{100!}{98!}$

   f) $\dfrac{16!}{10! \times 8!}$

12. Permutations (section 4.2) Evaluate.

   a) $_5P_3$

   b) $_7P_1$

   c) $P(6, 2)$

   d) $_9P_9$

   e) $_{100}P_1$

   f) $P(100, 2)$

13. Permutations (section 4.2) A baseball team has 13 members. If a batting line-up consists of 9 players, how many different batting line-ups are possible?

14. Permutations (section 4.2) What is the maximum number of three-digit area codes possible if the area codes cannot start with either 1 or 0?

15. Combinations (section 5.2) Evaluate these expressions.

   a) $_6C_3$

   b) $C(4, 3)$

   c) $_8C_8$

   d) $_{11}C_0$

   e) $\binom{6}{4} \times \binom{7}{5}$

   f) $\binom{100}{1}$

   g) $_{20}C_2$

   h) $_{20}C_{18}$

16. Combinations (section 5.2) A pizza shop has nine toppings available. How many different three-topping pizzas are possible if each topping is selected no more than once?

17. Combinations (section 5.3) A construction crew has 12 carpenters and 5 drywallers. How many different safety committees could they form if the members of this committee are

   a) any 5 of the crew?

   b) 3 carpenters and 2 drywallers?

18. Matrices (section 1.6) Identify any square matrices among the following. Also identify any column or row matrices.

   a) $\begin{bmatrix} 3 & 4 \\ 0 & 1 \end{bmatrix}$

   b) $[0.4 \ 0.3 \ 0.2]$

   c) $\begin{bmatrix} 1 & 0 \\ 0.5 & 0.5 \\ 0.8 & 0.6 \end{bmatrix}$

   d) $\begin{bmatrix} -2 & 3 & 9 \\ 0 & 11 & -4 \\ 3 & 6 & -1 \end{bmatrix}$

   e) $\begin{bmatrix} 49 & 63 \\ 25 & 14 \\ 72 & 9 \end{bmatrix}$

   f) $\begin{bmatrix} 8 \\ 16 \\ 32 \end{bmatrix}$

19. Matrices (section 1.7) Given $A = [0.3 \ 0.7]$ and $B = \begin{bmatrix} 0.4 & 0.6 \\ 0.55 & 0.45 \end{bmatrix}$, perform the following matrix operations, if possible. If the operation is not possible, explain why.

   a) $A \times B$

   b) $B \times A$

   c) B^2

   d) B^3

   e) A^2

   f) $A \times A^t$

Basic Probability Concepts

How likely is rain tomorrow? What are the chances that you will pass your driving test on the first attempt? What are the odds that the flight will be on time when you go to meet someone at the airport?

Probability is the branch of mathematics that attempts to predict answers to questions like these. As the word *probability* suggests, you can often predict only what *might* happen. However, you may be able to calculate how likely it is. For example, if the weather report forecasts a 90% chance of rain, there is still that slight possibility that sunny skies will prevail. While there are no sure answers, in this case it *probably* will rain.

Work with a partner. Have each partner take three identical slips of paper, number them 1, 2, and 3, and place them in a hat, bag, or other container. For each trial, both partners will randomly select one of their three slips of paper. Replace the slips after each trial. Score points as follows:

- If the product of the two numbers shown is less than the sum, Player A gets a point.
- If the product is greater than the sum, Player B gets a point.
- If the product and sum are equal, neither player gets a point.

1. Predict who has the advantage in this game. Explain why you think so.

2. Decide who will be Player A by flipping a coin or using the random number generator on a graphing calculator. Organize your results in a table like the one below.

Trial	1	2	3	4	5	6	7	8	9	10
Number drawn by A										
Number drawn by B										
Product										
Sum										
Point awarded to:										

3. **a)** Record the results for 10 trials. Total the points and determine the winner. Do the results confirm your prediction? Have you changed your opinion on who has the advantage? Explain.

   b) To estimate a probability for each player getting a point, divide the number of points each player earned by the total number of trials.

4. **a)** Perform 10 additional trials and record point totals for each player over all 20 trials. Estimate the probabilities for each player, as before.

   b) Are the results for 20 trials consistent with the results for 10 trials? Explain.

   c) Are your results consistent with those of your classmates? Comment on your findings.

5. Based on your results for 20 trials, predict how many points each player will have after 50 trials.

6. Describe how you could alter the game so that the other player has the advantage.

The investigation you have just completed is an example of a **probability experiment**. In probability, an experiment is a well-defined process consisting of a number of **trials** in which clearly distinguishable **outcomes**, or possible results, are observed.

The **sample space, S,** of an experiment is the set of all possible outcomes. For the sum/product game in the investigation, the outcomes are all the possible pairings of slips drawn by the two players. For example, if Player A draws 1 and Player B draws 2, you can label this outcome (1, 2). In this particular game, the result is the same for the outcomes (1, 2) and (2, 1), but with different rules it might be important who draws which number, so it makes sense to view the two outcomes as different.

Outcomes are often equally likely. In the sum/product game, each possible pairing of numbers is as likely as any other. Outcomes are often grouped into **events**. An example of an event is drawing slips for which the product is greater than the sum, and there are several outcomes in which this event happens. Different events often have different chances of occurring. Events are usually labelled with capital letters.

Example 1 Outcomes and Events

Let event A be a point awarded to Player A in the sum/product game.
List the outcomes that make up event A.

Solution

Player A earns a point if the sum of the two numbers is greater than the product. This event is sometimes written as event $A = \{\text{sum} > \text{product}\}$.
A useful technique in probability is to tabulate the possible outcomes.

Sums						Products				
		Player A						**Player A**		
		1	2	3				1	2	3
Player B	1	2	3	4		**Player B**	1	1	2	3
	2	3	4	5			2	2	4	6
	3	4	5	6			3	3	6	9

Use the tables shown to list the outcomes where the sum is greater than the product:
(1, 1), (1, 2), (1, 3), (2, 1), (3, 1)
These outcomes make up event A. Using this list, you can also write event A as
event A = {(1, 1), (1, 2), (1, 3), (2, 1), (3, 1)}

The **probability** of event A, **$P(A)$**, is a quantified measure of the likelihood that the event will occur. *The probability of an event is always a value between 0 and 1.* A probability of 0 indicates that the event is impossible, and 1 signifies that the event is a certainty. Most events in probability studies fall somewhere between these extreme values. Probabilities less than 0 or greater than 1 have no meaning. Probability can be expressed as fractions, decimals, or percents. Probabilities expressed as percents are always between 0% and 100%. For example, a 70% chance of rain tomorrow means the same as a probability of 0.7, or $\frac{7}{10}$, that it will rain.

The three basic types of probability are

• empirical probability, based on direct observation or experiment
• theoretical probability, based on mathematical analysis
• subjective probability, based on informed guesswork

The **empirical probability** of a particular event (also called **experimental** or **relative frequency probability**) is determined by dividing the number of times that the event actually occurs in an experiment by the number of trials. In the sum/product investigation, you were calculating empirical probabilities. For example, if you had found that in the first ten trials, the product was greater than the sum four times, then the empirical probability of this event would be

$$P(A) = \frac{4}{10}$$
$$= \frac{2}{5} \text{ or } 0.4$$

The **theoretical probability** of a particular event is deduced from analysis of the possible outcomes. Theoretical probability is also called **classical** or *a priori* probability. *A priori* is Latin for "from the preceding," meaning based on analysis rather than experiment.

For example, if all possible outcomes are equally likely, then

$$P(A) = \frac{n(A)}{n(S)}$$

Project Prep

You will need to determine theoretical probabilities to design and analyse your game in the probability project.

where $n(A)$ is the number of outcomes in which event A can occur, and $n(S)$ is the total number of possible outcomes. You used tables to list the outcomes for A in Example 1, and this technique allows you to find the theoretical probability $P(A)$ by counting $n(A) = 5$ and $n(S) = 9$. Another way to determine the values of $n(A)$ and $n(S)$ is by organizing the information in a tree diagram.

Example 2 Using a Tree Diagram to Calculate Probability

Determine the theoretical probabilities for each key event in the sum/product game.

Solution

The tree diagram shows the nine possible outcomes, each equally likely, for the sum/product game.

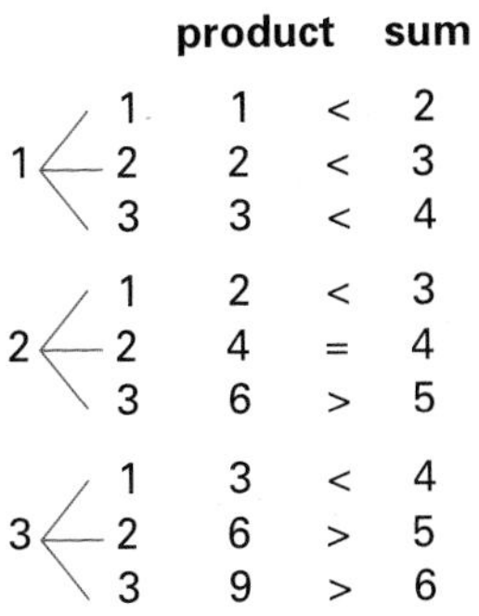

Let event A be a point for Player A, event B a point for Player B, and event C a tie between sum and product. From the tree diagram, five of the nine possible outcomes have the sum greater than the product. Therefore, the theoretical probability of this event is

$$P(A) = \frac{n(A)}{n(S)}$$
$$= \frac{5}{9}$$

		product		sum
1	1	1	<	2
	2	2	<	3
	3	3	<	4
2	1	2	<	3
	2	4	=	4
	3	6	>	5
3	1	3	<	4
	2	6	>	5
	3	9	>	6

Similarly,

$$P(B) = \frac{n(B)}{n(S)} \quad \text{and} \quad P(C) = \frac{n(C)}{n(S)}$$
$$= \frac{3}{9} \qquad\qquad = \frac{1}{9}$$

In Example 2, you know that one, and only one, of the three events will occur. The sum of the probabilities of all possible events always equals 1.

$$P(A) + P(B) + P(C) = \frac{5}{9} + \frac{3}{9} + \frac{1}{9}$$
$$= 1$$

Here, the numerator in each fraction represents the number of ways that each event can occur. The total of these numerators is the total number of possible outcomes, which is equal to the denominator.

Empirical probabilities may differ sharply from theoretical probabilities when only a few trials are made. Such **statistical fluctuation** can result in an event occurring more frequently or less frequently than theoretical probability suggests. Over a large number of trials, however, statistical fluctuations tend to cancel each other out, and empirical probabilities usually approach theoretical values. Statistical fluctuations often appear in sports, for example, where a team can enjoy a temporary winning streak that is not sustainable over an entire season.

In most problems, you will be determining theoretical probability. Therefore, from now on you may take the term *probability* to mean *theoretical probability* unless stated otherwise.

Example 3 Dice Probabilities

Many board games involve a roll of two six-sided dice to see how far you may move your pieces or counters. What is the probability of rolling a total of 7?

Solution

The table shows the totals for all possible rolls of two dice.

<table>
<tr><td colspan="2"></td><td colspan="6">First Die</td></tr>
<tr><td colspan="2"></td><td>1</td><td>2</td><td>3</td><td>4</td><td>5</td><td>6</td></tr>
<tr><td rowspan="6">Second Die</td><td>1</td><td>2</td><td>3</td><td>4</td><td>5</td><td>6</td><td>7</td></tr>
<tr><td>2</td><td>3</td><td>4</td><td>5</td><td>6</td><td>7</td><td>8</td></tr>
<tr><td>3</td><td>4</td><td>5</td><td>6</td><td>7</td><td>8</td><td>9</td></tr>
<tr><td>4</td><td>5</td><td>6</td><td>7</td><td>8</td><td>9</td><td>10</td></tr>
<tr><td>5</td><td>6</td><td>7</td><td>8</td><td>9</td><td>10</td><td>11</td></tr>
<tr><td>6</td><td>7</td><td>8</td><td>9</td><td>10</td><td>11</td><td>12</td></tr>
</table>

To calculate the probability of a particular total, count the number of times it appears in the table. For event $A = \{\text{rolling } 7\}$,

$$P(A) = \frac{n(A)}{n(S)}$$

$$= \frac{n(\text{rolls totalling } 7)}{n(\text{all possible rolls})}$$

$$= \frac{6}{36}$$

$$= \frac{1}{6}$$

The probability of rolling a total of 7 is $\frac{1}{6}$.

A useful and important concept in probability is the complement of an event. The **complement** of event A, A' or $\sim A$, is the event that "event A does *not* happen." Thus, whichever outcomes make up A, all the other outcomes make up A'. Because A and A' together include all possible outcomes, the sum of their probabilities must be 1. Thus,

$$P(A) + P(A') = 1 \quad \text{and} \quad P(A') = 1 - P(A)$$

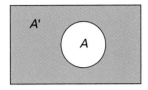

The event A' is usually called "A-prime," or sometimes "not-A"; $\sim A$ is called "tilde-A."

Example 4 The Complement of an Event

What is the probability that a randomly drawn integer between 1 and 40 is *not* a perfect square?

Solution

Let event $A = \{$a perfect square$\}$. Then, the complement of A is the event $A' = \{$*not* a perfect square$\}$. In this case, you need to calculate $P(A')$, but it is easier to do this by finding $P(A)$ first. There are six perfect squares between 1 and 40: 1, 4, 9, 16, 25, and 36. The probability of a perfect square is, therefore,

$$P(A) = \frac{n(A)}{n(S)}$$

$$= \frac{6}{40}$$

$$= \frac{3}{20}$$

Thus,

$$P(A') = 1 - P(A)$$

$$= 1 - \frac{3}{20}$$

$$= \frac{17}{20}$$

There is a $\frac{17}{20}$ or 85% chance that a random integer between 1 and 40 will not be a perfect square.

Subjective probability, the third basic type of probability, is an estimate of likelihood based on intuition and experience—an educated guess. For example, a well-prepared student may be 90% confident of passing the next data management test. Subjective probabilities often figure in everyday speech in expressions such as "I think the team has only a 10% chance of making the finals this year."

Example 5 Determining Subjective Probability

Estimate the probability that
a) the next pair of shoes you buy will be the same size as the last pair you bought
b) an expansion baseball team will win the World Series in their first season
c) the next person to enter a certain coffee shop will be male

Solution

a) There is a small chance that the size of your feet has changed significantly or that different styles of shoes may fit you differently, so 80–90% would be a reasonable subjective probability that your next pair of shoes will be the same size as your last pair.

b) Expansion teams rarely do well during their first season, and even strong teams have difficulty winning the World Series. The subjective probability of a brand-new team winning the World Series is close to zero.

WEB CONNECTION

www.mcgrawhill.ca/links/MDM12

For some interesting baseball statistics, visit the above web site and follow the links. Write a problem that could be solved using probabilities.

c) Without more information about the coffee shop in question, your best estimate is to assume that the shop's patrons are representative of the general population. This assumption gives a subjective probability of 50% that the next customer will be male.

Note that the answers in Example 5 contain estimates, assumptions, and, in some cases, probability *ranges*. While not as rigorous a measure as theoretical or empirical probability, subjective probabilities based on educated guesswork can still prove useful in some situations.

- A probability experiment is a well-defined process in which clearly identifiable outcomes are measured for each trial.

- An event is a collection of outcomes satisfying a particular condition. The probability of an event can range between 0 (impossible) and 1 or 100% (certain).

- The empirical probability of an event is the number of times the event occurs divided by the total number of trials.

- The theoretical probability of an event A is given by $P(A) = \dfrac{n(A)}{n(S)}$, where $n(A)$ is the number of outcomes making up A, $n(S)$ is the total number of outcomes in the sample space S, and all outcomes are equally likely to occur.

- A subjective probability is based on intuition and previous experience.

- If the probability of event A is given by $P(A)$, then the probability of the complement of A is given by $P(A') = 1 - P(A)$.

Communicate Your Understanding

1. Give two synonyms for the word *probability*.

2. a) Explain why $P(A) + P(A') = 1$.

 b) Explain why probabilities less than 0 or greater than 1 have no meaning.

3. Explain the difference between theoretical, empirical, and subjective probability. Give an example of how you would determine each type.

4. Describe three situations in which statistical fluctuations occur.

5. a) Describe a situation in which you might determine the probability of event A indirectly by calculating $P(A')$ first.

 b) Will this method always yield the same result as calculating $P(A)$ directly?

 c) Defend your answer to part b) using an explanation or proof, supported by an example.

Practise

1. Determine the probability of
   a) tossing heads with a single coin
   b) tossing two heads with two coins
   c) tossing at least one head with three coins
   d) rolling a composite number with one die
   e) not rolling a perfect square with two dice
   f) drawing a face card from a standard deck of cards

2. Estimate a subjective probability of each of the following events. Provide a rationale for each estimate.
   a) the sun rising tomorrow
   b) it never raining again
   c) your passing this course
   d) your getting the next job you apply for

3. Recall the sum/product game at the beginning of this section. Suppose that the game were altered so that the slips of paper showed the numbers 2, 3, and 4, instead of 1, 2, and 3.
   a) Identify all the outcomes that will produce each of the three possible events
      i) $p > s$ ii) $p < s$ iii) $p = s$
   b) Which player has the advantage in this situation?

Apply, Solve, Communicate

4. The town planning department surveyed residents of a town about home ownership. The table shows the results of the survey.

Residents	At Address Less Than 2 Years	At Address More Than 2 Years	Total for Category
Owners	2000	8000	10 000
Renters	4500	1500	6 000
Total	6500	9500	16 000

Determine the following probabilities.
   a) P(resident owns home)
   b) P(resident rents and has lived at present address less than two years)
   c) P(homeowner has lived at present address more than two years)

5. **Application** Suppose your school's basketball team is playing a four-game series against another school. So far this season, each team has won three of the six games in which they faced each other.
   a) Draw a tree diagram to illustrate all possible outcomes of the series.
   b) Use your tree diagram to determine the probability of your school winning exactly two games.
   c) What is the probability of your school sweeping the series (winning all four games)?
   d) Discuss any assumptions you made in the calculations in parts b) and c).

6. **Application** Suppose that a graphing calculator is programmed to generate a random natural number between 1 and 10 inclusive. What is the probability that the number will be prime?

7. **Communication**
   a) A game involves rolling two dice. Player A wins if the throw totals 5, 7, or 9. Player B wins if any other total is thrown. Which player has the advantage? Explain.
   b) Suppose the game is changed so that Player A wins if 5, 7, or doubles (both dice showing the same number) are thrown. Who has the advantage now? Explain.
   c) Design a similar game in which each player has an equal chance of winning.

8. a) Based on the randomly tagged sample, what is the empirical probability that a deer captured at random will be a doe?

b) If ten deer are captured at random, how many would you expect to be bucks?

C

9. Inquiry/Problem Solving Refer to the prime number experiment in question 6. What happens to the probability if you change the upper limit of the sample space? Use a graphing calculator or appropriate computer software to investigate this problem. Let A be the event that the random natural number will be a prime number. Let the random number be between 1 and n inclusive. Predict what you think will happen to $P(A)$ as n increases. Investigate $P(A)$ as a function of n, and reflect on your hypothesis. Did you observe what you expected? Why or why not?

10. Suppose that the Toronto Blue Jays face the New York Yankees in the division final. In this best-of-five series, the winner is the first team to win three games. The games are played in Toronto and in New York, with Toronto hosting the first, second, and if needed, fifth games. The consensus among experts is that Toronto has a 65% chance of winning at home and a 40% chance of winning in New York.

a) Construct a tree diagram to illustrate all the possible outcomes.

b) What is the chance of Toronto winning in three straight games?

c) For each outcome, add to your tree diagram the probability of that outcome.

d) Communication Explain how you found your answers to parts b) and c).

11. Communication Prior to a municipal election, a public-opinion poll determined that the probability of each of the four candidates winning was as follows:

Jonsson 10%

Trimble 32%

Yakamoto 21%

Audette 37%

a) How will these probabilities change if Jonsson withdraws from the race after ballots are cast?

b) How will these probabilities change if Jonsson withdraws from the race before ballots are cast?

c) Explain why your answers to a) and b) are different.

12. Inquiry/Problem Solving It is known from studying past tests that the correct answers to a certain university professor's multiple-choice tests exhibit the following pattern.

Correct Answer	Percent of Questions
A	15%
B	25%
C	30%
D	15%
E	15%

a) Devise a strategy for guessing that would maximize a student's chances for success, assuming that the student has no idea of the correct answers. Explain your method.

b) Suppose that the study of past tests revealed that the correct answer choice for any given question was the same as that of the immediately preceding question only 10% of the time. How would you use this information to adjust your strategy in part a)? Explain your reasoning.

Odds

Odds are another way to express a level of confidence about an outcome. Odds are commonly used in sports and other areas. Odds are often used when the probability of an event versus its complement is of interest, for example whether a sprinter will win or lose a race or whether a basketball team will make it to the finals.

INVESTIGATE & INQUIRE: Tennis Tournament

For an upcoming tennis tournament, a television commentator estimates that the top-seeded (highest-ranked) player has "a 25% probability of winning, but her odds of winning are only 1 to 3."

1. **a)** If event A is the top-seeded player winning the tournament, what is A'?
   b) Determine $P(A')$.

2. **a)** How are the odds of the top-seeded player winning related to $P(A)$ and $P(A')$?
   b) Should the television commentator be surprised that the odds were *only* 1 to 3? Why or why not?

3. **a)** What factors might the commentator consider when estimating the probability of the top-seeded player winning the tournament?

   b) How accurate do you think the commentator's estimate is likely to be? Would you consider such an estimate primarily a classical, an empirical, or a subjective probability? Explain.

WEB CONNECTION

www.mcgrawhill.ca/links/MDM12

For more information about tennis rankings and other tennis statistics, visit the above web site and follow the links. Locate some statistics about a tennis player of your choice. Use odds to describe these statistics.

The **odds in favour** of an event's occurring are given by the ratio of the probability that the event will occur to the probability that it will not occur.

$$\text{odds in favour of } A = \frac{P(A)}{P(A')}$$

Giving odds in favour of an event is a common way to express a probability.

Example 1 Determining Odds

A messy drawer contains three red socks, five white socks, and four black socks. What are the odds in favour of randomly drawing a red sock?

Solution

Let the event A be drawing a red sock. The probability of this event is

$$P(A) = \frac{3}{12}$$

$$= \frac{1}{4}$$

The probability of not drawing a red sock is
$$P(A') = 1 - P(A)$$

$$= \frac{3}{4}$$

Using the definition of odds,

$$\text{odds in favour of } A = \frac{P(A)}{P(A')}$$

$$= \frac{\frac{1}{4}}{\frac{3}{4}}$$

$$= \frac{1}{3}$$

Project Prep

A useful feature you could include in your probability project is a calculation of the odds of winning your game.

Therefore, the odds in favour of drawing a red sock are $\frac{1}{3}$, or less than 1. You are more likely *not* to draw a red sock. These odds are commonly written as 1:3, which is read as "one to three" or "one in three."

Notice in Example 1 that the ratio of red socks to other socks is 3:9, which is the same as the odds in favour of drawing a red sock. In fact, the odds in favour of an event A can also be found using

$$\text{odds in favour of } A = \frac{n(A)}{n(A')}$$

A common variation on the theme of odds is to express the odds *against* an event happening.

$$\text{odds against } A = \frac{P(A')}{P(A)}$$

Example 2 Odds Against an Event

If the chance of a snowstorm in Windsor, Ontario, in January is estimated at 0.4, what are the odds against Windsor's having a snowstorm next January? Is a January snowstorm more likely than not?

Solution

Let event A = {snowstorm in January}.
Since $P(A) + P(A') = 1$,

$$\begin{aligned}
\text{odds against } A &= \frac{P(A')}{P(A)} \\
&= \frac{1 - P(A)}{P(A)} \\
&= \frac{1 - 0.4}{0.4} \\
&= \frac{0.6}{0.4} \\
&= \frac{3}{2}
\end{aligned}$$

The odds against a snowstorm are 3:2, which is greater than 1:1. So a snowstorm is *less* likely to occur than not.

Sometimes, you might need to convert an expression of odds into a probability. You can do this conversion by expressing $P(A')$ in terms of $P(A)$.

Example 3 Probability From Odds

A university professor, in an effort to promote good attendance habits, states that the odds of passing her course are 8 to 1 when a student misses fewer than five classes. What is the probability that a student with good attendance will pass?

Solution

Let the event A be that a student with good attendance passes. Since

$$\text{odds in favour of } A = \frac{P(A)}{P(A')},$$

$$\frac{8}{1} = \frac{P(A)}{P(A')}$$

$$= \frac{P(A)}{1 - P(A)}$$

$$8 - 8P(A) = P(A)$$

$$8 = 9P(A)$$

$$P(A) = \frac{8}{9}$$

The probability that a student with good attendance will pass is $\frac{8}{9}$, or approximately 89%.

In general, it can be shown that if the odds in favour of $A = \frac{h}{k}$, then $P(A) = \frac{h}{h + k}$.

Example 4 Using the Odds-Probability Formula

The odds of Rico's hitting a home run are 2:7. What is the probability of Rico's hitting a home run?

Solution

Let A be the event that Rico hits a home run. Then, $h = 2$ and $k = 7$, and

$$P(A) = \frac{h}{h + k}$$

$$= \frac{2}{2 + 7}$$

$$= \frac{2}{9}$$

Rico has approximately a 22% chance of hitting a home run.

Key Concepts

- The odds in favour of A are given by the ratio $\frac{P(A)}{P(A')}$.

- The odds against A are given by the ratio $\frac{P(A')}{P(A)}$.

- If the odds in favour of A are $\frac{h}{k}$, then $P(A) = \frac{h}{h + k}$.

1. Explain why the terms *odds* and *probability* have different meanings. Give an example to illustrate your answer.

2. Would you prefer the odds in favour of passing your next data management test to be 1:3 or 3:1? Explain your choice.

3. Explain why odds can be greater than 1, but probabilities must be between 0 and 1.

Practise

1. Suppose the odds in favour of good weather tomorrow are 3:2.

   a) What are the odds against good weather tomorrow?

   b) What is the probability of good weather tomorrow?

2. The odds against the Toronto Argonauts winning the Grey Cup are estimated at 19:1. What is the probability that the Argos will win the cup?

3. Determine the odds in favour of rolling each of the following sums with a standard pair of dice.

   a) 12   **b)** 5 or less

   c) a prime number   **d)** 1

4. Calculate the odds in favour of each event.

   a) New Year's Day falling on a Friday

   b) tossing three tails with three coins

   c) not tossing exactly two heads with three coins

   d) randomly drawing a black 6 from a complete deck of 52 cards

   e) a random number from 1 to 9 inclusive being even

Apply, Solve, Communicate

5. Greta's T-shirt drawer contains three tank tops, six V-neck T-shirts, and two sleeveless shirts. If she randomly draws a shirt from the drawer, what are the odds that she will

   a) draw a V-neck T-shirt?

   b) not draw a tank top?

6. **Application** If the odds in favour of Boris beating Elena in a chess game are 5 to 4, what is the probability that Elena will win an upset victory in a best-of-five chess tournament?

7. **a)** Based on the randomly tagged sample, what are the odds in favour of a captured deer being a cross-hatched buck?

   b) What are the odds against capturing a doe?

Visit the above web site and follow the links for more information about Canadian wildlife.

8. The odds against A, by definition, are equivalent to the odds in favour of A'. Use this definition to show that the odds against A are equal to the reciprocal of the odds in favour of A.

9. Application Suppose the odds of the Toronto Maple Leafs winning the Stanley Cup are 1:5, while the odds of the Montréal Canadiens winning the Stanley Cup are 2:13. What are the odds in favour of either Toronto or Montréal winning the Stanley Cup?

10. What are the odds against drawing

   a) a face card from a standard deck?

   b) two face cards?

11. Mike has a loaded (or unfair) six-sided die. He rolls the die 200 times and determines the following probabilities for each score:

$P(1) = 0.11$
$P(2) = 0.02$
$P(3) = 0.18$
$P(4) = 0.21$
$P(5) = 0.40$

   a) What is $P(6)$?

   b) Mike claims that the odds in favour of tossing a prime number with this die are the same as with a fair die. Do you agree with his claim?

   c) Using Mike's die, devise a game with odds in Mike's favour that an unsuspecting person would be tempted to play. Use probabilities to show that the game is in Mike's favour. Explain why a person who does not realize that the die is loaded might be tempted by this game.

12. George estimates that there is a 30% chance of rain the next day if he waters the lawn, a 40% chance if he washes the car, and a 50% chance if he plans a trip to the beach. Assuming George's estimates are accurate, what are the odds

   a) in favour of rain tomorrow if he waters the lawn?

   b) in favour of rain tomorrow if he washes the car?

   c) against rain tomorrow if he plans a trip to the beach?

C

13. Communication A volleyball coach claims that at the next game, the odds of her team winning are 3:1, the odds against losing are 5:1, and the odds against a tie are 7:1. Are these odds possible? Explain your reasoning.

14. Inquiry/Problem Solving Aki is a participant on a trivia-based game show. He has an equal likelihood on any given trial of being asked a question from one of six categories: Hollywood, Strange Places, Number Fun, Who?, Having a Ball, and Write On! Aki feels that he has a 50/50 chance of getting Having a Ball or Strange Places questions correct, but thinks he has a 90% probability of getting any of the other questions right. If Aki has to get two of three questions correct, what are his odds of winning?

15. Inquiry/Problem Solving Use logic and mathematical reasoning to show that if the odds in favour of A are given by $\dfrac{h}{k}$, then $P(A) = \dfrac{h}{h + k}$. Support your reasoning with an example.

Probabilities Using Counting Techniques

How likely is it that, in a game of cards, you will be dealt just the hand that you need? Most card players accept this question as an unknown, enjoying the unpredictability of the game, but it can also be interesting to apply counting analysis to such problems.

In some situations, the possible outcomes are not easy or convenient to count individually. In many such cases, the counting techniques of permutations and combinations (see Chapters 4 and 5, respectively) can be helpful for calculating theoretical probabilities, or you can use a simulation to determine an empirical probability.

INVESTIGATE & INQUIRE: Fishing Simulation

Suppose a pond has only three types of fish: catfish, trout, and bass, in the ratio 5:2:3. There are 50 fish in total. Assuming you are allowed to catch only three fish before throwing them back, consider the following two events:

- event A = {catching three trout}
- event B = {catching the three types of fish, in alphabetical order}

1. Carry out the following probability experiment, independently or with a partner. You can use a hat or paper bag to represent the pond, and some differently coloured chips or markers to represent the fish. How many of each type of fish should you release into the pond? Count out the appropriate numbers and shake the container to simulate the fish swimming around.

2. Draw a tree diagram to illustrate the different possible outcomes of this experiment.

3. Catch three fish, one at a time, and record the results in a table. Replace all three fish and shake the container enough to ensure that they are randomly distributed. Repeat this process for a total of ten trials.

4. Based on these ten trials, determine the empirical probability of event A, catching three trout. How accurate do you think this value is? Compare your results with those of the rest of the class. How can you obtain a more accurate empirical probability?

5. Repeat step 4 for event B, which is to catch a bass, catfish, and trout in order.

6. Perform step 3 again for 10 new trials. Calculate the empirical probabilities of events A and B, based on your 20 trials. Do you think these probabilities are more accurate than those from 10 trials? Explain why or why not.

7. If you were to repeat the experiment for 50 or 100 trials, would your results be more accurate? Why or why not?

8. In this investigation, you knew exactly how many of each type of fish were in the pond because they were counted out at the beginning. Describe how you could use the techniques of this investigation to estimate the ratios of different species in a real pond.

This section examines methods for determining the theoretical probabilities of successive or multiple events.

Example 1 Using Permutations

Two brothers enter a race with five friends. The racers draw lots to determine their starting positions. What is the probability that the older brother will start in lane 1 with his brother beside him in lane 2?

Solution

A permutation $_nP_r$, or $P(n, r)$, is the number of ways to select r objects from a set of n objects, *in a certain order*. (See Chapter 4 for more about permutations.) The sample space is the total number of ways the first two lanes can be occupied. Thus,

$$n(S) = {_7}P_2$$

$$= \frac{7!}{(7-2)!}$$

$$= \frac{7!}{5!}$$

$$= \frac{7 \times 6 \times (5!)}{5!}$$

$$= 42$$

The specific outcome of the older brother starting in lane 1 and the younger brother starting in lane 2 can only happen one way, so $n(A) = 1$. Therefore,

$$P(A) = \frac{n(A)}{n(S)}$$

$$= \frac{1}{42}$$

The probability that the older brother will start in lane 1 next to his brother in lane 2 is $\frac{1}{42}$, or approximately 2.3%.

Example 2 Probability Using Combinations

A focus group of three members is to be randomly selected from a medical team consisting of five doctors and seven technicians.
a) What is the probability that the focus group will be comprised of doctors only?
b) What is the probability that the focus group will not be comprised of doctors only?

Solution

a) A combination $_nC_r$, also written $C(n, r)$ or $\binom{n}{r}$, is the number of ways to select r objects from a set of n objects, *in any order*. (See Chapter 5 for more about combinations.) Let event A be selecting three doctors to form the focus group. The number of possible ways to make this selection is

$$n(A) = {}_5C_3$$
$$= \frac{5!}{3!(5-3)!}$$
$$= \frac{5 \times 4 \times 3!}{3! \times 2!}$$
$$= \frac{20}{2}$$
$$= 10$$

However, the focus group can consist of any three people from the team of 12.
$$n(S) = {}_{12}C_3$$
$$= \frac{12!}{3!(12-3)!}$$
$$= \frac{12 \times 11 \times 10 \times 9!}{3! \times 9!}$$
$$= \frac{1320}{6}$$
$$= 220$$

```
GRAPHING CALCULATOR
 5 nCr 3
                    10
12 nCr 3
                   220
■
```

The probability of selecting a focus group of doctors only is
$$P(A) = \frac{n(A)}{n(S)}$$
$$= \frac{10}{220}$$
$$= \frac{1}{22}$$

The probability of selecting a focus group consisting of three doctors is $\frac{1}{22}$, or approximately 0.045.

b) Either the focus group is comprised of doctors only, or it is not. Therefore, the probability of the complement of A, $P(A')$, gives the desired result.

$$P(A') = 1 - P(A)$$
$$= 1 - \frac{1}{22}$$
$$= \frac{21}{22}$$

So, the probability of selecting a focus group not comprised of doctors only is $\frac{21}{22}$, or approximately 0.955.

Project Prep

When you determine the classical probabilities for your probability project, you may need to apply the counting techniques of permutations and combinations.

Example 3 Probability Using the Fundamental Counting Principle

What is the probability that two or more students out of a class of 24 will have the same birthday? Assume that no students were born on February 29.

Solution 1 Using Pencil and Paper

The simplest method is to find the probability of the complementary event that no two people in the class have the same birthday.

Pick two students at random. The second student has a different birthday than the first for 364 of the 365 possible birthdays. Thus, the probability that the two students have different birthdays is $\frac{364}{365}$. Now add a third student. Since there are 363 ways this person can have a different birthday from the other two students, the probability that all three students have different birthdays is $\frac{364}{365} \times \frac{363}{365}$. Continuing this process, the probability that none of the 24 people have the same birthday is

$$P(A') = \frac{n(A')}{n(S)}$$
$$= \frac{364}{365} \times \frac{363}{365} \times \frac{362}{365} \times \ldots \times \frac{342}{365}$$
$$\doteq 0.462$$

$$P(A) = 1 - P(A')$$
$$= 1 - 0.462$$
$$= 0.538$$

The probability that at least two people in the group have the same birthday is approximately 0.538.

Solution 2 Using a Graphing Calculator

Use the iterative functions of a graphing calculator to evaluate the formula above much more easily. The prod(function on the LIST MATH menu will find the product of a series of numbers. The seq(function on the LIST OPS menu generates a sequence for the range you specify. Combining these two functions allows you to calculate the probability in a single step.

Key Concepts

- In probability experiments with many possible outcomes, you can apply the fundamental counting principle and techniques using permutations and combinations.

- Permutations are useful when order is important in the outcomes; combinations are useful when order is not important.

Communicate Your Understanding

1. In the game of bridge, each player is dealt 13 cards out of the deck of 52. Explain how you would determine the probability of a player receiving

   a) all hearts   **b)** all hearts in ascending order

2. **a)** When should you apply permutations in solving probability problems, and when should you apply combinations?

   b) Provide an example of a situation where you would apply permutations to solve a probability problem, other than those in this section.

   c) Provide an example of a situation where you would apply combinations to solve a probability problem, other than those in this section.

Practise

1. Four friends, two females and two males, are playing contract bridge. Partners are randomly assigned for each game. What is the probability that the two females will be partners for the first game?

2. What is the probability that two out of a group of eight friends will have the same birthday?

3. A fruit basket contains five red apples and three green apples. Without looking, you randomly select two apples. What is the probability that

   a) you will select two red apples?

   b) you will not select two green apples?

4. Refer to Example 1. What is the probability that the two brothers will start beside each other in any pair of lanes?

Apply, Solve, Communicate

5. An athletic committee with three members is to be randomly selected from a group of six gymnasts, four weightlifters, and eight long-distance runners. Determine the probability that

  a) the committee is comprised entirely of runners

  b) the committee is represented by each of the three types of athletes

6. A messy drawer contains three black socks, five blue socks, and eight white socks, none of which are paired up. If the owner grabs two socks without looking, what is the probability that both will be white?

7. a) A family of nine has a tradition of drawing two names from a hat to see whom they will each buy presents for. If there are three sisters in the family, and the youngest sister is always allowed the first draw, determine the probability that the youngest sister will draw both of the other two sisters' names. If she draws her own name, she replaces it and draws another.

  b) Suppose that the tradition is modified one year, so that the first person whose name is drawn is to receive a "main" present, and the second a less expensive, "fun" present. Determine the probability that the youngest sister will give a main present to the middle sister and a fun present to the eldest sister.

8. Application

  a) Laura, Dave, Monique, Marcus, and Sarah are going to a party. What is the probability that two of the girls will arrive first?

  b) What is the probability that the friends will arrive in order of ascending age?

  c) What assumptions must be made in parts a) and b)?

9. A hockey team has two goalies, six defenders, eight wingers, and four centres. If the team randomly selects four players to attend a charity function, what is the likelihood that

  a) they are all wingers?

  b) no goalies or centres are selected?

10. Application A lottery promises to award ten grand-prize trips to Hawaii and sells 5 400 000 tickets.

  a) Determine the probability of winning a grand prize if you buy

  i) 1 ticket

  ii) 10 tickets

  iii) 100 tickets

  b) Communication How many tickets do you need to buy in order to have a 5% chance of winning a grand prize? Do you think this strategy is sensible? Why or why not?

  c) How many tickets do you need to ensure a 50% chance of winning?

11. Suki is enrolled in one data-management class at her school and Leo is in another. A school quiz team will have four volunteers, two randomly selected from each of the two classes. Suki is one of five volunteers from her class, and Leo is one of four volunteers from his. Calculate the probability of the two being on the team and explain the steps in your calculation.

12. a) Suppose 4 of the 22 tagged bucks are randomly chosen for a behaviour study. What is the probability that

   i) all four bucks have the cross-hatched antlers?

   ii) at least one buck has cross-hatched antlers?

b) If two of the seven cross-hatched males are randomly selected for a health study, what is the probability that the eldest of the seven will be selected first, followed by the second eldest?

ACHIEVEMENT CHECK

Knowledge/ Understanding	Thinking/Inquiry/ Problem Solving	Communication	Application

13. Suppose a bag contains the letters to spell *probability*.

a) How many four-letter arrangements are possible using these letters?

b) What is the probability that Barb chooses four letters from the bag in the order that spell her name?

c) Pick another four-letter arrangement and calculate the probability that it is chosen.

d) What four-letter arrangement would be most likely to be picked? Explain your reasoning.

14. Communication Refer to the fishing investigation at the beginning of this section.

a) Determine the theoretical probability of

   i) catching three trout

   ii) catching a bass, catfish, and trout in alphabetical order

b) How do these results compare with the empirical probabilities from the investigation? How do you account for any differences?

c) Could the random-number generator of a graphing calculator be used to simulate this investigation? If so, explain how. If not, explain why.

d) Outline the steps you would use to model this problem with software such as Fathom™ or a spreadsheet.

e) Is the assumption that the fish are randomly distributed likely to be completely correct? Explain. What other assumptions might affect the accuracy of the calculated probabilities?

15. A network of city streets forms square blocks as shown in the diagram.

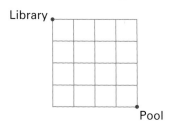

Jeanine leaves the library and walks toward the pool at the same time as Miguel leaves the pool and walks toward the library. Neither person follows a particular route, except that both are always moving toward their destination. What is the probability that they will meet if they both walk at the same rate?

16. Inquiry/Problem Solving A committee is formed by randomly selecting from eight nurses and two doctors. What is the minimum committee size that ensures at least a 90% probability that it will not be comprised of nurses only?

Dependent and Independent Events

If you have two examinations next Tuesday, what is the probability that you will pass both of them? How can you predict the risk that a critical network server and its backup will both fail? If you flip an ordinary coin repeatedly and get heads 99 times in a row, is the next toss almost certain to come up tails?

In such situations, you are dealing with **compound events** involving two or more separate events.

INVESTIGATE & INQUIRE: Getting Out of Jail in MONOPOLY®

While playing MONOPOLY® for the first time, Kenny finds himself in jail. To get out of jail, he needs to roll doubles on a pair of standard dice.

1. Determine the probability that Kenny will roll doubles on his first try.

2. Suppose that Kenny fails to roll doubles on his first two turns in jail. He reasons that on his next turn, his odds are now 50/50 that he will get out of jail. Explain how Kenny has reasoned this.

3. Do you agree or disagree with Kenny's reasoning? Explain.

4. What is the probability that Kenny will get out of jail on his third attempt?

5. After how many turns is Kenny certain to roll doubles? Explain.

6. Kenny's opponent, Roberta, explains to Kenny that each roll of the dice is an independent event and that, since the relatively low probability of rolling doubles never changes from trial to trial, Kenny may never get out of jail and may as well just forfeit the game. Explain the flaws in Roberta's analysis.

In some situations involving compound events, the occurrence of one event has no effect on the occurrence of another. In such cases, the events are **independent**.

Example 1 Simple Independent Events

a) A coin is flipped and turns up heads. What is the probability that the second flip will turn up heads?

b) A coin is flipped four times and turns up heads each time. What is the probability that the fifth trial will be heads?

c) Find the probability of tossing five heads in a row.

d) Comment on any difference between your answers to parts b) and c).

Solution

a) Because these events are independent, the outcome of the first toss has no effect on the outcome of the second toss. Therefore, the probability of tossing heads the second time is 0.5.

b) Although you might think "tails has to come up sometime," there is still a 50/50 chance on each independent toss. The coin has no memory of the past four trials! Therefore, the fifth toss still has just a 0.5 probability of coming up heads.

c) Construct a tree diagram to represent five tosses of the coin.

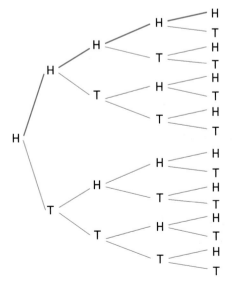

There is an equal number of outcomes in which the first flip turns up tails.

The number of outcomes doubles with each trial. After the fifth toss, there are 2^5 or 32 possible outcomes, only one of which is five heads in a row. So, the probability of five heads in a row, prior to any coin tosses, is $\frac{1}{32}$ or 0.031 25.

d) The probability in part c) is much less than in part b). In part b), you calculate only the probability for the fifth trial on its own. In part c), you are finding the probability that every one of five separate events actually happens.

Example 2 Probability of Two Different Independent Events

A coin is flipped while a die is rolled. What is the probability of flipping heads and rolling 5 in a single trial?

Solution

Here, two independent events occur in a single trial. Let A be the event of flipping heads, and B be the event of rolling 5. The notation $P(A$ and $B)$ represents the compound, or joint, probability that both events, A and B, will occur simultaneously. For independent events, the probabilities can simply be multiplied together.

$$P(A \text{ and } B) = P(A) \times P(B)$$
$$= \frac{1}{2} \times \frac{1}{6}$$
$$= \frac{1}{12}$$

The probability of simultaneously flipping heads while rolling 5 is $\frac{1}{12}$ or approximately 8.3%

In general, the compound probability of two independent events can be calculated using the **product rule for independent events**:

$$P(A \text{ and } B) = P(A) \times P(B)$$

From the example above, you can see that the product rule for independent events agrees with common sense. The product rule can also be derived mathematically from the fundamental counting principle (see Chapter 4).

Proof:

A and *B* are separate events and so they correspond to separate sample spaces, S_A and S_B.

Their probabilities are thus

$$P(A) = \frac{n(A)}{n(S_A)} \text{ and } P(B) = \frac{n(B)}{n(S_B)}.$$

Call the sample space for the compound event *S*, as usual.

You know that

$$P(A \text{ and } B) = \frac{n(A \text{ and } B)}{n(S)} \qquad (1)$$

Because *A* and *B* are independent, you can apply the fundamental counting principle to get an expression for $n(A \text{ and } B)$.

$$n(A \text{ and } B) = n(A) \times n(B) \qquad (2)$$

Similarly, you can also apply the fundamental counting principle to get an expression for $n(S)$.

$$n(S) = n(S_A) \times n(S_B) \qquad (3)$$

Substitute equations (2) and (3) into equation (1).

$$\begin{aligned} P(A \text{ and } B) &= \frac{n(A)n(B)}{n(S_A)n(S_B)} \\ &= \frac{n(A)}{n(S_A)} \times \frac{n(B)}{n(S_B)} \\ &= P(A) \times P(B) \end{aligned}$$

Example 3 Applying the Product Rule for Independent Events

Soo-Ling travels the same route to work every day. She has determined that there is a 0.7 probability that she will wait for at least one red light and that there is a 0.4 probability that she will hear her favourite new song on her way to work.

a) What is the probability that Soo-Ling will not have to wait at a red light and will hear her favourite song?

b) What are the odds in favour of Soo-Ling having to wait at a red light and not hearing her favourite song?

Solution

a) Let A be the event of Soo-Ling having to wait at a red light, and B be the event of hearing her favourite song. Assume A and B to be independent events. In this case, you are interested in the combination A' and B.

$$
\begin{aligned}
P(A' \text{ and } B) &= P(A') \times P(B) \\
&= (1 - P(A)) \times P(B) \\
&= (1 - 0.7) \times 0.4 \\
&= 0.12
\end{aligned}
$$

There is a 12% chance that Soo-Ling will hear her favourite song and not have to wait at a red light on her way to work.

b)
$$
\begin{aligned}
P(A \text{ and } B') &= P(A) \times P(B') \\
&= P(A) \times (1 - P(B)) \\
&= 0.7 \times (1 - 0.4) \\
&= 0.42
\end{aligned}
$$

The probability of Soo-Ling having to wait at a red light and not hearing her favourite song is 42%.

The odds in favour of this happening are

$$
\begin{aligned}
\text{odds in favour} &= \frac{P(A \text{ and } B')}{1 - P(A \text{ and } B')} \\
&= \frac{42\%}{100\% - 42\%} \\
&= \frac{42}{58} \\
&= \frac{21}{29}
\end{aligned}
$$

The odds in favour of Soo-Ling having to wait at a red light and not hearing her favourite song are 21:29.

In some cases, the probable outcome of an event, B, depends directly on the outcome of another event, A. When this happens, the events are said to be **dependent**. The **conditional probability** of B, **$P(B\,|\,A)$**, is the probability that B occurs, *given* that A has already occurred.

Example 4 Probability of Two Dependent Events

A professional hockey team has eight wingers. Three of these wingers are 30-goal scorers, or "snipers." Every fall the team plays an exhibition match with the club's farm team. In order to make the match more interesting for the fans, the coaches agree to select two wingers at random from the pro team to play for the farm team. What is the probability that two snipers will play for the farm team?

Solution

Let A = {first winger is a sniper} and B = {second winger is a sniper}. Three of the eight wingers are snipers, so the probability of the first winger selected being a sniper is

$$P(A) = \frac{3}{8}$$

If the first winger selected is a sniper, then there are seven remaining wingers to choose from, two of whom are snipers. Therefore,

$$P(B \mid A) = \frac{2}{7}$$

Applying the fundamental counting principle, the probability of randomly selecting two snipers for the farm team is the number of ways of selecting two snipers divided by the number of ways of selecting any two wingers.

$$P(A \text{ and } B) = \frac{3 \times 2}{8 \times 7}$$

$$= \frac{3}{28}$$

There is a $\frac{3}{28}$ or 10.7% probability that two professional snipers will play for the farm team in the exhibition game.

Notice in Example 4 that, when two events A and B are dependent, you can still multiply probabilities to find the probability that they both happen. However, you must use the conditional probability for the second event. Thus, the probability that both events will occur is given by the **product rule for dependent events**:

$$P(A \text{ and } B) = P(A) \times P(B \mid A)$$

This reads as: "The probability that both A and B will occur equals the probability of A times the probability of B given that A has occurred."

Project Prep

When designing your game for the probability project, you may decide to include situations involving independent or dependent events. If so, you will need to apply the appropriate product rule in order to determine classical probabilities.

Example 5 Conditional Probability From Compound Probability

Serena's computer sometimes crashes while she is trying to use her e-mail program, OutTake. When OutTake "hangs" (stops responding to commands), Serena is usually able to close OutTake without a system crash. In a computer magazine, she reads that the probability of OutTake hanging in any 15-min period is 2.5%, while the chance of OutTake and the operating system failing together in any 15-min period is 1%. If OutTake is hanging, what is the probability that the operating system will crash?

Solution

Let event A be OutTake hanging, and event B be an operating system failure. Since event A can trigger event B, the two events are dependent. In fact, you need to find the conditional probability $P(B \mid A)$. The data from the magazine tells you that $P(A) = 2.5\%$, and $P(A \text{ and } B) = 1\%$. Therefore,

$$P(A \text{ and } B) = P(A) \times P(B \mid A)$$
$$1\% = 2.5\% \times P(B \mid A)$$
$$P(B \mid A) = \frac{1\%}{2.5\%}$$
$$= 0.4$$

There is a 40% chance that the operating system will crash when OutTake is hanging.

Example 5 suggests a useful rearrangement of the product rule for dependent events.

$$P(B \mid A) = \frac{P(A \text{ and } B)}{P(A)}$$

This equation is sometimes used to define the conditional probability $P(B \mid A)$.

Key Concepts

- If A and B are independent events, then the probability of both occurring is given by $P(A \text{ and } B) = P(A) \times P(B)$.

- If event B is dependent on event A, then the conditional probability of B given A is $P(B \mid A)$. In this case, the probability of both events occurring is given by $P(A \text{ and } B) = P(A) \times P(B \mid A)$.

Communicate Your Understanding

1. Consider the probability of randomly drawing an ace from a standard deck of cards. Discuss whether or not successive trials of this experiment are independent or dependent events. Consider cases in which drawn cards are

   a) replaced after each trial

   b) not replaced after each trial

2. Suppose that for two particular events A and B, it is true that $P(B \mid A) = P(B)$. What does this imply about the two events? (*Hint*: Try substituting this equation into the product rule for dependent events.)

Practise

1. Classify each of the following as independent or dependent events.

	First Event	Second Event
a)	Attending a rock concert on Tuesday night	Passing a final examination the following Wednesday morning
b)	Eating chocolate	Winning at checkers
c)	Having blue eyes	Having poor hearing
d)	Attending an employee training session	Improving personal productivity
e)	Graduating from university	Running a marathon
f)	Going to a mall	Purchasing a new shirt

2. Amitesh estimates that he has a 70% chance of making the basketball team and a 20% chance of having failed his last geometry quiz. He defines a "really bad day" as one in which he gets cut from the team and fails his quiz. Assuming that Amitesh will receive both pieces of news tomorrow, how likely is it that he will have a really bad day?

3. In the popular dice game Yahtzee®, a Yahtzee occurs when five identical numbers turn up on a set of five standard dice. What is the probability of rolling a Yahtzee on one roll of the five dice?

Apply, Solve, Communicate

4. There are two tests for a particular antibody. Test A gives a correct result 95% of the time. Test B is accurate 89% of the time. If a patient is given both tests, find the probability that

   a) both tests give the correct result

   b) neither test gives the correct result

   c) at least one of the tests gives the correct result

5. **a)** Rocco and Biff are two koala bears participating in a series of animal behaviour tests. They each have 10 min to solve a maze. Rocco has an 85% probability of succeeding if he can smell the eucalyptus treat at the other end. He can smell the treat 60% of the time. Biff has a 70% chance of smelling the treat, but when he does, he can solve the maze only 75% of the time. Neither bear will try to solve the maze unless he smells the eucalyptus. Determine which koala bear is more likely to enjoy a tasty treat on any given trial.

   b) Communication Explain how you arrived at your conclusion.

6. Shy Tenzin's friends assure him that if he asks Mikala out on a date, there is an 85% chance that she will say yes. If there is a 60% chance that Tenzin will summon the courage to ask Mikala out to the dance next week, what are the odds that they will be seen at the dance together?

7. When Ume's hockey team uses a "rocket launch" breakout, she has a 55% likelihood of receiving a cross-ice pass while at full speed. When she receives such a pass, the probability of getting her slapshot away is $\frac{1}{3}$. Ume's slapshot scores 22% of the time. What is the probability of Ume scoring with her slapshot when her team tries a rocket launch?

8. **Inquiry/Problem Solving** Show that if A and B are dependent events, then the conditional probability $P(A \mid B)$ is given by

$$P(A \mid B) = \frac{P(A \text{ and } B)}{P(B)}.$$

9. A consultant's study found Megatran's call centre had a 5% chance of transferring a call about schedules to the lost articles department by mistake. The same study shows that, 1% of the time, customers calling for schedules have to wait on hold, only to discover that they have been mistakenly transferred to the lost articles department. What are the chances that a customer transferred to lost articles will be put on hold?

10. Pinder has examinations coming up in data management and biology. He estimates that his odds in favour of passing the data-management examination are 17:3 and his odds against passing the biology examination are 3:7. Assume these to be independent events.

   a) What is the probability that Pinder will pass both exams?

   b) What are the odds in favour of Pinder failing both exams?

   c) What factors could make these two events dependent?

11. **Inquiry/Problem Solving** How likely is it for a group of five friends to have the same birth month? State any assumptions you make for your calculation.

12. Determine the probability that a captured deer has the bald patch condition.

13. **Communication** Five different CD-ROM games, Garble, Trapster, Zoom!, Bungie, and Blast 'Em, are offered as a promotion by SugarRush cereals. One game is randomly included with each box of cereal.

   a) Determine the probability of getting all 5 games if 12 boxes are purchased.

   b) Explain the steps in your solution.

   c) Discuss any assumptions that you make in your analysis.

14. **Application** A critical circuit in a communication network relies on a set of eight identical relays. If any one of the relays fails, it will disrupt the entire network. The design engineer must ensure a 90% probability that the network will not fail over a five-year period. What is the maximum tolerable probability of failure for each relay?

15. a) Show that if a coin is tossed n times, the probability of tossing n heads is given by $P(A) = \left(\frac{1}{2}\right)^n$.

   b) What is the probability of getting at least one tail in seven tosses?

16. What is the probability of not throwing 7 or doubles for six consecutive throws with a pair of dice?

17. Laurie, an avid golfer, gives herself a 70% chance of breaking par (scoring less than 72 on a round of 18 holes) if the weather is calm, but only a 15% chance of breaking par on windy days. The weather forecast gives a 40% probability of high winds tomorrow. What is the likelihood that Laurie will break par tomorrow, assuming that she plays one round of golf?

18. **Application** The Tigers are leading the Storm one game to none in a best-of-five playoff series. After a playoff win, the probability of the Tigers winning the next game is 60%, while after a loss, their probability of winning the next game drops by 5%. The first team to win three games takes the series. Assume there are no ties. What is the probability of the Storm coming back to win the series?

6.5 Mutually Exclusive Events

The phone rings. Jacques is really hoping that it is one of his friends calling about either softball or band practice. Could the call be about both?

In such situations, more than one event could occur during a single trial. You need to compare the events in terms of the outcomes that make them up. What is the chance that at least one of the events happens? Is the situation "either/or," or can both events occur?

INVESTIGATE & INQUIRE: Baseball Pitches

Marie, at bat for the Coyotes, is facing Anton, who is pitching for the Power Trippers. Anton uses three pitches: a fastball, a curveball, and a slider. Marie feels she has a good chance of making a base hit, or better, if Anton throws either a fastball or a slider. The count is two strikes and three balls. In such full-count situations, Anton goes to his curveball one third of the time, his slider half as often, and his fastball the rest of the time.

1. Determine the probability of Anton throwing his
   a) curveball   b) slider   c) fastball

2. a) What is the probability that Marie will get the pitch she does not want?
   b) Explain how you can use this information to determine the probability that Marie will get a pitch she likes.

3. a) Show another method of determining this probability.
   b) Explain your method.

4. What do your answers to questions 2 and 3 suggest about the probabilities of events that cannot happen simultaneously?

The possible events in this investigation are said to be **mutually exclusive** (or **disjoint**) since they cannot occur at the same time. The pitch could not be both a fastball and a slider, for example. In this particular problem, you were interested in the probability of *either* of two favourable events. You can use the notation $P(A \text{ or } B)$ to stand for the probability of either A or B occurring.

Example 1 Probability of Mutually Exclusive Events

Teri attends a fundraiser at which 15 T-shirts are being given away as door prizes. Door prize winners are randomly given a shirt from a stock of 2 black shirts, 4 blue shirts, and 9 white shirts. Teri really likes the black and blue shirts, but is not too keen on the white ones. Assuming that Teri wins the first door prize, what is the probability that she will get a shirt that she likes?

Solution

Let A be the event that Teri wins a black shirt, and B be the event that she wins a blue shirt.

$$P(A) = \frac{2}{15} \quad \text{and} \quad P(B) = \frac{4}{15}$$

Teri would be happy if either A or B occurred.
There are $2 + 4 = 6$ non-white shirts, so

$$P(A \text{ or } B) = \frac{6}{15}$$

$$= \frac{2}{5}$$

The probability of Teri winning a shirt that she likes is $\frac{2}{5}$ or 40%. Notice that this probability is simply the sum of the probabilities of the two mutually exclusive events.

When events A and B are mutually exclusive, the probability that A or B will occur is given by the **addition rule for mutually exclusive events**:

$$P(A \text{ or } B) = P(A) + P(B)$$

A Venn diagram shows mutually exclusive events as non-overlapping, or disjoint. Thus, you can apply the additive counting principle (see Chapter 4) to prove this rule.

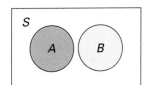

Proof:

If A and B are mutually exclusive events, then

$$P(A \text{ or } B) = \frac{n(A \text{ or } B)}{n(S)}$$

$$= \frac{n(A) + n(B)}{n(S)} \qquad \textit{A and B are disjoint sets, and thus share no elements.}$$

$$= \frac{n(A)}{n(S)} + \frac{n(B)}{n(S)}$$

$$= P(A) + P(B)$$

In some situations, events are **non-mutually exclusive**, which means that they can occur simultaneously. For example, consider a board game in which you need to roll either an 8 or doubles, using two dice.

Notice that in one outcome, rolling two fours, both events have occurred simultaneously. Hence, these events are not mutually exclusive. Counting the outcomes in the diagram shows that the probability of rolling either an 8 or doubles is $\frac{10}{36}$ or $\frac{5}{18}$. You need to take care not to count the (4, 4) outcome twice. You are applying the principle of inclusion and exclusion, which was explained in greater detail in Chapter 5.

	Second die					
	1	2	3	4	5	6
1	2	3	4	5	6	7
2	3	4	5	6	7	8
First die 3	4	5	6	7	8	9
4	5	6	7	8	9	10
5	6	7	8	9	10	11
6	7	8	9	10	11	12

Example 2 Probability of Non-Mutually Exclusive Events

A card is randomly selected from a standard deck of cards. What is the probability that either a heart or a face card (jack, queen, or king) is selected?

Solution

Let event A be that a heart is selected, and event B be that a face card is selected.

$$P(A) = \frac{13}{52} \text{ and } P(B) = \frac{12}{52}$$

If you add these probabilities, you get

$$P(A) + P(B) = \frac{13}{52} + \frac{12}{52}$$
$$= \frac{25}{52}$$

However, since the jack, queen, and king of hearts are in both A and B, the sum $P(A) + P(B)$ actually includes these outcomes *twice*.

A♣ 2♣ 3♣ 4♣ 5♣ 6♣ 7♣ 8♣ 9♣ 10♣ J♣ Q♣ K♣

A♦ 2♦ 3♦ 4♦ 5♦ 6♦ 7♦ 8♦ 9♦ 10♦ J♦ Q♦ K♦

A♥ 2♥ 3♥ 4♥ 5♥ 6♥ 7♥ 8♥ 9♥ 10♥ J♥ Q♥ K♥

A♠ 2♠ 3♠ 4♠ 5♠ 6♠ 7♠ 8♠ 9♠ 10♠ J♠ Q♠ K♠

Based on the diagram, the actual theoretical probability of drawing either a heart or a face card is $\frac{22}{52}$, or $\frac{11}{26}$. You can find the correct value by subtracting the probability of selecting the three elements that were counted twice.

$$P(A \text{ or } B) = \frac{13}{52} + \frac{12}{52} - \frac{3}{52}$$

$$= \frac{22}{52}$$

$$= \frac{11}{26}$$

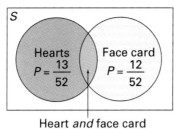

The probability that either a heart or a face card is selected is $\frac{11}{26}$.

When events A and B are non-mutually exclusive, the probability that A or B will occur is given by the **addition rule for non-mutually exclusive events**:

$$P(A \text{ or } B) = P(A) + P(B) - P(A \text{ and } B)$$

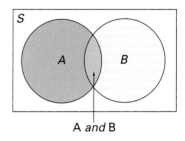

Example 3 Applying the Addition Rule for Non-Mutually Exclusive Events

An electronics manufacturer is testing a new product to see whether it requires a surge protector. The tests show that a voltage spike has a 0.2% probability of damaging the product's power supply, a 0.6% probability of damaging downstream components, and a 0.1% probability of damaging both the power supply and other components. Determine the probability that a voltage spike will damage the product.

Project Prep

When analysing the possible outcomes for your game in the probability project, you may need to consider mutually exclusive or non-mutually exclusive events. If so, you will need to apply the appropriate addition rule to determine theoretical probabilities.

Solution

Let A be damage to the power supply and C be damage to other components.

The overlapping region represents the probability that a voltage surge damages both the power supply and another component. The probability that either A or C occurs is given by

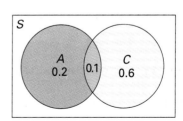

$$P(A \text{ or } C) = P(A) + P(C) - P(A \text{ and } C)$$

$$= 0.2\% + 0.6\% - 0.1\%$$

$$= 0.7\%$$

There is a 0.7% probability that a voltage spike will damage the product.

Practise

1. Classify each pair of events as mutually exclusive or non-mutually exclusive.

	Event A	Event B
a)	Randomly drawing a grey sock from a drawer	Randomly drawing a wool sock from a drawer
b)	Randomly selecting a student with brown eyes	Randomly selecting a student on the honour roll
c)	Having an even number of students in your class	Having an odd number of students in your class
d)	Rolling a six with a die	Rolling a prime number with a die
e)	Your birthday falling on a Saturday next year	Your birthday falling on a weekend next year
f)	Getting an A on the next test	Passing the next test
g)	Calm weather at noon tomorrow	Stormy weather at noon tomorrow
h)	Sunny weather next week	Rainy weather next week

2. Nine members of a baseball team are randomly assigned field positions. There are three outfielders, four infielders, a pitcher, and a catcher. Troy is happy to play any position except catcher or outfielder. Determine the probability that Troy will be assigned to play

   a) catcher

   b) outfielder

   c) a position he does not like

3. A car dealership analysed its customer database and discovered that in the last model year, 28% of its customers chose a 2-door model, 46% chose a 4-door model, 19% chose a minivan, and 7% chose a 4-by-4 vehicle. If a customer was selected randomly from this database, what is the probability that the customer

   a) bought a 4-by-4 vehicle?

   b) did not buy a minivan?

   c) bought a 2-door or a 4-door model?

   d) bought a minivan or a 4-by-4 vehicle?

Apply, Solve, Communicate

4. As a promotion, a resort has a draw for free family day-passes. The resort considers July, August, March, and December to be "vacation months."

  a) If the free passes are randomly dated, what is the probability that a day-pass will be dated within

   i) a vacation month?

   ii) June, July, or August

  b) Draw a Venn diagram of the events in part a).

5. A certain provincial park has 220 campsites. A total of 80 sites have electricity. Of the 52 sites on the lakeshore, 22 of them have electricity. If a site is selected at random, what is the probability that

  a) it will be on the lakeshore?

  b) it will have electricity?

  c) it will either have electricity or be on the lakeshore?

  d) it will be on the lakeshore and not have electricity?

6. A market-research firm monitored 1000 television viewers, consisting of 800 adults and 200 children, to evaluate a new comedy series that aired for the first time last week. Research indicated that 250 adults and 148 children viewed some or all of the program. If one of the 1000 viewers was selected, what is the probability that

  a) the viewer was an adult who did not watch the new program?

  b) the viewer was a child who watched the new program?

  c) the viewer was an adult or someone who watched the new program?

7. Application In an animal-behaviour study, hamsters were tested with a number of intelligence tasks, as shown in the table below.

Number of Tests	Number of Hamsters
0	10
1	6
2	4
3	3
4 or more	5

If a hamster is randomly chosen from this study group, what is the likelihood that the hamster has participated in

  a) exactly three tests?

  b) fewer than two tests?

  c) either one or two tests?

  d) no tests or more than three tests?

8. Communication

  a) Prove that, if A and B are non-mutually exclusive events, the probability of either A or B occurring is given by $P(A \text{ or } B) = P(A) + P(B) - P(A \text{ and } B)$.

  b) What can you conclude if $P(A \text{ and } B) = 0$? Give reasons for your conclusion.

9. Inquiry/Problem Solving Design a game in which the probability of drawing a winning card from a standard deck is between 55% and 60%.

10. Determine the probability that a captured deer has either cross-hatched antlers or bald patches. Are these events mutually exclusive? Why or why not?

11. The eight members of the debating club pose for a yearbook photograph. If they line up randomly, what is the probability that

  a) either Hania will be first in the row or Aaron will be last?

  b) Hania will be first and Aaron will not be last?

12. Consider a Stanley Cup playoff series in which the Toronto Maple Leafs hockey team faces the Ottawa Senators. Toronto hosts the first, second, and if needed, fifth and seventh games in this best-of-seven contest. The Leafs have a 65% chance of beating the Senators at home in the first game. After that, they have a 60% chance of a win at home if they won the previous game, but a 70% chance if they are bouncing back from a loss. Similarly, the Leafs' chances of victory in Ottawa are 40% after a win and 45% after a loss.

a) Construct a tree diagram to illustrate all the possible outcomes of the first three games.

b) Consider the following events:

A = {Leafs lose the first game but go on to win the series in the fifth game}

B = {Leafs win the series in the fifth game}

C = {Leafs lose the series in the fifth game}

Identify all the outcomes that make up each event, using strings of letters, such as *LLSLL*. Are any pairs from these three events mutually exclusive?

c) What is the probability of event A in part b)?

d) What is the chance of the Leafs winning in exactly five games?

e) Explain how you found your answers to parts c) and d).

13. A grade 12 student is selected at random to sit on a university liaison committee. Of the 120 students enrolled in the grade 12 university-preparation mathematics courses,

- 28 are enrolled in data management only
- 40 are enrolled in calculus only
- 15 are enrolled in geometry only
- 16 are enrolled in both data management and calculus
- 12 are enrolled in both calculus and geometry
- 6 are enrolled in both geometry and data management
- 3 are enrolled in all three of data management, calculus, and geometry

a) Draw a Venn diagram to illustrate this situation.

b) Determine the probability that the student selected will be enrolled in either data management or calculus.

c) Determine the probability that the student selected will be enrolled in only one of the three courses.

14. Application For a particular species of cat, the odds against a kitten being born with either blue eyes or white spots are 3:1. If the probability of a kitten exhibiting only one of these traits is equal and the probability of exhibiting both traits is 10%, what are the odds in favour of a kitten having blue eyes?

15. Communication

a) A standard deck of cards is shuffled and three cards are selected. What is the probability that the third card is either a red face card or a king if the king of diamonds and the king of spades are selected as the first two cards?

b) Does this probability change if the first two cards selected are the queen of diamonds and the king of spades? Explain.

16. Inquiry/Problem Solving The table below lists the degrees granted by Canadian universities from 1994 to 1998 in various fields of study.

a) If a Canadian university graduate from 1998 is chosen at random, what is the probability that the student is

i) a male?

ii) a graduate in mathematics and physical sciences?

iii) a male graduating in mathematics and physical sciences?

iv) not a male graduating in mathematics and physical sciences?

v) a male *or* a graduate in mathematics and physical sciences?

b) If a male graduate from 1996 is selected at random, what is the probability that he is graduating in mathematics and physical sciences?

c) If a mathematics and physical sciences graduate is selected at random from the period 1994 to 1996, what is the probability that the graduate is a male?

d) Do you think that being a male and graduating in mathematics and physical sciences are independent events? Give reasons for your hypothesis.

	1994	1995	1996	1997	1998
Canada	178 074	178 066	178 116	173 937	172 076
Male	76 470	76 022	75 106	73 041	71 949
Female	101 604	102 044	103 010	100 896	100 127
Social sciences	69 583	68 685	67 862	66 665	67 019
Male	30 700	29 741	29 029	28 421	27 993
Female	38 883	38 944	38 833	38 244	39 026
Education	30 369	30 643	29 792	27 807	25 956
Male	9093	9400	8693	8036	7565
Female	21 276	21 243	21 099	19 771	18 391
Humanities	23 071	22 511	22 357	21 373	20 816
Male	8427	8428	8277	8034	7589
Female	14 644	14 083	14 080	13 339	13 227
Health professions and occupations	12 183	12 473	12 895	13 073	12 658
Male	3475	3461	3517	3460	3514
Female	8708	9012	9378	9613	9144
Engineering and applied sciences	12 597	12 863	13 068	12 768	12 830
Male	10 285	10 284	10 446	10 125	10 121
Female	2312	2579	2622	2643	2709
Agriculture and biological sciences	10 087	10 501	11 400	11 775	12 209
Male	4309	4399	4756	4780	4779
Female	5778	6102	6644	6995	7430
Mathematics and physical sciences	9551	9879	9786	9738	9992
Male	6697	6941	6726	6749	6876
Female	2854	2938	3060	2989	3116
Fine and applied arts	5308	5240	5201	5206	5256
Male	1773	1740	1780	1706	1735
Female	3535	3500	3421	3500	3521
Arts and sciences	5325	5271	5755	5532	5340
Male	1711	1628	1882	1730	1777
Female	3614	3643	3873	3802	3563

Applying Matrices to Probability Problems

In some situations, the probability of an outcome depends on the outcome of the previous trial. Often this pattern appears in stock market trends, weather patterns, athletic performance, and consumer habits. Dependent probabilities can be calculated using Markov chains, a powerful probability model pioneered about a century ago by the Russian mathematician Andrei Markov.

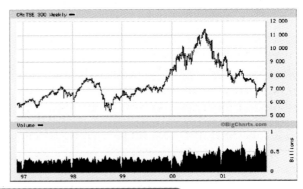

INVESTIGATE & INQUIRE: Running Late

Although Marla tries hard to be punctual, the demands of her home life and the challenges of commuting sometimes cause her to be late for work. When she is late, she tries especially hard to be punctual the next day. Suppose that the following pattern emerges: If Marla is punctual on any given day, then there is a 70% chance that she will be punctual the next day and a 30% chance that she will be late. On days she is late, however, there is a 90% chance that she will be punctual the next day and just a 10% chance that she will be late. Suppose Marla is punctual on the first day of the work week.

1. Create a tree diagram of the possible outcomes for the second and third days. Show the probability for each branch.

2. a) Describe two branches in which Marla is punctual on day 3.

   b) Use the product rule for dependent events on page 332 to calculate the compound probability of Marla being punctual on day 2 and on day 3.

   c) Find the probability of Marla being late on day 2 and punctual on day 3.

   d) Use the results from parts b) and c) to determine the probability that Marla will be punctual on day 3.

3. Repeat question 2 for the outcome of Marla being late on day 3.

4. a) Create a 1×2 matrix A in which the first element is the probability that Marla is punctual and the second element is the probability that she is late on day 1. Recall that Marla is punctual on day 1.

   b) Create a 2×2 matrix B in which the elements in each row represent *conditional* probabilities that Marla will be punctual and late. Let the first row be the probabilities after a day in which Marla was punctual, and the second row be the probabilities after a day in which she was late.

c) Evaluate $A \times B$ and $A \times B^2$.

d) Compare the results of part c) with your answers to questions 2 and 3. Explain what you notice.

e) What does the first row of the matrix B^2 represent?

The matrix model you have just developed is an example of a **Markov chain**, a probability model in which the outcome of any trial depends directly on the outcome of the previous trial. Using matrix operations can simplify probability calculations, especially in determining long-term trends.

The 1×2 matrix A in the investigation is an **initial probability vector, $S^{(0)}$**, and represents the probabilities of the initial state of a Markov chain. The 2×2 matrix B is a **transition matrix, P**, and represents the probabilities of moving from any initial state to a new state in any trial.

These matrices have been arranged such that the product $S^{(0)} \times P$ generates the row matrix that gives the probabilities of each state after one trial. This matrix is called the **first-step probability vector, $S^{(1)}$**. In general, the **nth-step probability vector, $S^{(n)}$**, can be obtained by repeatedly multiplying the probability vector by P. Sometimes these vectors are also called **first-state** and **nth-state vectors**, respectively.

Notice that each entry in a probability vector or a transition matrix is a probability and must therefore be between 0 and 1. The possible states in a Markov chain are always mutually exclusive events, one of which must occur at each stage. Therefore, the entries in a probability vector must sum to 1, as must the entries in each row of the transition matrix.

Example 1 Probability Vectors

Two video stores, Video Vic's and MovieMaster, have just opened in a new residential area. Initially, they each have half of the market for rented movies. A customer who rents from Video Vic's has a 60% probability of renting from Video Vic's the next time and a 40% chance of renting from MovieMaster. On the other hand, a customer initially renting from MovieMaster has only a 30% likelihood of renting from MovieMaster the next time and a 70% probability of renting from Video Vic's.

a) What is the initial probability vector?

b) What is the transition matrix?

c) What is the probability of a customer renting a movie from each store the second time?

d) What is the probability of a customer renting a movie from each store the third time?

e) What assumption are you making in part d)? How realistic is it?

Solution

a) Initially, each store has 50% of the market, so, the initial probability vector is

$$\begin{array}{cc} VV & MM \end{array}$$
$$S^{(0)} = [0.5 \quad 0.5]$$

b) The first row of the transition matrix represents the probabilities for the second rental by customers whose initial choice was Video Vic's. There is a 60% chance that the customer returns, so the first entry is 0.6. It is 40% likely that the customer will rent from MovieMaster, so the second entry is 0.4.

Similarly, the second row of the transition matrix represents the probabilities for the second rental by customers whose first choice was MovieMaster. There is a 30% chance that a customer will return on the next visit, and a 70% chance that the customer will try Video Vic's.

$$\begin{array}{cc} & VV \quad MM \end{array}$$
$$P = \begin{bmatrix} 0.6 & 0.4 \\ 0.7 & 0.3 \end{bmatrix} \begin{array}{c} VV \\ MM \end{array}$$

Regardless of which store the customer chooses the first time, you are assuming that there are only two choices for the next visit. Hence, the sum of the probabilities in each row equals one.

c) To find the probabilities of a customer renting from either store on the second visit, calculate the first-step probability vector, $S^{(1)}$:

$$S^{(1)} = S^{(0)}P$$
$$= [0.5 \quad 0.5] \begin{bmatrix} 0.6 & 0.4 \\ 0.7 & 0.3 \end{bmatrix}$$
$$= [0.65 \quad 0.35]$$

This new vector shows that there is a 65% probability that a customer will rent a movie from Video Vic's on the second visit to a video store and a 35% chance that the customer will rent from MovieMaster.

d) To determine the probabilities of which store a customer will pick on the third visit, calculate the second-step probability vector, $S^{(2)}$:

$$S^{(2)} = S^{(1)}P$$
$$= [0.65 \quad 0.35] \begin{bmatrix} 0.6 & 0.4 \\ 0.7 & 0.3 \end{bmatrix}$$
$$= [0.635 \quad 0.365]$$

So, on a third visit, a customer is 63.5% likely to rent from Video Vic's and 36.5% likely to rent from MovieMaster.

e) To calculate the second-step probabilities, you assume that the conditional transition probabilities do not change. This assumption might not be realistic since customers who are 70% likely to switch away from MovieMaster may not be as much as 40% likely to switch back, unless they forget why they switched in the first place. In other words, Markov chains have no long-term memory. They recall only the latest state in predicting the next one.

Note that the result in Example 1d) could be calculated in another way.

$$S^{(2)} = S^{(1)}P$$
$$= (S^{(0)}P)P$$
$$= S^{(0)}(PP) \quad \textit{since matrix multiplication is associative}$$
$$= S^{(0)}P^2$$

Similarly, $S^{(3)} = S^{(0)}P^3$, and so on. In general, the nth-step probability vector, $S^{(n)}$, is given by

$$S^{(n)} = S^{(0)}P^n$$

This result enables you to determine higher-state probability vectors easily using a graphing calculator or software.

Example 2 Long-Term Market Share

A marketing-research firm has tracked the sales of three brands of hockey sticks. Each year, on average,
- Player-One keeps 70% of its customers, but loses 20% to Slapshot and 10% to Extreme Styx
- Slapshot keeps 65% of its customers, but loses 10% to Extreme Styx and 25% to Player-One
- Extreme Styx keeps 55% of its customers, but loses 30% to Player-One and 15% to Slapshot

a) What is the transition matrix?

b) Assuming each brand begins with an equal market share, determine the market share of each brand after one, two, and three years.

c) Determine the long-range market share of each brand.

d) What assumption must you make to answer part c)?

Solution 1 Using Pencil and Paper

a) The transition matrix is

$$P = \begin{matrix} & \begin{matrix} P & S & E \end{matrix} \\ \begin{bmatrix} & 0.7 & 0.2 & 0.1 \\ & 0.25 & 0.65 & 0.1 \\ & 0.3 & 0.15 & 0.55 \end{bmatrix} & \begin{matrix} P \\ S \\ E \end{matrix} \end{matrix}$$

b) Assuming each brand begins with an equal market share, the initial probability vector is

$$S^{(0)} = \begin{bmatrix} \dfrac{1}{3} & \dfrac{1}{3} & \dfrac{1}{3} \end{bmatrix}$$

To determine the market shares of each brand after one year, compute the first-step probability vector.

$$S^{(1)} = S^{(0)}P$$

$$= \begin{bmatrix} \dfrac{1}{3} & \dfrac{1}{3} & \dfrac{1}{3} \end{bmatrix} \begin{bmatrix} 0.7 & 0.2 & 0.1 \\ 0.25 & 0.65 & 0.1 \\ 0.3 & 0.15 & 0.55 \end{bmatrix}$$

$$= [0.41\overline{6} \quad 0.\overline{3} \quad 0.25]$$

So, after one year Player-One will have a market share of approximately 42%, Slapshot will have 33%, and Extreme Styx will have 25%.

Similarly, you can predict the market shares after two years using

$$S^{(2)} = S^{(1)}P$$

$$= [0.41\overline{6} \quad 0.\overline{3} \quad 0.25] \begin{bmatrix} 0.7 & 0.2 & 0.1 \\ 0.25 & 0.65 & 0.1 \\ 0.3 & 0.15 & 0.55 \end{bmatrix}$$

$$= [0.45 \quad 0.3375 \quad 0.2125]$$

After two years, Player-One will have approximately 45% of the market, Slapshot will have 34%, and Extreme Styx will have 21%.

The probabilities after three years are given by

$$S^{(3)} = S^{(2)}P$$

$$= [0.45 \quad 0.3375 \quad 0.2125] \begin{bmatrix} 0.7 & 0.2 & 0.1 \\ 0.25 & 0.65 & 0.1 \\ 0.3 & 0.15 & 0.55 \end{bmatrix}$$

$$= [0.463 \quad 0.341 \quad 0.196]$$

After three years, Player-One will have approximately 46% of the market, Slapshot will have 34%, and Extreme Styx will have 20%.

c) The results from part b) suggest that the relative market shares may be converging to a steady state over a long period of time. You can test this hypothesis by calculating higher-state vectors and checking for stability.

For example,

$$S^{(10)} = S^{(9)}P \qquad\qquad S^{(11)} = S^{(10)}P$$
$$= [0.471 \ \ 0.347 \ \ 0.182] \qquad\qquad = [0.471 \ \ 0.347 \ \ 0.182]$$

The values of $S^{(10)}$ and $S^{(11)}$ are equal. It is easy to verify that they are equal to all higher orders of $S^{(n)}$ as well. The Markov chain has reached a **steady state**. A steady-state vector is a probability vector that remains unchanged when multiplied by the transition matrix. A steady state has been reached if

$$S^{(n)} = S^{(n)}P$$
$$= S^{(n+1)}$$

In this case, the steady state vector $[0.471 \ \ 0.347 \ \ 0.182]$ indicates that, over a long period of time, Player-One will have approximately 47% of the market for hockey sticks, while Slapshot and Extreme Styx will have 35% and 18%, respectively, based on current trends.

d) The assumption you make in part c) is that the transition matrix does not change, that is, the market trends stay the same over the long term.

Solution 2 Using a Graphing Calculator

a) Use the **MATRX EDIT** menu to enter and store a matrix for the transition matrix B.

b) Similarly, enter the initial probability vector as matrix A. Then, use the **MATRX EDIT** menu to enter the calculation $A \times B$ on the home screen. The resulting matrix shows the market shares after one year are 42%, 33%, and 25%, respectively.

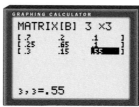

To find the second-step probability vector use the formula $S^{(2)} = S^{(0)}P^2$. Enter $A \times B^2$ using the **MATRX NAMES** menu and the $\boxed{x^2}$ key. After two years, therefore, the market shares are 45%, 34%, and 21%, respectively.

Similarly, enter $A \times B^3$ to find the third-step probability vector. After three years, the market shares are 46%, 34%, and 20%, respectively.

c) Higher-state probability vectors are easy to determine with a graphing calculator.

$$S^{(10)} = S^{(0)}P^{10}$$
$$= [0.471 \quad 0.347 \quad 0.182]$$

$$S^{(100)} = S^{(0)}P^{100}$$
$$= [0.471 \quad 0.347 \quad 0.182]$$

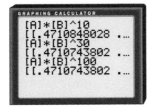

$S^{(10)}$ and $S^{(100)}$ are equal. The tiny difference between $S^{(10)}$ and $S^{(100)}$ is unimportant since the original data has only two significant digits. Thus, [0.471 0.347 0.182] is a steady-state vector, and the long-term market shares are predicted to be about 47%, 35%, and 18% for Player-One, Slapshot, and Extreme Styx, respectively.

Regular Markov chains always achieve a steady state. A Markov chain is regular if the transition matrix P or some power of P has no zero entries. Thus, regular Markov chains are fairly easy to identify. A regular Markov chain will reach the same steady state *regardless* of the initial probability vector.

Example 3 Steady State of a Regular Markov Chain

Suppose that Player-One and Slapshot initially split most of the market evenly between them, and that Extreme Styx, a relatively new company, starts with a 10% market share.
a) Determine each company's market share after one year.
b) Predict the long-term market shares.

Solution

a) The initial probability vector is
$$S^{(0)} = [0.45 \quad 0.45 \quad 0.1]$$

Using the same transition matrix as in Example 2,
$$S^{(1)} = S^{(0)}P$$
$$= [0.45 \quad 0.45 \quad 0.1] \begin{bmatrix} 0.7 & 0.2 & 0.1 \\ 0.25 & 0.65 & 0.1 \\ 0.3 & 0.15 & 0.55 \end{bmatrix}$$
$$= [0.4575 \quad 0.3975 \quad 0.145]$$

These market shares differ from those in Example 2, where $S^{(1)} = [0.41\overline{6} \quad 0.\overline{3} \quad 0.25]$.

Project Prep

In the probability project, you may need to use Markov chains to determine long-term probabilities.

b) $S^{(100)} = S^{(0)}P^{100}$

$$= [0.471 \ \ 0.347 \ \ 0.182]$$

In the long term, the steady state is the same as in Example 2. Notice that although the short-term results differ as seen in part a), the same steady state is achieved in the long term.

The steady state of a regular Markov chain can also be determined analytically.

Example 4 Analytic Determination of Steady State

The weather near a certain seaport follows this pattern: If it is a calm day, there is a 70% chance that the next day will be calm and a 30% chance that it will be stormy. If it is a stormy day, the chances are 50/50 that the next day will also be stormy. Determine the long-term probability for the weather at the port.

Solution

The transition matrix for this Markov chain is

$$P = \begin{matrix} & C & S \\ & \begin{bmatrix} 0.7 & 0.3 \\ 0.5 & 0.5 \end{bmatrix} & \begin{matrix} C \\ S \end{matrix} \end{matrix}$$

The steady-state vector will be a 1×2 matrix, $S^{(n)} = [p \ \ q]$.

The Markov chain will reach a steady state when $S^{(n)} = S^{(n)}P$, so

$$[p \ \ q] = [p \ \ q] \begin{bmatrix} 0.7 & 0.3 \\ 0.5 & 0.5 \end{bmatrix}$$

$$= [0.7p + 0.5q \ \ \ 0.3p + 0.5q]$$

Setting first elements equal and second elements equal gives two equations in two unknowns. These equations are dependent, so they define only one relationship between p and q.

$p = 0.7p + 0.5q$
$q = 0.3p + 0.5q$

Subtracting the second equation from the first gives

$p - q = 0.4p$
$q = 0.6p$

Now, use the fact that the sum of probabilities at any state must equal 1,

$$p + q = 1$$
$$p + 0.6p = 1$$
$$p = \frac{1}{1.6}$$
$$= 0.625$$
$$q = 1 - p$$
$$= 0.375$$

So, the steady-state vector for the weather is [0.625 0.375]. Over the long term, there will be a 62.5% probability of a calm day and 37.5% chance of a stormy day at the seaport.

Key Concepts

- The theory of Markov chains can be applied to probability models in which the outcome of one trial directly affects the outcome of the next trial.

- Regular Markov chains eventually reach a steady state, which can be used to make long-term predictions.

Communicate Your Understanding

1. Why must a transition matrix always be square?

2. Given an initial probability vector $S^{(0)} = [0.4\ 0.6]$ and a transition matrix $P = \begin{bmatrix} 0.5 & 0.5 \\ 0.3 & 0.7 \end{bmatrix}$, state which of the following equations is easier to use for determining the third-step probability vector:
   $$S^{(3)} = S^{(2)}P \quad \text{or} \quad S^{(3)} = S^{(0)}P^3$$
   Explain your choice.

3. Explain how you can determine whether a Markov chain has reached a steady state after k trials.

4. What property or properties must events A, B, and C have if they are the only possible different states of a Markov chain?

Practise

1. Which of the following cannot be an initial probability vector? Explain why.

a) [0.2 0.45 0.25]

b) [0.29 0.71]

c) $\begin{bmatrix} 0.4 \\ 0.6 \end{bmatrix}$

d) [0.4 −0.1 0.7]

e) [0.4 0.2 0.15 0.25]

2. Which of the following cannot be a transition matrix? Explain why.

a) $\begin{bmatrix} 0.3 & 0.3 & 0.4 \\ 0.1 & 0 & 0.9 \\ 0.2 & 0.3 & 0.4 \end{bmatrix}$

b) $\begin{bmatrix} 0.2 & 0.8 \\ 0.65 & 0.35 \end{bmatrix}$

c) $\begin{bmatrix} 0.5 & 0.1 & 0.4 \\ 0.3 & 0.22 & 0.48 \end{bmatrix}$

3. Two competing companies, ZapShot and E-pics, manufacture and sell digital cameras. Customer surveys suggest that the companies' market shares can be modelled using a Markov chain with the following initial probability vector $S^{(0)}$ and transition matrix P.

$$S^{(0)} = [0.67 \ 0.33] \qquad P = \begin{bmatrix} 0.6 & 0.4 \\ 0.5 & 0.5 \end{bmatrix}$$

Assume that the first element in the initial probability vector pertains to ZapShot. Explain the significance of

a) the elements in the initial probability vector

b) each element of the transition matrix

c) each element of the product $S^{(0)}P$

Apply, Solve, Communicate

4. Refer to question 3.

a) Which company do you think will increase its long-term market share, based on the information provided? Explain why you think so.

b) Calculate the steady-state vector for the Markov chain.

c) Which company increased its market share over the long term?

d) Compare this result with your answer to part a). Explain any differences.

5. For which of these transition matrices will the Markov chain be regular? In each case, explain why.

a) $\begin{bmatrix} 0.2 & 0.8 \\ 0.95 & 0.05 \end{bmatrix}$

b) $\begin{bmatrix} 1 & 0 \\ 0 & 1 \end{bmatrix}$

c) $\begin{bmatrix} 0.1 & 0.6 & 0.3 \\ 0.33 & 0.3 & 0.37 \\ 0.5 & 0 & 0.5 \end{bmatrix}$

6. Gina noticed that the performance of her baseball team seemed to depend on the outcome of their previous game. When her team won, there was a 70% chance that they would win the next game. If they lost, however, there was only a 40% chance that they would their next game.

a) What is the transition matrix of the Markov chain for this situation?

b) Following a loss, what is the probability that Gina's team will win two games later?

c) What is the steady-state vector for the Markov chain, and what does it mean?

7. **Application** Two popcorn manufacturers, Ready-Pop and ButterPlus, are competing for the same market. Trends indicate that 65% of consumers who purchase Ready-Pop will stay with Ready-Pop the next time, while 35% will try ButterPlus. Among those who purchase ButterPlus, 75% will buy ButterPlus again and 25% will switch to Ready-Pop. Each popcorn producer initially has 50% of the market.

   a) What is the initial probability vector?

   b) What is the transition matrix?

   c) Determine the first- and second-step probability vectors.

   d) What is the long-term probability that a customer will buy Ready-Pop?

8. **Inquiry/Problem Solving** The weather pattern for a certain region is as follows. On a sunny day, there is a 50% probability that the next day will be sunny, a 30% chance that the next day will be cloudy, and a 20% chance that the next day will be rainy. On a cloudy day, the probability that the next day will be cloudy is 35%, while it is 40% likely to be rainy and 25% likely to be sunny the next day. On a rainy day, there is a 45% chance that it will be rainy the next day, a 20% chance that the next day will be sunny, and a 35% chance that the next day will be cloudy.

   a) What is the transition matrix?

   b) If it is cloudy on Wednesday, what is the probability that it will be sunny on Saturday?

   c) What is the probability that it will be sunny four months from today, according to this model?

   d) What assumptions must you make in part c)? Are they realistic? Why or why not?

9. **Application** On any given day, the stock price for Bluebird Mutual may rise, fall, or remain unchanged. These states, R, F, and U, can be modelled by a Markov chain with the transition matrix:

$$\begin{array}{ccc} R & F & U \end{array}$$
$$\begin{bmatrix} 0.75 & 0.15 & 0.1 \\ 0.25 & 0.6 & 0.15 \\ 0.4 & 0.4 & 0.2 \end{bmatrix} \begin{array}{c} R \\ F \\ U \end{array}$$

   a) If, after a day of trading, the value of Bluebird's stock has fallen, what is the probability that it will rise the next day?

   b) If Bluebird's value has just risen, what is the likelihood that it will rise one week from now?

   c) Assuming that the behaviour of the Bluebird stock continues to follow this established pattern, would you consider Bluebird to be a safe investment? Explain your answer, and justify your reasoning with appropriate calculations.

10. Assume that each doe produces one female offspring. Let the two states be D, a normal doe, and B, a doe with bald patches. Determine

   a) the initial probability vector

   b) the transition matrix for each generation of offspring

   c) the long-term probability of a new-born doe developing bald patches

   d) Describe the assumptions which are inherent in this analysis. What other factors could affect the stability of this Markov chain?

11. When Mazemaster, the mouse, is placed in a maze like the one shown below, he will explore the maze by picking the doors at random to move from compartment to compartment. A transition takes place when Mazemaster moves through one of the doors into another compartment. Since all the doors lead to *other* compartments, the probability of moving from a compartment back to the same compartment in a single transition is zero.

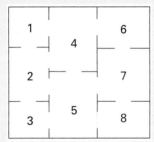

a) Construct the transition matrix, P, for the Markov chain.

b) Use technology to calculate P^2, P^3, and P^4.

c) If Mazemaster starts in compartment 1, what is the probability that he will be in compartment 4 after

   i) two transitions?

   ii) three transitions?

   iii) four transitions?

d) Predict where Mazemaster is most likely to be in the long run. Explain the reasoning for your prediction.

e) Calculate the steady-state vector. Does it support your prediction? If not, identify the error in your reasoning in part d).

12. Communication Refer to Example 4 on page 351.

a) Suppose that the probability of stormy weather on any day following a calm day increases by 0.1. Estimate the effect this change will have on the steady state of the Markov chain. Explain your prediction.

b) Calculate the new steady-state vector and compare the result with your prediction. Discuss any difference between your estimate and the calculated steady state.

c) Repeat parts a) and b) for the situation in which the probability of stormy weather following either a calm or a stormy day increases by 0.1, compared to the data in Example 4.

d) Discuss possible factors that might cause the mathematical model to be altered.

13. For each of the transition matrices below, decide whether the Markov chain is regular and whether it approaches a steady state. (*Hint*: An irregular Markov chain could still have a steady-state vector.)

a) $\begin{bmatrix} 0 & 1 \\ 1 & 0 \end{bmatrix}$ **b)** $\begin{bmatrix} 0 & 1 \\ 0.5 & 0.5 \end{bmatrix}$

c) $\begin{bmatrix} 1 & 0 \\ 0.5 & 0.5 \end{bmatrix}$

14. Refer to Example 2 on page 347.

a) Using a graphing calculator, find P^{100}. Describe this matrix.

b) Let $S^{(0)} = [a \ b \ c]$. Find an expression for the value of $S^{(0)}P^{100}$. Does this expression depend on $S^{(0)}$, P, or both?

c) What property of a regular Markov chain can you deduce from your answer to part b)?

15. Inquiry/Problem Solving The transition matrix for a Markov chain with steady-state vector of $\left[\begin{array}{cc} \dfrac{7}{13} & \dfrac{6}{13} \end{array}\right]$ is $\left[\begin{array}{cc} 0.4 & 0.6 \\ m & n \end{array}\right]$. Determine the unknown transition matrix elements, m and n.

Investment Broker

Many people use the services of an investment broker to help them invest their earnings. An investment broker provides advice to clients on how to invest their money, based on their individual goals, income, and risk tolerance, among other factors. An investment broker can work for a financial institution, such as a bank or trust company, or a brokerage, which is a company that specializes in investments. An investment broker typically buys, sells, and trades a variety of investment items, including stocks, bonds, mutual funds, and treasury bills.

An investment broker must be able to read and interpret a variety of financial data including periodicals and corporate reports. Based on experience and sound mathematical principles, the successful investment broker must be able to make reasonable predictions of uncertain outcomes.

Because of the nature of this industry, earnings often depend directly on performance. An investment broker typically earns a commission, similar to that for a sales representative. In the short term, the investment broker can expect some fluctuations in earnings. In the long term, strong performers can expect a very comfortable living, while weak performers are not likely to last long in the field.

Usually, an investment broker requires a minimum of a bachelor's degree in economics or business, although related work experience in investments or sales is sometimes an acceptable substitute. A broker must have a licence from the provincial securities commission and must pass specialized courses in order to trade in specific investment products such as securities, options, and futures contracts. The chartered financial analyst (CFA) designation is recommended for brokers wishing to enter the mutual-fund field or other financial-planning services.

WEB CONNECTION

www.mcgrawhill.ca/links/MDM12

Visit the above web site and follow the links to find out more about an investment broker and other careers related to mathematics.

Review of Key Concepts

6.1 Basic Probability Concepts
Refer to the Key Concepts on page 311.

1. A bag of marbles contains seven whites, five blacks, and eight cat's-eyes. Determine the probability that a randomly drawn marble is
   a) a white marble
   b) a marble that is not black

2. When a die was rolled 20 times, 4 came up five times.
   a) Determine the empirical probability of rolling a 4 with a die based on the 20 trials.
   b) Determine the theoretical probability of rolling a 4 with a die.
   c) How can you account for the difference between the results of parts a) and b)?

3. Estimate the subjective probability of each event and provide a rationale for your decision.
   a) All classes next week will be cancelled.
   b) At least one severe snow storm will occur in your area next winter.

6.2 Odds
Refer to the Key Concepts on page 317.

4. Determine the odds in favour of flipping three coins and having them all turn up heads.

5. A restaurant owner conducts a study that measures the frequency of customer visits in a given month. The results are recorded in the following table.

Number of Visits	Number of Customers
1	4
2	6
3	7
4 or more	3

Based on this survey, calculate
   a) the odds that a customer visited the restaurant exactly three times
   b) the odds in favour of a customer having visited the restaurant fewer than three times
   c) the odds against a customer having visited the restaurant more than three times

6.3 Probabilities Using Counting Techniques
Refer to the Key Concepts on page 324.

6. Suppose three marbles are selected at random from the bag of marbles in question 1.
   a) Draw a tree diagram to illustrate all possible outcomes.
   b) Are all possible outcomes equally likely? Explain.
   c) Determine the probability that all three selected marbles are cat's-eyes.
   d) Determine the probability that none of the marbles drawn are cat's-eyes.

7. The Sluggers baseball team has a starting line-up consisting of nine players, including Tyrone and his sister Amanda. If the batting order is randomly assigned, what is the probability that Tyrone will bat first, followed by Amanda?

8. A three-member athletics council is to be randomly chosen from ten students, five of whom are runners. The council positions are president, secretary, and treasurer. Determine the probability that
   a) the committee is comprised of all runners
   b) the committee is comprised of the three eldest runners
   c) the eldest runner is president, second eldest runner is secretary, and third eldest runner is treasurer

6.4 Dependent and Independent Events
Refer to the Key Concepts on page 333.

9. Classify each of the following pairs of events as independent or dependent.

	First Event	Second Event
a)	Hitting a home run while at bat	Catching a pop fly while in the field
b)	Staying up late	Sleeping in the next day
c)	Completing your calculus review	Passing your calculus exam
d)	Randomly selecting a shirt	Randomly selecting a tie

10. Bruno has just had job interviews with two separate firms: Golden Enterprises and Outer Orbit Manufacturing. He estimates that he has a 40% chance of receiving a job offer from Golden and a 75% chance of receiving an offer from Outer Orbit.

    a) What is the probability that Bruno will receive both job offers?

    b) Is Bruno applying the concept of theoretical, empirical, or subjective probability? Explain.

11. Karen and Klaus are the parents of James and twin girls Britta and Kate. Each family member has two shirts in the wash. If a shirt is pulled from the dryer at random, what is the probability that the shirt belongs to

    a) Klaus, if it is known that the shirt belongs to one of the parents?

    b) Britta, if it is known that the shirt is for a female?

    c) Kate, if it is known that the shirt belongs to one of the twins?

    d) Karen or James

12. During a marketing blitz, a telemarketer conducts phone solicitations continuously from 16 00 until 20 00. Suppose that you have a 20% probability of being called during this blitz. If you generally eat dinner between 18 00 and 18 30, how likely is it that the telemarketer will interrupt your dinner?

6.5 Mutually Exclusive Events
Refer to the Key Concepts on page 340.

13. Classify each pair of events as mutually exclusive or non-mutually exclusive.

	First Event	Second Event
a)	Randomly selecting a classical CD	Randomly selecting a rock CD
b)	Your next birthday occurring on a Wednesday	Your next birthday occurring on a weekend
c)	Ordering a hamburger with cheese	Ordering a hamburger with no onions
d)	Rolling a perfect square with a die	Rolling an even number with a die

14. a) Determine the probability of drawing either a 5 or a black face card from a standard deck of cards.

    b) Illustrate this situation with a Venn diagram.

15. In a data management class of 26 students, there are 9 with blonde hair, 7 with glasses, and 4 with blonde hair and glasses.

    a) Draw a Venn diagram to illustrate this situation.

    b) If a student is selected at random, what is the probability that the student will have either blonde hair or glasses?

16. Of 150 students at a school dance, 110 like pop songs and 70 like heavy-metal songs. A third of the students like both pop and heavy-metal songs.

   a) If a pop song is played, what are the odds in favour of a randomly selected student liking the song?

   b) What are the odds in favour of a student disliking both pop and heavy-metal songs?

   c) Discuss any assumptions which must be made in parts a) and b).

17. The four main blood types are A, B, AB, and O. The letters A and B indicate whether two factors (particular molecules on the surface of the blood cells) are present. Thus, type AB blood has both factors while type O has neither. Roughly 42% of the population have type A blood, 10% have type B, 3% have type AB, and 45% have type O. What is the probability that a person

   a) has blood factor A?

   b) does not have blood factor B?

6.6 Applying Matrices to Probability Problems

Refer to the Key Concepts on page 352.

18. Alysia, the star on her bowling team, tends to bowl better when her confidence is high. When Alysia bowls a strike, there is a 50% probability that she will bowl a strike in the next frame. If she does not bowl a strike, then she has a 35% probability of bowling a strike in the next frame. Assume that Alysia starts the first game with a strike.

   a) What is the initial probability vector for this situation?

   b) What is the transition matrix?

   c) Determine the probability that Alysia will bowl a strike in the second, third, and tenth frames.

   d) There are ten frames in a game of bowling. What is the probability that Alysia will bowl a strike in the first frame of the second game?

   e) What is the long-term probability that Alysia will bowl a strike?

   f) What assumptions must be made in parts d) and e)?

19. A year-long marketing study observed the following trends among consumers of three competing pen manufacturers.

   - 20% of Blip Pens customers switched to Stylo and 10% switched to Glyde-Wryte.
   - 15% of Stylo customers switched to Glyde-Wryte and 25% switched to Blip Pens.
   - 30% of Glyde-Wryte customers switched to Blip Pens and 5% switched to Stylo.

   a) What is the transition matrix for this situation?

   b) Determine the steady-state vector.

   c) What is the expected long-term market share of Glyde-Wryte if these trends continue?

Chapter Test

1. A coin is tossed three times.

   a) Draw a tree diagram to illustrate the possible outcomes.

   b) Determine the probability that heads will appear each time.

2. A jumbled desk drawer contains three pencils, four pens, and two markers. If you randomly pull out a writing instrument,

   a) what is the probability that it is not a pencil?

   b) what are the odds in favour of pulling out a pen?

3. A die is rolled ten times. What is the probability that a prime number will be rolled every time?

4. If Juanita bumps into Troy in the hallway between periods 2 and 3, there is a 25% chance that she will be late for class. If she does not bump into Troy, she will make it to class on time. If there is a 20% chance that Juanita will bump into Troy, how likely is it that she will be late on any given day?

5. Of 150 workers surveyed in an industrial community, 65 worked in the paper mill and 30 worked in the water-treatment plant.

   a) What is the probability that a worker surveyed at random works

   i) in either the paper mill or the water-treatment plant?

   ii) somewhere other than the paper mill or the water-treatment plant?

   b) What assumptions must you make in part a)?

6. The gene for blood type A is dominant over the one for type O blood. To have type O blood, a child must inherit type O genes from both parents. If the parents of a child both have one blood type A gene and one blood type O gene, what are the odds in favour of the child having type O blood?

7. Of the members of a track-and-field club, 42% entered track events at the most recent provincial meet, 32% entered field events, and 20% entered both track and field events.

   a) Illustrate the club's entries with a Venn diagram.

   b) What is the probability that a randomly selected member of the club

   i) entered either a track event or a field event at the provincial meet?

   ii) did not compete at the meet?

8. Five siblings, Paula, Mike, Stephanie, Kurt, and Emily, are randomly seated along one side of a long table. What is the probability that the children are seated

   a) with the three girls in the middle?

   b) in order of age?

9. Naomi, a fan of alternative music, has 12 CDs.

Band	Number of CDs
Nine Inch Nails	3
Soundgarden	4
Monster Magnet	2
Pretty & Twisted	1
Queensrÿche	2

If Naomi randomly loads her player with five CDs, what is the probability that it will hold

a) no Soundgarden CDs?

b) exactly one Monster Magnet CD?

c) three Nine Inch Nails CDs or three Soundgarden CDs?

d) one CD from each band?

10. Ursula, an electrical engineer in a quality-control department, checks silicon-controlled rectifiers (SCRs) for manufacturing defects. She has noticed that when a defective SCR turned up on the assembly line, there was a 0.07 probability that the next unit would also be defective. If, however, an SCR passed inspection, then there was just a 0.004 likelihood that the next unit would fail inspection.

a) Assuming that Ursula has just found a defective SCR, find

   i) the initial probability vector

   ii) the transition matrix

   iii) the probability that the next two SCRs will both fail inspection

b) Is this Markov chain regular? Explain why or why not.

c) What does your answer to part b) imply about the long-term probability of an SCR failing inspection? Quantify your answer.

ACHIEVEMENT CHECK

Knowledge/Understanding	Thinking/Inquiry/Problem Solving	Communication	Application

11. Candice owns a chocolate shop. One of her most popular products is a box of 40 assorted chocolates, 5 of which contain nuts.

a) If a person selects two chocolates at random from the box, what is the probability that

   i) both of the chocolates contain nuts?

   ii) at least one of the chocolates contains nuts?

   iii) only one of the chocolates contains nuts?

   iv) neither of the chocolates contain nuts?

b) Describe how you could simulate choosing the two chocolates. Outline a method using

   i) a manual technique

   ii) appropriate technology

c) Suppose you ran 100 trials with either of your simulations. Would you expect empirical probabilities based on the results of these trials to match the probabilities you computed in part a)? Why or why not?

d) Due to the popularity of the chocolates with nuts, Candice is planning to double the number of them in each box. She claims that having 10 of the 40 chocolates contain nuts will double the probability that one or both of two randomly selected chocolates will contain nuts. Do you agree with her claim? Support your answer with probability calculations.

Wrap-Up

Implementing Your Action Plan

1. Determine the probability of winning your game. If the game is simple enough, you can present both a theoretical and an empirical probability. For complex games, you may have to rely on empirical probability alone. If practical, use technology to simulate your game, and run enough trials to have confidence in your results.

2. Develop a winning strategy for your game. If no winning strategy is possible, explain why.

3. Prepare all the components for the game, such as a board, tokens, instruction sheets, or score cards.

4. Have other students, and perhaps your teacher, try your game. Note their comments and any difficulties they had with the game.

5. Keep track of the outcomes and analyse them using technology, where appropriate. Determine whether you need to make any adjustments to the rules or physical design of your game.

6. Record any problems that arise as you implement the plan, and outline how you deal with them.

Suggested Resources

- Toy stores
- Web sites for manufacturers and players' groups
- CD-ROM games
- Books on games of chance
- Statistical software
- Spreadsheets
- Graphing calculators

Evaluating Your Project

1. Assess your game in terms of
   - clarity of the rules
   - enjoyment by the players
   - originality
   - physical design, including attractiveness and ease of use

2. Consider the quality and extent of your mathematical analysis. Have you included all appropriate theoretical and empirical probabilities? Have you properly analysed the effectiveness of possible strategies for winning? Are there other mathematical investigations that you could apply to your game? Have you made appropriate use of technology? Have you documented all of your analysis?

3. If you were to do this project again, what would you do differently? Why?

4. Are there questions that arose from your game that warrant further investigation? How could you address these issues in a future project?

Presentation

1. Explain and demonstrate the game you have created.

2. Discuss comments made by the players who tested your game.

3. Outline the probabilities that apply to your game.

4. Present the outcomes as analysed and organized data.

5. Outline possible winning strategies and comment on their effectiveness.

6. Discuss the positive aspects of your project, its limitations, and how it could be extended or improved.

7. Listen to presentations by your classmates and ask for clarification or suggest improvements, where appropriate. Consider how you might be able to apply both the strengths and weaknesses of other presentations to improve your project.

Preparing for the Culminating Project

Applying Project Skills

In the course of this probability project, you will have developed skills that could be essential for your culminating project:

- developing and carrying out an action plan
- applying techniques for determining probabilities
- working with others to test your ideas
- evaluating your own work
- presenting results
- critiquing the work of other students

Keeping on Track

At this point, you should have implemented enough of the action plan for your culminating project to determine what data you need and how to analyse these data. You should have considered whether you need to refine or redefine your project. For example, you may have found that the data you require are not readily available or your method of analysis is not practical. In such cases, you should discuss your revised action plan with your teacher. Section 9.3 provides suggestions for developing and implementing your action plan.

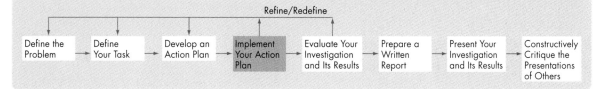

Refine/Redefine

Define the Problem → Define Your Task → Develop an Action Plan → Implement Your Action Plan → Evaluate Your Investigation and Its Results → Prepare a Written Report → Present Your Investigation and Its Results → Constructively Critique the Presentations of Others

1. Evaluate.

   a) $7!$ b) $_7P_1$ c) $_7C_1$

   d) $P(7, 7)$ e) $\binom{7}{2}$ f) $C(7, 2)$

2. Use the binomial theorem to

   a) expand $(3x - 2y)^5$

   b) factor $2a^4 - 8a^3b + 12a^2b^2 - 8ab^3 + 2b^4$

3. If upper-case and lower-case letters are considered as different letters, how many six-letter computer passwords are possible

   a) with no repeated letters?

   b) with at least one capital letter?

4. In how many ways can 12 different cars be parked in the front row of a used-car lot if the owner does not want the red one beside the orange one because the colours clash?

5. What is the probability that a random integer between 1 and 50, inclusive, is not a prime number?

6. A computer expert estimates that the odds of a chess grand master defeating the latest chess-playing computer are 4:5. What is the probability that the chess master will win a match against the computer?

7. a) How many divisors of 4725 are there?

   b) How many of these divisors are divisible by 5?

8. Eight friends, three of whom are left-handed, get together for a friendly game of volleyball. If they split into two teams randomly, what is the probability that one team is comprised of

   a) all right-handed players?

   b) two right-handed and two left-handed players?

9. A manager interviews in random order five candidates for a promotion. What is the likelihood that the most experienced candidate will be interviewed first, followed by the second most experienced candidate?

10. If four decks of cards are shuffled together, what is the probability of dealing a 13-card hand that includes exactly two black 3s?

11. At Inglis Park in Owen Sound, you can see adult salmon jumping over a series of logs as they swim upstream to spawn. The salmon have a 0.6 probability of a successful jump if they rest prior to the jump, but only a 0.3 probability immediately after jumping the previous log. If the fish are rested when they come to the first log, what is the probability that a salmon will clear

    a) both of the first two logs on the first try without resting?

    b) all of the first four logs on the first try if it rests after the second jump?

12. The weather forecast calls for a 12% chance of rain tomorrow, but it is twice as likely that it will snow. What is the probability that it will neither rain nor snow tomorrow?

13. Sasha and Pedro meet every Tuesday for a game of backgammon. They find that after winning a game, Sasha has a 65% probability of winning the next game. Similarly, Pedro has a 60% probability of winning after he has won a game. Pedro won the game last week.

    a) What are the probabilities of each player winning this week?

    b) What is the probability of Pedro winning the game two weeks later?

    c) If Pedro and Sasha play 100 games, how many games is each player likely to win?

Endangered Species

Background

The Canadian Nature Federation has identified 380 endangered species in Canada. Of these, 115 are near extinction. Worldwide, there are over 9485 endangered species. Some species, such as whooping cranes or giant pandas, have populations of only a few hundred.

There is often debate about the state of a particular species. In 2000, the Red List compiled by the World Conservation Union (IUCN) listed 81 species of whale that are classified as extinct, endangered, or vulnerable. Most nations have stopped all commercial whale hunting. However, Japan and Norway continue whaling, and Japan even increased the number of whale species that it hunts.

In southern Africa, elephant herds are often reduced by killing whole families of elephants. The governments involved claim that such culling keeps the herds within the limits of the available food supply and prevents the elephants from encroaching on farmers' fields. However, critics claim that the culling is done primarily for the profit derived from selling the elephants' ivory tusks and reduces the number of elephants dangerously. Thirty years ago, there were three million elephants in Africa. Estimates today place the number at less than 250 000.

In 2001, the government of British Columbia reintroduced hunting of grizzly bears, claiming that their numbers have rebounded enough that hunting will not endanger the species. The government places the number of adult grizzlies at over 13 000. The Canadian Nature Federation claims the number of grizzlies is at most 6000, and could be as low as 4000.

As you can see from these examples, obtaining accurate statistics is vital to the management of endangered species.

Your Task

Collect and analyse data about a species of your choice to determine whether it is endangered.

Developing an Action Plan

You will need to select a species that you believe may be endangered, find reliable sources of data about this species, and outline a method for analysing the data and drawing conclusions from them.

CHAPTER 7

Probability Distributions

Specific Expectations	Section
Identify examples of discrete random variables.	7.1
Construct a discrete probability distribution function by calculating the probabilities of a discrete random variable.	7.1
Calculate expected values and interpret them within applications as averages over a large number of trials.	7.1, 7.2, 7.3, 7.4
Determine probabilities, using the binomial distribution.	7.2
Interpret probability statements, including statements about odds, from a variety of sources.	7.1, 7.2, 7.3, 7.4
Identify the advantages of using simulations in contexts.	7.1, 7.2, 7.3, 7.4
Design and carry out simulations to estimate probabilities in situations for which the calculation of the theoretical probabilities is difficult or impossible.	7.1, 7.2, 7.3, 7.4
Assess the validity of some simulation results by comparing them with the theoretical probabilities, using the probability concepts developed in the course.	7.1, 7.2, 7.3, 7.4

FREE COLLECTOR CARD INSIDE!

KRAKKED KORN Cereal

The World's Greatest Mathematicians!

Pythagoras

Carl Gauss
•Fundamental Theorem of Algebra
•Prime Number Theorem
•Founded non-euclidean geometry

Pierre de Fermat
•Greatest amateur mathematician
•Fermat's Last Theorem

René Descartes
•Invented analytic geometry

Leonhard Euler
•Introduced modern mathematical notation
•Laid down foundations of graph theory
•Author of famous calculus text

Jacob Bernoulli
•One of the founders of calculus
•Proved Law of Large Numbers
•Bernoulli trials

Blaise Pascal
•Laid down principles for theory of probability
•Pascal's triangle

Chapter Problem

Collecting Cards

The Big K cereal company has randomly placed seven different collector cards of the world's great mathematicians into its Krakked Korn cereal boxes. Each box contains one card and each card is equally likely.

1. If you buy seven boxes of the cereal, what is the probability that you will get all seven different cards?

2. What is the probability that you will get seven copies of the same card?

3. Estimate the number of boxes you would have to buy to have a 50% chance of getting a complete set of the cards.

4. Design a simulation to estimate how many boxes of Krakked Korn you would have to buy to be reasonably sure of collecting an entire set of cards.

In this chapter, you will learn how to use probability distributions to calculate precise answers to questions like these.

Review of Prerequisite Skills

If you need help with any of the skills listed in purple below, refer to Appendix A.

1. **Order of operations** Evaluate.

   a) $\left(\dfrac{1}{3}\right)^3\left(\dfrac{2}{3}\right)$

   b) $\left(\dfrac{3}{4}\right)^2\left(\dfrac{1}{4}\right)$

   c) $\left(\dfrac{2}{5}\right)^2\left(\dfrac{3}{5}\right)^2$

   d) $1.2\left(\dfrac{1}{5}\right) + 3.1\left(\dfrac{2}{5}\right) + 2.4\left(\dfrac{3}{5}\right) + 4.2\left(\dfrac{4}{5}\right)$

   e) $0.2 + (0.8)(0.2) + (0.8)^2(0.2) + (0.8)^3(0.2)$

2. **Sigma notation** Write the following in sigma notation.

   a) $t_1 + t_2 + \dots + t_{12}$

   b) $(0)(_9C_0) + (1)(_9C_1) + (2)(_9C_2) + \dots + (9)(_9C_9)$

   c) $\dfrac{2}{3} + \dfrac{3}{4} + \dfrac{4}{5} + \dfrac{5}{6} + \dfrac{6}{7} + \dfrac{7}{8}$

   d) $\dfrac{a_0 + a_1 + a_2 + a_3 + a_4 + a_5}{6}$

3. **Sigma notation** Expand and simplify.

   a) $\displaystyle\sum_{k=1}^{6} k^2$

   b) $\displaystyle\sum_{m=1}^{15} b_{m-1}$

   c) $\displaystyle\sum_{i=0}^{7} {}_7C_i$

   d) $\displaystyle\sum_{x=0}^{8} (0.3)^x(0.7)$

4. **Binomial theorem (section 5.5)** Use the binomial theorem to expand and simplify.

   a) $(x + y)^6$

   b) $(0.4 + 0.6)^4$

   c) $\left(\dfrac{1}{3} + \dfrac{2}{3}\right)^5$

   d) $(p + q)^n$

5. **Probability (Chapter 6)** When rolling two dice,

   a) what is the probability of rolling a sum of 7?

   b) what is the probability of rolling a 3 and a 5?

   c) what is the probability of rolling a 3 or a 5?

   d) what is the probability of rolling a sum of 8?

   e) what is the probability of rolling doubles?

6. **Probability (Chapter 6)** In a family of four children, what is the probability that all four are girls?

7. **Probability (Chapter 6)** Three people each select a letter of the alphabet.

   a) What is the probability that they select the same letter?

   b) What is the probability that they select different letters?

Probability Distributions

In Chapter 6, the emphasis was on the probability of individual outcomes from experiments. This chapter develops models for distributions that show the probabilities of all possible outcomes of an experiment. The distributions can involve outcomes with equal or different likelihoods. Distribution models have applications in many fields including science, game theory, economics, telecommunications, and manufacturing.

INVESTIGATE & INQUIRE: Simulating a Probability Experiment

For project presentations, Mr. Fermat has divided the students in his class into five groups, designated A, B, C, D, and E. Mr. Fermat randomly selects the order in which the groups make their presentations. Develop a simulation to compare the probabilities of group A presenting their project first, second, third, fourth, or fifth.

Method 1: Selecting by Hand

1. Label five slips of paper as A, B, C, D, and E.

2. Randomly select the slips one by one. Set up a table to record the order of the slips and note the position of slip A in the sequence.

3. Repeat this process for a total of ten trials.

4. Combine your results with those from all of your classmates.

5. Describe the results and calculate an empirical probability for each of the five possible outcomes.

6. Reflect on the results. Do you think they accurately represent the situation? Why or why not?

Method 2: Selecting by Computer or Graphing Calculator

1. Use a computer or graphing calculator to generate random numbers between 1 and 5. The generator must be programmed to not repeat a number within a trial. Assign A = 1, B = 2, C = 3, D = 4, and E = 5.

2. Run a series of trials and tabulate the results. If you are skilled in programming, you can set the calculator or software to run a large number of trials and tabulate the results for you. If you run fewer than 100 trials, combine your results with those of your classmates.

See Appendix B for details on software and graphing calculator functions you could use in your simulation.

3. Calculate an empirical probability for each possible outcome.

4. Reflect on the results. Do you think they accurately represent the situation? Why or why not?

The methods in the investigation on page 369 can be adapted to simulate any type of probability distribution:

Step 1 Choose a suitable tool to simulate the random selection process. You could use software, a graphing calculator, or manual methods, such as dice, slips of paper, and playing cards. (See section 1.4.) Look for simple ways to model the selection process.

Step 2 Decide how many trials to run. Determine whether you need to simulate the full situation or if a sample will be sufficient. You may want several groups to perform the experiment simultaneously and then pool their results.

Step 3 Design each trial so that it simulates the actual situation. In particular, note whether you must simulate the selected items being *replaced* (independent outcomes) or *not replaced* (dependent outcomes).

Step 4 Set up a method to record the frequency of each outcome (such as a table, chart, or software function). Combine your results with those of your classmates, if necessary.

Step 5 Calculate empirical probabilities for the simulated outcomes. The sum of the probabilities in the distribution must equal 1.

Step 6 Reflect on the results and decide if they accurately represent the situation being simulated.

Many probability experiments have numerical outcomes—outcomes that can be counted or measured. A **random variable**, X, has a single value (denoted x) for each outcome in an experiment. For example, if X is the number rolled with a die, then x has a different value for each of the six possible outcomes. Random variables can be discrete or continuous. **Discrete variables** have values that are separate from each other, and the number of possible values can be small. **Continuous variables** have an infinite number of possible values in a continuous interval. This chapter describes distributions involving discrete random variables. These variables often have integer values.

Usually you select the property or attribute that you want to measure as the random variable when calculating probability distributions. The probability of a random variable having a particular value x is represented as $P(X = x)$, or $P(x)$ for short.

Project Prep

The difference between a discrete random variable and a continuous random variable will be important for your probability distributions project.

Example 1 Random Variables

Classify each of the following random variables as discrete or continuous.
a) the number of phone calls made by a salesperson
b) the length of time the salesperson spent on the telephone
c) a company's annual sales
d) the number of widgets sold by the company
e) the distance from Earth to the sun

Solution

a) Discrete: The number of phone calls must be an integer.

b) Continuous: The time spent can be measured to fractions of a second.

c) Discrete: The sales are a whole number of dollars and cents.

d) Discrete: Presumably the company sells only whole widgets.

e) Continuous: Earth's distance from the sun varies continuously since Earth moves in an elliptical orbit around the sun.

Example 2 Uniform Probability Distribution

Determine the probability distribution for the order of the group presentations simulated in the investigation on page 369.

Solution

Rather than considering the selection of a group to be first, second, and so on, think of each group randomly choosing its position in the order of presentations. Since each group would have an equal probability for choosing each of the five positions, each probability is $\frac{1}{5}$.

Random Variable, x	Probability, P(x)
Position 1	$\frac{1}{5}$
Position 2	$\frac{1}{5}$
Position 3	$\frac{1}{5}$
Position 4	$\frac{1}{5}$
Position 5	$\frac{1}{5}$

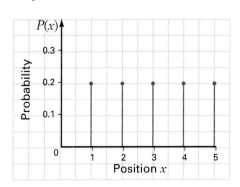

Observe that all outcomes in this distribution are equally likely in any single trial. A distribution with this property is a **uniform probability distribution**. The sum of the probabilities in this distribution is 1. In fact, all probability distributions must sum to 1 since they include all possible outcomes.

All outcomes in a uniform probability distribution are equally likely. So, for all values of x,

Probability in a Discrete Uniform Distribution

$$P(x) = \frac{1}{n},$$

where n is the number of possible outcomes in the experiment.

An **expectation** or **expected value**, $E(X)$, is the *predicted* average of all possible outcomes of a probability experiment. The expectation is equal to the sum of the products of each outcome (random variable $= x_i$) with its probability, $P(x_i)$.

Expectation for a Discrete Probability Distribution

$$E(X) = x_1 P(x_1) + x_2 P(x_2) + \ldots + x_n P(x_n)$$

$$= \sum_{i=1}^{n} x_i P(x_i)$$

Recall that the capital sigma, Σ, means "the sum of." The limits below and above the sigma show that the sum is from the first term ($i = 1$) to the nth term.

Example 3 Dice Game

Consider a simple game in which you roll a single die. If you roll an even number, you gain that number of points, and, if you roll an odd number, you lose that number of points.

a) Show the probability distribution of points in this game.

b) What is the expected number of points per roll?

c) Is this a fair game? Why?

Solution

a) Here the random variable is the number of points scored, not the number rolled.

Number on Upper Face	Points, x	Probability, $P(x)$
1	−1	$\frac{1}{6}$
2	2	$\frac{1}{6}$
3	−3	$\frac{1}{6}$
4	4	$\frac{1}{6}$
5	−5	$\frac{1}{6}$
6	6	$\frac{1}{6}$

b) Since each outcome occurs $\frac{1}{6}$ of the time, the expected number of points per roll is

$$E(X) = \left(-1 \times \frac{1}{6}\right) + \left(2 \times \frac{1}{6}\right) + \left(-3 \times \frac{1}{6}\right) + \left(4 \times \frac{1}{6}\right) + \left(-5 \times \frac{1}{6}\right) + \left(6 \times \frac{1}{6}\right)$$

$$= (-1 + 2 - 3 + 4 - 5 + 6) \times \frac{1}{6}$$

$$= 0.5$$

You would expect that the score in this game would average out to 0.5 points per roll.

c) The game is not fair because the points gained and lost are not equal. For a game to be **fair**, the expected outcome must be 0.

Example 4 Canoe Lengths

A summer camp has seven 4.6-m canoes, ten 5.0-m canoes, four 5.2-m canoes, and four 6.1-m canoes. Canoes are assigned randomly for campers going on a canoe trip.

a) Show the probability distribution for the length of an assigned canoe.

b) What is the expected length of an assigned canoe?

Solution

a) Here the random variable is the canoe length.

Length of Canoe (m), x	Probability, $P(x)$
4.6	$\frac{7}{25}$
5.0	$\frac{10}{25}$
5.2	$\frac{4}{25}$
6.1	$\frac{4}{25}$

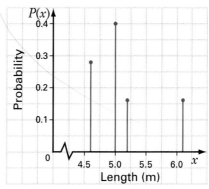

Observe that the sum of the probabilities is again 1, but the probabilities are not equal. This distribution is not uniform.

b) $$E(X) = (4.6)\left(\frac{7}{25}\right) + (5.0)\left(\frac{10}{25}\right) + (5.2)\left(\frac{4}{25}\right) + (6.1)\left(\frac{4}{25}\right)$$

$$= 5.1$$

The expected length of the canoe is 5.1 m.

- A random variable, X, has a single value for each outcome in the experiment. Discrete random variables have separated values while continuous random variables have an infinite number of outcomes along a continuous interval.

- A probability distribution shows the probabilities of all the possible outcomes of an experiment. The sum of the probabilities in any distribution is 1.

- Expectation, or the predicted average of all possible outcomes of a probability experiment, is

$$E(X) = x_1 P(x_1) + x_2 P(x_2) + \ldots + x_n P(x_n)$$
$$= \sum_{i=1}^{n} x_i P(x_i)$$

- The expected outcome in a fair game is 0.

- The outcomes of a uniform probability distribution all have the same probability, $P(x) = \dfrac{1}{n}$, where n is the number of possible outcomes in the experiment.

- You can simulate a probability distribution with manual methods, calculators, or computer software.

Communicate Your Understanding

1. Explain the principal differences between the graphs of the probability distributions in Example 2 and Example 4.

2. In the game of battleship, you select squares on a grid and your opponent tells you if you scored a "hit." Is this process a uniform distribution? What evidence can you provide to support your position?

Practise

1. Classify each of the following random variables as discrete or continuous.

 a) number of times you catch a ball in a baseball game

 b) length of time you play in a baseball game

 c) length of a car in centimetres

 d) number of red cars on the highway

 e) volume of water in a tank

 f) number of candies in a box

2. Explain whether each of the following experiments has a uniform probability distribution.

 a) selecting the winning number for a lottery

 b) selecting three people to attend a conference

 c) flipping a coin

d) generating a random number between 1 and 20 with a calculator

e) guessing a person's age

f) cutting a card from a well-shuffled deck

g) rolling a number with two dice

3. Given the following probability distributions, determine the expected values.

a)

x	P(x)
5	0.3
10	0.25
15	0.45

b)

x	P(x)
1 000	0.25
100 000	0.25
1 000 000	0.25
10 000 000	0.25

c)

x	P(x)
1	$\frac{1}{6}$
2	$\frac{1}{5}$
3	$\frac{1}{4}$
4	$\frac{1}{3}$
5	$\frac{1}{20}$

4. A spinner has eight equally-sized sectors, numbered 1 through 8.

a) What is the probability that the arrow on the spinner will stop on a prime number?

b) What is the expected outcome, to the nearest tenth?

Apply, Solve, Communicate

B

5. A survey company is randomly calling telephone numbers in your exchange.

a) Do these calls have a uniform distribution? Explain.

b) What is the probability that a particular telephone number will receive the next call?

c) What is the probability that the last four digits of the next number called will all be the same?

6. **a)** Determine the probability distribution for the sum rolled with two dice.

b) Determine the expected sum of two dice.

c) Repeat parts a) and b) for the sum of three dice.

7. There are only five perfectly symmetrical polyhedrons: the tetrahedron (4 faces), the cube (6 faces), the octahedron (8 faces), the dodecahedron (12 faces), and the icosahedron (20 faces). Calculate the expected value for dies made in each of these shapes.

8. A lottery has a $1 000 000 first prize, a $25 000 second prize, and five $1000 third prizes. A total of 2 000 000 tickets are sold.

a) What is the probability of winning a prize in this lottery?

b) If a ticket costs $2.00, what is the expected profit per ticket?

9. **Communication** A game consists of rolling a die. If an even number shows, you receive double the value of the upper face in points. If an odd number shows, you lose points equivalent to triple the value of the upper face.

a) What is the expectation?

b) Is this game fair? Explain.

10. **Application** In a lottery, there are 2 000 000 tickets to be sold. The prizes are as follows:

Prize ($)	Number of Prizes
1 000 000	1
50 000	5
1 000	10
50	50

What should the lottery operators charge per ticket in order to make a 40% profit?

11. In a family with two children, determine the probability distribution for the number of girls. What is the expected number of girls?

12. A computer has been programmed to draw a rectangle with perimeter of 24 cm. The program randomly chooses integer lengths. What is the expected area of the rectangle?

13. Suppose you are designing a board game with a rule that players who land on a particular square must roll two dice to determine where they move next. Players move ahead five squares for a roll with a sum of 7 and three squares for a sum of 4 or 10. Players move back *n* squares for any other roll.

a) Develop a simulation to determine the value of *n* for which the expected move is zero squares.

b) Use the probability distribution to verify that the value of *n* from your simulation does produce an expected move of zero squares.

14. **Inquiry/Problem Solving** Cheryl and Fatima each have two children. Cheryl's oldest child is a boy, and Fatima has at least one son.

a) Develop a simulation to determine whether Cheryl or Fatima has the greater probability of having two sons.

b) Use the techniques of this section to verify the results of your simulation.

15. Suppose you buy four boxes of the Krakked Korn cereal. Remember that each box has an equal probability of containing any one of the seven collector cards.

a) What is the probability of getting

i) four identical cards?

ii) three identical cards?

iii) two identical and two different cards?

iv) two pairs of identical cards?

v) four different cards?

b) Sketch a probability distribution for the number of different cards you might find in the four boxes of cereal.

c) Is the distribution in part b) uniform?

ACHIEVEMENT CHECK

Knowledge/ Understanding	Thinking/Inquiry/ Problem Solving	Communication	Application

16. A spinner with five regions is used in a game. The probabilities of the regions are

$P(1) = 0.3$
$P(2) = 0.2$
$P(3) = 0.1$
$P(4) = 0.1$
$P(5) = 0.3$

a) Sketch and label a spinner that will generate these probabilities.

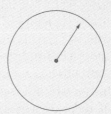

b) The rules of the game are as follows: If you spin and land on an even number, you receive double that number of points. If you land on an odd number, you lose that number of points. What is the expected number of points a player will win or lose?

c) Sketch a graph of the probability distribution for this game.

d) Show that this game is not fair. Explain in words.

e) Alter the game to make it fair. Prove mathematically that your version is fair.

17. Application The door prizes at a dance are gift certificates from local merchants. There are four $10 certificates, five $20 certificates, and three $50 certificates. The prize envelopes are mixed together in a bag and are drawn at random.

a) Use a tree diagram to illustrate the possible outcomes for selecting the first two prizes to be given out.

b) Determine the probability distribution for the number of $20 certificates in the first two prizes drawn.

c) What is the probability that exactly three of the first five prizes selected will be $10 certificates?

d) What is the expected number of $10 certificates among the first five prizes drawn?

18. Most casinos have roulette wheels. In North America, these wheels have 38 slots, numbered 1 to 36, 0, and 00. The 0 and 00 slots are coloured green. Half of the remaining slots are red and the other half are black. A ball rolls around the wheel and players bet on which slot the ball will stop in. If a player guesses correctly, the casino pays out according to the type of bet.

a) Calculate the house advantage, which is the casino's profit, as a percent of the total amount wagered for each of the following bets. Assume that players place their bets randomly.

  i) single number bet, payout ratio 35:1

  ii) red number bet, payout ratio 1:1

  iii) odd number bet, payout ratio 1:1

  iv) 6-number group, payout ratio 5:1

  v) 12-number group, payout ratio 2:1

b) Estimate the weekly profit that a roulette wheel could make for a casino. List the assumptions you have to make for your calculation.

c) European roulette wheels have only one zero. Describe how this difference would affect the house advantage.

19. Inquiry/Problem Solving Three concentric circles are drawn with radii of 8 cm, 12 cm, and 20 cm. If a dart lands randomly on this target, what are the probabilities of it landing in each region?

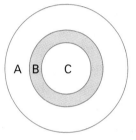

20. A die is a random device for which each possible value of the random variable has a probability of $\frac{1}{6}$. Design a random device with the probabilities listed below and determine the expectation for each device. Use a different type of device in parts a) and b).

a) $P(0) = \frac{1}{4}$

$P(1) = \frac{1}{6}$

$P(2) = P(3) = \frac{1}{8}$

$P(4) = P(5) = P(6) = P(7) = \frac{1}{12}$

b) $P(0) = \frac{1}{6}$

$P(1) = P(2) = \frac{1}{4}$

$P(3) = \frac{1}{3}$

21. Communication Explain how the population mean, μ, and the expectation, $E(X)$, are related.

Binomial Distributions

A manufacturing company needs to know the expected number of defective units among its products. A polling company wants to estimate how many people are in favour of a new environmental law. In both these cases, the companies can view the individual outcomes in terms of "success" or "failure." For the manufacturing company, a success is a product without defects; for the polling company, a success is an interview subject who supports the new law. Repeated independent trials measured in terms of such successes or failures are **Bernoulli trials**, named after Jacob Bernoulli (1654–1705), a Swiss mathematician who published important papers on logic, algebra, geometry, calculus, and probability.

WEB CONNECTION

www.mcgrawhill.ca/links/MDM12

To learn more about Bernoulli trials, visit the above web site and follow the links. Describe an application of Bernoulli trials that interests you.

INVESTIGATE & INQUIRE: Success/Failure Simulation

The Choco-Latie Candies company makes candy-coated chocolates, 40% of which are red. The production line mixes the candies randomly and packages ten per box. Develop a simulation to determine the expected number of red candies in a box.

1. Determine the key elements of the selection process and choose a random-number generator or other tool to simulate the process.

2. Decide whether each trial in your simulation must be independent. Does the colour of the first candy in a box affect the probability for the next one? Describe how you can set up your simulation to reflect the way in which the candies are packed into the boxes.

3. Run a set of ten trials to simulate filling one of the boxes and record the number of red candies. How could you measure this result in terms of successes and failures?

4. Simulate filling at least nine more boxes and record the number of successes for each box.

5. Review your results. Do you think the ten sets of trials are enough to give a reasonable estimate of the expected number of red candies? Explain your reasoning. If necessary, simulate additional sets of trials or pool your results with those of other students in your class.

6. Summarize the results by calculating empirical probabilities for each possible number of red candies in a box. The sum of these probabilities must equal 1. Calculate the expected number of red candies per box based on these results.

7. Reflect on the results. Do they accurately represent the expected number of red candies in a box?

8. Compare your simulation and its results with those of the other students in your class. Which methods produced the most reliable results?

The probabilities in the simulation above are an example of a **binomial distribution**. For such distributions, all the trials are *independent* and have only two possible outcomes, success or failure. The probability of success is the same in every trial—the outcome of one trial does not affect the probabilities of any of the later trials. The random variable is the number of successes in a given number of trials.

Example 1 Success/Failure Probabilities

A manufacturer of electronics components produces precision resistors designed to have a tolerance of ±1%. From quality-control testing, the manufacturer knows that about one resistor in six is actually within just 0.3% of its nominal value. A customer needs three of these more precise resistors. What is the probability of finding exactly three such resistors among the first five tested?

Solution

You can apply the concept of Bernoulli trials because the tolerances of the resistors are independent of each other. A success is finding a resistor with a tolerance of ±0.3% or less.

For each resistor, the probability of success is about $\frac{1}{6}$ and the probability of failure is about $\frac{5}{6}$.

You can choose three resistors from the batch of five in $_5C_3$ ways. Since the outcomes are independent events, you can apply the product rule for independent events. The probability of success with all three resistors in each of these combinations is the product of the probabilities for the individual resistors.

$$\frac{1}{6} \times \frac{1}{6} \times \frac{1}{6} = \left(\frac{1}{6}\right)^3$$

The probability for the three successful trials is equal to the number of ways they can occur multiplied by the probability for each way: ${}_5C_3\left(\dfrac{1}{6}\right)^3$.

Similarly, there are ${}_2C_2$ ways of choosing the other two resistors and the probability of a failure with both these resistors is $\left(\dfrac{5}{6}\right)^2$. Thus, the probability for the two failures is ${}_2C_2\left(\dfrac{5}{6}\right)^2 = \left(\dfrac{5}{6}\right)^2$ since ${}_nC_n = 1$.

Now, apply the product rule for independent events to find the probability of having three successes and two failures in the five trials.

$$P(x = 3) = {}_5C_3\left(\dfrac{5}{6}\right)^2 \times \left(\dfrac{5}{6}\right)^2$$
$$= 10\left(\dfrac{1}{216}\right)\left(\dfrac{25}{36}\right)$$
$$= 0.032150\ldots$$

The probability that exactly three of the five resistors will meet the customer's specification is approximately 0.032.

You can apply the method in Example 1 to show that the probability of x successes in n Bernoulli trials is

Probability in a Binomial Distribution

$P(x) = {}_nC_x\,p^x q^{n-x}$,
where p is the probability of success on any individual trial and $q = 1 - p$ is the probability of failure.

Since the probability for all trials is the same, the expectation for a success in any one trial is p. The expectation for n independent trials is

Expectation for a Binomial Distribution

$E(X) = np$

Example 2 Expectation in a Binomial Distribution

Tan's family moves to an area with a different telephone exchange, so they have to get a new telephone number. Telephone numbers in the new exchange start with 446, and all combinations for the four remaining digits are equally likely. Tan's favourite numbers are the prime numbers 2, 3, 5, and 7.

a) Calculate the probability distribution for the number of these prime digits in Tan's new telephone number.

b) What is the expected number of these prime digits in the new telephone number?

Solution 1   Using Pencil and Paper

a) The probability of an individual digit being one of Tan's favourite numbers is $\frac{4}{10} = 0.4$.

So,
$$p = 0.4 \quad \text{and} \quad q = 1 - 0.4$$
$$= 0.6$$

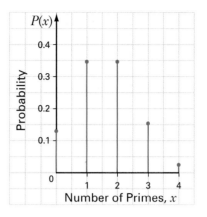

Number of Primes, x	Probability, P(x)
0	$_4C_0(0.4)^0(0.6)^4 = 0.1296$
1	$_4C_1(0.4)^1(0.6)^3 = 0.3456$
2	$_4C_2(0.4)^2(0.6)^2 = 0.3456$
3	$_4C_3(0.4)^3(0.6)^1 = 0.1536$
4	$_4C_4(0.4)^4(0.6)^0 = 0.0256$

b) You can calculate the expectation in two ways.

Using the equation for the expectation of any probability distribution,

$$E(X) = 0(0.1296) + 1(0.3456) + 2(0.3456) + 3(0.1536) + 4(0.0256)$$
$$= 1.6$$

Using the formula for a binomial expectation,

$$E(X) = np$$
$$= 4(0.4)$$
$$= 1.6$$

On average, there will be 1.6 of Tan's favourite digits in telephone numbers in his new exchange.

Solution 2   Using a Graphing Calculator

a) Check that lists L1, L2, and L3 are clear. Enter all the possible values for x into L1.

Use the binompdf(function from the DISTR menu to calculate the probabilities for each value of x. Binompdf stands for binomial probability density function.

Enter the formula binompdf(4,.4,L1) into L2 to find the values of $P(x)$.

b) In L3, calculate the value of $xP(x)$ using the formula L1 × L2. Then, use the
sum(function in the LIST OPS menu to calculate the expected value.

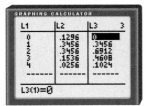 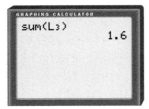

*For more details on software and graphing
calculator functions, see Appendix B.*

Solution 3 Using a Spreadsheet

a) Open a new spreadsheet. Create headings for x, $P(x)$, and $xP(x)$ in columns A to C.
Enter the possible values of the random variable x in column A.
Use the BINOMDIST function to calculate the probabilities for the different
values of x. The syntax for this function is

BINOMDIST(*number of successes, trials, probability of success, cumulative*)

Enter this formula in cell B3 and then copy the formula into cells B4 through B7.

Note that you must set the cumulative feature to 0 or FALSE, so that the function
calculates the probability of exactly 4 successes rather than the probability of up to
4 successes.

b) To calculate $xP(x)$ in column C, enter the formula A3*B3 in C3 and then copy cell C3
down to row 7. Use the SUM function to calculate the expected value.

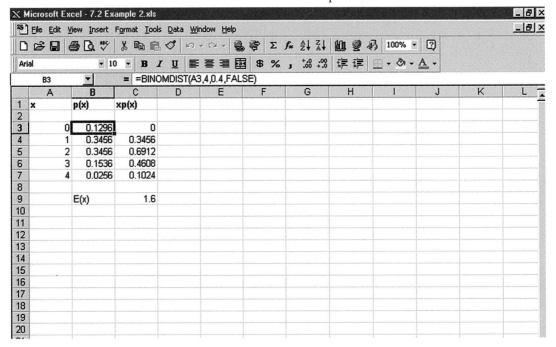

Solution 4   Using Fathom™

a) Open a new Fathom™ document. Drag a new **collection** box to the work area and name it Primes. Create five new cases.

Drag a new **case table** to the work area. Create three new attributes: x, px, and xpx. Enter the values from 0 to 4 in the x attribute column. Then, use the **binomialProbability function** to calculate the probabilities for the different values of x. The syntax of this function is

binomialProbability(*number of successes, trials, probability of success*)

Right-click on the px attribute and then select Edit Formula/Functions/ Distributions/Binomial to enter binomialProbability(x,4,.4).

b) You can calculate $xP(x)$ with the formula x*px.

Next, double-click on the **collection** box to open the **inspector**. Select the Measures tab, and name a new measure Ex. Right-click on Ex and use the **sum function**, located in the Functions/Statistical/One Attribute menu, to enter the formula sum(x*px).

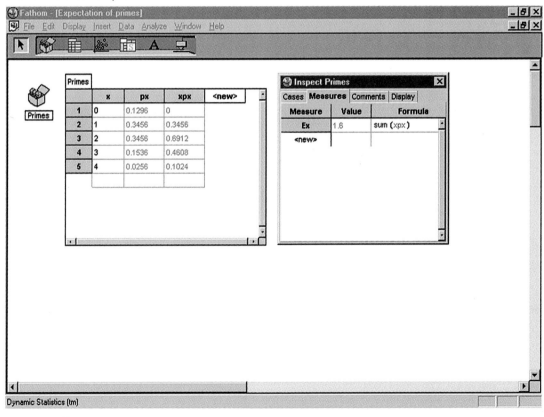

Example 3 Counting Candies

Consider the Choco-Latie candies described in the investigation on page 378.

a) What is the probability that at least three candies in a given box are red?

b) What is the expected number of red candies in a box?

Solution

a) A success is a candy being red, so $p = 0.4$ and $q = 1 - 0.4 = 0.6$. You could add the probabilities of having exactly three, four, five, ..., or ten red candies, but it is easier to use an indirect method.

$$P(\geq 3 \text{ red}) = 1 - P(<3)$$
$$= 1 - P(0 \text{ red}) - P(1 \text{ red}) - P(2 \text{ red})$$
$$= 1 - {}_{10}C_0(0.4)^0(0.6)^{10} - {}_{10}C_1(0.4)^1(0.6)^9 - {}_{10}C_2(0.4)^2(0.6)^8$$
$$\doteq 0.8327$$

The probability of at least three candies being red is approximately 0.8327.

b) $E(X) = np$
$$= 10(0.4)$$
$$= 4$$

The expected number of red candies in a box is 4.

Key Concepts

- A binomial distribution has a specified number of independent trials in which the outcome is either success or failure. The probability of a success is the same in each trial.

- The probability of x successes in n independent trials is
$$P(x) = {}_nC_x\, p^x\, q^{n-x},$$
where p is the probability of success on an individual trial and q is the probability of failure on that same individual trial ($p + q = 1$).

- The expectation for a binomial distribution is $E(X) = np$.

- To simulate a binomial experiment,
  - choose a simulation method that accurately reflects the probabilities in each trial
  - set up the simulation tool to ensure that each trial is independent
  - record the number of successes and failures in each experiment
  - summarize the results by calculating the probabilities for r successes in n trials (the sum of individual probabilities must equal 1)

1. Consider this question: If five cards are dealt from a standard deck, what is the probability that two of the cards are the ace and king of spades?

   a) Explain why the binomial distribution is not a suitable model for this scenario.

   b) How could you change the scenario so that it does fit a binomial distribution? What attributes of a binomial distribution would you use in your modelling?

2. Describe how the graph in Example 2 differs from the graph of a uniform distribution.

3. Compare your results from the simulation of the Choco-Latie candies at the beginning of this section with the calculated values in Example 3. Explain any similarities or differences.

Practise

A

1. Which of the following situations can be modelled by a binomial distribution? Justify your answers.

   a) A child rolls a die ten times and counts the number of 3s.

   b) The first player in a free-throw basketball competition has a free-throw success rate of 88.4%. A second player takes over when the first player misses the basket.

   c) A farmer gives 12 of the 200 cattle in a herd an antibiotic. The farmer then selects 10 cattle at random to test for infections to see if the antibiotic was effective.

   d) A factory producing electric motors has a 0.2% defect rate. A quality-control inspector needs to determine the expected number of motors that would fail in a day's production.

2. Prepare a table and a graph for a binomial distribution with

   a) $p = 0.2$, $n = 5$

   b) $p = 0.5$, $n = 8$

Apply, Solve, Communicate

B

3. Suppose that 5% of the first batch of engines off a new production line have flaws. An inspector randomly selects six engines for testing.

   a) Show the probability distribution for the number of flawed engines in the sample.

   b) What is the expected number of flawed engines in the sample?

4. **Application** Design a simulation to predict the expected number of 7s in Tan's new telephone number in Example 2.

5. The faces of a 12-sided die are numbered from 1 to 12. What is the probability of rolling 9 at least twice in ten tries?

6. **Application** A certain type of rocket has a failure rate of 1.5%.

   a) Design a simulation to illustrate the expected number of failures in 100 launches.

   b) Use the methods developed in this section to determine the probability of fewer than 4 failures in 100 launches.

   c) What is the expected number of failures in 100 launches of the rocket?

7. Suppose that 65% of the families in a town own computers. If eight families are surveyed at random,

a) what is the probability that at least four own computers?

b) what is the expected number of families with computers?

8. Inquiry/Problem Solving Ten percent of a country's population are left-handed.

a) What is the probability that 5 people in a group of 20 are left handed?

b) What is the expected number of left-handed people in a group of 20?

c) Design a simulation to show that the expectation calculated in part b) is accurate.

9. Inquiry/Problem Solving Suppose that Bayanisthol, a new drug, is effective in 65% of clinical trials. Design a problem involving this drug that would fit a binomial distribution. Then, provide a solution to your problem.

10. Pythag-Air-US Airlines has determined that 5% of its customers do not show up for their flights. If a passenger is bumped off a flight because of overbooking, the airline pays the customer $200. What is the expected payout by the airline, if it overbooks a 240-seat airplane by 5%?

11. A department-store promotion involves scratching four boxes on a card to reveal randomly printed letters from A to F. The discount is 10% for each A revealed, 5% for each B revealed, and 1% for the other four letters. What is the expected discount for this promotion?

12. a) Expand the following binomials.

  i) $(p + q)^6$ **ii)** $(0.2 + 0.8)^5$

b) Use the expansions to show how the binomial theorem is related to the binomial probability distribution.

ACHIEVEMENT CHECK

| Knowledge/ Understanding | Thinking/Inquiry/ Problem Solving | Communication | Application |

13. Your local newspaper publishes an Ultimate Trivia Contest with 12 extremely difficult questions, each having 4 possible answers. You have no idea what the correct answers are, so you make a guess for each question.

a) Explain why this situation can be modelled by a binomial distribution.

b) Use a simulation to predict the expected number of correct answers.

c) Verify your prediction mathematically.

d) What is the probability that you will get at least 6 answers correct?

e) What is the probability that you will get fewer than 2 answers correct?

f) Describe how the graph of this distribution would change if the number of possible answers for each question increases or decreases.

C

14. The French mathematician Simeon-Denis Poisson (1781–1840) developed what is now known as the *Poisson distribution*. This distribution can be used to approximate the binomial distribution if p is very small and n is very large. It uses the formula

$$P(x) = \frac{e^{-np}(np)^x}{x!},$$

where e is the irrational number 2.718 28 ... (the base for the natural logarithm).

Use the Poisson distribution to approximate the following situations. Compare the results to those found using the binomial distribution.

a) A certain drug is effective in 98% of cases. If 2000 patients are selected at random, what is the probability that the drug was ineffective in exactly 10 cases?

b) Insurance tables indicate that there is a probability of 0.01 that a driver of a specific model of car will have an accident requiring hospitalization within a one-year period. If the insurance company has 4500 policies, what is the probability of fewer than 5 claims for accidents requiring hospitalization?

c) On election day, only 3% of the population voted for the Environment Party. If 1000 voters were selected at random, what is the probability that fewer than 8 of them voted for the Environment Party?

15. Communication Suppose heads occurs 15 times in 20 tosses of a coin. Do you think the coin is fair? Explain your reasoning.

16. Inquiry/Problem Solving

a) Develop a formula for $P(x)$ in a "trinomial" distribution that has three possible outcomes with probabilities p, q, and r, respectively.

b) Use your formula to determine the probability of rolling a 3 twice and a 5 four times in ten trials with a standard die.

17. Communication A judge in a model-airplane contest says that the probability of a model landing without damage is 0.798, so there is only "one chance in five" that any of the seven models in the finals will be damaged. Discuss the accuracy of the judge's statement.

Career Connection

Actuary

Actuaries are statistics specialists who use business, analytical, and mathematical skills to apply mathematical models to insurance, pensions, and other areas of finance. Actuaries assemble and analyse data and develop probability models for the risks and costs of accidents, sickness, death, pensions, unemployment, and so on. Governments and private companies use such models to determine pension contributions and fair prices for insurance premiums. Actuaries may also be called upon to provide legal evidence on the value of future earnings of an accident victim.

Actuaries must keep up-to-date on social issues, economic trends, business issues, and the law. Most actuaries have a degree in actuarial science, statistics, or mathematics and have studied statistics, calculus, algebra, operations research, numerical analysis, and interest theory. A strong background in business or economics is also useful.

Actuaries work for insurance companies, pension-management firms, accounting firms, labour unions, consulting groups, and federal and provincial governments.

WEB CONNECTION

www.mcgrawhill.ca/links/MDM12

For more information about a career as an actuary, visit the above web site and follow the links.

Geometric Distributions

In some board games, you cannot move forward until you roll a specific number, which could take several tries. Manufacturers of products such as switches, relays, and hard drives need to know how many operations their products can perform before failing. In some sports competitions, the winner is the player who scores the most points before missing a shot. In each of these situations, the critical quantity is the **waiting time** or **waiting period**—the number of trials before a specific outcome occurs.

INVESTIGATE & INQUIRE: Simulating Waiting Times

To get out of jail in the game of MONOPOLY®, you must either roll doubles or pay the bank $50. Design a simulation to find the probabilities of getting out of jail in *x* rolls of the two dice.

1. Select a random-number generator to simulate the selection process.

2. Decide how to simplify the selection process. Decide, also, whether the full situation needs to be simulated or whether a proportion of the trials would be sufficient.

3. Design each trial so that it simulates the actual situation. Determine whether your simulation tool must be reset or replaced after each trial to properly correspond to rolling two dice.

4. Set up a method to record the frequency of each outcome. Record the number of failures before a success finally occurs.

5. Combine your results with those of your classmates, if necessary.

6. Use these results to calculate an empirical probability for each outcome and the expected waiting time before a success.

7. Reflect on the results. Do they accurately represent the expected number of failures before success?

8. Compare your simulation and its results with those of other students in your class. Which simulation do you think worked best? Explain the reasons for your choice.

The simulation above models a **geometric distribution**. Like binomial distributions, trials in a geometric distribution have only two possible outcomes, success or failure, whose probabilities do not change from one trial to the next. However, the random variable for a geometric distribution is the *waiting time*, the number of unsuccessful independent trials before success occurs. Having different random variables causes significant differences between binomial and geometric distributions.

Example 1 Getting out of Jail in MONOPOLY®

a) Calculate the probability distribution for getting out of jail in MONOPOLY® in x rolls of the dice.

b) Estimate the expected number of rolls before getting out of jail.

Solution 1 Using Pencil and Paper

a) The random variable is the number of unsuccessful rolls before you get out of jail. You can get out of jail by rolling doubles, and $P(\text{doubles}) = \dfrac{6}{36}$. So, for each independent roll,

$$p = \frac{6}{36} \quad \text{and} \quad q = 1 - \frac{1}{6}$$
$$= \frac{1}{6} \qquad\qquad = \frac{5}{6}$$

You can apply the product rule to find the probability of successive independent events (see section 6.3). Thus, each unsuccessful roll preceding the successful one adds a factor of $\dfrac{5}{6}$ to the probability.

Unsuccessful Rolls (Waiting Time), x	Probability, $P(x)$
0	$\dfrac{1}{6} = 0.166\,66\ \ldots$
1	$\left(\dfrac{5}{6}\right)\left(\dfrac{1}{6}\right) = 0.138\,88\ \ldots$
2	$\left(\dfrac{5}{6}\right)^2\left(\dfrac{1}{6}\right) = 0.115\,74\ \ldots$
3	$\left(\dfrac{5}{6}\right)^3\left(\dfrac{1}{6}\right) = 0.096\,45\ \ldots$
4	$\left(\dfrac{5}{6}\right)^4\left(\dfrac{1}{6}\right) = 0.080\,37\ \ldots$
...	...

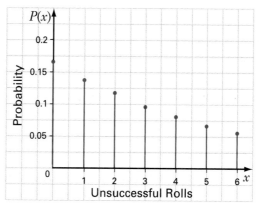

This distribution theoretically continues forever since one possible outcome is that the player never rolls doubles. However, the probability for a waiting time decreases markedly as the waiting time increases. Although this distribution is an infinite geometric series, its terms still sum to 1 since they represent the probabilities of all possible outcomes.

b) Calculate the first six terms. If these terms approach zero rapidly, the sum of these terms will give a rough first approximation for the expectation.

$$E(X) = \sum_{x=0}^{\infty} xP(x)$$
$$= (0)(0.16666...) + (1)(0.13888...) + (2)(0.11574...) + 3(0.09645...)$$
$$+ (4)(0.08037...) + (5)(0.06698...) + ...$$
$$> 0 + 0.13888 + 0.23148 + 0.28935 + 0.32150 + 0.33489 + ...$$
$$> 1.3$$

Clearly, the six terms are not approaching zero rapidly. All you can conclude is that the expected number of rolls before getting out of jail in MONOPOLY® is definitely more than 1.3.

Solution 2   Using a Graphing Calculator

a) You can use the calculator's lists to display the probabilities. Clear lists L1 to L4. For a start, enter the integers from 0 to 39 into L1. The seq(function in the LIST OPS menu provides a convenient way to enter these numbers.

seq(A,A,0,39)→L1

Next, use the geometric probability density function in the DISTR menu to calculate the probability of each value of x. The geometpdf(function has the syntax

geometpdf(*probability of success, number of trial on which first success occurs*)

Since x is the number of trials *before* success occurs, the number of the trial on which the first success occurs is $x + 1$. Therefore, enter geometpdf(1/6,L1+1)→L2.

Notice the dramatic decrease in probability for the higher values of x.

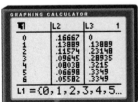 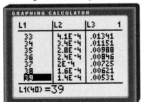

b) In L3, calculate the value of $xP(x)$ with the simple formula L1 × L2. You can sum these values to get a reasonable estimate for the expectation. To determine the accuracy of this estimate, it is helpful to look at the cumulative or running total of the $xP(x)$ values in L3.

To do this, in L4, select 6:cumSum(from the LIST OPS menu and type L3).

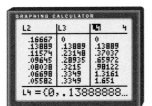 

Note how the running total of $xP(x)$ in L4 increases more and more slowly toward the end of the list, suggesting that the infinite series for the expectation will total to a little more than 4.97. Thus, 5.0 would be a reasonable estimate for the number of trials before rolling doubles. You can check this estimate by performing similar calculations for waiting times of 100 trials or more.

Solution 3   Using a Spreadsheet

a) Open a new spreadsheet. Enter the headings x, p(x), and xp(x) in columns A, B, and C. Use the Fill feature to enter a sequence of values of the random variable x in column A, starting with 0 and going up to 100. Calculate the probability $P(x)$ for each value of the random variable x by entering the formula $(5/6)^\text{^}A3*(1/6)$ in cell B3 and then copying it down the rest of the column.

b) You can calculate $xP(x)$ in column C by entering the formula A3*B3 in cell C3 and then copying it down the rest of the column. Next, calculate the cumulative expected values in column D using the SUM function (with absolute cell references for the first cell) in cell D3 and copying it down the column.

Note that for $x = 50$ the cumulative expected value is over 4.99. By $x = 100$, it has reached 4.999 999 and is increasing extremely slowly. Thus, 5.000 00 is an accurate estimate for the expected number of trials before rolling doubles.

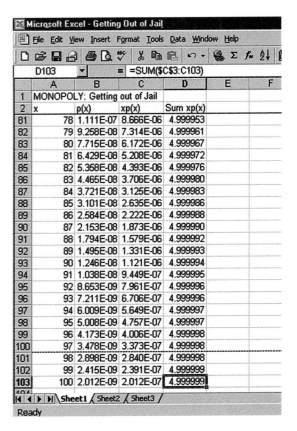

You can use the method in Example 1 to show that the probability of success after a waiting time of x failures is

Probability in a Geometric Distribution

$P(x) = q^x p,$

where p is the probability of success in each single trial and q is the probability of failure.

The expectation of a geometric distribution is the sum of an infinite series. Using calculus, it is possible to show that this expectation converges to a simple formula.

Expectation for a Geometric Distribution

$$E(X) = \sum_{x=0}^{\infty} xP(x)$$

$$= \frac{q}{p}$$

Project Prep

Techniques for calculating expected values will be useful for your probability distributions project.

Example 2 Expectation of Geometric Distribution

Use the formula for the expectation of a geometric distribution to evaluate the accuracy of the estimates in Example 1.

Solution

$E(X) = \dfrac{q}{p}$

$\quad\;\; = \dfrac{\frac{5}{6}}{\frac{1}{6}}$

$\quad\;\; = 5$

For this particular geometric distribution, the simple manual estimate in Example 1 is accurate only to an order of magnitude. However, the calculator and the spreadsheet estimates are much more accurate.

Example 3 Basketball Free Throws

Jamaal has a success rate of 68% for scoring on free throws in basketball. What is the expected waiting time before he misses the basket on a free throw?

Solution

Here, the random variable is the number of trials before Jamaal misses on a free throw. For calculating the waiting time, a success is Jamaal *failing* to score. Thus,

$q = 0.68$ and $p = 1 - 0.68$
$= 0.32$

Using the expectation formula for the geometric distribution,

$E(X) = \dfrac{q}{p}$

$= \dfrac{0.68}{0.32}$

$= 2.1$

The expectation is that Jamaal will score on 2.1 free throws before missing.

Example 4 Traffic Management

Suppose that an intersection you pass on your way to school has a traffic light that is green for 40 s and then amber or red for a total of 60 s.

a) What is the probability that the light will be green when you reach the intersection at least once a week?

b) What is the expected number of days before the light is green when you reach the intersection?

Solution

a) Each trial is independent with

$p = \dfrac{40}{100}$ and $q = 0.60$

$= 0.40$

There are five school days in a week. To get a green light on one of those five days, your waiting time must be four days or less.

$P(0 \le x \le 4) = 0.40 + (0.60)(0.40) + (0.60)^2(0.40) + (0.60)^3(0.40) + (0.60)^4(0.40)$
$= 0.92$

The probability of the light being green when you reach the intersection at least once a week is 0.92.

b)

$$E(X) = \frac{q}{p}$$

$$= \frac{0.60}{0.40}$$

$$= 1.5$$

The expected waiting time before catching a green light is 1.5 days.

Key Concepts

- A geometric distribution has a specified number of independent trials with two possible outcomes, success or failure. The random variable is the number of unsuccessful outcomes before a success occurs.

- The probability of success after a waiting time of x failures is $P(x) = q^x p$, where p is the probability of success in each single trial and q is the probability of failure.

- The expectation of a geometric distribution is $E(X) = \frac{q}{p}$.

- To simulate a geometric experiment, you must ensure that the probability on a single trial is accurate for the situation and that each trial is independent. Summarize the results by calculating probabilities and the expected waiting time.

Communicate Your Understanding

1. Describe how the graph in Example 1 differs from those of the uniform and binomial distributions.

2. Consider this question: What is the expected number of failures in 100 launches of a rocket that has a failure rate of 1.5%? Explain why this problem does not fit a geometric distribution and how it could be rewritten so that it does.

Practise

1. Which of the following situations is modelled by a geometric distribution? Explain your reasoning.

a) rolling a die until a 6 shows

b) counting the number of hearts when 13 cards are dealt from a deck

c) predicting the waiting time when standing in line at a bank

d) calculating the probability of a prize being won within the first 3 tries

e) predicting the number of successful launches of satellites this year

2. Prepare a table and a graph for six trials of a geometric distribution with

   a) $p = 0.2$ **b)** $p = 0.5$

Apply, Solve, Communicate

3. For a 12-sided die,

   a) what is the probability that the first 10 will be on the third roll?

   b) what is the expected waiting time until a 1 is rolled?

4. The odds in favour of a Pythag-Air-US Airlines flight being on time are 3:1.

   a) What is the probability that this airline's next eight flights will be on time?

   b) What is the expected waiting time before a flight delay?

5. **Communication** To finish a board game, Sarah needed to land on the last square by rolling a sum of 2 with two dice. She was dismayed that it took her eight tries. Should she have been surprised? Explain.

6. In a TV game show, the grand prize is randomly hidden behind one of three doors. On each show, the finalist gets to choose one of the doors. What is the probability that no finalists will win a grand prize on four consecutive shows?

7. **Application** A teacher provides pizza for his class if they earn an A-average on any test. The probability of the class getting an A-average on one of his tests is 8%.

   a) What is the probability that the class will earn a pizza on the fifth test?

   b) What is the probability that the class will not earn a pizza for the first seven tests?

   c) What is the expected waiting time before the class gets a pizza?

8. Minh has a summer job selling replacement windows by telephone. Of the people he calls, nine out of ten hang up before he can give a sales pitch.

   a) What is the probability that, on a given day, Minh's first sales pitch is on his 12th call?

   b) What is the expected number of hang-ups before Minh can do a sales pitch?

9. Despite its name, Zippy Pizza delivers only 40% of its pizzas on time.

   a) What is the probability that its first four deliveries will be late on any given day?

   b) What is the expected number of pizza deliveries before one is on time?

10. A poll indicated that 34% of the population agreed with a recent policy paper issued by the government.

    a) What is the probability that the pollster would have to interview five people before finding a supporter of the policy?

    b) What is the expected waiting time before the pollster interviews someone who agrees with the policy?

11. Suppose that 1 out of 50 cards in a scratch-and-win promotion gives a prize.

    a) What is the probability of winning on your fourth try?

    b) What is the probability of winning within your first four tries?

    c) What is the expected number of cards you would have to try before winning?

12. A top NHL hockey player scores on 93% of his shots in a shooting competition.

    a) What is the probability that the player will not miss the goal until his 20th try?

    b) What is the expected number of shots before he misses?

13. Application A computer manufacturer finds that 1.5% of its chips fail quality-control testing.

a) What is the probability that one of the first five chips off the line will be defective?

b) What is the expected waiting time for a defective chip?

14. Inquiry/Problem Solving Three friends of an avid golfer have each given her a package of 5 balls for her birthday. The three packages are different brands. The golfer keeps all 15 balls in her golf bag and picks one at random at the start of each round.

a) Design a simulation to determine the waiting time before the golfer has tried all three brands. Assume that the golfer does not lose any of the balls.

b) Use the methods described in this section to calculate the expected waiting time before the golfer has tried the three different brands.

15. The Big K cereal company has randomly placed one of a set of seven different collector cards in each box of its Krakked Korn cereal. Each card is equally likely.

a) Design a simulation to estimate how many boxes of Krakked Korn you would have to buy to get a complete set of cards.

b) Use the methods described in this section to calculate the average number of boxes of Krakked Korn you would have to buy to get a complete set of cards.

16. Inquiry/Problem Solving Consider a geometric distribution where the random variable is the number of the trial with the first success instead of the number of failures before a success. Develop formulas for the probabilities and expectation for this distribution.

17. Communication A manufacturer of computer parts lists a mean time before failure (MTBF) of 5.4 years for one of its hard drives. Explain why this specification is different from the expected waiting time of a geometric distribution. Could you use the MTBF to calculate the probability of the drive failing in any one-year period?

18. In a sequence of Bernoulli trials, what is the probability that the second success occurs on the fifth trial?

19. Communication The rack behind a coat-check counter collapses and 20 coats slip off their numbered hangers. When the first person comes to retrieve one of these coats, the clerk brings them out and holds them up one at a time for the customer to identify.

a) What is the probability that the clerk will find the customer's coat

i) on the first try?

ii) on the second try?

iii) on the third try?

iv) in fewer than 10 tries?

b) What is the expected number of coats the clerk will have to bring out before finding the customer's coat?

c) Explain why you cannot use a geometric distribution to calculate this waiting time.

Hypergeometric Distributions

When choosing the starting line-up for a game, a coach obviously has to choose a different player for each position. Similarly, when a union elects delegates for a convention or you deal cards from a standard deck, there can be no repetitions. In such situations, each selection reduces the number of items that could be selected in the next trial. Thus, the probabilities in these trials are dependent. Often we need to calculate the probability of a specific number of successes in a given number of dependent trials.

In Ontario, a citizen can be called for jury duty every three years. Although most juries have 12 members, those for civil trials in Ontario usually require only 6 members. Suppose a civil-court jury is being selected from a pool of 18 citizens, 8 of whom are men. Develop a simulation to determine the probability distribution for the number of women selected for this jury.

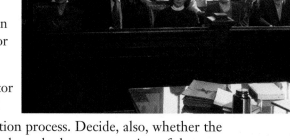

1. Select a random-number generator to simulate the selection process.

2. Decide how to simplify the selection process. Decide, also, whether the full situation needs to be simulated or whether a proportion of the trials would be sufficient.

3. Design each trial so that it simulates the actual situation. Ensure that each trial is dependent by setting the random-number generator so that there are no repetitions within each series of trials.

4. Set up a method to record the number of successes in each experiment. Pool your results with those of other students in your class, if necessary.

5. Use the results to estimate the probabilities of x successes (women) in r trials (selections of a juror).

6. Reflect on the results. Do they accurately represent the probability of x women being selected?

7. Compare your simulation and its results with those of your classmates. Which are the better simulations? Explain why.

Data in Action

The cost of running the criminal, civil, and family courts in Ontario was about $310 million for 2001. These courts have the equivalent of 3300 full-time employees.

The simulation in the investigation models a **hypergeometric distribution**. Such distributions involve a series of *dependent* trials, each with success or failure as the only possible outcomes. The probability of success changes as each trial is made. The random variable is the number of successful trials in an experiment. Calculations of probabilities in a hypergeometric distribution generally require formulas using combinations.

Example 1 Jury Selection

a) Determine the probability distribution for the number of women on a civil-court jury selected from a pool of 8 men and 10 women.

b) What is the expected number of women on the jury?

Solution 1 Using Pencil and Paper

a) The selection process involves dependent events since each person who is already chosen for the jury cannot be selected again. The total number of ways the 6 jurors can be selected from the pool of 18 is

$$n(S) = {}_{18}C_6$$
$$= 18\ 564$$

This combination could also be written as $C(18, 6)$ or $\binom{18}{6}$.

There can be from 0 to 6 women on the jury. The number of ways in which x women can be selected is ${}_{10}C_x$. The men can fill the remaining $6 - x$ positions on the jury in ${}_8C_{6-x}$ ways. Thus, the number of ways of selecting a jury with x women on it is ${}_{10}C_x \times {}_8C_{6-x}$ and the probability of a jury with x women is

$$P(x) = \frac{n(x)}{n(S)}$$
$$= \frac{{}_{10}C_x \times {}_8C_{6-x}}{{}_{18}C_6}$$

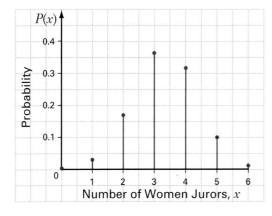

Number of Women, x	Probability, $P(x)$
0	$\dfrac{{}_{10}C_0 \times {}_8C_6}{{}_{18}C_6} \doteq 0.001\ 51$
1	$\dfrac{{}_{10}C_1 \times {}_8C_5}{{}_{18}C_6} \doteq 0.030\ 17$
2	$\dfrac{{}_{10}C_2 \times {}_8C_4}{{}_{18}C_6} \doteq 0.169\ 68$
3	$\dfrac{{}_{10}C_3 \times {}_8C_3}{{}_{18}C_6} \doteq 0.361\ 99$
4	$\dfrac{{}_{10}C_4 \times {}_8C_2}{{}_{18}C_6} \doteq 0.316\ 74$
5	$\dfrac{{}_{10}C_5 \times {}_8C_1}{{}_{18}C_6} \doteq 0.108\ 60$
6	$\dfrac{{}_{10}C_6 \times {}_8C_0}{{}_{18}C_6} \doteq 0.011\ 31$

b) $E(X) = \displaystyle\sum_{i=0}^{6} x_i P(x_i)$

$\doteq (0)(0.001\ 51) + (1)(0.030\ 17) + (2)(0.169\ 68) + (3)(0.361\ 99)$
$+ (4)(0.316\ 74) + (5)(0.108\ 60) + (6)(0.011\ 31)$
$\doteq 3.333\ 33$

The expected number of women on the jury is approximately 3.333.

Solution 2 Using a Graphing Calculator

a) Enter the possible values for x, 0 to 6, in L1. Then, enter the formula for $P(x)$ in L2:

(10 nCr L1) × (8 nCr (6–L1)) ÷ (18 nCr 6)

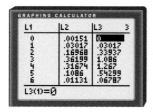

b) Calculate $xP(x)$ in L3 using the formula L1 × L2.

QUIT to the home screen. You can find the expected number of women by using the sum(function in the LIST MATH menu.

The expected number of women on the jury is approximately 3.333.

Solution 3 Using a Spreadsheet

a) Open a new spreadsheet. Create titles x, p(x), and xp(x) in columns A to C.

Enter the values of the random variable x in column A, ranging from 0 to 6. Next, use the combinations function to enter the formula for $P(x)$ in cell B3 and copy it to cells B4 through B9.

b) Calculate $xP(x)$ in column C by entering the formula A3*B3 in cell C3 and copying it to cells C4 through C9. Then, calculate the expected value using the SUM function.

The expected number of women on the jury is approximately 3.333.

	C11			=	=SUM(C3:C9)
	A	B	C	D	
1	x	p(x)	xp(x)		
2					
3		0 0.001508	0		
4		1 0.030166	0.030166		
5		2 0.169683	0.339367		
6		3 0.361991	1.085973		
7		4 0.316742	1.266968		
8		5 0.108597	0.542986		
9		6 0.011312	0.067873		
10					
11		E(x)	3.333333		
12					
13					
14					
15					
16					

Solution 4 • Using Fathom™

Open a new Fathom™ document. Drag a new **collection** box to the work area and name it **Number of Women Jurors**. Create seven new cases.

Drag a new **case table** to the work area. Create three new attributes: **x, px,** and **xpx.** Enter the values from 0 to 6 for the **x** attribute. Right-click on the **px** attribute, select **Edit Formula,** and enter

combinations(10,x)*combinations(8,6-x)/combinations(18,6)

Similarly, calculate $xP(x)$ using the formula **x*px.** Next, double-click on the **collection** box to open the **inspector**. Select the **Measures** tab, and name a new measure **Ex.** Right-click on **Ex** and use the **sum function** to enter the formula sum(x*px).

The expected number of women on the jury is approximately 3.333.

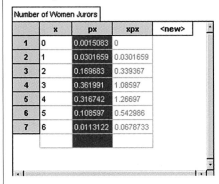

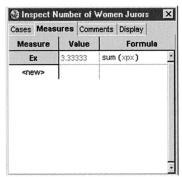

You can generalize the methods in Example 1 to show that for a hypergeometric distribution, the probability of x successes in r dependent trials is

Probability in a Hypergeometric Distribution

$$P(x) = \frac{{}_aC_x \times {}_{n-a}C_{r-x}}{{}_nC_r},$$

where a is the number of successful outcomes among a total of n possible outcomes.

Although the trials are dependent, you would expect the *average* probability of a success to be the same as the ratio of successes in the population, $\frac{a}{n}$. Thus, the expectation for r trials would be

Expectation for a Hypergeometric Distribution

$$E(X) = \frac{ra}{n}$$

This formula can be proven more rigorously by some challenging algebraic manipulation of the terms when $P(x) = \frac{{}_aC_x \times {}_{n-a}C_{r-x}}{{}_nC_r}$ is substituted into the equation for the expectation of any probability distribution, $E(X) = \sum_{i=1}^{n} x_i P(x_i)$.

Example 2 Applying the Expectation Formula

Calculate the expected number of women on the jury in Example 1.

Solution

$$E(X) = \frac{ra}{n}$$
$$= \frac{6 \times 10}{18}$$
$$= 3.3\overline{3}$$

The expected number of women jurors is $3.3\overline{3}$.

Example 3 Expectation of a Hypergeometric Distribution

A box contains seven yellow, three green, five purple, and six red candies jumbled together.

a) What is the expected number of red candies among five candies poured from the box?

b) Verify that the expectation formula for a hypergeometric distribution gives the same result as the general equation for the expectation of any probability distribution.

Solution

a) $n = 7 + 3 + 5 + 6$ $r = 5$ $a = 6$
$ = 21$

Using the expectation formula for the hypergeometric distribution,

$$E(X) = \frac{ra}{n}$$
$$ = \frac{5 \times 6}{21}$$
$$ = 1.4285\ldots$$

One would expect to have approximately 1.4 red candies among the 5 candies.

b) Using the general formula for expectation,

$$E(X) = \sum xP(x)$$
$$= (0)\frac{_6C_0 \times {}_{15}C_5}{_{21}C_5} + (1)\frac{_6C_1 \times {}_{15}C_4}{_{21}C_5} + (2)\frac{_6C_2 \times {}_{15}C_3}{_{21}C_5} + (3)\frac{_6C_3 \times {}_{15}C_2}{_{21}C_5} + (4)\frac{_6C_4 \times {}_{15}C_1}{_{21}C_5} + (5)\frac{_6C_5 \times {}_{15}C_0}{_{21}C_5}$$
$$= 1.4285\ldots$$

Again, the expected number of red candies is approximately 1.4.

Example 4 Wildlife Management

In the spring, the Ministry of the Environment caught and tagged 500 raccoons in a wilderness area. The raccoons were released after being vaccinated against rabies. To estimate the raccoon population in the area, the ministry caught 40 raccoons during the summer. Of these 15 had tags.

a) Determine whether this situation can be modelled with a hypergeometric distribution.

b) Estimate the raccoon population in the wilderness area.

WEB CONNECTION

www.mcgrawhill.ca/links/MDM12

To learn more about sampling and wildlife, visit the above web site and follow the links. Write a brief description of some of the sampling techniques that are used.

Solution

a) The 40 raccoons captured during the summer were all different from each other. In other words, there were no repetitions, so the trials were dependent. The raccoons were either tagged (a success) or not (a failure). Thus, the situation does have all the characteristics of a hypergeometric distribution.

b) Assume that the number of tagged raccoons caught during the summer is equal to the expectation for the hypergeometric distribution. You can substitute the known values in the expectation formula and then solve for the population size, n.

Here, the number of raccoons caught during the summer is the number of trials, so $r = 40$. The number of tagged raccoons is the number of successes in the population, so $a = 500$.

$$E(X) = \frac{ra}{n}, \quad \text{so} \quad 15 \doteq \frac{40 \times 500}{n}$$

$$n \doteq \frac{40 \times 500}{15}$$

$$n \doteq 1333.3$$

The raccoon population in the wilderness area is approximately 1333.

Alternatively, you could assume that the proportion of tagged raccoons among the sample captured during the summer corresponds to that in the whole population. Then, $\frac{15}{40} = \frac{500}{n}$, which gives the same estimate for n as the calculation shown above.

Key Concepts

- A hypergeometric distribution has a specified number of dependent trials having two possible outcomes, success or failure. The random variable is the number of successful outcomes in the specified number of trials. The individual outcomes cannot be repeated within these trials.

- The probability of x successes in r dependent trials is $P(x) = \dfrac{{}_aC_x \times {}_{n-a}C_{r-x}}{{}_nC_r}$,

  where n is the population size and a is the number of successes in the population.

- The expectation for a hypergeometric distribution is $E(X) = \dfrac{ra}{n}$.

- To simulate a hypergeometric experiment, ensure that the number of trials is representative of the situation and that each trial is dependent (no replacement or resetting between trials). Record the number of successes and summarize the results by calculating probabilities and expectation.

Communicate Your Understanding

1. Describe how the graph in Example 1 differs from the graphs of the uniform, binomial, and geometric distributions.

2. Consider this question: What is the probability that 5 people out of a group of 20 are left handed if 10% of the population is left-handed? Explain why this situation does not fit a hypergeometric model. Rewrite the question so that you can use a hypergeometric distribution.

Practise

1. Which of these random variables have a hypergeometric distribution? Explain why.

   a) the number of clubs dealt from a deck

   b) the number of attempts before rolling a six with a die

   c) the number of 3s produced by a random-number generator

   d) the number of defective screws in a random sample of 20 taken from a production line that has a 2% defect rate

   e) the number of male names on a page selected at random from a telephone book

   f) the number of left-handed people in a group selected from the general population

   g) the number of left-handed people selected from a group comprised equally of left-handed and right-handed people

2. Prepare a table and a graph of a hypergeometric distribution with

   a) $n = 6, r = 3, a = 3$ b) $n = 8, r = 3, a = 5$

Apply, Solve, Communicate

3. There are five cats and seven dogs in a pet shop. Four pets are chosen at random for a visit to a children's hospital.

   a) What is the probability that exactly two of the pets will be dogs?

   b) What is the expected number of dogs chosen?

4. **Communication** Earlier this year, 520 seals were caught and tagged. On a recent survey, 30 out of 125 seals had been tagged.

   a) Estimate the size of the seal population.

   b) Explain why you cannot calculate the exact size of the seal population.

5. Of the 60 grade-12 students at a school, 45 are taking English. Suppose that 8 grade-12 students are selected at random for a survey.

   a) Develop a simulation to determine the probability that 5 of the selected students are studying English.

   b) Use the formulas developed in this section to verify your simulation results.

6. **Inquiry/Problem Solving** In a study of Canada geese, 200 of a known population of 1200 geese were caught and tagged. Later, another 50 geese were caught.

   a) Develop a simulation to determine the expected number of tagged geese in the second sample.

   b) Use the formulas developed in this section to verify your simulation results.

7. **Application** In a mathematics class of 20 students, 5 are bilingual. If the class is randomly divided into 4 project teams,

   a) what is the probability that a team has fewer than 2 bilingual students?

   b) what is the expected number of bilingual students on a team?

8. In a swim meet, there are 16 competitors, 5 of whom are from the Eastern Swim Club.

   a) What is the probability that 2 of the 5 swimmers in the first heat are from the Eastern Swim Club?

   b) What is the expected number of Eastern Swim Club members in the first heat?

9. The door prizes at a dance are four $10 gift certificates, five $20 gift certificates, and three $50 gift certificates. The prize envelopes are mixed together in a bag, and five prizes are drawn at random.

   a) What is the probability that none of the prizes is a $10 gift certificate?

   b) What is the expected number of $20 gift certificates drawn?

10. A 12-member jury for a criminal case will be selected from a pool of 14 men and 11 women.

  a) What is the probability that the jury will have 6 men and 6 women?

  b) What is the probability that at least 3 jurors will be women?

  c) What is the expected number of women?

11. Seven cards are dealt from a standard deck.

  a) What is the probability that three of the seven cards are hearts?

  b) What is the expected number of hearts?

12. A bag contains two red, five black, and four green marbles. Four marbles are selected at random, without replacement. Calculate

  a) the probability that all four are black

  b) the probability that exactly two are green

  c) the probability that exactly two are green and none are red

  d) the expected numbers of red, black, and green marbles

13. A calculator manufacturer checks for defective products by testing 3 calculators out of every lot of 12. If a defective calculator is found, the lot is rejected.

  a) Suppose 2 calculators in a lot are defective. Outline two ways of calculating the probability that the lot will be rejected. Calculate this probability.

  b) The quality-control department wants to have at least a 30% chance of rejecting lots that contain only one defective calculator. Is testing 3 calculators in a lot of 12 sufficient? If not, how would you suggest they alter their quality-control techniques to achieve this standard? Support your answer with mathematical calculations.

14. Suppose you buy a lottery ticket for which you choose six different numbers between 1 and 40 inclusive. The order of the first five numbers is not important. The sixth number is a bonus number. To win first prize, all five regular numbers and the bonus number must match, respectively, the randomly generated winning numbers for the lottery. For the second prize, you must match the bonus number plus four of the regular numbers.

  a) What is the probability of winning first prize?

  b) What is the probability of winning second prize?

  c) What is the probability of not winning a prize if your first three regular numbers match winning numbers?

15. **Inquiry/Problem Solving** Under what conditions would a binomial distribution be a good approximation for a hypergeometric distribution?

16. **Inquiry/Problem Solving** You start at a corner five blocks south and five blocks west of your friend. You walk north and east while your friend walks south and west at the same speed. What is the probability that the two of you will meet on your travels?

17. A research company has 50 employees, 20 of whom are over 40 years old. Of the 22 scientists on the staff, 12 are over 40. Compare the expected numbers of older and younger scientists in a randomly selected focus group of 10 employees.

Review of Key Concepts

7.1 Probability Distributions

Refer to the Key Concepts on page 374.

1. Describe the key characteristics of a uniform distribution. Include an example with your description.

2. James has designed a board game that uses a spinner with ten equal sectors numbered 1 to 10. If the spinner stops on an odd number, a player moves forward double that number of squares. However, if the spinner stops on an even number, the player must move back half that number of squares.

   a) What is the expected move per spin?

   b) Is this rule "fair"? Explain why you might want an "unfair" spinner rule in a board game.

3. Suppose a lottery has sold 10 000 000 tickets at $5.00 each. The prizes are as follows:

Prize	Number of Prizes
$2 000 000	1
$1 000	500
$100	10 000
$5	100 000

   Determine the expected value of each ticket.

4. An environmental artist is planning to construct a rectangle with 36 m of fencing as part of an outdoor installation. If the length of the rectangle is a randomly chosen integral number of metres, what is the expected area of this enclosure?

5. Which die has the higher expectation?

   a) an 8-sided die with its faces numbered 3, 6, 9, and so on, up to 24

   b) a 12-sided die with its faces numbered 2, 4, 6, and so on, up to 24

7.2 Binomial Distributions

Refer to the Key Concepts on page 384.

6. Describe the key characteristics of a binomial distribution. Include an example with your description.

7. Cal's Coffee prints prize coupons under the rims of 20% of its paper cups. If you buy ten cups of coffee,

   a) what is the probability that you would win at least seven prizes?

   b) what is your expected number of prizes?

8. Use a table and a graph to display the probability distribution for the number of times 3 comes up in five rolls of a standard die.

9. A factory produces computer chips with a 0.9% defect rate. In a batch of 100 computer chips, what is the probability that

   a) only 1 is defective?

   b) at least 3 are defective?

10. A dart board contains 20 equally-sized sectors numbered 1 to 20. A dart is randomly tossed at the board 10 times.

    a) What is the probability that the dart lands in the sector labelled 20 a total of 5 times?

    b) What is the expected number of times the dart would land in a given sector?

11. Each question in a 15-question multiple-choice quiz has 5 possible answers. Suppose you guess randomly at each answer.

    a) Show the probability distribution for the number of correct answers.

    b) Verify the formula, $E(X) = np$, for the expectation of the number of correct answers.

7.3 Geometric Distributions

Refer to the Key Concepts on page 394.

12. Describe the key characteristics of a geometric distribution. Include an example with your description.

13. Your favourite TV station has ten minutes of commercials per hour. What is the expected number of times you could randomly select this channel without hitting a commercial?

14. A factory making printed-circuit boards has a defect rate of 2.4% on one of its production lines. An inspector tests randomly selected circuit boards from this production line.

a) What is the probability that the first defective circuit board will be the sixth one tested?

b) What is the probability that the first defective circuit board will be among the first six tested?

c) What is the expected waiting time until the first defective circuit board?

15. A computer has been programmed to generate a list of random numbers between 1 and 25.

a) What is the probability that the number 10 will not appear until the 6th number?

b) What is the expected number of trials until a 10 appears?

16. In order to win a particular board game, a player must roll, with two dice, the exact number of spaces remaining to reach the end of the board. Suppose a player is two spaces from the end of the board. Show the probability distribution for the number of rolls required to win, up to ten rolls.

7.4 Hypergeometric Distributions

Refer to the Key Concepts on page 403.

17. Describe the key characteristics of a hypergeometric distribution. Include an example with your description.

18. Of the 15 students who solved the challenge question in a mathematics contest, 8 were enrolled in mathematics of data management. Five of the solutions are selected at random for a display.

a) Prepare a table and graph of the probability distribution for the number of solutions in the display that were prepared by mathematics of data management students.

b) What is the expected number of solutions in the display that were prepared by mathematics of data management students?

19. Seven cards are randomly dealt from a standard deck. Show the probability distribution of the number of cards dealt that are either face cards or aces.

20. One summer, conservation officials caught and tagged 98 beavers in a river's flood plain. Later, 50 beavers were caught and 32 had been tagged. Estimate the size of the beaver population.

21. Suppose that 48 of the tagged beavers in question 20 were males.

a) Develop a simulation to estimate the probabilities for the number of tagged males among the 32 beavers captured a second time.

b) Verify the results of your simulation with mathematical calculations.

Chapter Test

ACHIEVEMENT CHART

Category	Knowledge/ Understanding	Thinking/Inquiry/ Problem Solving	Communication	Application
Questions	All	5, 10, 12	1, 12	3, 4, 6–8, 10–12

1. Determine if a uniform, binomial, geometric, or hypergeometric distribution would be the best model for each of the following experiments. Explain your reasoning.

   a) drawing names out of a hat without replacement and recording the number of names that begin with a consonant

   b) generating random numbers on a calculator until it displays a 5

   c) counting the number of hearts in a hand of five cards dealt from a well-shuffled deck

   d) asking all students in a class whether they prefer cola or ginger ale

   e) selecting the winning ticket in a lottery

   f) predicting the expected number of heads when flipping a coin 100 times

   g) predicting the number of boys among five children randomly selected from a group of eight boys and six girls

   h) determining the waiting time before picking a winning number in a lottery.

2. A lottery ticket costs $2.00 and a total of 4 500 000 tickets were sold. The prizes are as follows:

Prize	Number of Prizes
$500 000	1
$50 000	2
$5 000	5
$500	20
$50	100

   Determine the expected value of each ticket.

3. Of 25 people invited to a birthday party, 5 prefer vanilla ice cream, 8 prefer chocolate, and 4 prefer strawberry. The host surveys 6 of these people at random to determine how much ice cream to buy.

   a) What is the probability that at least 3 of the people surveyed prefer chocolate ice cream?

   b) What is the probability that none prefer vanilla?

   c) What is the expected number of people who prefer strawberry?

   d) What is the expected number of people who do not have a preference for any of the three flavours?

4. Suppose you randomly choose an integer n between 1 and 5, and then draw a circle with a radius of n centimetres. What is the expected area of this circle to the nearest hundredth of a square centimetre?

5. At the Statsville County Fair, the probability of winning a prize in the ring-toss game is 0.1.

   a) Show the probability distribution for the number of prizes won in 8 games.

   b) If the game will be played 500 times during the fair, how many prizes should the game operators keep in stock?

6. **a)** What is the probability that a triple will occur within the first five rolls of three dice?

   b) What is the expected waiting time before a triple?

7. In July of 2000, 38% of the population of Canada lived in Ontario. Design a simulation to estimate the expected number of residents of Ontario included in a random survey of 25 people in Canada.

8. A multiple-choice trivia quiz has ten questions, each with four possible answers. If someone simply guesses at each answer,

   a) what is the probability of only one or two correct guesses?

   b) what is the probability of getting more than half the questions right?

   c) what is the expected number of correct guesses?

9. In an experiment, a die is rolled repeatedly until all six faces have finally shown.

   a) What is the probability that it only takes six rolls for this event to occur?

   b) What is the expected waiting time for this event to occur?

10. The Burger Barn includes one of three different small toys with every Barn Burger. Each of the toys is equally likely to be included with a burger. Design a simulation to determine the number of Barn Burger purchases necessary for 3:1 odds of collecting all three different toys.

11. To determine the size of a bear population in a provincial park, 23 bears were caught and fitted with radio collars. One month later, 8 of 15 bears sighted had radio collars. What is the approximate size of the bear population?

ACHIEVEMENT CHECK

Knowledge/Understanding	Thinking/Inquiry/Problem Solving	Communication	Application

12. Louis inserts a 12-track CD into a CD player and presses the random play button. This CD player's random function chooses each track independently of any previously played tracks.

   a) What is the probability that the CD player will select Louis's favourite track first?

   b) What is the probability that the second selection will not be his favourite track?

   c) What is the expected waiting time before Louis hears his favourite track?

   d) Sketch a graph of the probability distribution for the waiting times.

   e) Explain how having a different number of tracks on the CD would affect the graph in part d).

   f) If Louis has two favourite tracks, what is the expected waiting time before he hears both tracks?

CHAPTER 8

The Normal Distribution

Specific Expectations	Section
Interpret one-variable statistics to describe the characteristics of a data set.	8.1, 8.2, 8.3
Organize and summarize data from secondary sources.	8.1, 8.2, 8.3
Identify situations that give rise to common distributions.	8.1, 8.2, 8.4, 8.5, 8.6
Interpret probability statements, including statements about odds, from a variety of sources.	8.1, 8.2, 8.3, 8.4, 8.5, 8.6
Assess the validity of some simulation results by comparing them with the theoretical probabilities, using the probability concepts developed in the course.	8.2
Describe the position of individual observations within a data set, using z-scores and percentiles.	8.2
Demonstrate an understanding of the properties of the normal distribution.	8.2, 8.3, 8.4, 8.5, 8.6
Make probability statements about normal distributions.	8.2, 8.3, 8.4, 8.5, 8.6
Illustrate sampling bias and variability by comparing the characteristics of a known population with the characteristics of samples taken repeatedly from that population, using different sampling techniques.	8.3
Assess the validity of conclusions made on the basis of statistical studies.	8.3, 8.5, 8.6
Determine probabilities, using the binomial distribution.	8.4, 8.5, 8.6

Chapter Problem

The Restless Earth

The table shows the number of major earthquakes around the world from 1900 through 1999.

1. Can you predict the number of earthquakes around the world next year?

2. Is it possible to quantify how accurate your prediction is likely to be?

In this chapter, you will develop the skills required to answer these questions as well as others involving statistical predictions. You will learn more about probability distributions, including continuous distributions. In particular, this chapter introduces the normal distribution, one of the most common and important probability distributions. Also, you will analyse statements about probabilities made in the media, statistical studies, and a wide variety of other applications.

Major Earthquakes (7.0 or Greater on the Richter Scale)					
Year	Frequency	Year	Frequency	Year	Frequency
1900	13	1934	22	1968	30
1901	14	1935	24	1969	27
1902	8	1936	21	1970	29
1903	10	1937	22	1971	23
1904	16	1938	26	1972	20
1905	26	1939	21	1973	16
1906	32	1940	23	1974	21
1907	27	1941	24	1975	21
1908	18	1942	27	1976	25
1909	32	1943	41	1977	16
1910	36	1944	31	1978	18
1911	24	1945	27	1979	15
1912	22	1946	35	1980	18
1913	23	1947	26	1981	14
1914	22	1948	28	1982	10
1915	18	1949	36	1983	15
1916	25	1950	39	1984	8
1917	21	1951	21	1985	15
1918	21	1952	17	1986	6
1919	14	1953	22	1987	11
1920	8	1954	17	1988	8
1921	11	1955	19	1989	7
1922	14	1956	15	1990	12
1923	23	1957	34	1991	11
1924	18	1958	10	1992	23
1925	17	1959	15	1993	16
1926	19	1960	22	1994	15
1927	20	1961	18	1995	25
1928	22	1962	15	1996	22
1929	19	1963	20	1997	20
1930	13	1964	15	1998	16
1931	26	1965	22	1999	23
1932	13	1966	19		
1933	14	1967	16		

Review of Prerequisite Skills

If you need help with any of the skills named in purple below, refer to Appendix A.

1. **Graphing exponential functions**

   a) Graph the following functions:

   i) $y = 10^{-x}$

   ii) $y = 10^{-2x}$

   iii) $y = 10^{-\frac{x}{2}}$

   b) Determine the y-intercepts of each function in part a).

   c) For $x > 0$, which function has

   i) the largest area under its curve?

   ii) the smallest area under its curve?

2. **Representing data (Chapter 2)** During July, a local theatre recorded the following numbers of patrons per day over a 30-day period.

102	116	113	132	128	117	156	182
183	171	160	140	154	160	122	187
185	158	112	145	168	187	117	108
171	171	156	163	168	182		

   a) Construct a histogram of these data.

   b) Determine the mean and standard deviation of these data.

   c) Construct a box-and-whisker plot of these data.

3. **Summary measures for data (Chapter 2)** Fifteen different cars were tested for stopping distances at a speed of 20 km/h. The results, in metres, are given below.

15	18	18	20	22	24	24	26
18	26	24	16	23	24	30	

   a) Find the mean, median, and mode of these data.

   b) Construct a box-and-whisker plot of these data.

   c) Find the standard deviation of these data.

4. **Summary measures for data (Chapter 2)** Analyse the following data, which represent the numbers of e-mail messages received by 30 executives on a Wednesday.

22	14	12	9	54	12	16	12	14
49	10	14	8	21	31	37	28	36
22	9	33	59	31	41	19	28	52
22	7	24						

5. **Summary measures for data (Chapter 2)** An insurance bureau listed the ratio of registered vehicles to cars stolen for selected towns and cities in Ontario. The results for a recent year were as follows:

50	38	53	56	69	90	94	88
58	68	78	89	89	52	50	70
83	98	91	90	90	84	80	70
83	89	79	75	78	73	92	105
100							

   Analyse these data and write a brief report of your findings.

6. **Scatter plots** The table below gives the depreciation (loss in value) of a new car for each year of ownership. Construct a scatter plot of the depreciation of this car.

Year of Ownership	Depreciation by End of Year
1	30%
2	20%
3	18%
4	15%
5	15%
6–9	10% per year
10	5%

7. Scatter plots The table below gives data on the planets in our solar system.

Planet	Mean Distance From the Sun (AU)	Time for One Revolution (years)
Mercury	0.387	0.241
Venus	0.723	0.615
Earth	1.000	1.000
Mars	1.523	1.881
Jupiter	5.203	11.861
Saturn	9.541	29.457
Uranus	19.190	84.008
Neptune	30.086	164.784
Pluto	39.507	248.350

a) Construct a scatter plot of these data.

b) Construct a line of best fit for these data.

c) Explore whether a curve would be a better approximation for this relationship.

d) Find out what Kepler's third law states, and investigate it using the data provided.

8. Z-scores (section 2.6) The mean age of the viewers of a popular quiz show is 38.3 years with a standard deviation of 12.71 years.

a) What is the z-score of a 25-year-old viewer?

b) What is the z-score of a 70-year-old viewer?

c) What is the z-score of a 40-year-old viewer?

d) What age range is within 1 standard deviation of the mean?

e) What age range is within 2 standard deviations of the mean?

f) What age range is within 1.3 standard deviations of the mean?

9. Z-scores (section 2.6) On a recent grade-11 mathematics contest, the mean score was 57.9 with a standard deviation of 11.6. On the grade-10 mathematics contest written at the same time, the mean score was 61.2 with a standard deviation of 11.9. Gavin scored 84.3 on the grade-11 contest and his sister, Patricia, scored 86.2 on the grade-10 contest. Explain why Gavin's results could be considered better than his sister's.

10. Binomial distribution (Chapter 7) According to Statistics Canada, about 85% of all Canadian households own a VCR. Twelve people were selected at random.

a) What is the probability that exactly 8 of them own a VCR?

b) What is the probability that no more than 8 of them own a VCR?

c) What is the probability that at least 8 of them own a VCR?

d) How many of them would you expect to own a VCR?

11. Geometric distribution (section 7.3) If you program a calculator to generate random integers between –4 and 10, what is

a) the probability that the first zero will not occur until the 6th number generated?

b) the probability that the first odd number will be generated within the first 5 numbers?

c) the expected waiting time before a negative number is generated?

Continuous Probability Distributions

Distributions like the binomial probability distribution and the hypergeometric distribution deal with discrete data. The possible values of the random variable are natural numbers, because they arise from counting processes (usually successful or unsuccessful trials). Many characteristics of a population such as the heights of human adults are continuous in nature, and have fractional or decimal values. Just as with discrete data, however, these continuous variables have statistical distributions. For some discrete quantities such as the earthquake data on page 411 a smooth, continuous model of their variation may be more useful than a bar graph. Continuous probability distributions, by contrast with those you studied in Chapter 7, allow fractional values and can be graphed as smooth curves.

INVESTIGATE & INQUIRE: Modelling Failure Rates

Manufacturers often compile reliability data to help them predict demand for repair services and to improve their products. The table below gives the failure rates for a model of computer printer during its first four years of use.

Age of Printer (Months)	Failure Rate (%)
0–6	4.5
6–12	2.4
12–18	1.2
18–24	1.5
24–30	2.2
30–36	3.1
36–42	4.0
42–48	5.8

1. Construct a scatter plot of these data. Use the midpoint of each interval. Sketch a smooth curve that is a good fit to the data.

2. Describe the resulting probability distribution. Is this distribution symmetric?

3. Why do you think printer failure rates would have this shape of distribution?

4. Calculate the mean and standard deviation of the failure rates. How useful are these summary measures in describing this distribution? Explain.

In the investigation above, you modelled failure rates as a smooth curve, which allowed you to describe some of the features of the distribution. A distribution which is not symmetric may be **positively skewed** (tail pulled to the right) or **negatively skewed** (tail pulled to the left). For example, the number of children in a Canadian family has a positively-skewed distribution, because there is a relatively low modal value of two and an extended tail that represents a small number of significantly larger families.

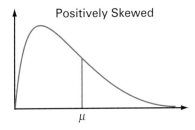

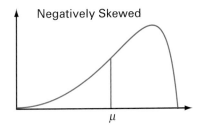

Both kinds of skewed distributions are **unimodal**. The single "hump" is similar to the mode of a set of discrete values, which you studied in Chapter 2. A distribution with two "humps" is called **bimodal**. This distribution may occur when a population consists of two groups with different attributes. For example, the distribution of adult shoe sizes is bimodal because men tend to have larger feet than women do.

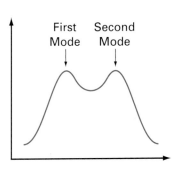

Modelling Distributions With Equations

Often you want to find the probability that a variable falls in a particular range of values. This kind of probability can be determined from the area under the distribution curve. The curve itself represents the **probability density**, the probability per unit of the continuous variable.

Many distribution curves can be modelled with equations that allow the areas to be calculated rather than estimated. The simplest such curve is the uniform distribution.

Example 1 Probabilities in a Uniform Distribution

The driving time between Toronto and North Bay is found to range evenly between 195 and 240 min. What is the probability that the drive will take less than 210 min?

Solution

The time distribution is uniform. This means that every time in the range is equally likely. The graph of this distribution will be a horizontal straight line. The total area under this line must equal 1 because all the possible driving times lie in the range 195 to 240 min. So, the height of the line is $\dfrac{1}{240 - 195} = 0.022$.

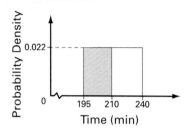

The probability that the drive will take less than 210 min will be the area under the probability graph to the left of 210. This area is a rectangle. So,
$$P(\text{driving time} \le 210) = 0.022 \times (210 - 195)$$
$$= 0.33$$

The **exponential distribution** predicts the waiting times between consecutive events in any random sequence of events. The equation for this distribution is

$$y = ke^{-kx}$$

where $k = \dfrac{1}{\mu}$ is the number of events per unit time and $e \doteq 2.71828$.

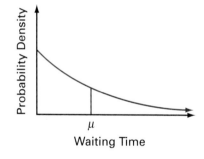

The longer the average wait, the smaller the value of k, and the more gradually the graph slopes downward.

Notice that the smallest waiting times are the most likely. This distribution is similar to the discrete geometric distribution in section 7.3. Recall that the geometric distribution models the number of trials before a success. If you think of the event "receiving a phone call in a given minute" as successive trials, you can see that the exponential distribution is the continuous equivalent of the geometric distribution.

Example 2 Exponential Distribution

The average time between phone calls to a company switchboard is $\mu = 2$ min.
a) Simulate this process by generating random arrival times for an 8-h business day. Group the waiting times in intervals and plot the relative frequencies of the intervals.
b) Draw the graph of $y = ke^{-kx}$ on your relative frequency plot. Comment on the fit of this curve to the data.

c) Calculate the probability that the time between two consecutive calls is less than 3 min.

Solution

a) Your random simulation should have a mean time between calls of 2 min. Over 8 h, you would expect about 240 calls. Use the RAND function of a spreadsheet to generate 240 random numbers between 0 and 480 in column A. These numbers simulate the times (in minutes) at which calls come in to the switchboard. Copy these numbers as values into column B so that you can sort them. Use the Sort feature to sort column B, then calculate the difference between each pair of consecutive numbers to find the waiting times between calls.

Next, copy the values for the waiting times into column D and sort them. In column F, use the COUNTIF function to cumulatively group the data into 1-min intervals. In column G, calculate the frequencies for the intervals by subtracting the cumulative frequency for each interval from that for the following interval. In column H, divide the frequencies by 239 to find the relative frequencies. Use the Chart feature to plot the relative frequencies.

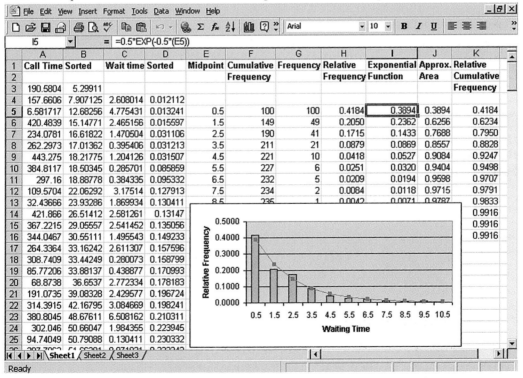

Formula bar: I5 = =0.5*EXP(-0.5*(E5))

	A	B	C	D	E	F	G	H	I	J	K
1	Call Time	Sorted	Wait time	Sorted	Midpoint	Cumulative Frequency	Frequency	Relative Frequency	Exponential Function	Approx. Area	Relative Cumulative Frequency
3	190.5804	5.29911									
4	157.6606	7.907125	2.608014	0.012112							
5	6.581717	12.68256	4.775431	0.013241	0.5	100	100	0.4184	0.3894	0.3894	0.4184
6	420.4839	15.14771	2.465156	0.015597	1.5	149	49	0.2050	0.2362	0.6256	0.6234
7	234.0781	16.61822	1.470504	0.031106	2.5	190	41	0.1715	0.1433	0.7688	0.7950
8	262.2973	17.01362	0.395406	0.031213	3.5	211	21	0.0879	0.0869	0.8557	0.8828
9	443.275	18.21775	1.204126	0.031507	4.5	221	10	0.0418	0.0527	0.9084	0.9247
10	384.8117	18.50345	0.285701	0.085859	5.5	227	6	0.0251	0.0320	0.9404	0.9498
11	297.16	18.88778	0.384335	0.095332	6.5	232	5	0.0209	0.0194	0.9598	0.9707
12	109.5704	22.06292	3.17514	0.127913	7.5	234	2	0.0084	0.0118	0.9715	0.9791
13	32.43666	23.93286	1.869934	0.130411	8.5	235	1	0.0042	0.0071	0.9787	0.9833
14	421.866	26.51412	2.581261	0.13147							0.9916
15	367.2215	29.05557	2.541452	0.135056							0.9916
16	344.0467	30.55111	1.495543	0.149233							0.9916
17	264.3364	33.16242	2.611307	0.157596							
18	308.7409	33.44249	0.280073	0.158799							
19	85.77206	33.88137	0.438877	0.170993							
20	68.8738	36.6537	2.772334	0.178183							
21	191.0735	39.08328	2.429577	0.196724							
22	314.3915	42.16795	3.084669	0.198241							
23	380.8045	48.67611	6.508162	0.210311							
24	302.046	50.66047	1.984355	0.223945							
25	94.74049	50.79088	0.130411	0.230332							

b) Since $\mu = 2$ min, the exponential equation is $P(X < x) = 0.5e^{-0.5x}$. You can use the EXP function to calculate values for the midpoint of each interval, and then plot these data with the Chart feature.

The exponential model fits the simulation data reasonably well. The sample size is small enough that statistical fluctuations could account for some intervals not fitting the model as closely as the others do.

c) You can estimate the probability of waiting times of various lengths from the cumulative relative frequencies for the simulation. Calculate these values in column K by dividing the cumulative frequencies by 239, the number of data. Cell K7 shows the relative cumulative frequency for the third interval, which gives an estimate of about 0.8 for $P(X < 3)$.

Alternatively, you can calculate this probability from the area under the probability distribution curve, as you did with the uniform distribution. The area will be approximately equal to the sum of the values of the exponential function at the midpoints of the first four intervals times the interval width. Use the SUM function to calculate this sum in column J. As shown in cell J7, this method also gives an estimate of about 0.8.

The exponential distribution is not symmetric, and its mode is always zero. The statistical measure you need to know, however, is the mean, which cannot easily be seen from the shape of the curve, but which can be found easily if you know the equation of the distribution, since $\mu = \dfrac{1}{k}$.

Notice also that the y-intercept of the curve is equal to k, so you could easily estimate the mean from the graph.

What would a symmetric version of an exponential distribution look like? Many quantities and characteristics have such a distribution, which is sometimes called a "bell curve" because of its shape. The photograph and curve on page 414 suggest one common example, people's heights. In fact, the bell curve is the most frequently observed probability distribution, and you will explore its mathematical formulation throughout this chapter.

Key Concepts

- Continuous probability distributions allow for fractional or decimal values of the random variable.

- Distributions can be represented by a relative-frequency table, a graph, or an equation.

- Probabilities can be computed by finding the area under the curve within the appropriate interval.

1. Draw or generate a relative-frequency graph for each supplier. Using a graphing calculator or Fathom™, plot the relative frequencies versus the mid-interval values for each supplier. Use the same axis scales for each data set.

2. Fit a normal curve to each data set. Use parameters $\mu = 975$ and $\sigma = 1$ to start or use your own choice of μ and σ. If using a graphing calculator, enter the normalpdf(function in the Y= editor as Y1=normalpdf(X,975,1). Using Fathom™, open the Graph menu, select Plot Function, and use the normalDensity() function to enter normalDensity(x,975,1). Adjust the parameters μ and σ to improve the fit as much as possible.

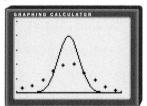

3. a) Using the techniques of section 2.1, create a table like the one below for each data set. Calculate the sample mean amount $\bar{x}$ of molybdenum for each supplier.

Supplier A				
Molybdenum (parts per 10 000)	Frequency (f)	Mid-Interval Value (x)	f × x	f × x²
970–971	3	970.5	2912	2.826×10^6
971–972				
...				
Total	$n = \sum f$ $= 100$		$\sum fx$	$\sum fx^2$

b) Compare the sample mean values $\bar{x}_A$ and $\bar{x}_B$ with the means μ_A and μ_B of the normal curves you fitted to each data set.

4. a) Calculate the sample standard deviation for each data set, using the

formula $s = \sqrt{\dfrac{\sum fx^2 - n(\bar{x})^2}{n-1}}$, the stdDev(function of the graphing

calculator, or the standard deviation function in Fathom™.

b) Compare these standard deviations with the standard deviations of the normal curves you fitted to each data set.

5. The product-development team at DynaJet Ltd. decide that alloy will be used only if its molybdenum content falls in the range 972.5 to 979.5 parts per 10 000. Based on your curve-fitting analysis, which supplier would you advise DynaJet to consider first? Which supplier would you expect to be more expensive?

Project Prep

You will need an understanding of the normal distribution to complete your probability distributions project.

A population that follows a normal distribution can be completely described by its mean, μ, and standard deviation, σ. The symmetric, unimodal form of a normal distribution makes both the mode and median equal to the mean. As you saw in the investigation, the smaller the value of σ, the more the data cluster about the mean, so the narrower the bell shape. Larger values of σ correspond to more dispersion and a wider bell shape.

Example 1 Predictions From a Normal Model

Giselle is 168 cm tall. In her high school, boys' heights are normally distributed with a mean of 174 cm and a standard deviation of 6 cm. What is the probability that the first boy Giselle meets at school tomorrow will be taller than she is?

Solution

You need to find the probability that the height of the first boy Giselle meets falls in a certain range. The normal distribution is a continuous curve, so the probability is the area under the appropriate part of the probability-distribution curve.

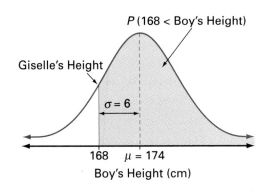

You need to find $P(168 <$ boy's height$)$. For this distribution, $168 = \mu - \sigma$. So, you can use the fact that, for *any* normally distributed random variable X, $P(\mu - \sigma < X < \mu - \sigma) \doteq 68\%$.

Break $P(168 <$ boy's height$)$ up as

$P(168 <$ boy's height$) = P(168 <$ boy's height $< 174) + P(174 >$ boy's height$)$

Since the heights of boys are normally distributed with mean $\mu = 174$ cm and standard deviation $\sigma = 6$ cm, 68% of the boys' heights lie within 1σ (6 cm) of the mean. So, the range 168 cm to 180 cm will contain 68% of the data. Because the normal distribution is symmetric, the bottom half of this range, 168 cm to 174 cm (the mean value), contains $\dfrac{68\%}{2} = 34\%$ of the data.

Therefore,

$$P(168 < \text{boy's height}) = P(168 < \text{boy's height} < 174) + P(174 > \text{boy's height})$$
$$\doteq 34\% + 50\%$$
$$= 84\%$$

So, the probability that Giselle will meet a boy taller than 168 cm is approximately 84%.

Your work on normal distributions so far should suggest a pattern of highly predictable properties. For example, although different normal distributions have different standard deviations, the value of σ scales the distribution curve in a simple, regular way. Doubling the value of σ, for example, is equivalent to a stretch of factor 2 in the horizontal direction about the line $x = \mu$, and a stretch of factor 0.5 in the vertical direction about the x-axis. The area under the curve between the values $\mu - \sigma$ and $\mu + \sigma$ is multiplied by $2 \times 0.5 = 1$, and therefore remains at approximately 68%. This fact was used in the solution of Example 1. These normal curves show how the areas are the same between each corresponding set of verticals.

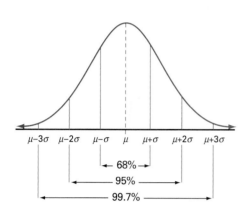 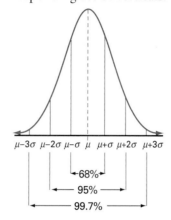

Notice that, for any normal distribution X,

• approximately 68% of the data values of X will lie within the range $\mu - \sigma$ and $\mu + \sigma$

• approximately 95% of the data values of X will lie within the range $\mu - 2\sigma$ and $\mu + 2\sigma$

• approximately 99.7% of the data values of X (almost all of them) will lie within the range $\mu - 3\sigma$ and $\mu + 3\sigma$

An understanding of these properties is vital for correctly interpreting statistics. For example, designers and administrators of psychometric tests need to understand the structure of normal distributions.

WEB CONNECTION
www.mcgrawhill.ca/links/MDM12

To learn more about the normal distribution, visit the above web site and follow the links. Write a brief description of how changes to the mean and standard deviation affect the graph of the normal distribution.

Equations and Probabilities for Normal Distributions

The curve of a normal distribution with mean μ and standard deviation σ is given by the equation

$$f(x) = \frac{1}{\sigma\sqrt{2\pi}}\, e^{-\frac{1}{2}\left(\frac{x-\mu}{\sigma}\right)^2}$$

A graphing calculator uses this equation when you select the normalpdf(function from the DISTR menu. The initials *pdf* stand for **probability density function**. Statisticians use this term for the "curve equation" of any continuous probability distribution.

As with the exponential distributions you studied in section 8.1, the family resemblance of all normal distributions is reflected in the form of the equations for these distributions. The bracketed expression $\left(\dfrac{x - \mu}{\sigma} \right)$ creates a peak at $x = \mu$, the centre of the characteristic bell shape. This exponent also scales the function horizontally by a factor of σ. Using calculus, it can be shown that the vertical scaling factor of $\dfrac{1}{\sigma\sqrt{2\pi}}$ ensures that the area under the whole curve is equal to 1.

Probabilities for normally distributed populations, as for other continuous distributions, are equal to areas under the distribution curve. The area under the curve from $x = a$ to $x = b$ gives the probability $P(a < X < b)$ that a data value X will lie between the values a and b.

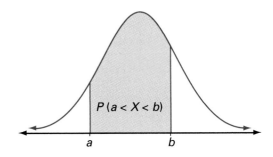

No simple formula exists for the areas under normal distribution curves. Instead, these areas have to be calculated using a more accurate version of "counting grid squares." However, you can simplify such calculations considerably by using z-scores.

Recall from section 2.6 that the z-score for a value of the random variable is $z = \dfrac{x - \mu}{\sigma}$. The distribution of the z-scores of a normally distributed variable is a normal distribution with mean 0 and standard deviation 1. This particular distribution is often called the **standard normal distribution**. Areas under this normal curve are known to a very high degree of accuracy and can be printed in table format, for easy reference. A table of these areas can be found on pages 606 and 607.

Example 2 Normal Probabilities

All That Glitters, a sparkly cosmetic powder, is machine-packaged in a process that puts approximately 50 g of powder in each package. The actual masses have a normal distribution with mean 50.5 and standard deviation 0.6. The manufacturers want to ensure that each package contains at least 49.5 g of powder. What percent of packages do not contain this much powder?

Solution 1: Using a Normal Distribution Table

Let X be the mass of powder in a package. The probability required is $P(X < 49.5)$. This probability is equal to the shaded area under the normal curve shown at the right.

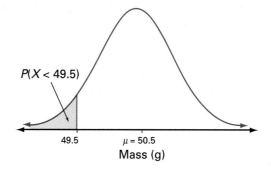

To calculate $P(X < 49.5)$, first find the corresponding z-score.

$$z = \frac{x - \mu}{\sigma}$$
$$= \frac{49.5 - 50.5}{0.6}$$
$$= -1.67$$

Use the table of Areas Under the Normal Distribution Curve on pages 606 and 607 to find the probability that a standard normal variable is less than this z-score. The table gives an area of 0.0475 for $z = -1.67$, so

$$P(X < 49.5) = P(Z < -1.67)$$
$$\doteq 0.0475$$

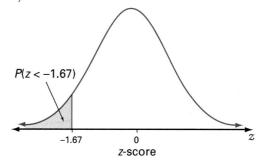

So, 4.8% of packages contain less than 49.5 g of powder.

Solution 2: Using a Graphing Calculator

If X is the mass of powder in a package, the required probability is $P(X < 49.5)$. A probability of this form, $P(X < x)$, is called a **cumulative probability**. From the DISTR menu, select the normalcdf(function. The initials *cdf* stand for cumulative density function.

The syntax for this function is normalcdf(*lower bound, upper bound, μ, σ*).

You need all the area to the left of the line $x = 49.5$, so the lower bound should be as far to the left along the x-axis as your calculator will go. Therefore, enter normalcdf(⁻1E99, 49.5, 50.5, 0.6).

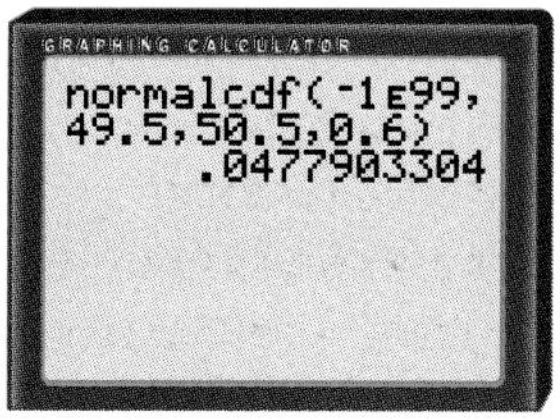

So, 4.8% of packages contain less than 49.5 g of powder.

As described in section 2.6, a z-score is the number of standard deviations a value lies above the mean. Negative z-scores represent values that lie below the mean. You can use z-scores to rank any set of data, using the standard deviation as a unit of measure. Thus, in Example 2, the mass 49.5 g has a z-score of −1.67, or 1.67 standard deviations below the mean. As well as allowing the probability to be read from the normal distribution table, converting 49.5 g to a z-score gives you a useful measure of where this value lies in the distribution.

Example 3 Standardized Test Scores

To qualify for a special program at university, Sharma had to write a standardized test. The test had a maximum score of 750, with a mean score of 540 and a standard deviation of 70. Scores on this test were normally distributed. Only those applicants scoring above the third quartile (the top 25%) are admitted to the program. Sharma scored 655 on this test. Will she be admitted to the program?

Solution 1: Using Z-Scores

One way to answer this question is to calculate Sharma's z-score.

$$z = \frac{x - \mu}{\sigma}$$

$$= \frac{655 - 540}{70}$$

$$\doteq 1.64$$

So, Sharma's score lies 1.64 standard deviations above the mean. The percent of area lying more than one standard deviation above or below the mean is 100% − 68% = 32%. Half of this area, or 16%, lies *above* $\mu + \sigma$. Therefore, Sharma's z-score of 1.64 places her well above the third quartile.

Solution 2: Using Percentiles

Just like a raw data set, a probability distribution can be analysed in terms of percentiles. The 80th percentile of a normal distribution of X, for example, is the value of x such that

$$P(X \le x) = 80\%$$

To discover whether Sharma's score is in the top 25%, calculate $P(X < 655)$ using the normalcdf(function on a graphing calculator or $P(Z < 1.64)$ using the table of Areas Under the Normal Distribution Curve on pages 606 and 607.

$P(X < 655) \doteq 0.9498$.

Sharma's score is, therefore, just below the 95th percentile for this test. At least 94% of all those who wrote this test scored lower than Sharma.

Solution 3: Using a Cut-Off Score

Find the score which represents the lower limit of the top 25% of all scores (the third quartile). This means you need to find a raw score, Q_3, for which the cumulative probability is

$$P(X < Q_3) = 0.75$$

You can use the invNorm(function in the DISTR menu on a graphing calculator to find Q_3.

So, a score of 587 or above qualifies as third quartile (top 25%). Sharma will be accepted to the program, since her score of 655 easily exceeds 587.

Key Concepts

- The normal distribution models many quantities that vary symmetrically about a mean value. This distribution has a characteristic bell shape.

- The standard normal distribution is a normal distribution with mean $\mu = 0$ and standard deviation $\sigma = 1$. Probabilities for this distribution are given in the table of Areas Under the Normal Distribution Curve on pages 606 and 607 and can be used if a graphing calculator is not available. The table gives probabilities of the form $P(Z < a)$.

- A z-score, $z = \dfrac{x - \mu}{\sigma}$, indicates the number of standard deviations a value lies from the mean. A z-score also converts a particular normal distribution to the standard normal distribution, so z-scores can be used with the areas under the normal distribution curve to find probabilities.

- Z-scores, percentiles, and cut-off scores are all useful techniques for analysing normal distributions.

1. Explain how to find the probability that the next student through the classroom door will be at most 172 cm tall, if heights for your class are normally distributed with a mean of 167 cm and a standard deviation of 2.5 cm.

2. Ron works in quality control for a company that manufactures steel washers. The diameter of a washer is normally distributed with a mean of 8 mm and a standard deviation of 1 mm. Ron checked a washer and found that its diameter was 9.5 mm. Should Ron be surprised by this result? Explain your answer to Ron.

3. Suppose the probability for a certain outcome based on a test score is 0.1. What information do you need in order to find the raw score for this outcome? Explain how to find the raw score.

Practise

1. Copy and complete the chart below, assuming a normal distribution for each situation.

Mean, μ	Standard Deviation, σ	Probability
12	3	$P(X < 9) =$
30	5	$P(X < 25) =$
5	2.2	$P(X > 6) =$
245	18	$P(233 < X < 242) =$

Apply, Solve, Communicate

2. Michael is 190 cm tall. In his high school, heights are normally distributed with a mean of 165 cm and a standard deviation of 20 cm. What is the probability that Michael's best friend is shorter than he is?

3. **Application** Testing has shown that new CD players have a mean lifetime of 6.2 years. Lifetimes for these CD players are normally distributed and have a standard deviation of 1.08 years. If the company offers a 5-year warranty on parts and labour, what percent of CD players will fail before the end of the warranty period?

4. A class of 135 students took a final examination in mathematics. The mean score on the examination was 68% with a standard deviation of 8.5%. Determine the percentile rank of each of the following people:

   a) Joey, who scored 78%

   b) Shaheed, who got 55%

   c) Michelle, who was very happy with her mark of 89%

5. **Communication** There have been some outstanding hitters in baseball. In 1911, Ty Cobb's batting average was 0.420. In 1941, Ted Williams batted 0.406. George Brett's 0.390 average in 1980 was one of the highest since Ted Williams. Batting averages have historically been approximately normally distributed, with means and standard deviations as shown below.

Decade	Mean	Standard Deviation
1910s	0.266	0.0371
1940s	0.267	0.0326
1970s–1980s	0.261	0.0317

Compute z-scores for each of these three outstanding hitters. Can you rank the three hitters? Explain your answer.

6. To run without causing damage, diameters of engine crankshafts for a certain car model must fall between 223.92 mm and 224.08 mm. Crankshaft diameters are normally distributed, with a mean of 224 mm and a standard deviation of 0.03 mm. What percent of these crankshafts are likely to cause damage?

7. The daily discharge of lead from a mine's tailings is normally distributed, with a mean level of 27 mg/L and a standard deviation of 14 mg/L. On what proportion of days will the daily discharge exceed 50 mg/L?

8. **Application** The success rate of shots on goal in an amateur hockey league's games is normally distributed, with a mean of 56.3% and a standard deviation of 8.1%.

   a) In what percent of the games will fewer than 40% of shots on goal score?

   b) In what percent of the games will the success rate be between 50% and 60%?

   c) What is the probability that a game will have a success rate of more than 66% of shots?

9. The daily sales by Hank's Hot Dogs have a mean of $572.50 and a standard deviation of $26.10.

   a) What percent of the time will the daily sales be greater than $564?

   b) What percent of the time will the daily sales be greater than $600?

10. **Inquiry/Problem Solving** The average heights of teenage girls are normally distributed, with a mean of 157 cm and a standard deviation of 7 cm.

   a) What is the probability that a teenage girl's height is greater than 170 cm?

   b) What range of heights would occur about 90% of the time?

11. **Inquiry/Problem Solving** The results of a university examination had a mean of 55 and a standard deviation of 13. The professor wishes to "bell" the marks by converting all the marks so that the mean is now 70, with a standard deviation of 10.

   a) Develop a process or formula to perform this conversion.

   b) What would happen to a mark of 80 after the conversion?

   c) What would happen to a mark of 40 after the conversion?

12. The heights of 16-month-old oak seedlings are normally distributed with a mean of 31.5 cm and a standard deviation of 10 cm. What is the range of heights between which 75% of the seedlings will grow?

13. The coach of a track team can send only the top 5% of her runners to a regional track meet. For the members of her team, times for a 1-km run are normally distributed with a mean of 5.6 min and a standard deviation of 0.76 min. What is the cut-off time to determine which members of the team qualify for the regional meet?

Normal Sampling and Modelling

Many statistical studies take sample data from an underlying normal population. As you saw in the investigation on page 422, the distribution of the sample data reflects the underlying distribution, with most values clustered about the mean in an approximate bell shape. Therefore, if a population is believed or expected to be normally distributed, predictions can be made from a sample taken from that population. As you will see, this predictive process is most reliable when the sample size is large.

● Example 1 Investment Returns

The annual returns from a particular mutual fund are believed to be normally distributed. Erin is considering investing in this mutual fund. She obtained a sample of 20 years of historic returns, which are listed in the table below.

Year	Return (%)	Year	Return (%)
1	7.2	11	6.4
2	12.3	12	27.0
3	17.1	13	14.5
4	17.9	14	25.2
5	10.8	15	−0.5
6	19.3	16	2.4
7	12.2	17	16.7
8	−13.1	18	12.8
9	20.2	19	2.9
10	18.6	20	18.8

a) Determine the mean and standard deviation of these data.

b) Assuming the data are normally distributed, what is the probability that an annual return will be

  i) at least 9%?

  ii) negative?

c) Out of the next ten years, how many years should Erin expect to show returns greater that 6%? What assumptions are necessary to answer this question?

Solution 1: Using a Normal Distribution Table

a) Using the formulas for the sample mean and sample standard deviation,

$$\bar{x} = \frac{\sum x}{n}$$
$$= \frac{248.7}{20}$$
$$= 12.435$$

$$s = \sqrt{\frac{\sum x^2 - n(\bar{x})^2}{n-1}}$$
$$= \sqrt{\frac{4805.37 - 20 \times (12.435)^2}{19}}$$
$$= 9.49$$

The mean of the data is $\bar{x} \doteq 12.4$ and the standard deviation is $s = 9.49$.

b) i) Find the z-score of 9.

$$z = \frac{x - \bar{x}}{s}$$
$$= \frac{9 - 12.4}{9.49}$$
$$= -0.36$$

Then, use the table of Areas Under the Normal Distribution Curve on pages 606 and 607 to find the probability.

$$P(X \geq 9) = P(Z \geq -0.36)$$
$$= 1 - P(Z \leq +0.36)$$
$$= 1 - 0.3594$$
$$\doteq 0.64$$

The probability of at least a 9% return is 0.64, or 64%.

ii) $P(X < 0) = P\left(Z < \dfrac{0 - 12.4}{9.49}\right)$
$$= P(Z < -1.31)$$
$$= 0.0951$$

The probability of a negative return is approximately 10%.

c) First, find the probability of a return greater than 6%.

$$P(X > 6) = P\left(Z > \frac{6 - 12.4}{9.5}\right)$$
$$= P(Z > -0.674)$$
$$= 1 - P(Z < -0.674)$$
$$\doteq 1 - 0.25$$
$$= 0.75$$

In any given year, there is a 75% probability of a return greater than 6%. Therefore, Erin can expect such a return in seven or eight years out of the next ten years. This prediction depends on the assumptions that the return data are normally distributed, and that this distribution does not change over the next ten years.

Solution 2: Using a Graphing Calculator

a) To find the mean and standard deviation, enter the returns in L1.

Use the 1-Var Stats command from the **STAT CALC** menu to obtain the following information.

From the calculator, the mean is $\bar{x} \doteq 12.4$ and the standard deviation is $s \doteq 9.49$.

Recall that, since the data is a sample, you should use the value of Sx rather than σx.

b) Since the underlying population is normally distributed, use a normal distribution with a mean of 12.4 and a standard deviation of 9.49 to make predictions about the population.

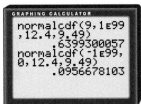

  i) $P(X \geq 9)$ is the area under the normal curve to the right of $x = 9$.

  Therefore, use the normalcdf(function as shown on the screen on the right.

  This screen shows the probability of a return of at least 9% as 0.64, or 64%.

  ii) For the area to the left of $x = 0$, use the normalcdf(function as shown on the screen on the right.

  The probability of a negative return is approximately 10%.

c) You can use the normalcdf(function to find the probability of a return greater than 6% and then proceed as in Solution 1.

Solution 3: Using a Spreadsheet

a) Copy the table into a spreadsheet starting at cell A1 and ending at cell B21. In cells E2 and E3, respectively, calculate the mean and standard deviation using the AVERAGE function and the STDEV function in Microsoft® Excel or by selecting Tools/Numeric Tools/Analysis…/Descriptive Statistics in Corel® Quattro® Pro.

b) **i)** You can use the NORMDIST function to find the cumulative probability for a result up to a given value. Subtract this probability from 1 to find the probability of an annual return of at least 9%:

E6: =NORMDIST(9,E2,E3,TRUE)
E7: =1-E6

From cell E7, you can see that $P(X \geq 9) \doteq 0.64$.

ii) Copy the NORMDIST function and change the value for X to 0 to find that there is about a 10% probability that next year's returns will be negative.

c) Copy the formula again and change the value for X to 6. The NORMDIST function will calculate the probability of an annual return of up to 6%. Subtracting this probability from 1 gives the probability of an annual return of greater than 6% (see cell G7).

$$P(X \geq 6) = 1 - P(X < 6)$$
$$\doteq 0.75$$

So, Erin should expect returns greater than 6% in seven or eight out of the next ten years.

	A	B	C	D	E	F	G	H
	Year	Return						
1		(%)						
2	1	7.2		mean	12.435			
3	2	12.3		Sx	9.494557			
4	3	17.1						
5	4	17.9		x	9	0	6	
6	5	10.8		P(X<x)	0.358756	0.095149	0.248962891	
7	6	19.3		P(X>x)	0.641244		0.751037109	
8	7	12.2						
9	8	-13.1						
10	9	20.2						
11	10	18.6						
12	11	6.4						
13	12	27						
14	13	14.5						
15	14	25.2						
16	15	-0.5						
17	16	2.4						
18	17	16.7						
19	18	12.8						
20	19	2.9						
21	20	18.8						
22								

Sheet1 / Sheet2 / Sheet3 /

Ready

Normal Models for Discrete Data

All the examples of normal distributions you have seen so far have modelled continuous data. There are many situations, however, where discrete data can also be modelled as normal distributions. For instance, the earthquake data presented in the Chapter Problem are discrete, but a statistician might well try a normal model for them. If the data set is reasonably large, and the data fall into a symmetric, unimodal bell shape, it makes sense to try fitting a smooth normal curve to them. Just as with the continuous investment data in Example 1, the normal model can then be used to make predictions.

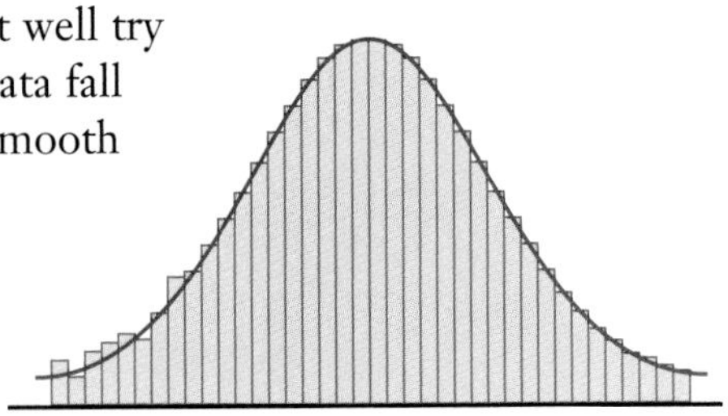

Example 2 Candy Boxes

A company produces boxes of candy-coated chocolate pieces. The number of pieces in each box is assumed to be normally distributed with a mean of 48 pieces and a standard deviation of 4.3 pieces. Quality control will reject any box with fewer than 44 pieces. Boxes with 55 or more pieces will result in excess costs to the company.

a) What is the probability that a box selected at random contains exactly 50 pieces?

b) What percent of the production will be rejected by quality control as containing too few pieces?

c) Each filling machine produces 130 000 boxes per shift. How many of these will lie within the acceptable range?

d) If you owned this company, what conclusions might you reach about your current production process?

Solution 1: Using a Normal Distribution Table

a) For a continuous distribution, the probabilities are for ranges of values. For example, all probabilities listed in the table of Areas Under the Normal Distribution Curve on pages 606 and 607 are of the form $P(Z < z)$, not $P(Z = z)$. Since a normal model is being used, discrete values such as "50 chocolates" have to be treated as though they were continuous. The simplest way is to calculate the value $P(49.5 < X < 50.5)$, treating a value of 50 chocolates as "between 49.5 and 50.5 chocolates." This technique, called **continuity correction**, enables predictions to be made about discrete quantities using a normal model.

$$P(49.5 < X < 50.5) = P\left(\frac{49.5 - 48}{4.3} < Z < \frac{50.5 - 48}{4.3}\right)$$
$$= P(0.35 < Z < 0.58)$$
$$= P(Z < 0.58) - P(Z < 0.35)$$
$$= 0.7190 - 0.6368$$
$$= 0.082$$

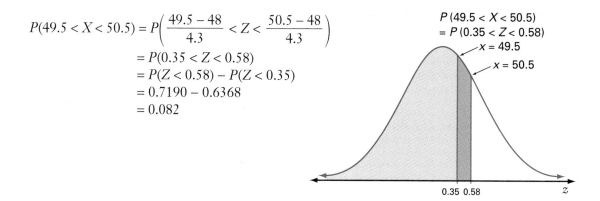

$P(49.5 < X < 50.5)$
$= P(0.35 < Z < 0.58)$
$x = 49.5$
$x = 50.5$

0.35 0.58

The probability that a box selected at random contains exactly 50 pieces is 0.082, or 8.2%.

b) A box is rejected by quality control if it has fewer than 44 pieces. A box with exactly 44 pieces is accepted, a box with exactly 43 pieces is not. With continuity correction, therefore, the probability required is $P(X < 43.5)$.

$$P(X < 43.5) = P\left(Z < \frac{43.5 - 48}{4.3}\right)$$
$$= P(Z < -1.05)$$
$$= 0.147$$

Approximately 14.7% of the production will be rejected by quality control as containing too few pieces.

c) The probability of a box being in the acceptable range of 44 to 54 pieces inclusive is

$$P(43.5 < X < 54.5) = P\left(\frac{43.5 - 48}{4.3} < Z < \frac{54.5 - 48}{4.3}\right)$$
$$= P(-1.05 < Z < +1.51)$$
$$= P(Z < +1.51) - P(Z < -1.05)$$
$$= 0.9345 - 0.1469$$
$$= 0.788$$

Thus, out of 130 000 boxes, approximately 130 000 × 0.788 or 102 000 boxes, to the nearest thousand, will be within the acceptable range.

d) Clearly there are too many rejects with the current process. The packaging process should be adjusted to reduce the standard deviation and get a more consistent number of pieces in each box. If such improvements are not possible, you might have to raise the price of each box to cover the cost of the high number of rejected boxes.

Solution 2: Using a Graphing Calculator

a) To find $P(X = 50)$ using a graphing calculator, apply continuity correction and calculate $P(49.5 < X < 50.5)$ using the normalcdf(function. Thus, the probability of a box containing exactly 50 pieces is approximately 0.083, or 8.3%.

b) You need to find $P(X < 43.5)$. Again use the normalcdf(function. Approximately 14.8% of the production will be rejected for having too few candies.

c) From the calculator, $P(43.5 < X < 54.5) = 0.787$. So, out of 130 000 boxes, 130 000 × 0.787 or 102 000 boxes, to the nearest thousand, will lie within the acceptable range.

d) See Solution 1.

Apply, Solve, Communicate

Use appropriate technology for these problems.
Assume that all the data are normally distributed.

1. A police radar unit measured the speeds, in
   kilometres per hour, of 70 cars travelling
   along a straight stretch of highway in Ontario.
   The speed limit on this highway is 100 km/h.
   The speeds of the 70 cars are listed below.

115	95	95	103	91	105	124	92
111	128	112	128	113	103	105	114
116	120	107	108	118	103	113	110
108	119	114	111	94	92	118	111
103	118	104	103	118	114	115	95
126	106	92	120	122	112	100	129
120	130	115	96	111	97	98	115
141	114	118	117	104	105	107	103
122	98	117	110	113	95		

   a) Calculate the mean and standard
      deviation of these data.

   b) What is the probability that a car
      travelling along this stretch of highway
      is speeding?

2. **Application** A university surveyed 50
   graduates from its engineering program to
   determine entry-level salaries. The results
   are listed below.

$30 400	$31 458	$31 338	$30 950	$33 560
$33 378	$32 250	$32 254	$32 000	$29 547
$32 228	$31 050	$29 074	$36 943	$33 830
$29 549	$30 838	$29 746	$31 116	$30 477
$39 708	$28 730	$34 802	$29 522	$33 582
$40 728	$33 570	$35 495	$36 416	$33 627
$29 639	$28 525	$34 169	$30 965	$33 912
$27 485	$34 299	$33 500	$30 477	$27 028
$40 829	$33 294	$28 528	$32 428	$31 526
$38 953	$36 246	$37 239	$28 469	$27 385

   a) Calculate the mean and standard
      deviation of these data.

   b) What is the probability that a graduate
      of this program will have an entry-level
      salary below $30 000?

3. **Communication** A local grocery store wants
   to obtain a profile of its typical customer. As
   part of this profile, the dollar values of
   purchases for 30 shoppers were recorded.
   The results are listed below.

$65.53	$57.11	$75.45	$53.73	$32.44
$68.85	$85.48	$65.60	$73.67	$73.11
$73.06	$56.51	$44.70	$101.77	$82.25
$45.30	$93.25	$62.47	$39.98	$68.45
$69.79	$56.90	$53.16	$65.09	$81.70
$88.95	$52.63	$68.22	$101.63	$64.45

   a) Calculate the mean and standard
      deviation of these data.

   b) What is the probability that a typical
      shopper's purchase is more than $60?

   c) What is the probability that a typical
      shopper's purchase is less than $50?

   d) Does the grocery store need to collect
      more data? Give reasons for your answer.

For questions 4 through 7 you will need access
to the E-STAT database.

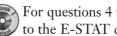

www.mcgrawhill.ca/links/MDM12

To connect to E-STAT, go to the above web site and
follow the links.

4. In E-STAT, access the People/Labour/Job
   Search section.

   a) Download the monthly help wanted
      index data from 1991–2001 for Canada
      and Ontario.

   b) Make a histogram for the data for
      Canada. Do these data appear to be
      normally distributed?

c) Calculate the mean and standard deviation of the data for Canada and the data for Ontario.

d) Do your calculations show that it was easier to find a job in Ontario than in the rest of Canada during this period?

5. From E-STAT, access the <u>Inflation</u> data table.

a) Download the table into a spreadsheet or Fathom™.

b) Calculate the mean and standard deviation of the data.

c) What is the probability that the inflation rate in a year was less than 3%?

6. From E-STAT, access the <u>Greenhouse Gas Emissions</u> data table.

a) Download the table into a spreadsheet or Fathom™.

b) Calculate the mean and standard deviation of the data.

c) Use these data to formulate and solve two questions involving probability.

7. Inquiry/Problem Solving From E-STAT, access a data table on an area of interest to you.

a) Download the table into a spreadsheet or Fathom™.

b) Use these data to formulate and solve two questions involving probability.

8. Babe Ruth played for the New York Yankees from 1920 to 1934. The list below gives the number of home runs he hit each year during that time.

| 54 | 59 | 35 | 41 | 46 | 25 | 47 | 60 |
| 54 | 46 | 49 | 46 | 41 | 34 | 22 | |

a) Calculate the mean and standard deviation of these data.

b) Estimate the probability that he would have hit more than 46 home runs if he had played another season for the Yankees.

9. Application The weekly demand for laser printer cartridges at Office Oasis is normally distributed with a mean of 350 cartridges and a standard deviation of 10 cartridges. The store has a policy of avoiding stockouts (having no product on hand). The manager decides that she wants the chance of a stockout in any given week to be at most 5%. How many cartridges should the store carry each week to meet this policy?

10. Application The table gives estimates of wolf population densities and population growth rates for the wolf population in Algonquin Park.

Year	Wolves/ 100 km²	Population Growth Rate
1988–89	4.91	
1989–90	2.47	−0.67
1990–91	2.80	0.12
1991–92	3.62	0.26
1992–93	2.53	−0.36
1993–94	2.23	−0.13
1994–95	2.82	0.24
1995–96	2.75	−0.02
1996–97	2.33	−0.17
1997–98	3.04	0.27
1998–99	1.59	−0.65

a) Group the population densities into intervals and make a frequency diagram. Do these data appear to be normally distributed?

b) Use the same method to determine whether the growth rate data appear to be normally distributed.

c) Is it possible that you would change your answer to part b) if you had a larger set of data? Explain why or why not.

WEB CONNECTION

www.mcgrawhill.ca/links/MDM12

To learn more about the decline in the wolf population in Algonquin Park, visit the above web site and follow the links.

11. Suppose the earthquake data given on page 411 are approximately normally distributed. Estimate the probability that the number of earthquakes in a given year will be greater than 30. What assumptions do you have to make for your estimate?

ACHIEVEMENT CHECK

Knowledge/ Understanding	Thinking/Inquiry/ Problem Solving	Communication	Application

12. A soft-drink manufacturer runs a bottle-filling machine, which is designed to pour 355 mL of soft drink into each can it fills. Overfilling costs money, but underfilling may result in unhappy consumers and lost sales. The quality-control inspector measured the volume of soft drink in 25 cans randomly selected from the filling machine. The results are shown below.

351.82	349.52	354.15	351.57	347.91
350.08	357.55	351.43	350.24	354.58
351.18	354.86	350.76	349.11	360.16
353.08	347.60	356.41	350.62	349.50
352.12	349.80	348.86	345.07	353.60

a) Calculate the mean and standard deviation of these data.

b) What is the probability that a can holds between 352 mL and 356 mL of soft drink?

c) Should the manufacturer adjust the filling machine? Justify your answer.

13. Inquiry/Problem Solving Given a chronological sequence of data, statistical fluctuations from day to day or year to year are sometimes reduced if you group or combine the data into longer periods.

a) Copy and complete the following table, using the data from Example 1 on page 432. Explain how each entry in the third column is calculated.

Year	Return (%)	Five-Year Return (%)
1	7.2	
2	12.3	
3	17.1	
4	17.9	
5	10.8	
6	19.3	
7	...	

b) Find the sample mean and standard deviation of the data in the third column. Compare these with the sample mean and standard deviation you found for the yearly returns in Example 1. Are the 5-year returns normally distributed Is there an advantage to longer-term investment in this fund?

c) Make a similar study of the earthquake data on page 411.

Normal Probability Plots

If it is not known whether the underlying population is normally distributed, you can use a graphing calculator or software to construct a normal probability plot of the sample data. A **normal probability plot** graphs the data according to the probabilities you would expect if the data are normal, using z-scores. If the plot is approximately linear (a straight line), the underlying population can be assumed to be normally distributed.

Using a Graphing Calculator

A toy tricycle comes with this label: "Easy-To-Assemble. An adult can complete this assembly in 20 min or less." Thirty-six adults were asked to complete the assembly of a tricycle, and record their times. Here are the results:

16	10	20	22	19	14	30	22	12	24	28	11	17	13	18	19	17	21
29	22	16	28	21	15	26	23	24	20	8		17	21	32	18	25	22	20

1. Using a graphing calculator, enter these data in L1. Find the mean and standard deviation of the data.

2. Make a normal probability plot of the data. Using STAT PLOT, select 1:Plot1, and the settings shown below.

   Based on the plot, are assembly times normally distributed?

3. **a)** What is the probability that an adult can complete this assembly in 20 min or less?

   b) What proportion of adults should complete this assembly within 15 to 30 min?

Using Fathom™

4. **a)** Open Fathom™, and open a new document if necessary. Drag a new **collection** box to the workspace. Rename the collection Assembly Times, and create 36 new cases.

   b) Drag a new **case table** to the workspace. Name the first column Times, the second column zTimes, and the third column Quantiles.

c) i) Enter the time data in the first column. **Sort** it in ascending order.

ii) Edit the formula in the second column to **zScore**(Times). This will calculate the z-scores for the data.

iii) Edit the formula in the third column to **normalQuantile**((uniqueRank(Times) − 0.5)/36, 0, 1).
This formula will calculate the z-scores of the quantiles corresponding to each entry in the **Times** column. The **uniqueRank( )** function returns the "row number" of the *sorted* data. Note that most of the quantile z-scores in the screen below are different from the z-scores for the corresponding data.

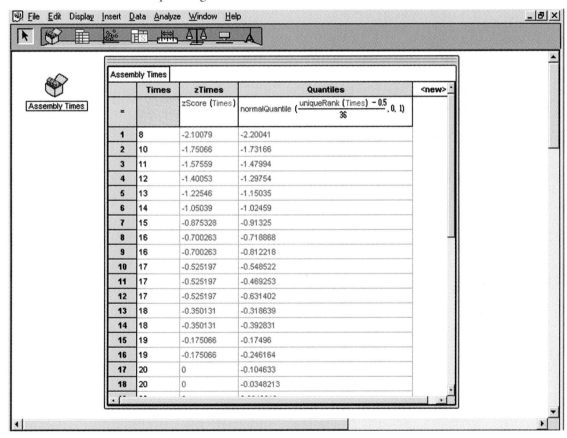

d) Drag a new graph to the workspace. Drag the **Times** title to the horizontal axis, and the **Quantiles** title to the vertical axis to generate a normal probability plot. Calculate the linear correlation coefficient for **Times** and **Quantiles** and comment on how near to linear this graph is. Are the data normally distributed?

e) Double click on the collection to open the **inspector**. Choose the **Measures** tab. Create four measures: **Mean, StdDev, P20orLess,** and **P15to30.** Use the **mean, standard deviation,** and **normalCumulative** functions to calculate the mean, the standard deviation, and the answers to question 3.

5. For each of the questions 1 to 6, 8, and 12 of section 8.3, pages 439 to 441, use a normal probability plot to determine how close to a normal distribution each data set is.

6. Let $x_1, x_2, \ldots x_n$ be a set of data, ranked in increasing order so that $x_1 \le x_2 \le \ldots \le x_n$.

For $i = 1, 2, \ldots, n$, define the **quantile** z_i by $P(Z < z_i) = \dfrac{(i - 0.5)}{n}$,

where Z is a standard normal distribution (mean 0, standard deviation 1).

a) For a data set of your choice, plot a graph of z_i against x_i. Remember to sort the x-values into increasing order. Use the invNorm(function on your graphing calculator, or the table of Areas Under the Normal Distribution Curve on pages 606 and 607, to calculate the z-values. Notice that these quantile z-values are different from the z-scores in earlier sections.

b) Compare this graph with the normal probability plot for the data set. Explain your findings.

c) Explain why, if the data are normally distributed, a graph of z_i against x_i should be close to a straight line.

Normal Approximation to the Binomial Distribution

The normal distribution is a continuous distribution. Many real-life situations involve discrete data, such as surveys of people or testing of units produced on an assembly line. As you saw in section 8.3, these situations can often be modelled by a normal distribution. If the discrete data have a binomial probability distribution and certain simple conditions are met, the normal distribution makes a very good approximation. This approximation allows the probabilities of value ranges to be calculated more easily than with the binomial formulas.

INVESTIGATE & INQUIRE: Approximating a Binomial Distribution

1. On your graphing calculator, enter the integers from 0 to 12 in L1. This list represents the number of successes in 12 trials.

2. With the cursor on L2, enter the binompdf(function. Use 12 for the number of trials and 0.5 for the probability of success. The calculator will place into L2 the binomial probabilities for each number of successes from L1.

3. Use STAT PLOT to construct a histogram using L1 as Xlist and L2 as Freq. Your result should look like the screen on the right.

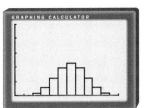

4. Now, construct a normal distribution approximation for this binomial distribution. From the DISTR menu, select the ShadeNorm(function and use 0, 0, 6, and $\sqrt{3}$ as the parameters. The normal approximation should now appear, superimposed on the binomial histogram.

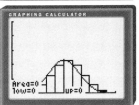

5. Investigate the effect of changing the binomial probability of success, p. Use p values of 0.3, 0.1, 0.95, and 0.7. Keep the number of trials at 12 and repeat steps 2 through 4 for each value of p. In step 4, use the ShadeNorm(function and enter 0, 0, 12p, $\sqrt{12pq}$ for each value of p. You will have to adjust the window settings for some of these situations.

6. For each different p value in step 5, make a subjective estimate of how good an approximation the normal distribution is for the underlying binomial distribution. Summarize your findings.

7. Statistical theory states that for a binomial distribution with n trials and probability of success p, a normal distribution with $\mu = np$ and $\sigma = \sqrt{npq}$ is a reasonable approximation as long as both np and nq are greater than 5. Recall that $q = 1 - p$ is the probability of failure. Do your results meet these criteria?

Example 1 Bank Loans

A bank found that 24% of its loans to new small businesses become delinquent. If 200 small businesses are selected randomly from the bank's files, what is the probability that at least 60 of them are delinquent? Compare the results from the normal approximation with the results from the calculations using a binomial distribution.

Solution 1: Using a Normal Distribution Table or Graphing Calculator

Here, $np = (200)(0.24)$ and $nq = (200)(0.76)$
$\qquad = 48 \qquad\qquad\qquad = 152$
$\qquad > 5 \qquad\qquad\qquad\quad\; > 5$

Therefore, the normal approximation should be reasonable.

$\mu = np \qquad\qquad\qquad \sigma = \sqrt{npq}$
$\;\; = (200)(0.24) \qquad\qquad = \sqrt{(200)(0.24)(0.76)}$
$\;\; = 48 \qquad\qquad\qquad\quad = 6.04$

Using the normal approximation with continuity correction, and referring to the table of Areas Under the Normal Distribution Curve on pages 606 and 607, the required probability is

$$P(X > 59.5) = P\left(Z > \frac{59.5 - 48}{6.04}\right)$$
$$= P(Z > 1.90)$$
$$= 1 - P(Z < 1.90)$$
$$= 0.029$$

Using a graphing calculator, this normal approximation probability can be calculated using the function normalcdf(59.5,1ᴇ99,48,6.04). Alternatively, the binomial probability can be calculated as 1−binomcdf(200,.24,59). The calculator screen shows the results of the two methods.

The normal approximation gives a result of 0.0285. The binomial cumulative density function gives 0.0307. The results of the two methods are close enough to be equal to the nearest percent. So, the probability that at least 60 of the 200 loans are delinquent is approximately 3%.

Solution 2: Using Fathom™

Open Fathom, and open a new document if necessary. Drag a new **collection** box to the workspace. Rename it Loans.

Double click on the collection to open the **inspector**. Choose the Measures tab.
Create two measures: Binomial and Normal.
Edit the Binomial formula to 1 − binomialCumulative(59,200,.24).
Edit the Normal formula to 1 − normalCumulative(59.5,48,6.04).
These functions give the same probability values as the equivalent graphing calculator functions.

Example 2 Market Share

QuenCola, a soft-drink company, knows that it has a 42% market share in one region of the province. QuenCola's marketing department conducts a blind taste test of 70 people at the local mall.

a) What is the probability that fewer than 25 people will choose QuenCola?

b) What is the probability that exactly 25 people will choose QuenCola?

Solution 1 Using a Normal Distribution Table

a) For this binomial distribution,

$$np = 70(0.42) \quad \text{and} \quad nq = 70(0.58)$$
$$= 29.4 \qquad\qquad\quad = 40.6$$
$$> 5 \qquad\qquad\qquad\quad > 5$$

Therefore, you can use the normal approximation.

$$\mu = np \qquad\qquad\quad \sigma = \sqrt{npq}$$
$$= (70)(0.42) \qquad\quad = \sqrt{(70)(0.42)(0.58)}$$
$$= 29.4 \qquad\qquad\quad = 4.13$$

Using the normal approximation with continuity correction,

$$P(X < 24.5) = P(Z < -1.19)$$
$$= 0.117$$

So, there is a 12% probability that fewer than 25 of the people surveyed will choose QuenCola.

b) The probability of exactly 25 people choosing QuenCola can also be calculated using the normal approximation.

$$P(24.5 < X < 25.5) = P(-1.19 < Z < -0.94)$$
$$= P(Z < -0.94) - P(Z < -1.19)$$
$$= 0.1736 - 0.1170$$
$$= 0.057$$

The probability that exactly 25 people will choose QuenCola is approximately 5.7%.

Solution 2: Using a Graphing Calculator

a) To find the probability that 24 or fewer people will choose QuenCola, you need to use a parameter of 24.5 for the normal approximation, but 24 for the cumulative binomial function. The normalcdf(and binomcdf(functions differ by less than 0.001.

$$P(X_{normal} < 24.5) = 0.118$$
$$P(X_{binomial} \leq 24) = 0.117$$

There is a 12% probability that fewer than 25 of the people surveyed will choose this company's product.

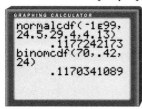

b) The normalcdf(and binompdf(functions return the following values.

$$P(24.5 < X_{normal} < 25.5) = 0.0548$$
$$P(X_{binomial} = 25) = 0.0556$$

Here, the normal approximation is a bit smaller than the value calculated using the actual binomial distribution. The manual calculation gave a slightly higher value because of rounding. The probability that exactly 25 of the people surveyed will choose this company's product is 5.6%.

Key Concepts

- A discrete binomial probability distribution can be approximated with a continuous normal distribution as long as np and nq are both greater than 5.

- To approximate the mean and standard deviation, the values $\mu = np$ and $\sigma = \sqrt{npq}$ are used.

- As with other discrete data, continuity correction should be applied when approximating a binomial distribution by a normal distribution.

1. The probability of an airline flight arriving on time is 88%. Explain how to use the normal approximation to find the probability that at least 350 of a random sample of 400 flights arrived on time.

2. Construct an example of a binomial distribution for which np is less than 5, and for which the normal distribution is *not* a good approximation. Then, show that the condition $np > 5$ and $nq > 5$ is needed.

3. Has technology reduced the usefulness of the normal approximation to a binomial distribution? Justify your answer.

Practise

1. For which of the binomial distributions listed below is the normal distribution a reasonable approximation?

   a) $n = 60$, $p = 0.4$ **b)** $n = 45$, $p = 0.1$

   c) $n = 80$, $p = 0.1$ **d)** $n = 30$, $p = 0.8$

2. Copy the table. Use the normal approximation to complete the table.

Sample Size, n	Probability of Success, p	μ	σ	Probability
60	0.4			$P(X < 22) =$
200	0.7			$P(X < 160) =$
75	0.6			$P(X > 50) =$
250	0.2			$P(X > 48) =$
1000	0.8			$P(780 < X < 840) =$
90 $\checkmark$	0.65			$P(52 < X < 62) =$
100	0.36			$P(X = 40) =$
3000	0.52			$P(X = 1650) =$

Apply, Solve, Communicate

Use the normal approximation to the binomial distribution, unless otherwise indicated.

3. It is estimated that 62% of television viewers "channel surf" during commercials. A market-research firm surveyed 1500 television viewers. What is the probability that at least 950 of them were channel surfing?

4. **Application** Salespeople sometimes advertise their products by telephoning strangers. Only about 1.5% of these "cold calls" result in a sale. Toni makes cold calls 8 h per day for 5 days. The average time for a cold call is 90 s. What is the probability that Toni gets at least 30 new customers for the week?

5. A theatre found that 7% of people who purchase tickets for a play do not show up. If the theatre's capacity is 250 people, what is the probability that there are fewer than 20 "no shows" for a sold-out performance?

6. A magazine reported that 18% of car drivers use a cellular phone while driving. In a survey of 200 drivers, what is the probability that exactly 40 of them will use a cellular phone while driving? Compare the results of using the binomial distribution and the normal approximation.

7. The human-resources manager at a company knows that 34% of the workforce belong to a union. If she randomly surveys 50 employees, what is the probability that exactly 30 of them do not belong to a union? Compare the results of using the binomial distribution with the results of using the normal approximation.

8. Application A recent survey of a gas-station's customers showed that 68% paid with credit cards, 29% used debit cards, and only 3% paid with cash. During her eight-hour shift as cashier at this gas station, Serena had a total of 223 customers.

a) What is the probability that

   i) at least 142 customers used a credit card?

   ii) fewer than 220 customers paid with credit or debit cards?

b) What is the expected number of customers who paid Serena with cash?

9. A computer-chip manufacturer knows that 72% of the chips produced are defective. Suppose 3000 chips are produced every hour.

a) What is the probability that

   i) at least 800 chips are acceptable?

   ii) exactly 800 chips are acceptable?

b) Compare the results of using the binomial distribution with those found using the normal approximation.

10. Calculate the probability that 200 rolls of two dice rolls will include

a) more than 30 sums of 5

b) between 30 and 40, inclusive, sums of 5

11. On some busy streets, diamond lanes are reserved for taxis, buses, and cars with three or more passengers. It is estimated that 20% of cars travelling in a certain diamond lane have fewer than three passengers. Sixty cars are selected at random.

a) Use the normal approximation to find the probability that

   i) fewer than 10 cars have fewer than three passengers

   ii) at least 15 cars have fewer than three passengers

b) Compare these results with those found using the binomial distribution.

c) How would the results compare if 600 cars were selected?

ACHIEVEMENT CHECK			
Knowledge/ Understanding	Thinking/Inquiry/ Problem Solving	Communication	Application

12. The probability of winning a large plush animal in the ring-toss game at the Statsville School Fair is 8%.

a) Find the probability of winning in at least 10% of 300 games, using

   i) a binomial distribution

   ii) a normal distribution

b) Predict how the probabilities of winning at least 50 times in 500 games will differ from the answers in part a). Explain your prediction.

c) Verify your prediction in part b) by calculating the probabilities using both distributions. Do your calculations support your predictions?

d) When designing the game, one student claims that having $np > 3$ and $nq > 3$ is a sufficient test for the normal approximation. Another student claims that np and nq both need to be over 10. Whom would you agree with and why?

C

13. Inquiry/Problem Solving

a) A newspaper knows that 64% of the households in a town are subscribers. If 50 households are surveyed randomly, how many of these households should the newspaper expect to be subscribers?

b) The marketing manager for the newspaper has asked you for an upper and lower limit for the number of subscribers likely to be in this sample. Find upper and lower bounds of a range that has a 90% probability of including this number.

Repeated Sampling and Hypothesis Testing

Repeated Sampling

When you draw a sample from a population, you often use the sample mean, $\bar{x}$, as an estimate of the population mean, μ, and the sample standard deviation, s, as an estimate of the population standard deviation, σ. However, the statistics for a single sample may differ radically from those of the underlying population. Statisticians try to address this problem by repeated sampling. Do additional samples improve the accuracy of the estimate?

INVESTIGATE & INQUIRE: Simulating Repeated Sampling

Simulate drawing samples of size 100 from a normally distributed population with mean $\mu = 10$ and standard deviation $\sigma = 5$. After 20 samples, examine the mean of the sample means.

The steps below outline a method using a graphing calculator. However, you can also simulate repeated sampling with a spreadsheet or statistical software such as Fathom™. See section 1.4 and the technology appendix descriptions of the software functions you could use.

1. Use the mode settings to set the number of decimal places to 2. Using the STAT EDIT menu, check that L1 and L2 are clear.

2. Place the cursor on L1. From the MATH PRB menu, select the randNorm(function and enter 10 as the mean, 5 as the standard deviation, and 100 as the number of trials. From the STAT CALC menu, select the 1-Var Stats command to find the mean of the 100 random values in L1. Enter this mean in L2.

3. Repeat step 2 twenty times. You should then have 20 entries in L2. Each of these entries is the sample mean, $\bar{x}$, of a random sample of 100 drawn from a population with a mean $\mu = 10$ and a standard deviation $\sigma = 5$.

4. Use the 1-Var Stats command to find the mean and standard deviation of L2.

5. Construct a histogram for L2. What does the shape of the histogram tell you about the distribution of the sample means?

You can construct a hypothesis test to investigate the 45-g net-weight claim, using these steps.

Project Prep

You will be asked to conduct a hypothesis test when you complete your probability distributions project.

- *Step 1* State the hypothesis being challenged, **null hypothesis** (null means no change in this context). The null hypothesis is usually denoted H_0. So, for the chocolate bars, H_0: $\mu = 45$.

- *Step 2* State the alternative hypothesis H_1 (sometimes called H_a). You suspect that the mean mass may be lower than 45 g; so, H_1: $\mu < 45$.

- *Step 3* Establish a decision rule. How strong must the evidence be to reject the null hypothesis? If, for a normal distribution with a population mean of 45 g, the probability of a sample with a mean of 44.5 g is *very small*, then getting such a sample would be strong evidence that the actual population mean is not 45 g. The **significance level**, α, is the probability threshold that you choose for deciding whether the observed results are rare enough to justify rejecting H_0. For example, if $\alpha = 0.05$, you are willing to be wrong 5% of the time.

- *Step 4* Conduct an experiment. For the chocolate bars, you weigh the sample of 30 bars.

- *Step 5* Assume H_0 is true. Calculate the probability of obtaining the results of the experiment given this assumption. If the standard deviation for chocolate bar weights is 2 g, then

$$\sigma_{\bar{x}} = \frac{\sigma}{\sqrt{n}} \qquad \text{and} \qquad P(\bar{x} < 44.5) = P\left(Z < \frac{44.5 - 45}{0.365}\right)$$

$$= \frac{2}{\sqrt{30}} \qquad\qquad\qquad\qquad \doteq P(Z < -1.37)$$

$$= 0.365 \qquad\qquad\qquad\qquad\quad = 0.0853$$

The probability of a sample mean of 44.5 or less, given a sample of size 30 from an underlying normal distribution with a mean of 45, is 8.5%.

- *Step 6* Compare this probability to the significance level, α: 8.5% is greater than 5%.

- *Step 7* Accept H_0 if the probability is greater than the significance level. Such probabilities show that the sample result is not sufficiently rare to support an alternative hypothesis. If the probability is less than the significance level reject H_0 and instead accept H_1. So, for the chocolate bars, you would accept H_0.

- *Step 8* Draw a conclusion. How reliable is the company's claim? In this case, the statistical evidence is slightly too weak to refute the company's claim.

Communicate Your Understanding

1. A researcher performed a hypothesis test, getting a result of $P(X < x) = 0.08$.

   a) Should the researcher accept or reject H_0 if $\alpha = 10\%$? Explain your answer.

   b) Should the researcher accept or reject H_0 if $\alpha = 5\%$? Explain your answer.

2. Outline the steps in a hypothesis test to determine whether getting 13 heads in 20 coin tosses is sufficient evidence to show that the coin is biased.

3. List at least four situations where hypothesis tests might be used. State H_0 and H_1 for each situation.

Practice

1. Copy and complete the table below.

Poplulation Mean	Population Standard Deviation	Sample Size	Mean of Sample Means	Standard Deviation of Sample Means
20	6	49		
12	4	25		
5	2	36		
40	8	100		
8.4	3.2	68		
17.6	10.4	87		
73.9	21.4	250		

2. For each situation, test the significance of the experimental results, given H_0 and H_1.

	Sample Size	Number of Successes	α	H_0	H_1
a)	50	23	10%	$p = 0.4$	$p > 0.4$
b)	200	55	5%	$p = 0.3$	$p < 0.3$
c)	250	175	1%	$p = 0.68$	$p > 0.68$
d)	40	8	10%	$p = 0.15$	$p > 0.15$
e)	400	80	1%	$p = 0.15$	$p > 0.15$

Apply , Solve, Communicate

3. **Application** A machine makes steel bearings with a mean diameter of 39 mm and a standard deviation of 3 mm. The bearing diameters are normally distributed. A quality-control technician found that in a sample of 50 bearings the mean diameter was 44 mm. Test the significance of this result with a significance level of 10%. Decide whether the machine needs to be adjusted.

4. **Application** A newspaper stated that 70% of the population supported a particular candidate's position on health care. In a random survey of 50 people, 31 agreed with the candidate's position. Test the significance of this result with a confidence level of 90%. Should the newspaper print a correction?

5. **Communication** A new drug will not be considered for acceptance by Health Canada unless it causes serious side effects in less than 0.01% of the population. In a trial with 80 000 people, 9 suffered serious side effects. Test the significance of this result with $\alpha = 0.01$. Do you recommend that this drug be accepted by Health Canada? Explain your answer.

6. A certain soft-drink manufacturer claims that its product holds 28% of the market. In a blind taste test, 13 out of 60 people chose this product. Does this test support or refute the soft drink manufacturer's claim? Choose a significance level you feel is appropriate for this situation.

7. Inquiry/Problem Solving An insurance company claims that 38% of automobile accidents occur within 5 km of home. The company examined 400 recent accidents and found that 120 occurred within 5 km of the driver's home. Does this result support or refute the company's claim? Choose a significance level and justify your choice.

8. Inquiry/Problem Solving A student-loan program claims that the average loan per student per year is $7500. Dana investigated this statement by asking 50 students about this year's student loan. The mean of the results was $5800. What additional information does Dana need to test the significance of this result? What significance level would be appropriate here? Why?

9. In finance, the *strong form* of the efficient-market hypothesis states that studying financial information about stocks is a waste of time, since all public and private information that might affect the stock's price is already reflected in the price of the stock. However, a study of 450 stocks found that only about 8% had price movements that could be accounted for in this way. At what significance level could you accept the strong form of the efficient-market hypothesis?

10. Does advertising influence behaviour? Before a recent advertising campaign, a children's breakfast cereal held 8% of the market. After the campaign, 18 families out of a sample of 200 families indicated they purchased the cereal. Was the advertising campaign a success? Select a confidence level you feel is appropriate for this situation.

11. Researchers often use repeated samples to test a hypothesis. Why do you think they use this method? Outline some of its advantages and disadvantages.

✓	ACHIEVEMENT CHECK		
Knowledge/ Understanding	Thinking/Inquiry/ Problem Solving	Communication	Application

12. A new medication is designed to lower cholesterol. The cholesterol level in a group of patients is normally distributed with a mean of 6.15 mmol/L and a standard deviation of 1.35 mmol/L. A sample of 40 people used the medication for 30 days, after which their mean cholesterol level was 5.87 mmol/L. The drug company wants to have a 95% confidence level that the drug is effective before releasing it.

a) Should the company release the drug?

b) The drug appears to have more effect if used for longer periods of time. What mean cholesterol level would the test group have to reach for the company to be confident about releasing the drug? Support your answer with mathematical calculations.

C

13. With a graphing calculator, you can use the Z-Test instruction to test a value for the mean of a normal distribution, based on a sample data set. Enter the first 20 years of the earthquake data on page 411. Use the Z-Test instruction. Switch the input to Data (as opposed to Stats). Choose and enter test values of μ_0 and σ, and also choose an alternative hypothesis. What information does this test give you?

8.6 Confidence Intervals

Governments often commission polls to gauge support for new initiatives. The polling organization surveys a small number of people and estimates support in the entire population based on the sample results. Opinion polls printed in newspapers often include a note such as "These results are accurate to within ±3%, 19 times in 20." In a statement like this, the figure ±3% is a margin of error, and the phrase "19 times in 20" is a confidence level of 0.95 or 95%. The statement means that, if 43% of the sample supported an initiative, there is a 95% probability (you can be 95% confident) that between 40% and 46% of the population supports the initiative. The range 40% to 46% is an example of a confidence interval. The probability of error for this finding is $1 - 0.95 = 0.05$ or 5%.

In such surveys, you do not know the population mean, μ. However, you can determine **confidence intervals**, ranges of values within which μ is likely to fall. These intervals are centered on the sample mean, $\bar{x}$, and their widths depend on the confidence level, $1 - \alpha$. For example, a 95% confidence interval has a 0.95 probability of including μ. In a normal distribution, μ is as likely to lie above the confidence interval as below it, so

$$P\left(\bar{x} - z_{0.975} \frac{\sigma}{\sqrt{n}} < \mu < \bar{x} + z_{0.975} \frac{\sigma}{\sqrt{n}}\right) = 0.95,$$

where $P(Z < z_{0.975}) = 0.975$
$= 97.5\%$

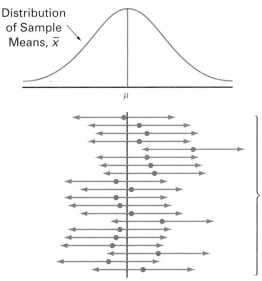

Distribution of Sample Means, $\bar{x}$

95% Confidence Intervals for Different Samples

A $(1 - \alpha)$ or $(1 - \alpha) \times 100\%$ confidence interval for μ, given a population standard deviation σ and a sample of size n with sample mean $\bar{x}$, represents the range of values

$$\bar{x} - z_{\frac{\alpha}{2}}\frac{\sigma}{\sqrt{n}} < \mu < \bar{x} + z_{\frac{\alpha}{2}}\frac{\sigma}{\sqrt{n}}$$

Project Prep

You will need to construct a confidence interval when you complete your probability distributions project.

The table gives a list of common confidence levels and their associated z-scores.

Confidence Level	Tail size, $\frac{\alpha}{2}$	z-score, $z_{\frac{\alpha}{2}}$
90%	0.05	1.645
95%	0.025	1.960
99%	0.005	2.576

Example 1 Drying Times

A paint manufacturer knows from experience that drying times for latex paints have a standard deviation of $\sigma = 10.5$ min. The manufacturer wants to use the slogan: "Dries in T minutes" on its advertising. Twenty test areas of equal size are painted and the mean drying time, $\bar{x}$, is found to be 75.4 min. Find a 95% confidence level for the actual mean drying time of the paint. What would be a reasonable value for T?

Solution 1: Using Pencil and Paper

For a 95% confidence level, the acceptable probability of error, or significance level, is $\alpha = 5\%$, so $z_{\frac{\alpha}{2}} = z_{0.975} = 1.960$. Substituting into the formula gives

$$\bar{x} - z_{\frac{\alpha}{2}}\frac{\sigma}{\sqrt{n}} < \mu < \bar{x} + z_{\frac{\alpha}{2}}\frac{\sigma}{\sqrt{n}}$$

$$75.4 - (1.960)\left(\frac{10.5}{\sqrt{20}}\right) < \mu < 75.4 + (1.960)\left(\frac{10.5}{\sqrt{20}}\right)$$

$$70.8 < \mu < 80.0$$

So, the manufacturer can be 95% confident that the actual mean drying time for the paint lies between 70.8 min and 80.0 min. It would be reasonable to advertise: "Dries in 80 minutes."

Solution 2: Using a Graphing Calculator

Use the Z-Interval instruction in the STAT TESTS menu. Make sure the input is set to Stats. Enter the population parameter $\sigma = 10.5$ and the sample parameters $\bar{x} = 75.4$ and $n = 20$. Set the confidence level to 0.95. Select Calculate and press ENTER to find the required interval.

Often, you want to know the *proportion* of a population that have a particular opinion or characteristic. This proportion is simply p, the probability of success in the binomial distribution. When data are expressed in terms of proportions, the confidence interval formula becomes

$$\hat{p} - z_{\frac{\alpha}{2}} \frac{\sqrt{pq}}{\sqrt{n}} < p < \hat{p} + z_{\frac{\alpha}{2}} \frac{\sqrt{pq}}{\sqrt{n}},$$

where $\hat{p}$ is the proportion in the sample. This sample proportion is an estimate of the population proportion just as the sample mean, $\bar{x}$, is an estimate of the population mean, μ.

For many polls, the population proportion is not known. In fact, the purpose of the polls is to estimate this parameter. Since $p \doteq \hat{p}$ and $q \doteq 1 - \hat{p}$, you can *estimate* a confidence interval using the formula

$$\hat{p} - z_{\frac{\alpha}{2}} \frac{\sqrt{\hat{p}(1 - \hat{p})}}{\sqrt{n}} < p < \hat{p} + z_{\frac{\alpha}{2}} \frac{\sqrt{\hat{p}(1 - \hat{p})}}{\sqrt{n}}$$

Example 2 Municipal Elections

Voter turnout in municipal elections is often very low. In a recent election, the mayor got 53% of the voters, but only about 1500 voters turned out.

a) Construct a 90% confidence interval for the proportion of people who support the mayor.

b) Comment on any assumptions you have to make for your calculation.

Solution

a) In this case, you want to find a confidence interval for a *proportion* of the population in a binomial distribution. Here, p is the proportion of the population who support the mayor, so you can use the election results to estimate this proportion:

$$p \doteq \hat{p} \qquad \text{and} \qquad q = 1 - p$$
$$= 0.53 \qquad\qquad\qquad \doteq 1 - 0.53$$
$$= 0.47$$

These estimated values gives a 90% confidence interval of

$$\hat{p} - z_{\frac{\alpha}{2}} \frac{\sqrt{pq}}{\sqrt{n}} < p < \hat{p} + z_{\frac{\alpha}{2}} \frac{\sqrt{pq}}{\sqrt{n}}$$

$$0.53 - 1.645 \frac{\sqrt{0.53(0.47)}}{\sqrt{1500}} < p < 0.53 + 1.645 \frac{\sqrt{0.53(0.47)}}{\sqrt{1500}}$$

$$0.51 < p < 0.55$$

So the mayor can be 90% confident of having the support of between 51% and 55% of the population.

b) You have to assume that the people who voted are representative of whole population. This assumption might not be valid because the people who take the trouble to vote are likely to be the ones most interested in municipal affairs.

Sample Sizes and Margin of Error

Given a sample of size n, from a normal population with standard deviation σ, you can use the sample mean to construct a confidence interval. You can express this confidence interval in terms of its central value and width. For example, suppose a sample of bolts has a 95% confidence interval of 7.51 mm $< \mu <$ 7.55 mm for the diameters. You can express this interval as a 95% confident estimate of 7.53 mm ± 0.02 mm for the mean diameter μ. Sometimes, however, a statistician might first decide on the confidence interval width, or **margin of error**, required, and then use this value to calculate the minimum sample size necessary to achieve this width. Opinion polls and other surveys are constructed in this way.

The width of the confidence interval $\bar{x} \pm z_{\frac{\alpha}{2}} \dfrac{\sigma}{\sqrt{n}}$ is $w = 2 z_{\frac{\alpha}{2}} \dfrac{\sigma}{\sqrt{n}}$. Solving this equation for the sample size, n, gives

$$n = \left(\frac{2 z_{\frac{\alpha}{2}} \sigma}{w} \right)^2$$

If the pollsters know (or have a good estimate of) the population standard deviation, σ, they can use this formula to find the sample size they require for a survey to have a specified margin of error.

Example 3 Sample Size for Quality-Control Testing

Suppose the diameters of the bolts mentioned above have a standard deviation of $\sigma = 0.1$. How large a sample would you need to be 90% confident that the mean diameter is 7.53 mm ± 0.01 mm?

Solution

For a 90% confidence level, $z_{\frac{\alpha}{2}} = 1.645$. Substituting the known values into the equation for n gives

$$n = \left(\frac{2z_{\frac{\alpha}{2}}\sigma}{w}\right)^2$$

$$= \left(\frac{2(1.645)(0.1)}{(0.02)}\right)^2$$

$$\doteq 271$$

You would need a sample of about 270 bolts.

You can use a similar method to find the sample sizes required for surveys involving population proportions. The margin of error for a proportion is

$$w = 2z_{\frac{\alpha}{2}}\frac{\sqrt{pq}}{\sqrt{n}}, \text{ so } n = \left(\frac{2z_{\frac{\alpha}{2}}\sqrt{pq}}{w}\right)^2. \text{ This formula simplifies to}$$

$$n = 4pq\left(\frac{z_{\frac{\alpha}{2}}}{w}\right)^2$$

Example 4 Sample Size for a Poll

A recent survey indicated that 82% of secondary-school students graduate within five years of entering grade 9. This result is considered accurate within plus or minus 3%, 19 times in 20. Estimate the sample size in this survey.

Solution

The result describes a confidence interval with a margin of error $w = 6\%$, and the confidence level is "19 times out of 20" or 95%, giving $\alpha = 0.05$.

Here, as in Example 2, you have a binomial distribution with data expressed as proportions. You can use the survey results to estimate p. Since 82% of the students in the survey graduated,

$$p \doteq \hat{p} \quad \text{and} \quad q = 1 - p$$
$$= 0.82 \qquad\qquad \doteq 1 - 0.82$$
$$= 0.18$$

WEB CONNECTION

www.mcgrawhill.ca/links/MDM12

To learn more about confidence intervals, visit the above web site and follow the links.

Substituting into the formula for n,

$$n = 4pq\left(\frac{z_{\frac{\alpha}{2}}}{w}\right)^2$$

$$\doteq 4(0.82)(0.18)\left(\frac{1.960}{0.06}\right)^2$$

$$\doteq 630$$

So, to obtain the stated level of accuracy and confidence, approximately 630 people would need to be surveyed.

Communicate Your Understanding

1. How does the population distribution affect the distribution of the sample means?

2. a) Why is the z-score for 97.5% used to construct a 95% confidence interval? Support your answer with a sketched distribution.

   b) Given that $z_{0.975} = 1.96$, approximately how many standard deviations wide is a 95% confidence interval?

3. To obtain the desired margin of error, an investigator must sample 2000 people. List at least three possible problems the investigator may encounter.

4. Interpret the following headline using confidence intervals: "4 out of every 13 Canadians think that the government should subsidize professional sports teams in Canada. These results are considered accurate within plus or minus 4%, nine times out of ten."

Practice

1. Construct the following confidence intervals.

Confidence Level	$\bar{x}$	σ	Sample Size
a) 90%	15	3	36
b) 95%	30	10	75
c) 99%	6.4	2.5	60
d) 90%	30.6	8.7	120
e) 95%	41.8	12.6	325
f) 99%	4.25	0.86	44

2. Interpret each of the following statements using confidence intervals.

   a) In a recent survey, 42% of high school graduates indicated that they expected to earn over $100 000 per year by the time they retire. This survey is considered accurate within plus or minus 3%, 19 times in 20.

   b) A survey done by the incumbent MP indicated that 48% of decided voters said they would vote for him again in the next election. The result is considered accurate within plus or minus 5%, nine times in ten.

c) According to a market research firm, 28% of teenagers will purchase the latest CD by the rock band Drench. The result is considered accurate within ±4%, 11 out of 15 times.

Apply, Solve, Communicate

3. Application A large water pipeline is being constructed to link a town with a fresh water aquifer. A construction supervisor measured the diameters of 40 pipe segments and found that the mean diameter was 25.5 cm. In the past, pipe manufactured by the same company have had a standard deviation of 7 mm. Determine a 95% confidence interval for the mean diameter of the pipe segments.

4. Application A study of 55 patients with low-back pain reported that the mean duration of the pain was 17.6 months, with a standard deviation of 5.1 months. Assuming that the duration of this problem is normally distributed in the population, determine a 99% confidence interval for the mean duration of low-back pain in the population.

5. The Statsville school board surveyed 70 parents on the question: "Should school uniforms be instituted at your school?" 28% of the respondents answered "Yes." Construct a 90% confidence interval for the proportion of Statsville parents who want school uniforms.

6. The most popular name for pleasure boats is "Serenity." A survey of 200 000 boat owners found that 12% of their boats were named "Serenity." Determine a 90% confidence interval for the proportion of pleasure boats carrying this name.

7. The football coach wants an estimate of the physical-fitness level of 44 players trying out for the varsity team. He counted the number of sit-ups done in 2 min by each player. Here are the results:

38	95	86	63	68	73	26	43
90	30	71	100	92	57	71	67
47	56	68	61	61	92	83	50
66	51	87	64	80	58	60	103
14	39	88	75	60	87	70	66
95	26	75	61				

Construct a 90% confidence interval for the mean number of sit-ups that varsity football players are capable of performing.

8. The students in a school environment club are concerned that recycling efforts are failing in smaller communities. The amount of waste recycled in 33 towns with under 5000 households is given below:

12%	10%	12%	11%	13%	12%	18%
10%	33%	3%	10%	15%	12%	18%
20%	24%	18%	5%	13%	12%	14%
25%	17%	22%	11%	12%	26%	20%
17%	22%	11%	20%	30%		

a) Construct a 95% confidence interval for the mean percent of waste recycled in towns with under 5000 households.

b) Communication Write a letter to the mayors of towns with under 5000 households outlining the results.

9. A market-research firm asked 300 people about their shampoo-purchasing habits. Fifty-five people said they bought S'No Flakes. Determine a 95% confidence interval for the percent of people who purchase S'No Flakes.

Review of Key Concepts

8.1 Continuous Probability Distributions
Refer to the Key Concepts on page 418.

1. Suppose the commuting time from Georgetown to downtown Toronto varies uniformly from 30 to 55 min, depending on traffic and weather conditions. Construct a graph of this distribution and use the graph to find

   a) the probability that a trip takes 45 min or less

   b) the probability that a trip takes more than 48 min

2. The lifetime of a critical component in microwave ovens is exponentially distributed with $k = 0.16$.

   a) Sketch a graph of this distribution. Identify the distribution by name.

   b) Calculate the approximate probability that this critical component will require replacement in less than five years.

3. Many people invest in the stock market by buying stocks recommended in investment newsletters. The table gives the annual returns after one year for 105 stocks recommended in investment newsletters.

   a) Construct a graph of these data.

   b) Describe the shape of the graph. Use terms such as symmetric, skewed, or bimodal.

   c) Calculate the mean and standard deviation for these data.

Return (%)	Number of Stocks
−15	3
−12	1
−9	1
−6	1
−3	2
0	6
+3	10
+6	16
+9	22
+12	18
+15	9
+18	9
+21	4
+24	1
+27	1
+30	1

8.2 Properties of the Normal Distribution
Refer to the Key Concepts on page 429.

4. An electrician is testing the accuracy of resistors that have a nominal resistance of 15 Ω (ohms). He finds that the distribution of resistances is approximately normal with a mean of 15.08 Ω and a standard deviation of 1.52 Ω. What is the probability that

   a) a resistor selected randomly has a resistance less than 13 Ω?

   b) a resistor selected randomly has a resistance greater than 14.5 Ω?

   c) a resistor selected randomly has a resistance between 13.8 Ω and 16.2 Ω?

5. The results of a blood test at a medical laboratory are normally distributed with $\mu = 60$ and $\sigma = 15$.

   a) What is the probability that a blood test chosen randomly from these data has a score greater than 90?

   b) What percent of these blood tests will have results between 50 and 80?

   c) How low must a score be to lie in the lowest 5% of the results?

6. The lifetimes of a certain brand of photographic light are normally distributed with a mean of 210 h and a standard deviation of 50 h.

   a) What is the probability that a particular light will last more than 250 h?

   b) What percent of lights will need to be replaced within 235 h?

   c) Out of 2000 lights, how many will have a lifetime between 200 h and 400 h?

c) According to a market research firm, 28% of teenagers will purchase the latest CD by the rock band Drench. The result is considered accurate within ±4%, 11 out of 15 times.

Apply, Solve, Communicate

3. **Application** A large water pipeline is being constructed to link a town with a fresh water aquifer. A construction supervisor measured the diameters of 40 pipe segments and found that the mean diameter was 25.5 cm. In the past, pipe manufactured by the same company have had a standard deviation of 7 mm. Determine a 95% confidence interval for the mean diameter of the pipe segments.

4. **Application** A study of 55 patients with low-back pain reported that the mean duration of the pain was 17.6 months, with a standard deviation of 5.1 months. Assuming that the duration of this problem is normally distributed in the population, determine a 99% confidence interval for the mean duration of low-back pain in the population.

5. The Statsville school board surveyed 70 parents on the question: "Should school uniforms be instituted at your school?" 28% of the respondents answered "Yes." Construct a 90% confidence interval for the proportion of Statsville parents who want school uniforms.

6. The most popular name for pleasure boats is "Serenity." A survey of 200 000 boat owners found that 12% of their boats were named "Serenity." Determine a 90% confidence interval for the proportion of pleasure boats carrying this name.

7. The football coach wants an estimate of the physical-fitness level of 44 players trying out for the varsity team. He counted the number of sit-ups done in 2 min by each player. Here are the results:

38	95	86	63	68	73	26	43
90	30	71	100	92	57	71	67
47	56	68	61	61	92	83	50
66	51	87	64	80	58	60	103
14	39	88	75	60	87	70	66
95	26	75	61				

Construct a 90% confidence interval for the mean number of sit-ups that varsity football players are capable of performing.

8. The students in a school environment club are concerned that recycling efforts are failing in smaller communities. The amount of waste recycled in 33 towns with under 5000 households is given below:

12%	10%	12%	11%	13%	12%	18%
10%	33%	3%	10%	15%	12%	18%
20%	24%	18%	5%	13%	12%	14%
25%	17%	22%	11%	12%	26%	20%
17%	22%	11%	20%	30%		

a) Construct a 95% confidence interval for the mean percent of waste recycled in towns with under 5000 households.

b) **Communication** Write a letter to the mayors of towns with under 5000 households outlining the results.

9. A market-research firm asked 300 people about their shampoo-purchasing habits. Fifty-five people said they bought S'No Flakes. Determine a 95% confidence interval for the percent of people who purchase S'No Flakes.

10. A manufacturing company wants to estimate the average number of sick days its employees take per year. A pilot study found the standard deviation to be 2.5 days. How large a sample must be taken to obtain an estimate with a maximum error of 0.5 day and a 90% confidence level?

11. An industrial-safety inspector wishes to estimate the average noise level, in decibels (dB), on a factory floor. She knows that the standard deviation is 8 dB. She wants to be 90% confident that the estimate is correct to within ±2 dB. How many noise-level measurements should she take?

12. An ergonomics advisor wants to estimate the percent of computer workers who experience carpal-tunnel syndrome. An initial survey of 50 workers found three cases of the syndrome. To be 99% confident of an accuracy of ±2%, how many workers must the advisor survey?

13. a) Obtain a survey result from your local newspaper that contains accuracy information. Determine the sample size. State explicitly any assumptions you needed to make.

b) Try to find a survey result from your local newspaper that gives the sample size. Use this information to estimate the standard deviation for the population surveyed.

14. Take 12 random samples of ten data from the earthquake table on page 411. Use these samples to estimate a 90% confidence interval for the mean number of major earthquakes you should expect this year.

15. Inquiry/Problem Solving The table below gives a sample of population growth rates of wolves in Algonquin Park.

Year	Population Growth Rate
1989–90	−0.67
1990–91	0.12
1991–92	0.26
1992–93	−0.36
1993–94	−0.13
1994–95	0.24
1995–96	−0.02
1996–97	−0.17
1997–98	0.27
1998–99	−0.65

a) Use these samples to estimate the mean and standard deviation for the growth rate of the wolf population in Algonquin Park. Explain your results.

b) Assuming that the growth rates are normally distributed, estimate the probability that the growth rate for the wolf population is less than zero.

c) A population is in danger of extinction if its population growth rate is −0.05 or less. A study based on these samples claimed that there is a 71% probability that this wolf population is in danger of extinction. Is the study correct?

d) Construct a 90% confidence interval for the true population growth rate of wolves in Algonquin Park.

16. A social scientist wants to estimate the average salary of office managers in a large city. She wants to be 95% confident that her estimate is correct. Assume that the salaries are normally distributed and that $\sigma = \$1050$. How large a sample must she take to obtain the desired information and be accurate within $200?

17. Communication An opinion pollster determines that a sample of 1500 people should give a margin of error of 3% at the 95% confidence level 19 times in 20. The pollster decides that an efficient way to find a representative sample of 1500 people is to conduct the poll at Pearson International Airport. Discuss, in terms of techniques such as stratified sampling, how representative the poll will be.

ACHIEVEMENT CHECK

Knowledge/ Understanding	Thinking/Inquiry/ Problem Solving	Communication	Application

18. Emilio has played ten rounds of golf at the Statsville course this season. His mean score is 80 and the standard deviation is 4. Assume Emilio's golf scores are normally distributed.

a) Find the 95% confidence interval of Emilio's mean golf score.

b) Predict how the confidence interval would change if the standard deviation of the golf scores was 8 instead of 4. Explain your reasoning.

c) Find the 95% confidence interval of Emilio's mean golf score if the standard deviation was 8 instead of 4. Does the answer support your prediction? Explain.

d) Emilio's most recent golf score at the course is 75. He claims that his game has improved and this latest score should determine whether he qualifies for entry in the Statsville tournament. Should the tournament organizers accept his claim? Justify your answer mathematically.

C

19. Given a sample of size n with mean $\bar{x}$, the population mean μ can be estimated via the 95% confidence interval defined by the probability

$$P\left(\bar{x} - z_{0.975}\frac{\sigma}{\sqrt{n}} < \mu < \bar{x} + z_{0.975}\frac{\sigma}{\sqrt{n}}\right) = 0.95 \quad (1)$$

If a value μ_0 is assumed for μ, a 5% significance level hypothesis test H_0: $\mu = \mu_0$: against H_1: $\mu \neq \mu_0$ can be performed on the sample mean $\bar{x}$, using the probability

$$P\left(\mu_0 - z_{0.975}\frac{\sigma}{\sqrt{n}} < \bar{x} < \mu_0 + z_{0.975}\frac{\sigma}{\sqrt{n}}\right) = 0.95 \quad (2)$$

a) Show that the probability in equation (1) can be rearranged as

$$P\left(|\bar{x} - \mu| < z_{0.975}\frac{\sigma}{\sqrt{n}}\right)$$

b) Show that the probability in equation (2) can also be rearranged into a similar form.

c) Use parts a) and b) to prove that, if μ_0 is in the 95% confidence interval defined by equation (1), then H_0 will be accepted at the 5% significance level, but if μ_0 is not in this confidence interval, H_0 will be rejected and H_1 accepted.

20. Suppose you are designing a poll on a subject for which you have no information on what people's opinions are likely to be. What sample size should you use to ensure a 90% confidence level that your results are accurate to plus or minus 2%? Explain your reasoning.

Review of Key Concepts

8.1 Continuous Probability Distributions
Refer to the Key Concepts on page 418.

1. Suppose the commuting time from Georgetown to downtown Toronto varies uniformly from 30 to 55 min, depending on traffic and weather conditions. Construct a graph of this distribution and use the graph to find

   a) the probability that a trip takes 45 min or less

   b) the probability that a trip takes more than 48 min

2. The lifetime of a critical component in microwave ovens is exponentially distributed with $k = 0.16$.

   a) Sketch a graph of this distribution. Identify the distribution by name.

   b) Calculate the approximate probability that this critical component will require replacement in less than five years.

3. Many people invest in the stock market by buying stocks recommended in investment newsletters. The table gives the annual returns after one year for 105 stocks recommended in investment newsletters.

Return (%)	Number of Stocks
−15	3
−12	1
−9	1
−6	1
−3	2
0	6
+3	10
+6	16
+9	22
+12	18
+15	9
+18	9
+21	4
+24	1
+27	1
+30	1

   a) Construct a graph of these data.

   b) Describe the shape of the graph. Use terms such as symmetric, skewed, or bimodal.

   c) Calculate the mean and standard deviation for these data.

8.2 Properties of the Normal Distribution
Refer to the Key Concepts on page 429.

4. An electrician is testing the accuracy of resistors that have a nominal resistance of 15 Ω (ohms). He finds that the distribution of resistances is approximately normal with a mean of 15.08 Ω and a standard deviation of 1.52 Ω. What is the probability that

   a) a resistor selected randomly has a resistance less than 13 Ω?

   b) a resistor selected randomly has a resistance greater than 14.5 Ω?

   c) a resistor selected randomly has a resistance between 13.8 Ω and 16.2 Ω?

5. The results of a blood test at a medical laboratory are normally distributed with $\mu = 60$ and $\sigma = 15$.

   a) What is the probability that a blood test chosen randomly from these data has a score greater than 90?

   b) What percent of these blood tests will have results between 50 and 80?

   c) How low must a score be to lie in the lowest 5% of the results?

6. The lifetimes of a certain brand of photographic light are normally distributed with a mean of 210 h and a standard deviation of 50 h.

   a) What is the probability that a particular light will last more than 250 h?

   b) What percent of lights will need to be replaced within 235 h?

   c) Out of 2000 lights, how many will have a lifetime between 200 h and 400 h?

8.3 Normal Sampling and Modelling

Refer to the Key Concepts on page 438.

7. The list below gives the age in months of 30 deer tagged in an Ontario provincial park last fall.

47	28	31	41	39	25	21	29	26	23
34	25	33	37	28	45	18	36	54	40
33	47	42	29	37	22	42	37	48	64

a) Use a method of your choice to assess whether these data are normally distributed. Explain your conclusion.

b) Find the mean and standard deviation of these data.

c) Determine the probability that a deer selected randomly from this sample is at least 30 months old. State and justify any assumptions you have made.

8. A quality-control inspector chose 20 bolt housings randomly from an assembly line. The interior diameters of these housings are listed in centimetres below.

2.29	2.23	2.48	2.24	2.40	2.23	2.37
2.33	2.37	2.31	2.31	2.26	2.26	2.18
2.33	2.31	2.30	2.30	2.24	2.34	

a) Find the mean and standard deviation for these data.

b) Assume the data are normally distributed. If the minimum acceptable diameter of these bolt housings is 2.25 cm, what proportion of the housings would be rejected as below this minimum?

9. Last year, Satsville High School ran nine classes of mathematics of data management, each with the same number of students. The class-average scores at year-end were as follows:

80.4	70.5	68.9	72.7	83.1	78.6	76.6
74.4	75.8					

a) Find the mean and standard deviation of these class-average scores.

b) How is the standard deviation of the class-average scores related to the underlying distribution of individual scores?

c) A study of the scores for all of the data-management students found that the scores were approximately normally distributed, with a mean of 75.7 and a standard deviation of 24.3. Given that all nine classes were the same size, find the most likely value for this class size. Explain your answer.

10. Use a normal approximation to find the probability that in a given year there will be more than 20 major earthquakes.

Chapter Problem

8.4 Normal Approximation to the Binomial Distribution
Refer to the Key Concepts on page 448.

11. A manufacturer of pencils has 60 dozen pencils randomly chosen from each day's production and checked for defects. A defect rate of 10% is considered acceptable.

 a) Assuming that 10% of all the manufacturer's pencils are actually defective, what is the probability of finding 80 or more defective pencils in this sample?

 b) If 110 pencils are found to be defective in today's sample, is it likely that the manufacturing process needs improvement? Explain your conclusion.

12. A store manager believes that 42% of her customers are repeat business (have visited her store within the last two weeks). Assuming she is correct, what is the probability that out of the next 500 customers, between 200 and 250 customers are repeat business?

8.5 Repeated Sampling and Hypothesis Testing
Refer to the Key Concepts on page 456.

13. An auto body repair shop plans its billing based on an average of 0.9 h to paint a car. The owner recently checked times for the last 50 cars painted and found that the average time for these cars was 1.2 h. She knows that the standard deviation is 0.4 h. Test the significance of this result at $\alpha = 0.10$.

14. A basketball coach claims that the average cost of basketball shoes is less than $80. He surveyed the costs of 36 pairs of basketball shoes in local stores and found the following prices:

$60	$50	$120	$110	$75	$110	$70
$40	$90	$65	$60	$85	$75	$80
$75	$80	$90	$45	$55	$70	$85
$85	$90	$90	$80	$50	$80	$85
$60	$70	$55	$95	$60	$45	$95
$70						

Test the coach's hypothesis at a 5% significance level.

15. A perfume company's long-term market share is estimated to be 6%. After an extensive advertising campaign, 11 out of 150 consumers surveyed claim to have purchased this company's perfume recently. Was the advertising campaign a success? Justify your assessment.

8.6 Confidence Intervals
Refer to the Key Concepts on page 464.

16. A study found that the average time it took for a university graduate to find a job was 5.4 months, with a standard deviation of 0.8 months. If a sample of 64 graduates were surveyed, determine a 95% confidence interval for the mean time to find a job.

17. In a survey of 200 households, 72 had central air-conditioning. Find a 90% confidence interval for the proportion of homes with central air-conditioning.

18. Here are the numbers of employees at 40 selected corporations in southern Ontario.

7685	11 778	11 370	9953	6200
900	2100	1270	1960	887
3100	7300	5400	3114	348
1650	400	873	195	173
725	3472	1570	256	895
120	347	40	2290	4236
850	540	164	285	12
390	60	713	175	213

a) Find a 90% confidence interval for the average number of employees at corporations in southern Ontario.

b) Comment on your findings. Does your confidence interval describe the sample data realistically? What problems exist with constructing a confidence interval for data of this sort?

19. A regional planner has been asked to estimate the average income of businesses in her region. She wants to be 90% confident of her conclusion. She sampled 40 companies and listed their net incomes (in thousands of dollars) as shown below:

84	49	3	133	85	4240	461	60
28	97	14	252	18	16	24	345
254	29	254	5	72	31	23	225
70	8	61	366	77	8	26	10
55	137	158	834	123	47	2	21

Estimate the average income of businesses in this region, with a 90% confidence level.

20. A survey of reading habits found that 63% of those surveyed said they regularly read at least part of a daily newspaper. The results are considered accurate within plus or minus 5%, 19 times in 20. Determine the number of people surveyed.

21. A market-research company found that 14.5% of those surveyed used Gleemodent toothpaste. The company states that the survey is accurate within ±4%, nine times out of ten. How many people did they survey?

22. A city's transportation department surveyed 50 students, 70 city residents, and 30 cyclists concerning their opinion on how well the city supported bicycling as an alternative means of transportation. The table below summarizes the survey results:

Rating	Students	Community	Cyclists
Excellent	0	4	0
Very Good	3	12	0
Good	23	17	8
Not So Good	15	19	8
Poor	7	11	11
Very Poor	2	7	3

a) Construct a 95% confidence interval for the proportion of students who feel that city's support is good or better.

b) Identify three other proportions which you feel are important. Construct confidence intervals for these proportions. Report your findings, together with reasons why these proportions may be important to the city's transportation department.

Chapter Test

1. Give a real-life example of data which could have the following probability distributions. Explain your answers.

   a) normal distribution

   b) uniform distribution

   c) exponential distribution

   d) bimodal distribution

2. The volume of orange juice in 2-L containers is normally distributed with a mean of 1.95 L and a standard deviation of 0.15 L.

   a) What is the probability that a container chosen at random has a volume of juice between 1.88 L and 2.15 L?

   b) If containers with less than 1.75 L are considered below standard, what proportion of juice containers would be rejected?

   c) Out of 500 containers, how many have a volume greater than 2.2 L?

3. According to its label, a soft-drink can contains 500 mL. Currently, the filling machine is set so that the volume per can is normally distributed, with a mean of 502 mL and a standard deviation of 1.5 mL. If too many cans contain less than 500 mL, the company will lose sales. If many cans contain more than 504 mL, the company will incur excess costs. Does the company need to recalibrate its filling machine?

4. A farmer finds the mass of the yields of 20 trees in an orchard. Here are the results in kilograms.

16.89	7.77	7.26	14.05	10.85	15.69
12.95	7.92	16.12	9.06	5.71	6.11
5.95	9.25	8.09	8.02	10.43	9.42
9.19	6.86				

   a) Find the mean and standard deviation of these data.

   b) What is the probability that a tree selected randomly from this orchard has a yield greater than 10 kg? State any assumptions you make in this calculation.

5. A manufacturer of mixed nuts promises: "At least 20% cashews in every can." A consumer-research agency tests 150 cans of nuts and finds a mean of 22% cashews with a standard deviation of 1.5%. The proportions of cashews are normally distributed.

   a) What is the probability that the population mean is less than 20% cashews?

   b) Based on the sample, what proportion of cans have between 15% and 30% cashews?

   c) The company must stop making their claim if more than 3% of the cans contain less than 20% cashews. Write a brief report to the company outlining whether they need a new motto.

   d) Suggest a better motto for the company.

6. Approximately 85% of applicants get their G-1 driver's licence the first time they try the test. If 80 applicants try the test, what is the probability that more than 10 applicants will need to retake the test?

7. A car assembly line produces 1920 cars per shift. A defect rate of 3% is considered acceptable. From the production of one recent shift, 65 cars were found to be defective.

a) Find the probability of this occurrence.

b) Explain to the shift supervisor, who has not studied probability theory, what your answer means and whether changes will need to be made in the production process to reduce the number of defective cars.

8. Find a 95% confidence interval for the percent of voters who are likely to vote "Yes" in a referendum if, in a sample of 125 voters, 55 said they would vote "Yes."

9. A politician asks a polling firm to determine the likelihood that he will be re-elected. The polling firm reports that of decided voters, 48% indicated they would vote for him if an election were called today. The result is accurate within plus or minus 5%, 19 times in 20.

a) How many decided voters were polled?

b) Should the politician be worried about his chances of re-election? Justify your answer.

c) The polling company found a large number of undecided voters. Should this fact influence the conclusion? What action, if any, should the politician take regarding this large pool of undecided voters?

ACHIEVEMENT CHECK

Knowledge/Understanding	Thinking/Inquiry/Problem Solving	Communication	Application

10. Students in the first-year statistics course at Statsville College wrote a 100-point examination in which the grades were normally distributed. The mean is 60 and the standard deviation is 12. Students in the first-year calculus course also wrote a 100-point examination in which the grades were normally distributed. For the calculus examination, the mean is 68 and the standard deviation is 8.

a) Betty has a mark of 85% in statistics and Brianna has a mark of 85% in Calculus. Who has the higher standing relative to her classmates? Justify your answer.

b) The statistics professor has decided to bell the marks in order to match the mean and standard deviation of the calculus class. Find Betty's new mark.

c) Suppose a subgroup of the students in this school, those on the school's honour roll, were selected. Would you still expect the distribution of their examination scores to be normally distributed? Would the distribution have the same mean and standard deviation? Explain your reasoning.

Wrap-Up

Implementing Your Action Plan

1. Determine the criteria for classifying your chosen species as endangered.

2. Collect sample data. You will need at least three estimates for the population of the species. More estimates would be useful.

3. Formulate a hypothesis concerning the status of the species.

4. Perform a hypothesis test to decide whether the data indicate that the species is endangered.

5. Determine a confidence interval for the estimated population of the species.

6. Draw a conclusion about the status of the species. Does your analysis lead to any further conclusions? For example, can you determine whether a total ban on commercial exploitation of the species is justified?

7. Present your results, using appropriate technology.

Evaluating Your Project

1. Identify factors that could affect the validity of your conclusion, such as possible measurement errors or bias in your data.

2. Are there improvements you could make to the methods you used for this project?

3. If you were to update your analysis a year from now, do you think your conclusions would be different? Justify your answer.

4. Could you apply your data collection and analysis techniques to other endangered species?

5. Did your research or analysis suggest related topics that you would like to explore?

Suggested Resources

- Canadian Nature Federation
- Committee on the Status of Wildlife in Canada
- Sheldrick Wildlife Trust
- World Conservation Union (IUCN)

WEB CONNECTION

www.mcgrawhill.ca/links/MDM12

Visit the above web site and follow the links to learn more about endangered species.

Presentation

Choose the most appropriate method for presenting your findings. Your could use one or a combination of the following forms:

- a written report
- an oral presentation
- a computer presentation (using software such as Corel® Presentations™ or Microsoft® PowerPoint®)
- a web page
- a display board

See section 9.5 and Appendix D for ideas on how to prepare a presentation. Be sure to document the sources for your data.

Preparing for the Culminating Project

Applying Project Skills

In this probability distributions project, you have learned new skills and further developed some of the skills used in the earlier projects. Many of these skills will be vital for your culminating project:

- carrying out an action plan
- formulating a hypothesis
- using technology to collect data
- using sample data to test a hypothesis
- formulating and interpreting a confidence interval
- evaluating the quality of data
- critiquing research methodology
- presenting results using appropriate technology

Keeping on Track

At this time, your data analysis should be complete and you should have determined what conclusions you can draw from it. You should be ready to evaluate your culminating project and prepare a presentation of your results. Section 9.4 details questions you can use to guide your evaluation of your own work. Appendix D outlines techniques for presentations.

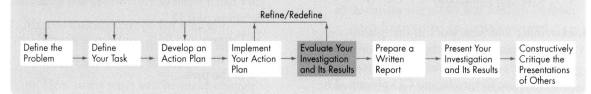

11. A group of data-entry clerks had the following results in a keyboarding test:

Speed (words/min)	44	62	57	28	46	71	50
Number of Errors	4	3	3	5	11	2	4

  a) Create a scatter plot and classify the linear correlation.

  b) Determine the correlation coefficient and the equation of the line of best fit.

  c) Identify the outlier and repeat part b) without it. Use calculations and a graph to show how this affects the strength of the linear correlation and the line of best fit.

  d) What does your analysis tell you about the relationship between speed and accuracy for this group of data-entry clerks?

12. A shopper observes that whenever the price of butter goes up, the price of cheese goes up also. Can the shopper conclude that the price of butter causes the increase in the price of cheese? If not, how would you account for the correlation in prices?

13. A market research company has a contract to determine the percent of adults in Ontario who want speed limits on expressways to be increased. The poll's results must be accurate within ±4%, 19 times in 20.

  a) If a small initial sample finds opinions almost equally divided, how many people should the company survey?

  b) Suggest a sampling method that would give reliable results.

14. How many ways can a bank of six jumpers on a circuit board be set if

  a) each jumper can be either on or off?

  b) each jumper can be off or connect either pins 1 and 2 or pins 2 and 3?

15. How many arrangements of the letters in the word *mathematics* begin with a vowel and end with a letter other than *h*?

16. In 2001, 78 books were nominated for the $25 000 Giller Award for Canadian fiction. How many different shortlists of 6 finalists could the jury select?

17. A marketing survey of consumers' soft-drink preferences collected the following data:
- 75 liked cola, 65 liked ginger ale, and 32 liked spring water.
- 43 liked cola and ginger ale.
- 13 liked cola and spring water.
- 15 liked ginger ale and spring water.
- 7 liked all three and 12 liked none of them.

  a) How many people were surveyed?

  b) How many liked only ginger ale?

  c) How many liked only spring water and ginger ale?

  d) How many liked only one of the choices?

  e) How many liked exactly two of the choices?

18. In how many ways can a box of 18 different chocolates be evenly distributed among three people?

19. Use Pascal's triangle to

  a) expand $\left(\dfrac{x}{3} + 3y\right)^5$

  b) develop a formula expressing the sum of the first n natural numbers in terms of combinations

20. Naomi's favourite cereal includes a free mini-puck with the emblem of one of the 30 NHL hockey teams. If equal numbers of the different pucks are randomly distributed in the cereal boxes, what is the probability that Naomi will get a mini-puck for one of the 6 original NHL teams in any given box of cereal?

8.2 Properties of the Normal Distribution

Many physical quantities like height and mass are distributed symmetrically and unimodally about the mean. Statisticians observe this bell curve so often that its mathematical model is known as the **normal distribution**.

Perhaps the most remarkable thing about the normal distribution is that a single mathematical formulation turns out to be the best model for statistical data from so many diverse sources. Like the exponential distribution in section 8.1, all normal curves can be described with a single form of equation. This equation can be used to calculate probabilities in a wide range of contexts. For example, in manufacturing, normal-distribution theory is used to design quality-control processes. The physical, social, and psychological sciences all make extensive use of normal models.

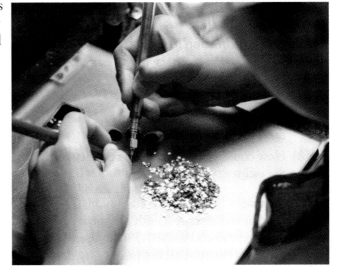

INVESTIGATE & INQUIRE: Plotting the Bell Curve

Rene 41® nickel superalloy, used to make components of jet and rocket engines, is a mixture of several metals, including molybdenum. The nominal percent of molybdenum by mass for this alloy is 9.75%. Dynajet Ltd. is considering two alloy suppliers for its line of jet-engine components. An analysis of the molybdenum content in 100 samples from supplier A and 75 samples from supplier B produced the following results:

Molybdenum (parts per 10 000)	Supplier A Frequency	Relative Frequency	Supplier B Frequency	Relative Frequency
970–971	3	0.030	1	0.013
971–972	5	0.050	2	0.027
972–973	9	0.090	9	0.120
973–974	15	0.150	15	0.200
974–975	19	0.190	20	0.267
975–976	20	0.200	16	0.213
976–977	14	0.140	8	0.107
977–978	9	0.090	3	0.040
978–979	4	0.040	1	0.013
979–980	2	0.020	0	0

Communicate Your Understanding

1. Explain the terms *discrete*, *continuous*, *symmetric*, *positively skewed*, and *negatively skewed*.

2. Give at least two examples of data that might result in
   a) a bimodal distribution
   b) an exponential distribution
   c) a positively-skewed distribution

3. Are summary measures such as mean and standard deviation always a good indicator of the probability distribution for a set of data? Justify your answer.

Apply, Solve, Communicate

A

1. Match the following distribution curves to the random variables. Give reasons for your choices.
   a) waiting times between arrivals at a pizza outlet during lunchtime
   b) collar sizes in the adult population
   c) hours worked per week

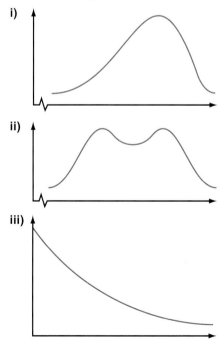

B

Use appropriate technology, wherever possible.

2. **Communication** The graph below shows a relative-frequency distribution for reading-test scores of grade 7 students in a school district.

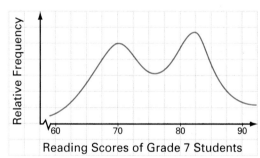

   a) Estimate the mean for these data.
   b) Give a possible explanation for the shape of this distribution.

11. A group of data-entry clerks had the following results in a keyboarding test:

Speed (words/min)	44	62	57	28	46	71	50
Number of Errors	4	3	3	5	11	2	4

a) Create a scatter plot and classify the linear correlation.

b) Determine the correlation coefficient and the equation of the line of best fit.

c) Identify the outlier and repeat part b) without it. Use calculations and a graph to show how this affects the strength of the linear correlation and the line of best fit.

d) What does your analysis tell you about the relationship between speed and accuracy for this group of data-entry clerks?

12. A shopper observes that whenever the price of butter goes up, the price of cheese goes up also. Can the shopper conclude that the price of butter causes the increase in the price of cheese? If not, how would you account for the correlation in prices?

13. A market research company has a contract to determine the percent of adults in Ontario who want speed limits on expressways to be increased. The poll's results must be accurate within ±4%, 19 times in 20.

a) If a small initial sample finds opinions almost equally divided, how many people should the company survey?

b) Suggest a sampling method that would give reliable results.

14. How many ways can a bank of six jumpers on a circuit board be set if

a) each jumper can be either on or off?

b) each jumper can be off or connect either pins 1 and 2 or pins 2 and 3?

15. How many arrangements of the letters in the word *mathematics* begin with a vowel and end with a letter other than *h*?

16. In 2001, 78 books were nominated for the $25 000 Giller Award for Canadian fiction. How many different shortlists of 6 finalists could the jury select?

17. A marketing survey of consumers' soft-drink preferences collected the following data:
- 75 liked cola, 65 liked ginger ale, and 32 liked spring water.
- 43 liked cola and ginger ale.
- 13 liked cola and spring water.
- 15 liked ginger ale and spring water.
- 7 liked all three and 12 liked none of them.

a) How many people were surveyed?

b) How many liked only ginger ale?

c) How many liked only spring water and ginger ale?

d) How many liked only one of the choices?

e) How many liked exactly two of the choices?

18. In how many ways can a box of 18 different chocolates be evenly distributed among three people?

19. Use Pascal's triangle to

a) expand $\left(\dfrac{x}{3} + 3y\right)^5$

b) develop a formula expressing the sum of the first n natural numbers in terms of combinations

20. Naomi's favourite cereal includes a free mini-puck with the emblem of one of the 30 NHL hockey teams. If equal numbers of the different pucks are randomly distributed in the cereal boxes, what is the probability that Naomi will get a mini-puck for one of the 6 original NHL teams in any given box of cereal?

Course Review

1. If the probability of rain tomorrow is 40%, what are the odds

   a) that it will rain?

   b) that it will not rain?

2. Describe each field you would include in a database for scheduling deliveries by an appliance dealer.

3. **a)** List the vertices of odd-degree in the following network.

   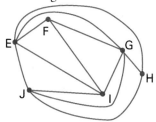

   b) Is this network

   i) complete?   **ii)** traceable?   **iii)** planar?

4. At Burger Barn, a combination meal consists of a hamburger and any two side orders, which can be soup, French fries, onion rings, coleslaw, pie, or ice cream. How many combination meals are possible?

5. Make a table of the probability distribution for the days of the month in a leap year.

6. **a)** What is a Bernoulli trial?

   b) What is the key difference between trials in a geometric probability distribution and those in a hypergeometric probability distribution?

7. Marc is making change for a $5 bill.

   a) List the ways he can provide change using only $1 and $2 coins.

   b) How many ways can he make change using only $1 coins and quarters?

   c) Would the number of ways be the same if he used only $2 coins and quarters? Explain your reasoning.

   d) How many ways can he make change using only $1 coins, $2 coins, and quarters?

8. Give an example of each of the following sampling techniques and list an advantage of using each type.

   a) stratified sample

   b) simple random sample

   c) systematic sample

9. In the fairy tale Rumplestiltskin, the queen must guess the name *Rumplestiltskin*.

   a) If she knows only what letters are in the name, how many different guesses could she make?

   b) At one guess per minute, how long would it take to try all the possible arrangements?

10. A survey asked randomly chosen movie fans how many videos they rented in the last month.

Number of Videos	Frequency
0–1	2
2–3	5
4–5	17
6–7	24
8–9	21
10–11	9
12–13	1

   a) Estimate the mean and median number of videos rented by those surveyed.

   b) Draw a histogram and relative frequency polygon for the data.

   c) Discuss any sources of inaccuracy in your calculations.

1. A lottery has sold 5 000 000 tickets at $1.00 each. The prizes are shown in the table. Determine the expected value per ticket.

Prize	Number of Prizes
$1 000 000	1
$50 000	10
$500	100
$10	1000

2. The speed limit in a school zone is 40 km/h. A survey of cars passing the school shows that their speeds are normally distributed with a mean of 38 km/h and a standard deviation of 6 km/h.

   a) What percent of cars passing the school are speeding?

   b) If drivers receive speeding tickets for exceeding the posted speed limit by 10%, what is the probability that a driver passing the school will receive a ticket?

3. Determine the probability distribution for the number of heads that you could get if you flipped a coin seven times. Show your results with a table and a graph.

4. Suppose that 82.5% of university students use a personal computer for their studies. If ten students are selected at random, what is

   a) the probability that exactly five use a personal computer?

   b) the probability that at least six use a personal computer?

   c) the expected number of students who use a personal computer?

5. Harvinder and Sean work in the quality-control department of a large electronics manufacturer that is having problems with its assembly line for producing CD players. The defective rate on this assembly line has gone up to 12%, and the department head wants to know the probability that a skid of 50 CD players will contain at least 4 defective units. Harvinder uses the binomial distribution to answer this question, while Sean uses the normal approximation. By what percent will Harvinder's answer exceed Sean's?

6. A box contains 15 red, 13 green, and 16 blue light bulbs. Bulbs are randomly selected from this box to replace all the bulbs in a string of 15 lights.

   a) Design a simulation to estimate the expected number of each colour of light bulb in the string.

   b) Calculate the theoretical probability of having exactly 5 red bulbs in the string.

   c) What is the expected number of blue bulbs?

   d) Would you expect your simulation to produce the same probability as you calculated in part b)? Why or why not?

7. A newspaper poll indicated that 70% of Canadian adults were in favour of the anti-terrorist legislation introduced in 2001. The poll is accurate within ± 4%, 19 times in 20. Estimate the number of people polled for this survey. Describe any assumptions you make about the sampling procedure.

8. The Ministry of Natural Resources conducted aerial surveys to estimate the number of wolves in Algonquin Park. Aerial surveys of 50 randomly selected 100-km² sections of the park had a mean of 1.67 wolves and a standard deviation of 0.32.

   a) Determine a 95% confidence interval for the mean number of wolves per 100 km² in the park.

   a) Describe any assumptions you made for your calculation in part a).

Presentation

Choose the most appropriate method for presenting your findings. Your could use one or a combination of the following forms:

- a written report
- an oral presentation
- a computer presentation (using software such as Corel® Presentations™ or Microsoft® PowerPoint®)
- a web page
- a display board

See section 9.5 and Appendix D for ideas on how to prepare a presentation. Be sure to document the sources for your data.

Applying Project Skills

In this probability distributions project, you have learned new skills and further developed some of the skills used in the earlier projects. Many of these skills will be vital for your culminating project:

- carrying out an action plan
- formulating a hypothesis
- using technology to collect data
- using sample data to test a hypothesis
- formulating and interpreting a confidence interval
- evaluating the quality of data
- critiquing research methodology
- presenting results using appropriate technology

Keeping on Track

At this time, your data analysis should be complete and you should have determined what conclusions you can draw from it. You should be ready to evaluate your culminating project and prepare a presentation of your results. Section 9.4 details questions you can use to guide your evaluation of your own work. Appendix D outlines techniques for presentations.

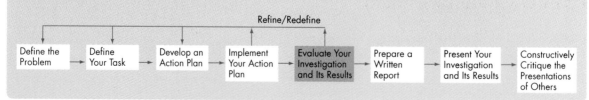

Refine/Redefine

Define the Problem → Define Your Task → Develop an Action Plan → Implement Your Action Plan → Evaluate Your Investigation and Its Results → Prepare a Written Report → Present Your Investigation and Its Results → Constructively Critique the Presentations of Others

21. Steve has a bag containing five red, three green, six orange, and ten black jelly beans. Steve's favourites are the black licorice ones. He randomly selects eight jelly beans.

  a) What is the probability that he will have at least four black ones?

  b) What is the expected number of black jelly beans?

22. Of the 24 guests invited to Hannah's party, 12 are male and 15 have dark hair. If 7 of the females have dark hair, what is the probability that the first guest to arrive will either have dark hair or be a male?

23. When testing its new insect repellant, a company found that 115 people out of an experimental group of 200 got fewer than the mean number of mosquito bites reported by the 100 people in the control group. Is this evidence sufficient for the company to claim that its spray is effective at a

  a) 5% significance level?

  b) 99% confidence level?

24. A particular car dealer's records show that 16.5% of the cars it sold were red.

  a) What is the probability that the first red car sold at the dealership on a given day will be the fifth car sold that day?

  b) What is the probability that the first red car sold will be among the first five cars sold?

  c) What is the expected waiting time before a red car is sold?

25. A town has three barbeque-chicken restaurants. In the past year, UltraChicken lost 20% of its customers to Churrasqueira Champion and 15% to Mac's Chicken, Churrasqueira Champion lost 10% of its customers to each of its two competitors, and Mac's Chicken lost 25% of its customers to UltraChicken and 30% to Churrasqueira Champion.

  a) Predict the long-term market share for each of these three restaurants.

  b) What assumptions must you make for your solution to part a)?

26. A battery maker finds that its production line has a 2.1% rate of defects.

  a) What is the probability that the first defect found will be in the 20th battery tested?

  b) What is the probability that there are no defects in the first 6 batteries tested?

  c) What is the expected waiting time until a defective battery is tested?

27. Explain the difference between a leading question and a loaded question.

28. During the winter, 42% of the patients of a walk-in clinic come because of symptoms of the common cold or flu.

  a) What is the probability that, of the 32 patients on one winter morning, exactly 10 had symptoms of the common cold or flu?

  b) What is the expected number of patients who have symptoms of the common cold or flu?

29. The Ministry of Natural Resources is concerned that hunters are killing a large number of wolves that leave the park to follow deer. For this reason, the Ministry is considering a permanent ban on wolf hunting in the area bordering the park. Outline the data and statistical analysis that you would require to determine whether such a ban is justified.

CHAPTER 9

Culminating Project: Integration of the Techniques of Data Management

Specific Expectations	Section
Pose a significant problem whose solution would require the organization and analysis of a large amount of data.	9.1, 9.2
Select and apply the tools of the course to design and carry out a study of the problem.	9.3
Compile a clear, well-organized, and fully justified report of the investigation and its findings.	9.3, 9.4
Create a summary of a project to present within a restricted length of time, using communications technology effectively.	9.4, 9.5
Answer questions about a project, fully justifying mathematical reasoning.	9.4, 9.5
Critique the mathematical work of others in a constructive fashion.	9.5

Overview of the Process

This chapter is designed to help you formulate and carry out a culminating project. The presentation of your project and its results may constitute a significant portion of your final assessment in this course. You will be required to integrate the skills and knowledge you have learned throughout the course to investigate a problem which is significant to you personally and also mathematically.

The smaller projects you may have completed throughout this course were designed to introduce you to the concepts you will need for the culminating project, and to give you practice in these skills. You may have already prepared a list of questions, examples, or ideas from this text that could be the basis for your culminating project. You may even consider expanding one of the smaller projects into your culminating project.

This chapter will provide a framework for developing your culminating project and its presentation. It will provide techniques, organizational ideas, suggestions, and timelines to make your culminating project a successful and rewarding experience.

As you work through your project, you may need to refine or even redefine your task. This may be due to limited or insufficient data for your original problem. You may also need to refine or redefine your task if your original problem is too large or has too much data to analyse. Even after you have implemented your plan, it may be necessary to refine or redefine your task based on the conclusions of your original study.

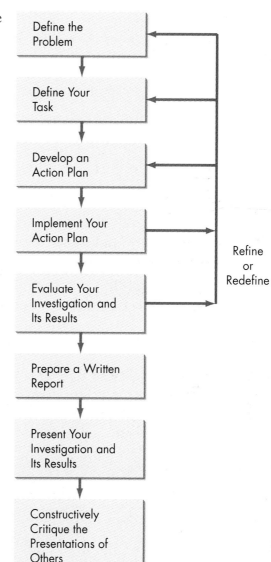

Defining the Problem

Your first task is to identify a topic for your project. Your topic needs to be of interest to you personally. It must also involve sufficient mathematical content and analysis to constitute a reasonable culminating project. Conversely, the problem cannot be so large that a reasonable analysis using the skills and knowledge from this course is impossible.

There are many ways to get started on your topic search. Some possible sources for ideas include:

- A social issue of interest to you
- A sport or hobby that you enjoy
- An issue from another course you would like to investigate
- An interesting article from a newspaper
- A question or issue from this textbook you want to investigate further
- A smaller project from this textbook which could be expanded
- An interesting issue from the Internet or other media
- An issue arising from employment or a possible future career
- An issue generated through brainstorming with others

Using Mind Maps to Identify or Refine a Problem

A useful tool for generating and organizing related topics is called a **mind map**. A mind map begins with a broad, general topic and generates sub-topics related to the main theme.

Example 1 First Level Mind Map

For the main topic ECOLOGY, construct a mind map showing sub-topics Pollution, Global Warming, Endangered Species, Rain Forest, Recycling, Nuclear Waste, Strip Mining, and Garbage Disposal.

Solution

A mind map for this topic is shown below.

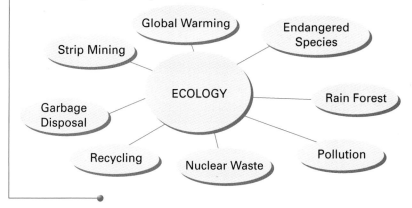

Mind maps typically have several levels. Each level flows from the level above it. The magnitude of the topics at each level is smaller than the previous level.

Example 2 Extended Mind Map

Extend the ECOLOGY mind map along the Pollution branch, with the sub-topics Smog, Pesticides, Public Transit, Industrial Development, and three other related sub-topics that your group has brainstormed.

Solution

The extended mind map is shown below. Add three ovals for the sub-topics that your group has brainstormed.

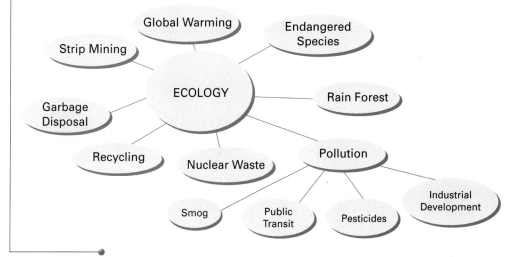

WEB CONNECTION

www.mcgrawhill.ca/links/MDM12

To learn more about creating mind maps, visit the above web site and follow the links. Write a brief description of how to create mind maps.

Posing a Problem

Once you have narrowed down your topic, you will need to pose a problem that you plan to investigate. Use the following checklists to evaluate potential problems.

The problem should satisfy *all* of the following:

- Be a significant problem of interest to you
- Involve the collection of a large amount of data
- Involve the organization of a large amount of data
- Involve the analysis of a large amount of data
- Allow the use of technology
- Allow the use of diagrams

The problem should use *some* of the following:

- One-variable statistics tools
- Two-variable statistics tools
- Matrices as tools
- Permutations and combinations
- Probability
- Probability distributions
- Simulations
- Hypothesis testing

Generating Questions

One way of posing a problem is to generate questions from data. For example, once a topic has been identified, do a preliminary data search. The type and quantity of available data may indicate some possible questions. Data from print sources, the Internet, and E-STAT are some resources that may be used.

WEB CONNECTION

www.mcgrawhill.ca/links/MDM12

For some links to interesting data sets visit the above web site. You might be able to find a data set that would help you decide on a topic for your culminating project

Practise

Work in small groups to complete questions 1 to 5.

1. Construct a mind map for each of the following topics. Brainstorm additional sub-topics to add to your group's mind map.

   a) FINANCE

   Sub-topics: tuition fees, student loans, interest rates, inflation, home ownership, stock market, mutual funds, and so on.

   b) MARKETING

   Sub-topics: market research, consumer behaviour, advertising, commercials, television, magazines, new products, taste tests, and so on.

   c) WEATHER

   Sub-topics: hurricanes, thunderstorms, tornadoes, temperature, seasons, prediction, and so on.

   d) GEOGRAPHY

   Sub-topics: demographics, crops, climate, currency, and so on.

2. Select four of the topics below. Then, brainstorm sub-topics and construct a mind map for each of your four topics. Try to include several levels for your mind maps.

AUTOMOBILES
MUSIC
MOVIES
SPORTS
DANCE
FASHION
TRAVEL
OCCUPATIONS
RISK

3. Often topics in the same sub-level of a mind map are related. These related topics can be joined with dashed lines to show their relationship. This type of mind map is known as a **mind web**. The following diagram shows a partial mind web for ECOLOGY, with dashed lines to show connected topics.

a) In your notebook, complete the ECOLOGY mind web.

b) Construct mind webs for the topics your group selected in question 2.

4. In your group, select a major topic for a mind web.

5. Refer to the data you collected for your Tools for Data Management project. Generate at least three questions from these data. The questions should be different from the ones you answered in the project.

6. The following table lists data related to vehicle collisions in Ontario in one year. In your group, brainstorm at least five questions that could be investigated using these data. Remember that to actually answer your questions, additional research would be necessary. Do not do any additional research at this time.

Vehicle Collisions in Ontario			
Age	Licensed Drivers	Number in Collisions	% of Drivers in Age Group in Collision
16	85 050	1 725	2.0
17	105 076	7 641	7.3
18	114 056	9 359	8.2
19	122 461	9 524	7.8
20	123 677	9 320	7.5
21–24	519 131	36 024	6.9
25–34	1 576 673	90 101	5.7
35–44	1 895 323	90 813	4.8
45–54	1 475 588	60 576	4.1
55–64	907 235	31 660	3.5
65–74	639 463	17 598	2.8
75 & older	354 581	9 732	2.7
Total	**7 918 314**	**374 073**	**4.7** (average)

7. Refer to the earthquake data on page 411. Use these data as a starting point to generate at least three questions to investigate. These questions may involve the earthquake data or a related topic.

8. Conduct a preliminary data search on the topic TOP GROSSING MOVIES. Generate at least three questions that could be investigated using your data.

9. Conduct a preliminary data search for a topic from the possible sources identified at the beginning of this section. Generate at least three questions from your data.

Moving Your Project Forward

Pose the question you will investigate in your culminating project.

Defining Your Task

Once you have posed a problem that you want to investigate, you need to clearly define exactly what your task will be. This definition will usually involve developing a **hypothesis** or thesis statement. You will then develop and implement an action plan to test your hypothesis.

Example 1 Tuition Fees

Suppose you plan to investigate the effect of increased tuition fees on university accessibility. State two possible hypotheses you could investigate on this topic.

Solution

One hypothesis might be: "Increases in tuition fees over the last ten years have forced large numbers of middle class students to give up on a university education."

Another hypothesis could be: "Increases in tuition fees over the last ten years have allowed universities to broaden the range of specialty programs offered."

Example 2 Drive Clean Program

Is there a connection between the number of defective vehicles identified through the Drive Clean program and the number of car accidents in a particular region? State at least two possible hypotheses.

Solution

Three possible hypotheses are:
1. There is a direct connection between the number of car accidents and the number of defective vehicles identified through the Drive Clean program.
2. The number of car accidents in a region has no connection to the number of defective vehicles identified through the Drive Clean program.
3. The number of defective vehicles identified through the Drive Clean program in a region has reduced the number of accidents.

Notice that in the examples you are not initially stating that your hypothesis is true. You are taking a position, which you will test by collecting and analysing data. Your hypothesis will determine what data you need to collect, and will drive your action plan.

Practise

For each question, state a hypothesis that could be tested.

1. Should car insurance premiums be related to driver age?

2. How is performance on standardized tests related to intelligence?

3. Is it possible for a major league sports team to make money given the large salaries of professional athletes?

4. Can future performance of the stock market be predicted with reasonable accuracy?

5. Should the colour of your car influence your car insurance rates?

6. Does it rain more often on the weekend?

7. Is income related to level of education?

8. Should we recycle 100% of our garbage?

9. Is the likelihood that a criminal will reoffend related to the severity of sentences for crimes?

10. Is the retail price of gasoline related to the price of crude oil?

11. Is life expectancy affected by celebrity?

12. Does the age at which a child learns to read affect his or her probability of going to college or university?

13. Can a model be devised to efficiently optimize school bus schedules, garbage collection, or recycling pick-up?

Moving Your Project Forward

Refer to the culminating project question you posed in section 9.1. State one or more hypotheses for your culminating project. Be sure to present your hypotheses to your teacher.

Developing and Implementing an Action Plan

Your action plan is a logical sequence of specific steps that must be carried out to test your hypothesis. Typical steps in an action plan include:

1. State your hypothesis.
2. Determine what data needs to be collected.
3. Decide how these data should be organized.
4. Decide how these data should be illustrated.
5. Determine what analysis needs to be done on these data.
6. Draw a conclusion based on your analysis.
7. Evaluate the quality of your investigation.
8. Write a report of your investigation and its results.
9. Develop the presentation of your investigation.
10. Establish time lines for each step of your plan.

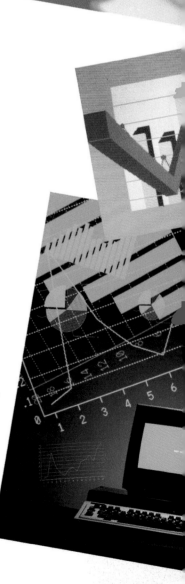

Example 1 Drafting an Action Plan

Outline an action plan for investigating the relationship between driver age and number of accidents.

Solution

1. **Hypothesis** The graduated licence system in Ontario has resulted in a dramatic decrease in the number of accidents involving teenage drivers.

2. **Data Collection** The table in section 9.1, question 6 on Vehicle Collisions in Ontario is a starting point for data collection. However, the data given in the table is only for one year. You will need accident data for other years, including the years before and after the introduction of graduated licences. You will need the data separated by age. You may also want the data separated by gender, or by region. You may want to consider other variables, such as driving for pleasure/work, accidents by time of day, accidents involving impaired drivers, accidents by type of vehicle, and so on. **Appendix C: Research Skills** on page 594 will help you with your data collection.

3. **Data Organization** The data should be organized to allow you to test your hypothesis. This means you will need to isolate the effects of the graduated licence system from other factors such as population that may have changed over the years you are examining.

4. **Data Presentation** When presenting the data, you need to keep in mind that you will be presenting these data both in your written report and in your class presentation. You may want to use bar graphs, circle graphs, line graphs, or tables to present your data. You also want to choose ways of presenting your data that help address your hypothesis. Note that this does not mean distorting the data by using inappropriate scales or ignoring outliers and data that does not support your hypothesis. You are testing the validity of your hypothesis, not trying to convince your audience of the correctness of your claim.

5. **Data Analysis** Use the tools you have learned in this course to analyse your data. Summary statistics, measures of dispersion, analysis of outliers, regression, and other techniques may be appropriate. It may be possible to perform a formal hypothesis test on your data. You should be able to relate your results to probability theory and probability distributions, where appropriate. You may be able to use a simulation to test your results. The choice of data analysis tools will depend on the hypothesis that you are testing.

6. **Conclusion** You should be able to state whether your results support or refute your hypothesis. You should also be able to indicate how strong your evidence is.

7. **Evaluation** Reflect on your conclusion as well as your entire investigation. Section 9.4 gives more detail on evaluating projects.

8. **Written Report** Your report should outline in detail your investigation and its conclusions. It should include all the parts of your action plan, the raw data (usually in an appendix), footnotes or endnotes, and bibliography. Section 9.5 gives more detail on writing reports.

9. **Presentation** Your presentation should be a summary of your investigation and its results. Section 9.5 and **Appendix D: Oral Presentation Skills** on page 598 will help you develop your presentation.

10. **Time Lines** Although listed last, time lines must be established and met for each stage of your project. It will not be possible to complete a project of this size in a short, restricted time. The following diagram may help you to establish time lines for this project.

Once your action plan has been developed, it must be implemented. At the implementation stage, it may be necessary to refine or even redefine your problem. Possible reasons for this include:

- Data is insufficient to test your hypothesis.
- Data is contradictory making it impossible to test your hypothesis.
- Question is too broad resulting in too much data to organize efficiently.
- Available data is of the wrong type for testing your hypothesis.
- Analysis requires techniques beyond the scope of this course.

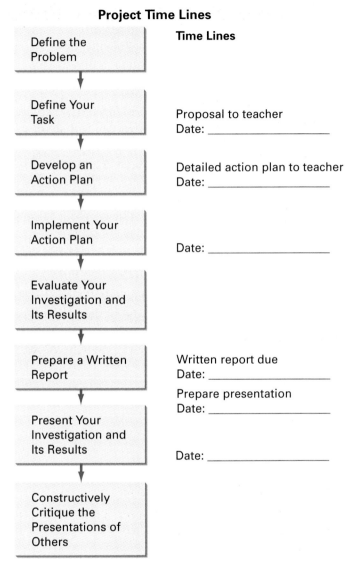

Project Time Lines

Time Lines

Define the Problem

Define Your Task
Proposal to teacher
Date: _____

Develop an Action Plan
Detailed action plan to teacher
Date: _____

Implement Your Action Plan
Date: _____

Evaluate Your Investigation and Its Results

Prepare a Written Report
Written report due
Date: _____
Prepare presentation
Date: _____

Present Your Investigation and Its Results
Date: _____

Constructively Critique the Presentations of Others

Practise

1. Select three of the hypotheses you developed in section 9.2. For each hypothesis, develop an action plan.

2. Review the action plans of at least two other members of your group or class. Make constructive suggestions to help your classmates improve their action plans.

Moving Your Project Forward

Develop a detailed action plan for your culminating project.

9.4 Evaluating Your Own Project

After completing your analysis and drawing a conclusion, you need to evaluate your investigation and the results. It is important that you look at the quality of your investigation and results. Otherwise, you may draw incorrect conclusions. For example, insufficient data may result in biased results. Flawed data collection techniques by the primary investigator may lead you to draw incorrect conclusions.

Your evaluation should include:

- reflection on your conclusion
- reflection on your investigation methodology
- reflection on what you learned from this project

Reflection on Your Conclusion

- Does the conclusion you reached support or refute your hypothesis?
- How strong is your evidence?
- Are there any limitations on your conclusion?
- In what ways could this conclusion be extended beyond the scope of your hypothesis?

Reflection on Your Investigation Methodology

- Were there any problems with data collection (sufficiency, availability, ease of access)?
- Why did you choose to organize the data the way you did?
- Are there other ways to organize the data? Why did you reject these methods?
- Were there any problems with the data analysis? How did you solve these?

- Was it necessary to revise/refine/redefine your problem as you progressed through your investigation? Outline what/when/why your problem was revised. Was it necessary to go back and collect more data to accurately judge your hypothesis?

- Did you examine your problem from various points of view (such as hypothesis testing, probability theory, or simulation)? If so, explain. If not, why not?

- Did you employ multiple tools from this course (one- or two-variable statistics, matrices, permutations, and so on)? If so, explain. If not, why not?

Reflection on Your Learning

- Do you feel that this project gave you a better understanding of how statisticians solve problems? Explain.

- Do you feel that this project gave you a better understanding of how the data-management tools from this course are used? Explain.

- If you were to do a similar project in the future, what would you do differently and why?

- Describe how you feel about this project. Did you enjoy it? Why or why not?

- Outline what you learned about presentations from this project.

Practise

1. Work alone or in a group to develop a rubric and weighting for scoring one of the other projects you completed in this course.

2. Prepare a "top ten" list of omissions from projects you have seen or completed in the past.

> **Moving Your Project Forward**
>
> Use the guidelines stated above to evaluate your culminating project.

Reporting, Presenting, and Critiquing Projects

Project Report

You are expected to compile a clear, well-organized, and fully justified report of your investigation and its results.

Your report should include:

- title that indicates the purpose of your project
- statement of your hypothesis
- background of your problem
- procedure and use of technology
- raw data such as graphs, charts, or tables, indicating sources
- summary of data in tables, graphs, and summary statistics
- analysis of your data, including calculations, and graphs
- results
- conclusions
- evaluation of your conclusions and of your investigation
- footnotes or endnotes and bibliography

Project Presentation

You are expected to present your culminating project in a clear and coherent manner. The presentation must be done within a restricted length of time and it must use communications technology effectively. The presentation should outline your investigation and its results, and should include a critical evaluation of your methodology and conclusions.

Knowing that you must present your project to others may influence the problem you decide to investigate, and how you decide to organize and analyse your data. As you progress through your investigation, you need to keep in mind that you will be presenting your findings.

Refer to **Appendix D: Oral Presentation Skills** on page 598. This appendix outlines in detail the important elements of a presentation. You will need to keep these in mind as you work through your project. If possible, rehearse your presentation with a classmate. Ask for feedback to improve your presentation before presenting it to the class.

Project Critiquing

Critiquing a project involves:

- identifying the strengths of the presentation
- identifying any problems or concerns with the presentation
- making suggestions to the presenters to help them improve future presentations

There are two distinct aspects to critiquing a presentation. The first is an evaluation of the presenter. Here, such things as audibility and clarity are evaluated from the perspective of the audience.

The second aspect to critiquing a presentation in this course is a critique of the mathematics involved in the project.

Maintaining the Quality in Your Presentation and Written Report

Use the following checklist to ensure that your written report and your presentation are high quality mathematical products. You can use this checklist to quickly identify strengths and weaknesses in your own presentation and in the presentations of others.

Mathematics Content Checklist

✓ Mathematical terminology is used correctly throughout.

✓ Mathematical terminology is used consistently throughout.

✓ Mathematical notation is used correctly throughout.

✓ Mathematical notation is used consistently throughout.

✓ Mathematical content is correct.

✓ Mathematical content is complete.

✓ Mathematical development is logical.

✓ All important steps in mathematical development are included.

✓ Any assumptions used in your analysis are explicitly stated.

✓ Any limitations of your analysis are identified.

✓ Any limitations of your analysis are explicitly stated.

✓ Possible extensions of your analysis are identified.

✓ Possible extensions of your analysis are discussed.

✓ Conclusions follow logically from your analysis.

✓ Possible audience questions are identified in advance.

✓ Answers to possible audience questions can be justified mathematically.

Practise

1. Think of an excellent presentation you have experienced. List five reasons why it was so good.

2. Make a list of common problems that are encountered when students explain mathematical reasoning to the class.

3. Presentation technology has many features that can enhance a report. List features that you find effective or ineffective based on presentations you have seen.

4. How can you distinguish between the style and content of a presentation while it is in progress?

Moving Your Project Forward

1. Present your culminating project.

2. Critique the presentations of other students in your class.

Appendix A Review of Prerequisite Skills

Evaluating expressions

To evaluate the expression $2x^2 - 7xy + 5$ for $x = 2$ and $y = -3$, substitute 2 for x and -3 for y in the expression. Then, simplify using the order of operations.

$$2x^2 - 7xy + 5 = 2(2)^2 - 7(2)(-3) + 5$$
$$= 8 + 42 + 5$$
$$= 55$$

1. Evaluate for $x = -2$, $y = 5$, and $z = 4$.

a) $x^2 + 5x - y^2$

b) $(2x + y)(3z - 2y)$

c) $8xy + 3y^3 - 6z$

d) $-5y - 4x^2y^2 + 3$

e) $2z - y(3x^2 - 4y)$

f) $8 + 6yx - 7y^2$

g) $xyz - xy - xz - yz$

h) $5x^2 - 9z^2y + 1$

i) $(xz - xy)(xz + xy)$

j) $\dfrac{(x - y)(x + z)}{(y + z)^2}$

k) $\dfrac{(x + 1)(y - 3)}{(x - 6)(z + 1)}$

l) $\dfrac{(x - 5)^2(y - 3)^2(z + 1)^2}{(x - y + z + 4)^3}$

Exponent laws

The exponent laws are used to simplify the following expression.

The exponent 3 is shared by all parts inside the brackets:

$$\dfrac{(3x^2y)^3(8x^5y^7)}{2x^4y^9}$$

Multiply the numeric coefficients and add exponents on like variables:

$$= \dfrac{(27x^6y^3)(8x^5y^7)}{2x^4y^9}$$

Divide the numeric coefficients and subtract exponents on like variables:

$$= \dfrac{216x^{11}y^{10}}{2x^4y^9}$$

$$= 108x^7y$$

1. Use the exponent laws to simplify each of the following.

a) $(4x^2)^5$

b) $(-3x^4y^5)^2$

c) $5(2x^3y)(5x^2y^2)^3$

d) $\dfrac{1}{9}\left(\dfrac{2}{5}\right)^2\left(\dfrac{3}{4}\right)^3$

e) $\left(\dfrac{a^2}{b}\right)^5\left(\dfrac{b^2}{a}\right)^3$

f) $\dfrac{(9x^2y^4)(-15x^5y^7)}{(3xy)(5x^2y^6)}$

g) $\left(\dfrac{2}{x^2}\right)^5(3x^2y^3)(4x^3)^2$

h) $\dfrac{(-2x^3y^3)^3(5x^2y)^4}{(10xy^3)^2}$

Fractions, percents, decimals

The following table shows how equivalent fractions, percents, and decimals can be expressed.

Fraction	Percent	Decimal
$\dfrac{63}{100}$	63%	0.63
$\dfrac{8}{100} = \dfrac{2}{25}$	8%	0.08
$\dfrac{0.5}{100} = \dfrac{1}{200}$	0.5%	0.005
$\dfrac{150}{100} = \dfrac{3}{2}$	150%	1.5

1. Copy and complete the following table. Express all fractions in their simplest form.

	Fraction	Percent	Decimal
a)	$\dfrac{75}{100}$		
b)	$\dfrac{1}{2}$		
c)	$8\dfrac{2}{5}$		
d)		34%	
e)		0.03%	
f)		5.6%	
g)			0.45
h)			0.03
i)			2.68

2. Find the percent change when the price of gasoline jumps from $0.699/L to $0.799/L.

3. A $25-shirt is on sale for 20% off. What is the total cost of the shirt, including 8% PST and 7% GST?

Graphing data

The table gives the ages of 80 cars sold at a used car lot.

Age (years)	Frequency	Percent
1	3	3.75
2	2	2.5
3	7	8.75
4	15	18.75
5	12	15
6	18	22.5
7	10	12.5
8	6	7.5
9	3	3.75
10	3	3.75
11	0	0
12	1	1.25

The percents were calculated by dividing each frequency by 80 and multiplying by 100.

The most frequent age of cars sold was 6 years old, at 22.5% of the total.

The least frequent age was 11 years old, at 0% of the total.

These data can be graphed using a bar graph or a circle graph as shown below.

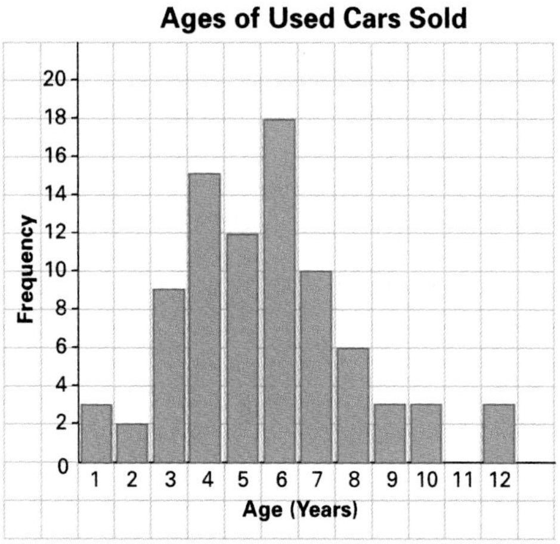

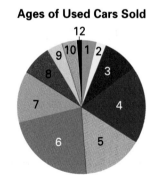

Ages of Used Cars Sold

1. The table shows the number of tickets sold for a two-week theatre production of *Little Shop of Horrors*.

   a) Construct a line graph of the ticket sales.

   b) Describe the trend in ticket sales.

   c) What types of graphs would not be suitable for these data?

Day	Tickets Sold
1	2350
2	2350
3	2350
4	2350
5	2189
6	2012
7	1850
8	1878
9	1504
10	920
11	1267
12	1422
13	998
14	835

2. Grade 12 students were asked to provide the number of universities to which they were considering applying. The following results were obtained.

   a) How many students were surveyed?

   b) Construct a bar graph.

   c) Determine the percent of total responses for each number.

   d) Construct a circle graph.

   e) What type of graph would not be suitable for these data?

Number of Universities	Frequency
0	35
1	25
2	56
3	27
4	18
5	12
6	3

Graphing exponential functions

In the exponential function $y = 0.3(2)^x$, the base is 2, the numerical coefficient is 0.3, and the exponent is x. Complete a table of values by finding the value of y for each value of x.

Plot the points on a grid and draw a smooth curve through the points.

x	y
−2	0.075
−1	0.15
0	0.3
1	0.6
2	1.2
3	2.4
4	4.8
5	9.6

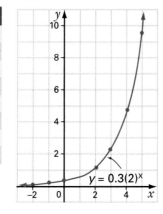

$y = 0.3(2)^x$

1. Graph each of the following functions.

a) $y = 2^x$ **b)** $y = 2^{-x}$

c) $y = 1.5(2)^x$ **d)** $y = 3(0.5)^x$

e) $y = 0.1(4)^x$ **f)** $y = 81\left(\dfrac{1}{3}\right)^x$

Graphing linear equations

To graph the line $y = \dfrac{2}{3}x - 1$ using the slope and y-intercept form $y = mx + b$, identify the slope as $m = \dfrac{2}{3}$ and the y-intercept as $b = -1$. Plot the y-intercept first. Then, plot a second point by moving 3 to the right (run) and 2 upward (rise) to the point $(3, 2)$.

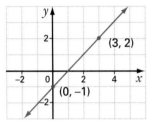

1. Graph each of the following using the slope and y-intercept.

a) $y = \dfrac{5}{3}x + 2$ **b)** $y = 4x - 5$

c) $y = -5x + 3$ **d)** $y = -\dfrac{1}{2}x + 2$

e) $y = -x - 2$ **f)** $y = 6x - 7$

g) $2x + y = 8$ **h)** $3x + 4y = 10$

i) $5x - 6y = 9$

Graphing quadratic functions

To graph the function $y = 2x^2 - 12x + 7$, you must first complete the square.

$$y = 2x^2 - 12x + 7$$

Factor 2 from the x^2 and x terms: $\quad y = 2(x^2 - 6x) + 7$

Add and subtract the square of half of the coefficient of x: $\quad y = 2(x^2 - 6x + 3^2 - 3^2) + 7$

$$y = 2(x^2 - 6x + 9) - 2(9) + 7$$

Complete the square: $\quad y = 2(x - 3)^2 - 11$

To graph $y = 2(x - 3)^2 - 11$, plot the vertex at $(3, -11)$.

Sketch the parabola opening upward, stretched vertically a factor of 2.

To find the y-intercept, substitute $x = 0$ and evaluate $y = 7$.

To find the x-intercepts, substitute $y = 0$ and evaluate $x \doteq 5.3$ or $x \doteq 0.7$.

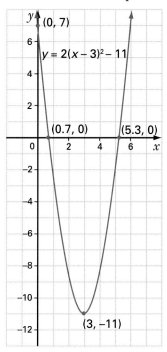

1. Graph the following functions and estimate any x- and y-intercepts.

   a) $y = 3x^2$ **b)** $y = -3x^2$ **c)** $y = 2(x - 1)^2 + 3$

   d) $y = 3x^2 + 6x - 5$ **e)** $y = -2x^2 + 8x + 10$ **f)** $y = 5x^2 - 8x + 2$

   g) $y = x^2 + 6x - 9$ **h)** $y = -x^2 + 5x + 3$ **i)** $y = 4x^2 + 6x + 1$

Mean, median, mode

The following marks were scored on a test marked out of 50.

45 38 26 44 45 20 38 32 29 18 32 33 33 41 45 50 31 27 25 24
38 36 25 35 42 30 38 31 32 29 30 39 38 39 25 26 21 49 12 34

The marks can be organized into a stem-and-leaf plot.

Test Marks	
1	2, 8
2	0, 1, 4, 5, 5, 5, 6, 6, 7, 9, 9
3	0, 0, 1, 1, 2, 2, 2, 3, 3, 4, 5, 6, 8, 8, 8, 8, 8, 9, 9
4	1, 2, 4, 5, 5, 5, 9
5	0

The mean is the sum of all the measures, divided by the number of measures.

$$\text{mean} = \frac{1325}{40}$$
$$= 33.125$$

The median is the middle measure when all the measures are placed in order from least to greatest.

$$\frac{1}{2} \text{ of } 40 = 20$$

The median is the midpoint of the 20th and 21st measures.

$$\frac{(32 + 33)}{2} = 32.5$$

The mode is the most frequent measure.
mode = 38

1. Calculate the mean, median, and mode for each of the following sets of data.

a) Student ages in years:
18 16 13 15 16 18 18 18 15 17 17 18 14 15 16 18 19 20 19
16 14 17 17 18 15 16 18

b) Prices, in dollars, of the Mario Lemieux rookie card at various stores:
89 58 79 79 47 99 88 125 79 89 64 79 78 90 95 75 89 79 79

Number patterns

A pattern can be described by identifying the operation that is needed to find successive terms.

44, 33, 22, . . . is a sequence of terms found by subtracting 11.

The next three terms would be 11, 0, −11.

1. Describe each of the following patterns. Find the next three terms.

a) 15, 12, 9 **b)** 4, 12, 36 **c)** 1, 4, 9

d) 2, 4, 6 **e)** 9, −3, 1 **f)** 80, −40, 20

g) 1, −1, 1, −1 **h)** 15, 5, −5 **i)** 1, 2, 4, 8, 16

j) $\frac{3}{4}, \frac{1}{2}, \frac{1}{3}$ **k)** p, pq, pqq, pqqq **l)** a, ab, abb, abbbb

m)

n)

Order of operations

To evaluate this expression, use the order of operations to multiply first and then simplify by subtracting.

$$(-2)(-6) - (5)(-3)^2 = 12 - (5)(9)$$
$$= 12 - 45$$
$$= -33$$

1. Evaluate each expression.

a) $(-3)(5) + (-6)(-8)$

b) $(12)(11) - (-9)(7)$

c) $-10(3)(2) + (-12)(7)(-2)$

d) $0.3(5.5)^2 - (-6.7)(2.1)^3 + 4.2(-1.1)^5$

e) $3.2\left(\dfrac{1}{2}\right) - 2.5\left(\dfrac{3}{5}\right) + 1.4\left(\dfrac{2}{7}\right)$

f) $\left(\dfrac{3}{4}\right)^2\left(\dfrac{8}{3}\right)$

g) $\dfrac{5}{2} - \dfrac{2}{5}\left(\dfrac{4}{3}\right)^3$

h) $\dfrac{2}{3} - \dfrac{1}{2}\left(\dfrac{5}{6} - \dfrac{7}{4}\right)$

Ratios of areas

One rectangle has dimensions of 6 cm by 7 cm. A second rectangle has dimensions that are 3 times those of the first. The ratio of their areas can be found in two ways.

Method 1: The area of the first rectangle is $6 \times 7 = 42$ cm².
The area of the second rectangle is $18 \times 21 = 378$ cm².
The ratio of their areas is 42:378 = 1:9.

Method 2: The dimensions have a ratio of 1:3.
The ratio of the areas is the square of the ratio of the dimensions.
So, the ratio is $1^2:3^2 = 1:9$.

1. a) A triangle has base of 15 mm and height of 8 mm. The dimensions of a second triangle are double those of the first. What is the ratio of their areas?

b) The dimensions of a square are triple those of another square. What is the ratio of the areas of the two squares?

c) The length and the width of a parallelogram are halved to create a second parallelogram. What is the ratio of the areas of the two parallelograms?

d) A pyramid has edges that are five times as long as the edges of another pyramid. What is the ratio of the volumes of the two pyramids?

Scatter plots

The measurements of the neck and wrist circumferences, in centimetres, of fifteen 17-year-old students, are recorded in the following table.

Neck	32	31	44	36	29	34	40	38	42	32	33	35	40	39	30
Wrist	15	15	19	17	15	16	21	17	20	16	17	17	18	17	12

Plot the data in a scatter plot, showing the linear relationship between the two measurements. Construct a line of best fit through the data.

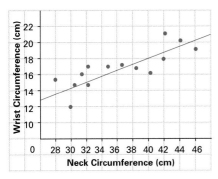

Refer to Appendix B: Technology for help in constructing the scatter plot and line of best fit using technology.

The slope of the line of best fit is positive.

Half of the points are above the line of best fit and the other half of the points are below the line.

The line can be extrapolated to find the wrist circumference of a 17-year-old person with a neck circumference of greater than 44 cm or less than 29 cm. A person with a neck circumference of 47 cm would expect to have a wrist circumference of about 21 cm. A person with a neck circumference of 28 cm would expect to have a wrist circumference of about 14 cm.

1. The following table shows the temperature of Earth's atmosphere at various altitudes on a particular day.

   a) Construct a scatter plot, with Altitude along the horizontal axis and Temperature along the vertical axis.

   b) Describe the relationship between altitude and temperature.

   c) Draw a line of best fit through the data. Describe your steps.

   d) What is the slope of your line of best fit? What does the slope represent?

   e) Write an equation relating temperature to altitude.

   f) What would the temperature be at an altitude of 12 km?

Altitude (km)	Temperature (°C)
0	13
1	0
2	−1
4	−13
5	−20
6	−27
8	−41
10	−55

2. In the city of Marktown, the annual cost, in dollars, of heating a house relative to its floor space, in square metres, is shown in the following table.

Floor Space (m²)	Heating Cost ($)
140	1071
160	1279
180	1452
200	1599
220	1821
240	2002
260	2242
280	2440

a) Construct a scatter plot, with Floor Space along the horizontal axis and Heating Cost along the vertical axis.

b) Describe the relationship between floor space and heating cost.

c) Draw a line of best fit through the data. Describe your steps.

d) What is the slope of your line of best fit? What does it represent?

e) Write an equation relating floor space and heating cost.

f) What would the heating cost be for a floor space of 350 m²?

Sigma notation

The following series can be written in sigma notation by writing a formula representing the terms and by writing the lower and upper limits of the variable, i, as 1 and 12, respectively.

$$1^2 + 2^2 + 3^2 + \ldots + 12^2 = \sum_{i=1}^{12} i^2$$

1. Express the following series in sigma notation.

a) $2 + 4 + 8 + 16 + 32 + 64$

b) $3! + 4! + 5! + \ldots + 15!$

c) $\dfrac{1}{2} + \dfrac{2}{3} + \dfrac{3}{4} + \dfrac{4}{5} + \dfrac{5}{6} + \dfrac{6}{7} + \dfrac{7}{8}$

d) $5x + 6x + 7x + 8x + 9x + 10x + 11x$

2. Expand each of the following. Do not simplify.

a) $\displaystyle\sum_{i=1}^{10} 3i$

b) $\displaystyle\sum_{m=1}^{8} 5(2)^m$

c) $\displaystyle\sum_{i=4}^{15} (2i - 1)$

d) $\displaystyle\sum_{k=3}^{6} \dfrac{k+1}{k^2}$

Similar triangles

In similar triangles, corresponding angles are equal and corresponding sides are proportional.
$\triangle ABC \sim \triangle DEF$ because $AB{:}DE = BC{:}EF = CA{:}FD = 1{:}2$.

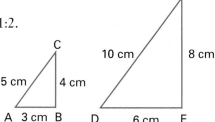

1. Determine which of the following triangles are similar. Explain.

a)
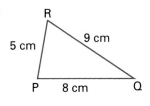

b)
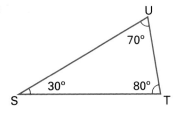

c)
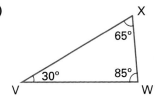

d)
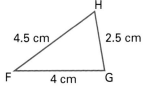

e)
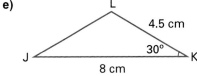

Simplifying expressions

The expression $(x - y)(2x - 3y)^2$ can be simplified by expanding the brackets and then collecting like terms.

$$(x - y)(2x - 3y)^2 = (x - y)(4x^2 - 12xy + 9y^2)$$
$$= 4x^3 - 12x^2y + 9xy^2 - 4x^2y + 12xy^2 - 9y^3$$
$$= 4x^3 - 16x^2y + 21xy^2 - 9y^3$$

1. Expand and simplify.

a) $(x + 2y)^2$ **b)** $(2x - 5)^2$ **c)** $(x^2 + y)^2$

d) $(5x - 2y)^2$ **e)** $(3x + 4)(2x + 1)^2$ **f)** $(5y - 2)^2(3y + 4)$

g) $(k - 3)^2(k + 4)$ **h)** $(5m - n)^2(5m + 2n)$ **i)** $(2y - 4x)(5y + 4x)^2$

Solving equations

To solve this equation, expand the brackets, then collect like terms on one side of the equation.
$$5(x + 3) = 4(x + 7)$$
$$5x + 15 = 4x + 28$$
$$5x - 4x = 28 - 15$$
$$x = 13$$

To solve an equation involving exponents, take the root of each side.
$$x^2 = 64$$
$$x = \pm\sqrt{64}$$
$$x = \pm 8$$

1. Solve for x.

a) $4x - 5 = 3x - 9$ **b)** $4x + 3 = 2x - 9$ **c)** $7x + 6 = 2x - 9$

d) $8(x - 3) = 3(2x + 4)$ **e)** $\dfrac{3}{4} = \dfrac{c}{10}$ **f)** $\dfrac{4.8}{1.2} = \dfrac{6.3}{y}$

g) $\dfrac{3x - 2}{5} = \dfrac{2x + 1}{3}$ **h)** $\dfrac{5x - 2}{4} = \dfrac{3x + 1}{2}$ **i)** $x^2 = 36$

j) $x^2 = 144$ **k)** $x^3 = 8$ **l)** $x^3 = 216$

Substituting into equations

To simplify $f(g(x))$, substitute $g(x)$ for x in $f(x)$, expand each set of brackets and simplify by collecting like terms.

Given $f(x) = 2x^2 - 3x + 1$ and $g(x) = 5x + 4$.

$$
\begin{aligned}
f(g(x)) &= 2(5x + 4)^2 - 3(5x + 4) + 1 \\
&= 2(25x^2 + 40x + 16) - 15x - 12 + 1 \\
&= 50x^2 + 80x + 32 - 15x - 12 + 1 \\
&= 50x^2 + 65x + 21
\end{aligned}
$$

1. Given $f(x) = 3x^2 - 6$, $g(x) = x^2 + 2$, and $h(x) = 2x + 1$. Substitute and simplify.

a) $f(-1)$ **b)** $g(5)$ **c)** $h(-10)$ **d)** $g(f(x))$ **e)** $g(f(-1))$

f) $f(g(x))$ **g)** $f(g(4))$ **h)** $h(f(x))$ **i)** $f(f(x))$ **j)** $f(h(2))$

k) $g(g(x))$ **l)** $g(g(10))$ **m)** $h(g(-1))$ **n)** $f(g(h(3)))$

2. In general, are $f(g(x))$ and $g(f(x))$ equal? Explain.

Tree diagrams

The tree diagram on the right illustrates the number of ways to pick an outfit from a green, blue, or grey shirt, black or brown pants, and a white or tan jacket. Each new decision branches out from the previous one to build a tree diagram.

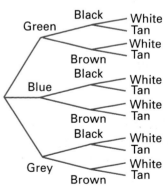

1. A spinner has four colours—red, green, orange, and purple. Draw a tree diagram to illustrate the possible outcomes for two spins.

2. A hockey team played three games in a tournament. The results could be a win, loss, or tie. Draw a tree diagram to illustrate the possible outcomes in the tournament.

3. A family is planning a vacation. They could fly, drive, take the train, or take the bus from Toronto to Sudbury. They could drive or take the bus from Sudbury to Timmins. They could drive, take the train, or fly from Timmins to Winnipeg. Draw a tree diagram to illustrate the possible ways of taking their vacation.

Contents   Page

Technology Tool Cross-Reference Table

Use the terms listed in the following table to help you determine the best technology tool to use for your calculations. For details on how to use each of these terms, refer to the corresponding entry in the appropriate section of Appendix B.

Topic	Graphing Calculator	Spreadsheets	Fathom™
Binomial Distribution	binomcdf(function binompdf(function	BINOMDIST function	binomialCumulative() function binomialProbability() function
Box-and-Whisker Plot	STAT PLOT		graph icon
Combinations	nCr function	combinations function	combinations function
Confidence Intervals	ZInterval instruction		
Correlation Coefficient	DiagnosticOn/Off LinReg instruction STAT PLOT	Chart feature CORREL function	correlation coefficient scatter plot
Factorials	! function	FACT(n) function	
Geometric Distribution	geometpdf(function		
Graphing Data	STAT PLOT TRACE instruction window settings Y= editor	Chart feature	graph icon scatter plot
Hypothesis Testing	Z-Test instruction		
Linear Regression	LinReg instruction STAT PLOT	line of best fit	linear regression
Line of Best Fit	LinReg instruction STAT PLOT	line of best fit	linear regression

Topic	Graphing Calculator	Spreadsheets	Fathom™
Matrix Operations	copy matrices multiply matrices store matrices	Matrices: addition and subtraction inverse multiplication scalar multiplication storing transpose	
Measures of Central Tendency	mean(function median(function 1-Var Stats command	average mean median mode	mean median mode
Measures of Spread	1-Var Stats command standard deviation	standard deviation	standard deviation
Non-linear Regression	Non-linear regression: CubicReg instruction ExpReg instruction QuadReg instruction		
Normal Distribution	invNorm(function normalcdf(function normalpdf(function ShadeNorm(function	NORMDIST function	normalCumulative function normalQuantile function
Organizing Data	augment(function cumSum(function prod(function seq(function SortA(function sum(function	COUNTIF function Fill feature filtered search MAX function search Sort feature SUM function	caseIndex function case table collection count function filter inspector sort sum function uniqueRank() function
Permutations	nPr function	permutations function	
Quartiles	interquartile range semi-interquartile range 1-Var Stats command		interquartile range quartiles semi-interquartile range
Random Numbers	randInt (function randNorm(function	random integers random real numbers	random function randomInteger function randomNormal function
Rounding Numbers	round(function	INT function ROUND function	
Scatter Plots	STAT PLOT	Chart feature	graph icon scatter plot
Standard Deviation	standard deviation 1-Var Stats	standard deviation	standard deviation
Variance	1-Var Stats	variance	variance
Z-scores			zScore function

Graphing Calculator

Note: Unless otherwise stated, all keystrokes are for the TI-83 Plus or TI-83 graphing calculator.

Function or Task	Keystroke(s), Menu, or Screen
augment(function augment(listA,listB)	The augment(function found under the LIST OPS menu is used to join together the elements of list A and list B. **Example:** Select 1:Edit… from the STAT EDIT menu to create lists L1 and L2 as shown: 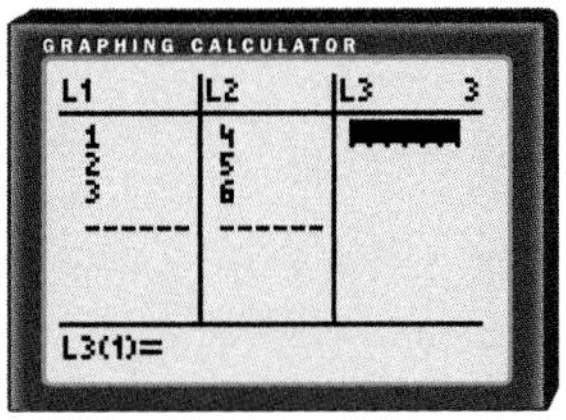 Press (2nd) (MODE) to QUIT to the home screen. Press (2nd) (STAT) (▸) to display the LIST OPS menu. Select 9:augment(and type L1 (,) L2 (). Press (STO▸) L3. 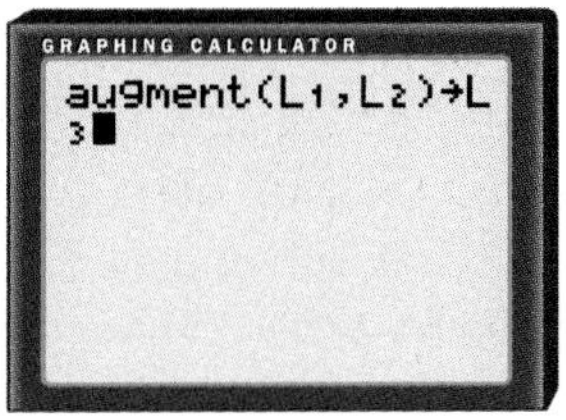 Press (ENTER). You can inspect L3 by selecting 1:Edit… from the STAT EDIT menu. 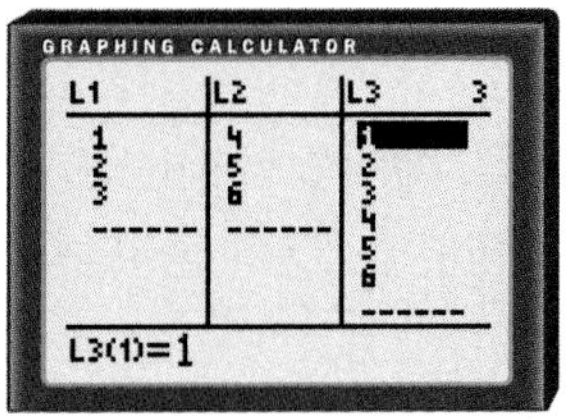

Function or Task	Keystroke(s), Menu, or Screen
binomcdf(function binomcdf(numtrials,p,x)	The binomcdf(function allows you to calculate the probability that an experiment whose only possible outcomes are success or failure, with a probability of success given by p, achieves x or fewer successes in the number of trials given by numtrials. The value for x can also be a list of numbers. If x is not specified, then a list of values from x = 0 to x = numtrials is generated.

Example 1:

A die is rolled ten times. What is the probability of getting five or fewer 2s? Press `2nd` `VARS` to display the DISTR menu. Scroll down the screen and select A:binomcdf(. Type 10 `,` 1 `÷` 6 `,` 5 `)` and press `ENTER`.

The probability is approximately 0.998.

Example 2:

To find all of the cumulative probabilities for Example 1, above, from x = 0 to x = 10 and store them in list L1, retrieve the binomcdf(function as above, but leave out the parameter x. Then, press `STO▸` followed by `2nd` L1 `ENTER`.

You can scroll through the list of probabilities using the right arrow key. You can also inspect list L1 by selecting 1:EDIT... from the STAT EDIT menu.

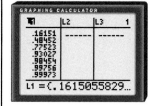

GRAPHING CALCULATOR

Example 3:

If you want only a single probability, for example, that you get exactly five 2s in ten rolls of a die, you need to subtract the cumulative probability that x = 4 from the cumulative probability that x = 5, as shown below:

```
GRAPHING CALCULATOR
binomcdf(10,1/6,
5)-binomcdf(10,1
/6,4)
        .0130238102
```

Note: You can more easily calculate a single probability using the binompdf(function.

Example 4:

Suppose that you want the cumulative probabilities of getting 3, 4, or 5 twos. This can be done as follows:

```
GRAPHING CALCULATOR
binomcdf(10,1/6,
{3,4,5})
{.9302721577 .9…
■
```

Note: Brace brackets {} are required for the list of x values.

---

binompdf(function

binompdf(numtrials,p,x)

The binompdf(function allows you to calculate the probability that an experiment whose only possible outcomes are success or failure, with a probability of success given by p, achieves x successes in the number of trials given by numtrials. x may also be a list of numbers. If x is not specified, then a list of values from x = 0 to x = numtrials is generated. This list can be stored in one of the graphing calculator's lists.

▼

Function or Task | **Keystroke(s), Menu, or Screen**

Example 1:

A die is rolled ten times. What is the probability of rolling exactly five 2s? Press (2nd) (VARS) to display the DISTR menu. Scroll down the screen and select 0:binompdf(. Type 10 (,) 1 (÷) 6 (,) 5 ()) and press (ENTER).

The probability is approximately 0.013.

Example 2:

To calculate all of the probabilities for Example 1, above, from x = 0 to x = 10 and store them in list L1, retrieve the binompdf(function as above, but leave out the parameter x. Then, press (STO▸) followed by (2nd) L1 (ENTER).

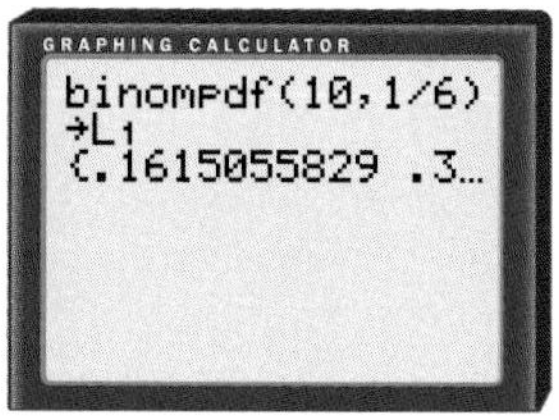

You can scroll through the list of probabilities using the right arrow key. You can also inspect list L1 by selecting 1:EDIT… from the STAT EDIT menu.

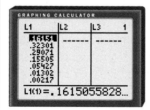

Once the probabilities are stored in the list, you can graph them using STAT PLOT.

Function or Task	Keystroke(s), Menu, or Screen

	Example 3:
	Suppose that you want only the probabilities of getting 3, 4, or 5 twos. This can be done as follows: 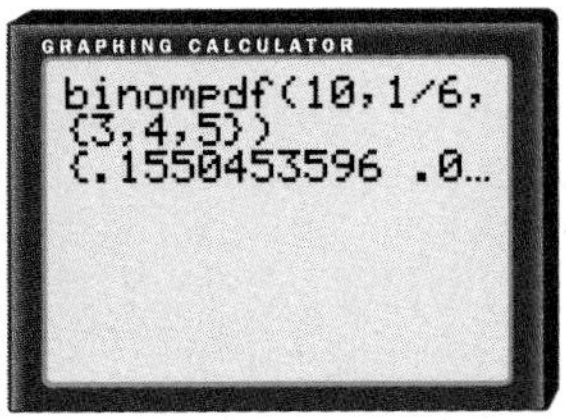
	
	GRAPHING CALCULATOR
	`binompdf(10,1/6,`
	`{3,4,5})`
	`{.1550453596 .0…`
	
	Note: Brace brackets {} are required for the list of x values.

ClrAllLists command	To clear all lists at once, press `2nd` `+` to access the MEMORY menu. 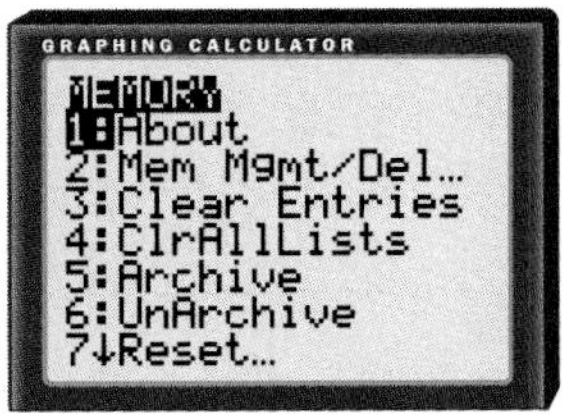 GRAPHING CALCULATOR **MEMORY** **1**:About 2:Mem Mgmt/Del… 3:Clear Entries 4:ClrAllLists 5:Archive 6:UnArchive 7↓Reset… Select 4:ClrAllLists and press `ENTER`. A **Done** message will indicate that all lists have been cleared. You can check the lists by selecting 1:EDIT… from the STAT EDIT menu.
ClrList command ClrList listname1, listname2,…	The ClrList command found under the STAT EDIT menu is used to clear the entries in one or more lists. It also removes any formula associated with the list name. For example, to clear lists L1 and L2, select 4:ClrList from the STAT EDIT menu and type L1 `,` L2. 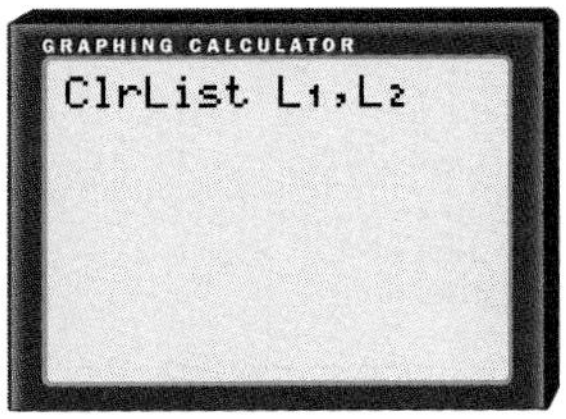 GRAPHING CALCULATOR `ClrList L₁,L₂` Press `ENTER`. *Note*: If you want to clear all of the lists at once, it is faster to use the ClrAllLists command.

Function or Task	Keystroke(s), Menu, or Screen
copy matrices	Suppose that you want to copy the elements of matrix [A] to matrix [B]. Enter matrix [A] into the TI-83 Plus as described under store matrices. $$A = \begin{bmatrix} 14 & 10 & 12 \\ 12 & 14 & 10 \\ 8 & 7 & 5 \\ 18 & 15 & 14 \end{bmatrix}$$ On the TI-83 Plus, press [2nd] [x^{-1}] 1 [STO▸] [2nd] [x^{-1}] 2 [ENTER]. This will copy matrix [A] to matrix [B]. On the TI-83, press [MATRX] 1 [STO▸] [MATRX] 2 [ENTER]. This will copy matrix [A] to matrix [B]. *Note*: On the TI-83 Plus, the MATRX menu is accessed by pressing [2nd] [x^{-1}]. On the TI-83, the MATRX menu is accessed by pressing the [MATRX] key.
cumSum(function cumSum(listname)	The cumSum(function returns the cumulative sum of the elements in a list. It is useful for calculating cumulative frequencies for a distribution. **Example:** Enter the numbers 1, 3, 5, 7, 9, and 11 in L1 by selecting 1:Edit... from the STAT EDIT menu. Move the cursor on top of the list name for L2. Press [2nd] [STAT] [▸] to display the LIST OPS menu. Select 6:cumSum(and type L1 [ )]. Press [ENTER]. 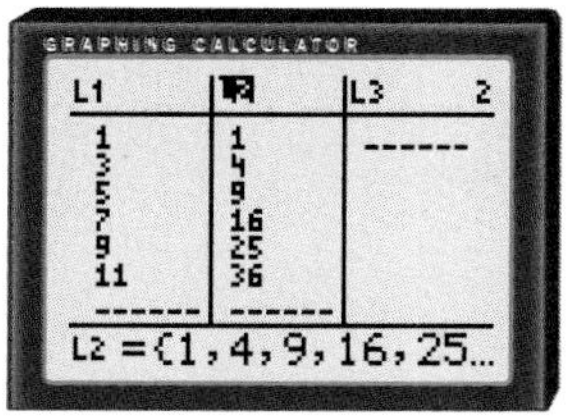

Function or Task	Keystroke(s), Menu, or Screen
DiagnosticOn	When you use one of the regression functions to generate a curve of best fit, the calculator will calculate correlation coefficients that allow you to judge how good the fit was. However, you must first turn on the diagnostic mode. Press (2nd) 0 to access the **CATALOG** menu. Scroll down the list until you line up the black arrow on the screen with **DiagnosticOn**. Press the (ENTER) key to select **DiagnosticOn**. Press (ENTER) again to turn on the diagnostic mode. Try one of the examples of linear regression or non-linear regression in this appendix to see the coefficients displayed on the screen. In a similar manner, you can turn off the diagnostic mode by accessing the **CATALOG** menu and selecting **DiagnosticOff**.
! function value !	The TI-83/TI-83 Plus does not have a factorial key. To calculate a factorial, use the ! function found on the **MATH PRB** menu. **Example:** To evaluate 8!, press 8 (MATH) (▶) (▶) (▶) (or (MATH) (◀)) to display the **MATH PRB** menu. Select 4:! and press (ENTER).  GRAPHING CALCULATOR 8! 　　　　40320 ■ *Note*: The TI-83/TI-83 Plus has the same maximum 69! limit that most scientific calculators have. Most spreadsheets have higher limits.

Function or Task	Keystroke(s), Menu, or Screen
>Frac function value >Frac	The >Frac function found under the MATH menu will display the results of a calculation in fractional form. **Example:** To add $\frac{1}{2} + \frac{1}{3}$, and display the result as a fraction rather than as a decimal, type 1 (÷) 2 (+) 1 (÷) 3 (MATH) 1 (ENTER). 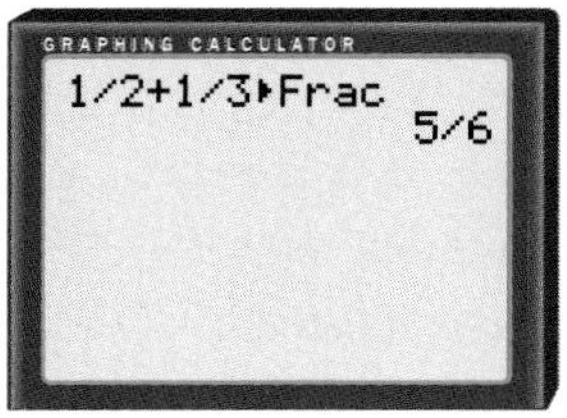 GRAPHING CALCULATOR 1/2+1/3▶Frac 5/6
geometpdf(function geometpdf(p,x)	The geometpdf(function calculates the probability that the first success of an event will occur on trial x, given a probability of success p. **Example:** Calculate the probability that the first roll of doubles on a pair of dice occurs on the fourth roll. In this case, $p = \frac{1}{6}$ and x = 4. Press (2nd) (VARS) to display the DISTR menu. Select D:geometpdf(and type 1 (÷) 6 (,) 4 ( )) and press (ENTER).  GRAPHING CALCULATOR geometpdf(1/6,4) .0964506173 ■ There is a probability of about 0.096 that the first doubles will occur on the fourth roll of the dice. This value is also the probability for a waiting time of three trials.

GRAPHING CALCULATOR

Function or Task	Keystroke(s), Menu, or Screen
invNorm(function invNorm(p,mean,standard deviation)	The invNorm(function is the opposite of the normalcdf(function. It allows you to calculate the distribution function which gives a probability specified by p using a normal distribution with a given **mean** and **standard deviation**. The mean and standard deviation are optional. If they are not specified, then the mean is assumed to be zero, and the standard deviation is assumed to be one. **Example:** A particular IQ test has a mean of 100 and a standard deviation of 10. There is a probability of 0.454 that a given participant scored at or below a particular IQ, and you wish to find this IQ. Press (2nd) (VARS) to display the DISTR menu. Select 3:invNorm(and type .454 (,) 100 (,) 10 ( )) and press (ENTER).  ``` GRAPHING CALCULATOR invNorm(.454,100 ,10) 98.84438412 ■ ``` The correct IQ is approximately 98.84.
interquartile range	The interquartile range of the elements of a list may be determined by carrying out the 1-Var Stats command on a list of data as described on page 529. The results of the 1-Var Stats command include the first quartile Q1 and the third quartile Q3. The interquartile range is calculated by subtracting Q3 − Q1.
Linear regression	See LinReg instruction

Function or Task	Keystroke(s), Menu, or Screen
LinReg instruction LinReg(ax+b) Xlist, Ylist, Function	You can use the LinReg method of regression if it looks like your scatter plot resembles a linear function. **Example:** Clear all functions in the Y= editor. Clear all lists using the ClrAllLists command. Press STAT PLOT and turn off all plots except **Plot1**. Ensure that you are set for a scatter plot, that Xlist is L1, and that Ylist is L2. Use the **STAT EDIT** menu to enter the integers 0, 1, 2, 3, and 4 into L1 and to enter the numbers 1.2, 4.3, 6.5, 10.8, and 12.5 into L2. Press the (ZOOM) key and select **9:ZoomStat** to fit the axes to the data. 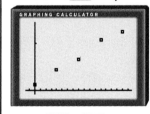 Press (STAT) (▸) to display the **STAT CALC** menu. Select 4:LinReg(ax+b) and type L1 (,) L2 (,) Y1. (To display Y1 press (VARS) (▸). Select 1:Function. Select 1:Y1.) Press (ENTER).  Press (GRAPH). 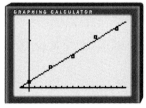 The regression equation is stored in the Y= editor. If you press (Y=), you will see the equation generated by the calculator.

Function or Task	Keystroke(s), Menu, or Screen
mean(function mean(listname)	The **mean(** function located under the **LIST MATH** menu returns the mean of the list specified by listname. **Example:** Select 1:Edit… from the **STAT EDIT** menu to enter the numbers 1, 2, 3, and 4 into L1. Press [2nd] [MODE] to QUIT to the home screen. Press [2nd] [STAT] [▶] [▶] to display the **LIST MATH** menu. Select 3:mean(and type L1 [ )]. Press [ENTER]. 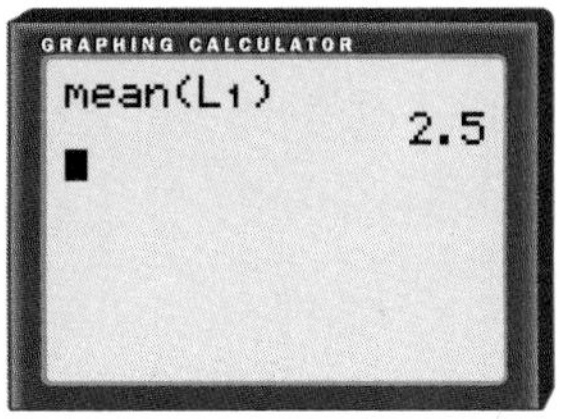 GRAPHING CALCULATOR mean(L₁) 2.5 ■
median(function median(listname)	The **median(** function located under the **LIST MATH** menu returns the median of the list specified by listname. **Example:** Select 1:Edit… from the **STAT EDIT** menu to enter the numbers 1, 2, 3, 4, 5, 6, and 7 into L1. Press [2nd] [MODE] to QUIT to the home screen. Press [2nd] [STAT] [▶] [▶] to display the **LIST MATH** menu. Select 4:median(and type L1 [ )]. Press [ENTER]. 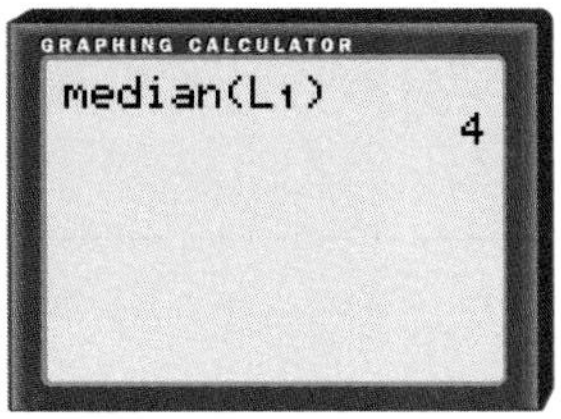 GRAPHING CALCULATOR median(L₁) 4

Function or Task	Keystroke(s), Menu, or Screen
mode settings	If you press the $\boxed{\text{MODE}}$ key, you will see a number of **mode settings** that affect the way the TI-83/TI-83 Plus displays and interprets numbers and graphs. 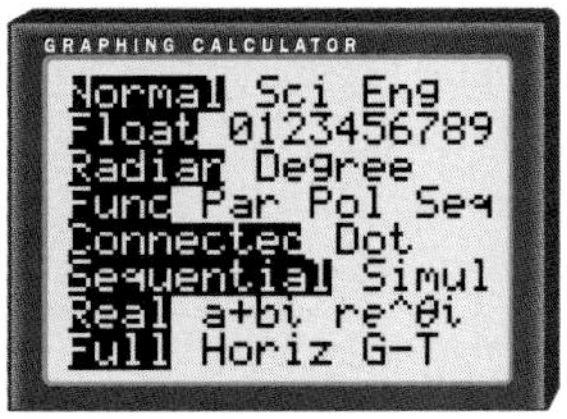

a) You have a choice of normal, scientific, or engineering format for real numbers.

b) You may choose a fixed number of decimal points for floating point numbers from 0 to 9.

c) You may measure angles in radians or degrees.

d) You may choose your graph plotting as **Func** (y as a function of x), **Par** (x and y are functions of a parameter t), **Pol** (polar coordinates r as a function of θ), or **Seq** (to plot sequences).

e) You may choose to connect or not to connect the dots plotted for functions.

f) You may plot your functions sequentially or simultaneously.

g) You may display numbers as **Real** (real numbers), **a+bi** (complex numbers in vector form), or **re^θi** (complex numbers in polar form).

h) You may plot your graph **Full** (Screen), as the top half of the screen with text at the bottom in **Horiz** mode or in the left half of the screen with the corresponding table in the right half of the screen in **G-T** mode.

Function or Task	Keystroke(s), Menu, or Screen

multiply matrices	To multiply matrix [A] by matrix [B], store the matrices [A] and [B] using the method described in store matrices. $$A = \begin{bmatrix} 5 & 1 & -2 \\ 4 & -2 & 0 \end{bmatrix} \quad B = \begin{bmatrix} 7 & 0 \\ -4 & 3 \\ 1 & -6 \end{bmatrix}$$ Using the TI-83 Plus, multiply the matrices by pressing `2nd` `x⁻¹` 1 `×` `2nd` `x⁻¹` 2 `STO▸` `2nd` `x⁻¹` 3 `ENTER`. Using the TI-83, multiply the matrices by pressing `MATRX` 1 `×` `MATRX` 2 `STO▸` `MATRX` 3 `ENTER`. These keystrokes will multiply [A] by [B] and store the result in [C]. The elements of [C] will be displayed on the screen. 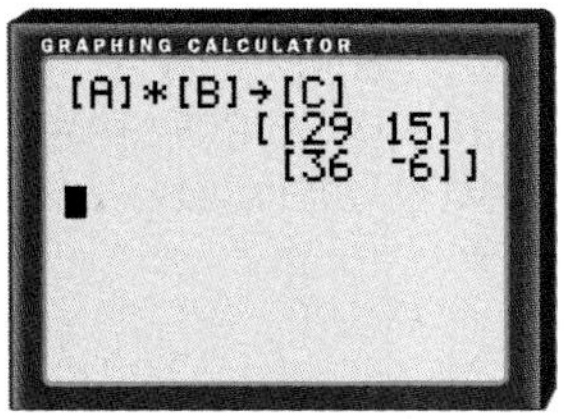
	
	 GRAPHING CALCULATOR
	`[A]*[B]→[C]`
	`       [[29 15]`
	`        [36 -6]]`
	`■`
	
	 Note: On the TI-83 Plus, the **MATRX** menu is accessed by pressing `2nd` `x⁻¹`. On the TI-83, the **MATRX** menu is accessed by pressing the `MATRX` key.

nCr function
 	
 value1 nCr value2
	To calculate a combination, use the nCr function located under the **MATH PRB** menu.
	
	 Example:
	 Evaluate the number of subsets of 10 objects taken 7 at a time, or 10 choose 7.
	 Type 10. Press `MATH` `▶` `▶` `▶` (or `MATH` `◀`) to display the **MATH PRB** menu. Select 3:nCr and type 7. Press `ENTER`. 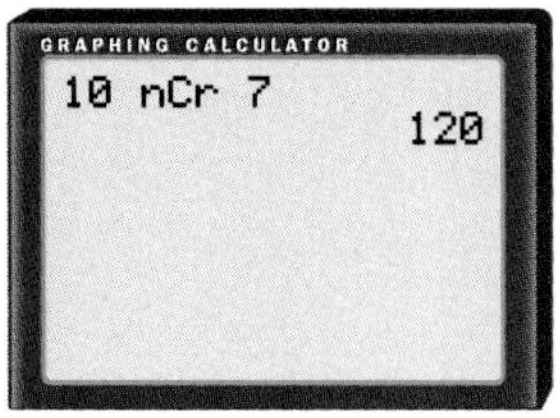
	
	 GRAPHING CALCULATOR
	`  10 nCr 7`
	`              120`

Function or Task	Keystroke(s), Menu, or Screen
Non-linear regression: CubicReg instruction CubicReg Xlist, Ylist, Function	You can use the CubicReg method of regression if it looks like your scatter plot resembles a cubic function, as shown below: 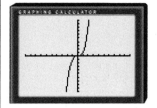 **Example:** Clear all functions in the Y= editor. Clear all lists using the ClrAllLists command. Press STAT PLOT and turn off all plots except Plot1. Ensure that you are set for a scatter plot, that Xlist is L1, and that Ylist is L2. Select 1:Edit... from the STAT EDIT menu to enter the integers 0, 1, 2, 3, and 4 into L1 and to enter the numbers 1.9, 2.4, 3.1, 4.5, and 8.9 into L2. Press the (ZOOM) key and select 9:ZoomStat to fit the axes to the data. 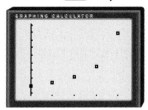 Press (STAT) (▶) to display the STAT CALC menu. Select 6:CubicReg and type L1 (,) L2 (,) Y1 (To display Y1, press (VARS) (▶). Select 1:Function. Select 1:Y1.) Press (ENTER). 

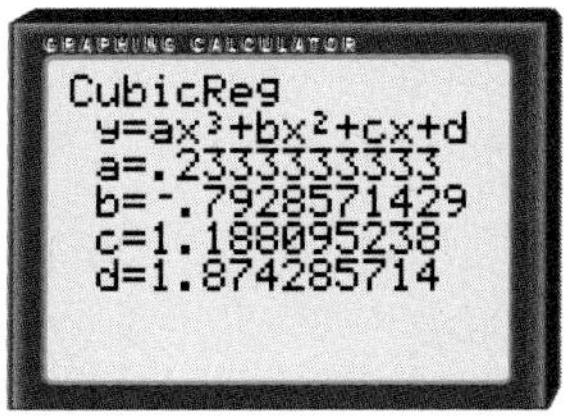

Function or Task	Keystroke(s), Menu, or Screen
	Press GRAPH. 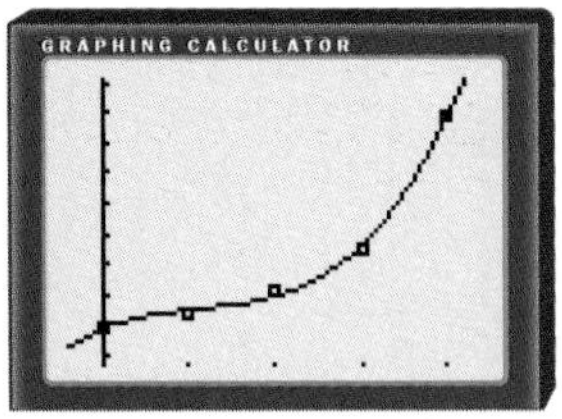 The regression equation is stored in the Y= editor. If you press (Y=), you will see the equation generated by the calculator.
Non-linear regression: ExpReg instruction ExpReg Xlist, Ylist, Function	You can use the **ExpReg** method of regression if it looks like your scatter plot resembles an exponential function, as shown below: 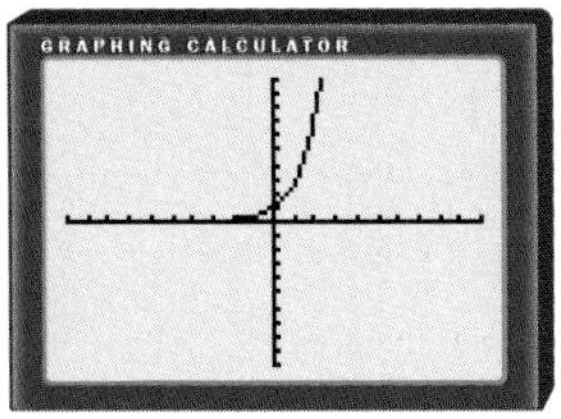 **Example:** Clear all functions in the Y= editor. Clear all lists using the ClrAllLists command. Press STAT PLOT and turn off all plots except Plot1. Ensure that you are set for a scatter plot, that Xlist is L1, and that Ylist is L2. Select 1:Edit... from the STAT EDIT menu to enter the integers 0, 1, 2, 3, and 4 into L1 and to enter the numbers 0.11, 0.25, 0.42, 0.85 and 1.55 into L2. Press the ZOOM key and select 9:ZoomStat to fit the axes to the data. 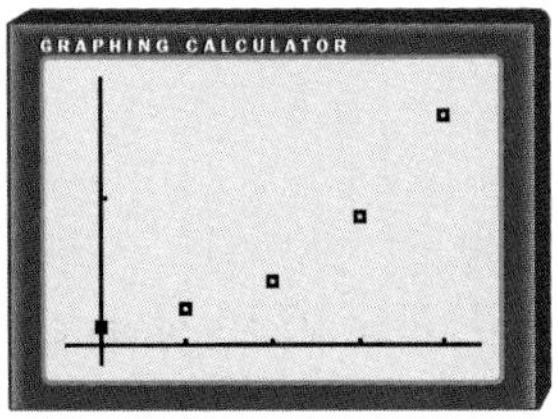

Function or Task	Keystroke(s), Menu, or Screen
	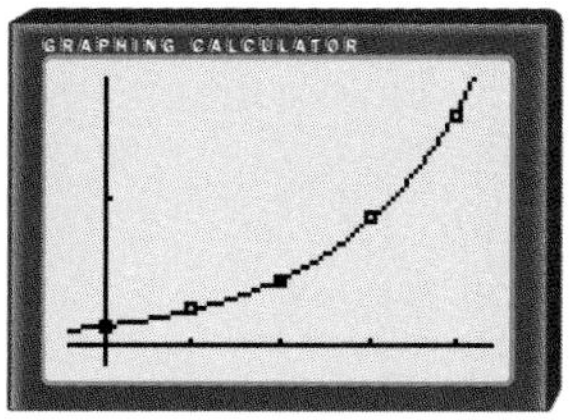 Press (STAT) (▶) to display the **STAT CALC** menu. Select **0:ExpReg** and type L1 (**,**) L2 (**,**) Y1. (To display Y1, press (VARS) (▶). Select 1:**Function**. Select 1:Y1.) Press (ENTER). 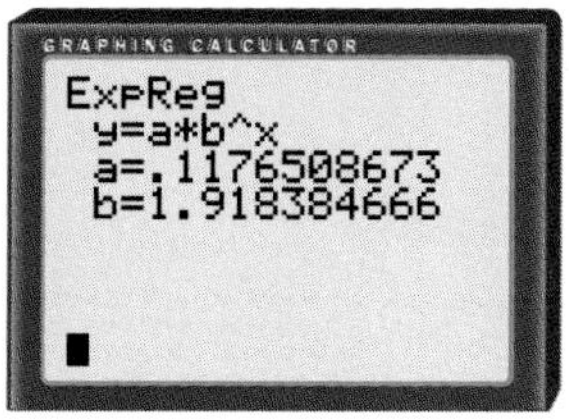 ![Graphing calculator screen showing ExpReg, y=a*b^x, a=.1176508673, b=1.918384666] Press (GRAPH). ![Graphing calculator screen showing an exponential curve through scatter plot points] The regression equation is stored in the Y= editor. If you press (Y=), you will see the equation generated by the calculator.
Non-linear regression: QuadReg instruction QuadReg Xlist, Ylist, Function	You can use the QuadReg method of regression if it looks like your scatter plot resembles a quadratic function, as shown below: 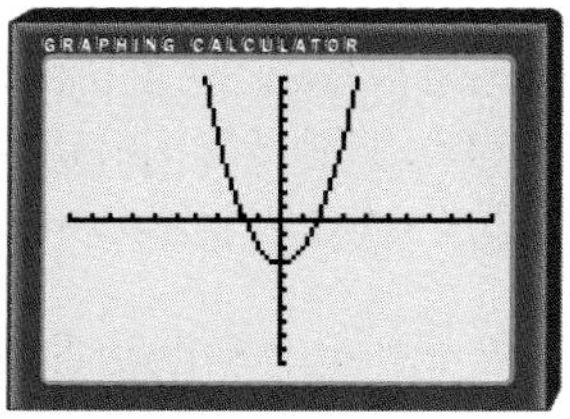 ![Graphing calculator screen showing a parabola] **Example:** Clear all functions in the Y= editor. Clear all lists using the ClrAllLists command. Press STAT PLOT and turn off all plots except Plot1. Ensure that you are set for a scatter plot, that Xlist is L1, and that Ylist is L2. Select 1:Edit… from the **STAT EDIT** menu to enter the integers 0, 1, 2, 3, and 4 into L1 and to enter the numbers 0.9, 1.3, 1.9, 2.7, and 4.1 into L2.

Function or Task	Keystroke(s), Menu, or Screen
	Press the (ZOOM) key and select **9:ZoomStat** to fit the axes to the data.

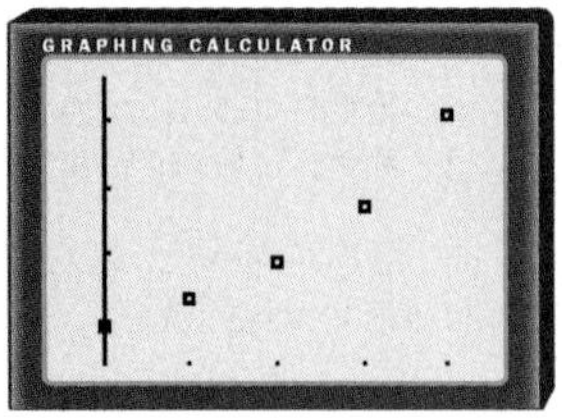

Press (STAT) (▶) to display the **STAT CALC** menu. Select **5:QuadReg** and type L1 (,) L2 (,) Y1. (To display Y1, press (VARS) (▶). Select 1:**Function**. Select 1:Y1.) Press (ENTER).

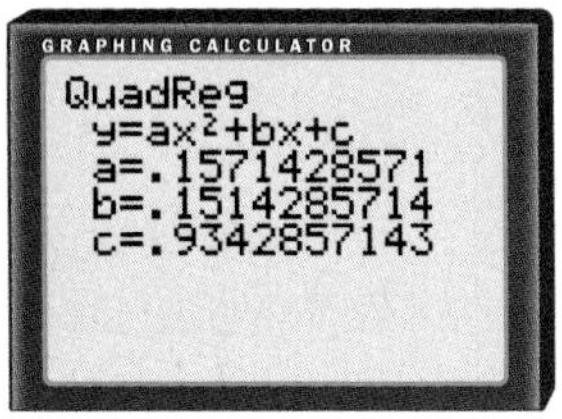

Press (GRAPH).

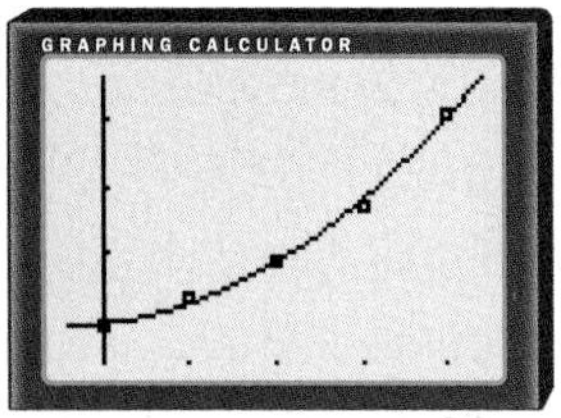

The regression equation is stored in the Y= editor. If you press (Y=), you will see the equation generated by the calculator.

| **normalcdf(function**

normalcdf(lowerbound, upperbound, mean, standard deviation) | The normalcdf(function allows you to calculate the probability that a given data point lies between **lowerbound** and **upperbound** using a normal distribution with a given **mean** and **standard deviation**. The mean and standard deviation are optional. If they are not specified, then the mean is assumed to be zero, and the standard deviation is assumed to be one. |

Function or Task	Keystroke(s), Menu, or Screen

Example 1:

A particular IQ test has a mean of 100 and a standard deviation of 10. Determine the probability that a given participant scored between 90 and 115.

Press (2nd) (VARS) to display the DISTR menu.

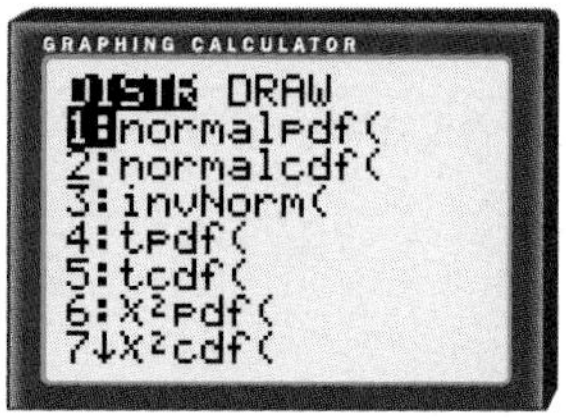

```
GRAPHING CALCULATOR
DISTR DRAW
1:normalpdf(
2:normalcdf(
3:invNorm(
4:tpdf(
5:tcdf(
6:X²pdf(
7↓X²cdf(
```

Select 2:normalcdf(and type 90 (,) 115 (,) 100 (,) 10 ( )). Press (ENTER).

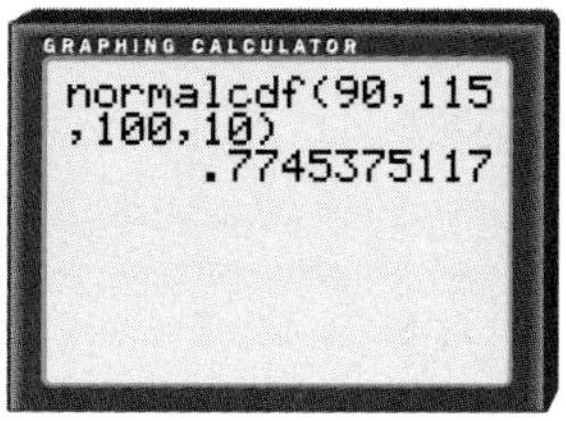

```
GRAPHING CALCULATOR
normalcdf(90,115
,100,10)
     .7745375117
```

The probability that a given participant scored between 90 and 115 is approximately 0.7745.

Example 2:

Calculate the probability that a participant scored 115 or less.
In this case, the value of **lowerbound** is $-\infty$. You can approximate $-\infty$ using a negative number like -1×10^{99}. Select the normalcdf(function as above. Type -1 and press (2nd) (,) to access the EE (Enter Exponent) function. Type 99 (,) 115 (,) 100 (,) 10 ( )) and press (ENTER).

```
GRAPHING CALCULATOR
normalcdf(-1E99,
115,100,10)
     .9331927713
```

The probability of scoring 115 or less is approximately 0.933.

Function or Task	Keystroke(s), Menu, or Screen

normalpdf(function
 	
normalpdf(x,mean,standard deviation)
	The normalpdf(function calculates the probability density function at a specified value for variable x using a normal distribution with a given mean and standard deviation. The mean and standard deviation are optional. If they are not specified, the mean is assumed to be zero, and the standard deviation is assumed to be one.
	
	Example:
	A particular IQ test has a mean of 100 and a standard deviation of 10. Determine the probability density function at a value of $x = 110$.
	Press 2nd VARS to select the DISTR menu. Select 1:normalpdf(and type 110 , 100 , 10). Press ENTER.
	
	GRAPHING CALCULATOR
	normalpdf(110,10
	0,10)
	      .0241970725
	
	The probability density is approximately 0.024.
	This function can also be used to plot the probability distribution. Change your window settings to:
	
	GRAPHING CALCULATOR
	WINDOW
	  Xmin=70
	  Xmax=130
	  Xscl=10
	  Ymin=0
	  Ymax=.1
	  Yscl=.02
	  Xres=1
	
	Press the Y= key to display the Y= editor, then press 2nd VARS to display the the DISTR menu. Select 1:normalpdf(. Press X,T,θ,n , 100 , 10). Press GRAPH.
	
	GRAPHING CALCULATOR
	Plot1 Plot2 Plot3
	\Y1=normalpdf(X,
	100,10)
	\Y2=
	\Y3=
	\Y4=
	\Y5=
	\Y6=
	
	You can then use the TRACE instruction to inspect the graph.

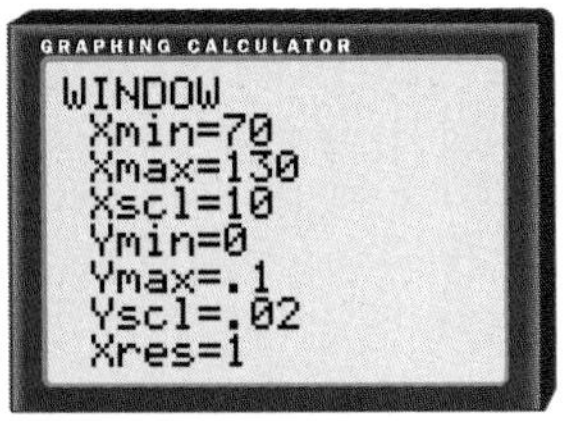

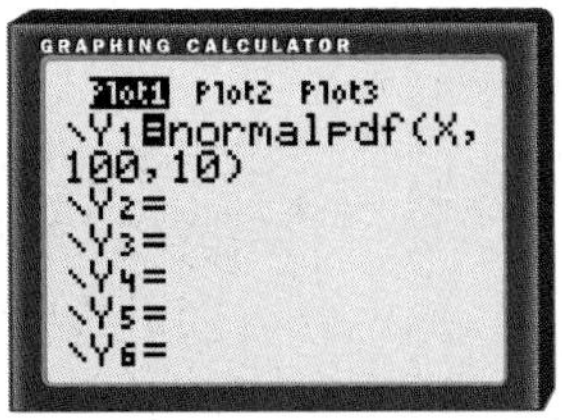

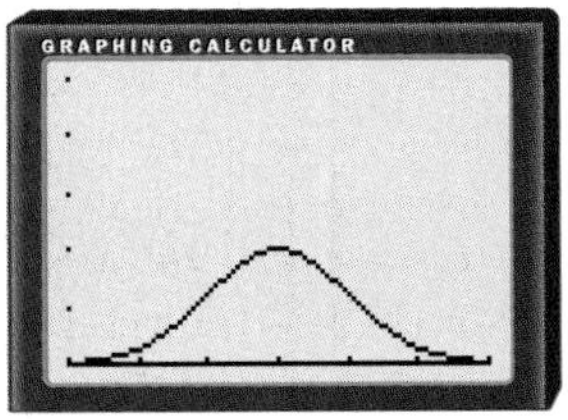

Function or Task	Keystroke(s), Menu, or Screen
nPr function value1 nPr value2	To calculate a permutation, use the nPr function available located under the **MATH PRB** menu. **Example:** Evaluate the number of arrangements of 10 objects taken 7 at a time. Type 10. Press (MATH) (▶) (▶) (▶) (or (MATH) (◀)) to display the **MATH PRB** menu. Select 2:nPr and type 7. Press (ENTER).  GRAPHING CALCULATOR 10 nPr 7 $\qquad$ 604800 ■
1-Var Stats command 1-Var Stats Xlist, Freqlist	The TI-83/TI-83 Plus can calculate various statistical variables for a list of numbers specified by **Xlist**. Similar variables for grouped data can be calculated by adding the **Freqlist**. **Example 1:** Ten automobiles were tested for fuel economy, and were found to burn the following amounts of fuel, measured in litres per 100 km. 8.4, 5.0, 4.8, 5.9, 7.3, 8.2, 6.4, 8.1, 9.5, 4.2 Use the ClrAllLists command to clear the lists in your calculator if necessary. Select 1:Edit… from the **STAT EDIT** menu to enter the above numbers into L1. Press (STAT) (▶) to display the **STAT CALC** menu. Select 1:1-Var Stats. Press (2nd) 1 to type L1 and press (ENTER). You can scroll down to see more statistics: 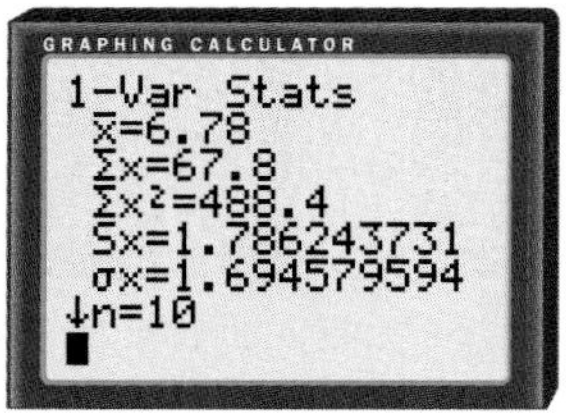 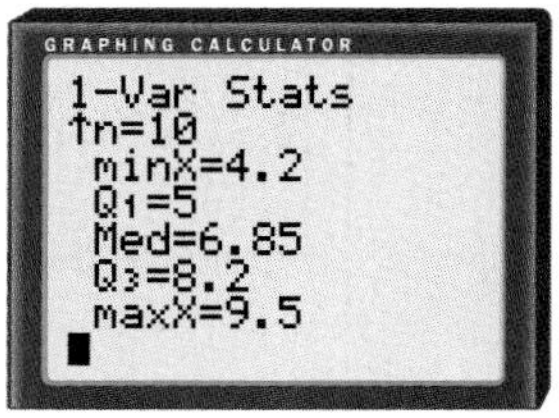 GRAPHING CALCULATOR 1-Var Stats x̄=6.78 Σx=67.8 Σx²=488.4 Sx=1.786243731 σx=1.694579594 ↓n=10 ■ GRAPHING CALCULATOR 1-Var Stats ↑n=10 minX=4.2 Q₁=5 Med=6.85 Q₃=8.2 maxX=9.5 ■

Function or Task	Keystroke(s), Menu, or Screen

The meanings are:
$\bar{x}$ is the mean
Σx is the sum of all the values
Σx^2 is the sum of the squares of the values
Sx is the sample standard deviation
σx is the population standard deviation
n is the number of values in the list
minX is the lowest value
Q_1 is the first quartile
Med is the median
Q_3 is the third quartile
maxX is the highest value

Example 2:

Twenty people were asked to write down the amount of cash they were carrying. The data were arranged into intervals. The frequency of occurrence in each interval was noted. The results were as follows:

Midpoint($)	Frequency
5	3
15	2
25	3
35	2
45	4
55	6
65	5
75	2
85	1
95	2

Use the ClrAllLists command to clear all the lists, if necessary. Enter these data into L1 and L2, respectively. Select the 1-Var Stats command as described in Example 1, but this time type L1 (,) L2. Press (ENTER).

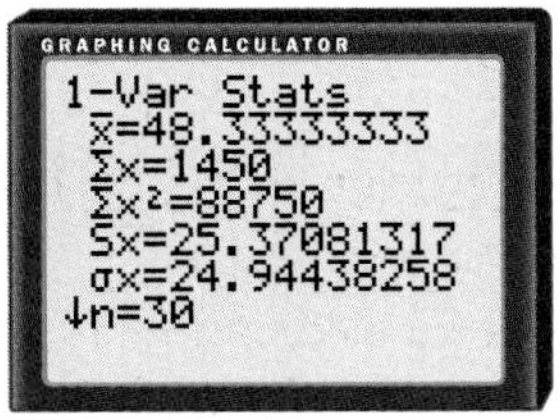

GRAPHING CALCULATOR
1-Var Stats
x̄=48.33333333
Σx=1450
Σx²=88750
Sx=25.37081317
σx=24.94438258
↓n=30

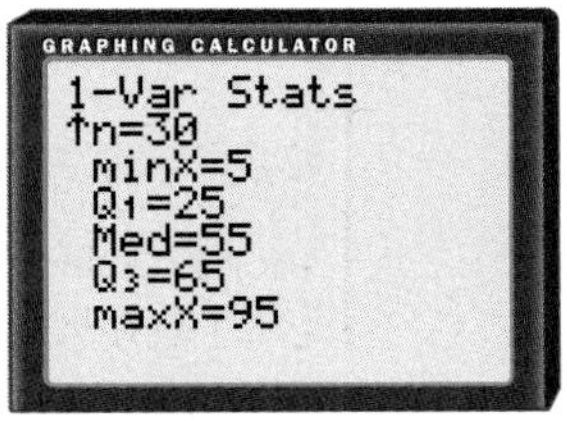

GRAPHING CALCULATOR
1-Var Stats
↑n=30
minX=5
Q1=25
Med=55
Q3=65
maxX=95

Function or Task	Keystroke(s), Menu, or Screen
prod(function prod(list,start,end)	The prod(function is used to find the product of the elements of a list beginning with element start and finishing with element end. If start and end are not specified, then the entire list is used. **Example:** Select 1:Edit... from the STAT EDIT menu to enter 1, 2, 3, 4, and 5 in L1. Press 2nd MODE to QUIT to the home screen. Press 2nd STAT ▶ ▶ to display the LIST MATH menu. Select 6:prod(and type L1 , 2 , 4) . Press ENTER. 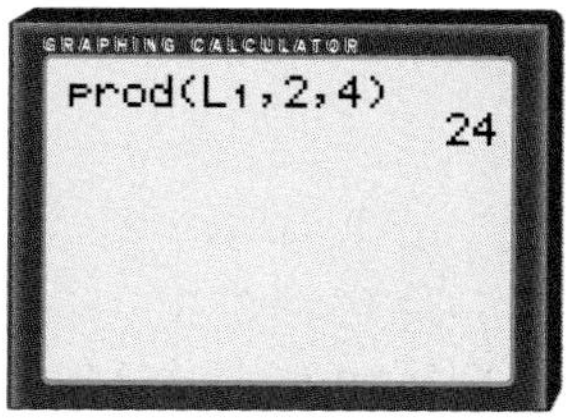 GRAPHING CALCULATOR prod(L1,2,4) 24
quartiles	The quartiles of the elements of a list may be determined by carrying out the 1-Var Stats command on the list, as described on page 529. The results of the 1-Var Stats command include the first quartile Q_1 and the third quartile Q_3.
randInt(function randInt(lowerbound, upperbound, numtrials)	When simulating probability problems, it is useful to be able to generate random integers. This can be done using the randInt(function located under the MATH PRB menu. The function is followed by a lowerbound, an upperbound, and an optional numtrials. **Example 1:** Simulate one roll of one die. The lowerbound is 1, the upperbound is 6, and you do not need to enter the numtrials, since the default value is assumed to be 1. Press MATH ▶ ▶ ▶ (or MATH ◀) to display the MATH PRB menu. Select 5:randInt(and type 1 , 6) . Press ENTER. You will get a random integer between 1 and 6 as shown in the first calculation of the screen shot following Example 3.

Function or Task	Keystroke(s), Menu, or Screen

Example 2:

If you want three rolls of the die, press the same keystrokes to select 5:randInt(again, but this time type 1 (,) 6 (,) 3 ()). Press (ENTER). You will get a list of three random rolls of the die as shown in the second calculation of the screen shot following Example 3.

Example 3:

You can use the function twice to get the sum of two dice rolled independently, as shown in the third calculation of the following screen shot.

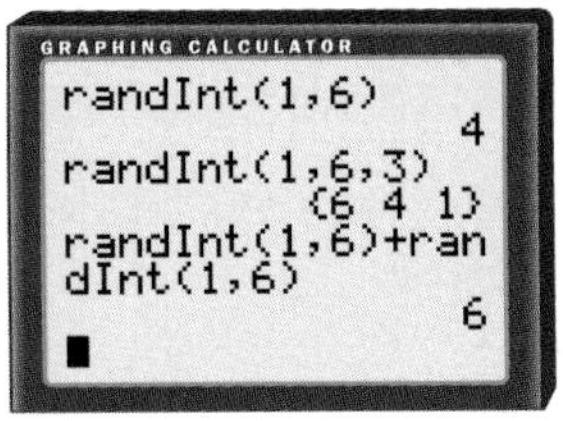

GRAPHING CALCULATOR
randInt(1,6)
4
randInt(1,6,3)
{6 4 1}
randInt(1,6)+randInt(1,6)
6

A Note About Seeds: Whenever you use the random integer function, you will generate the same series of random integers. The start of the series is controlled by the value of the variable **rand** which is stored internally in the TI-83 Plus, and is set to zero by default. If you change the default value to something else, you can generate a different series. For example, you can change the default to 1 using the keystrokes 1 (STO▸) (MATH) (◂) 1 (ENTER).

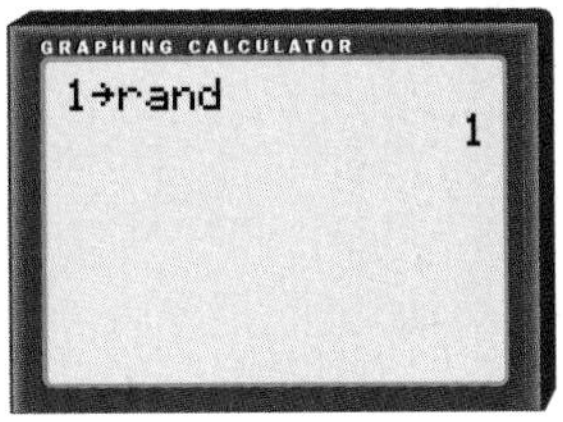

GRAPHING CALCULATOR
1→rand
1

Function or Task	Keystroke(s), Menu, or Screen
randNorm(function randNorm(mean, standard deviation, numtrials)	You can use the randNorm(function to select a random number from a normal distribution with a given **mean** and **standard deviation**. If numtrials is not specified, you get one random number. If you want more one random number, set the value for **numtrials**. **Example 1:** A particular IQ test has a mean of 100 and a standard deviation of 10. Find three random values assuming a normal distribution. Press (MATH) (▶) (▶) (▶) (or (MATH) (◀)) to display the MATH PRB menu. Select 6:randNorm(and type 100 (,) 10 (,) 3 ( )). Press (ENTER). You will get three random IQs from the distribution similar to the following screen:  Use the right arrow key to scroll through the other values. You can store these results in a list if you wish by adding (STO▸) (2nd) 1 (to use L1) to the randNorm(function as shown in the following screen:  *A Note About Seeds:* The random number seed discussed in the section on the randInt(function also applies to the randNorm(function.

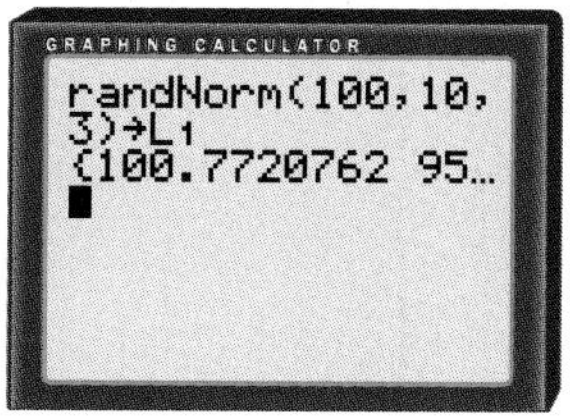

Function or Task	Keystroke(s), Menu, or Screen
round(function round(operand,#decimals)	The **round(** function located under the **MATH NUM** menu will return the **operand** correctly rounded to the number of decimal places specified in **#decimals**. The **operand** can be a number, an expression, a list name or a matrix name. In the case of a list or a matrix, the function will round all of the elements. **Example:** Evaluate the fraction $\frac{3}{7}$ rounded correctly to four decimal places. Press [MATH] [▶] to display the **MATH NUM** menu. Select **2:round(** and type 3 [÷] 7 [,] 4 [)]. Press [ENTER]. 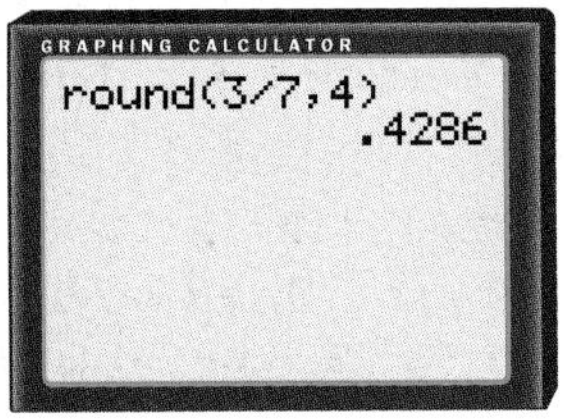 GRAPHING CALCULATOR round(3/7,4) .4286
semi-interquartile range	The **semi-interquartile range** is one half of the **interquartile range**. See interquartile range.
seq(function seq(expression, variable, begin, end, increment)	You can use the **seq(** function to create a list of numbers with various properties. If the value of **increment** is not specified, it is assumed to be one. **Example 1:** Enter a list of the squares of the odd numbers from 5 to 11 inclusive. The value of **begin** is 5, **end** is 11, and **increment** is 2. Press the [2nd] [STAT] [▶] to display the **LIST OPS** menu. Select **5:seq(** and type [ALPHA] A and press [x^2]. Type [,] [ALPHA] A [,] 5 [,] 11 [,] 2 [)]. Press [STO▸] [2nd] 1 to store the result in L1. Press [ENTER].  GRAPHING CALCULATOR seq(A²,A,5,11,2) →L₁ {25 49 81 121} ■

Function or Task	Keystroke(s), Menu, or Screen
	Notice the list of odd number squares, as expected. These numbers have also been stored in list L1. You can inspect L1 by selecting 1:Edit… from the STAT EDIT menu. 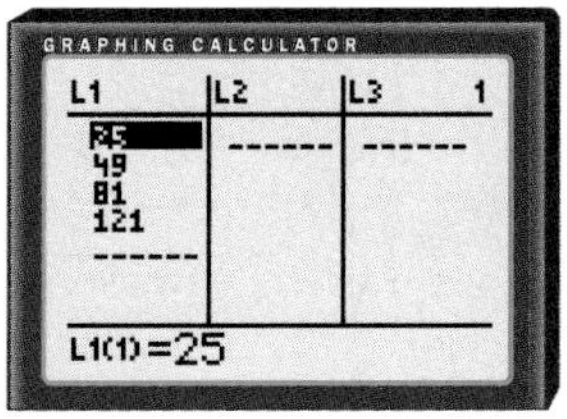 GRAPHING CALCULATOR L1　　L2　　L3　　1 25 49 81 121 ------- L1(1)=25
SortA(function SortA(listname)	The SortA(function located under the LIST OPS menu will sort the list specified by listname into ascending order. **Example:** Select 1:Edit… from the STAT EDIT menu to enter 1, 3, 4, and 2 into L1. Press (2nd) (STAT) (▶) display the LIST OPS menu. Select 1:SortA(and type L1 ( )). Press (ENTER). Press (2nd) 1 to display list L1. 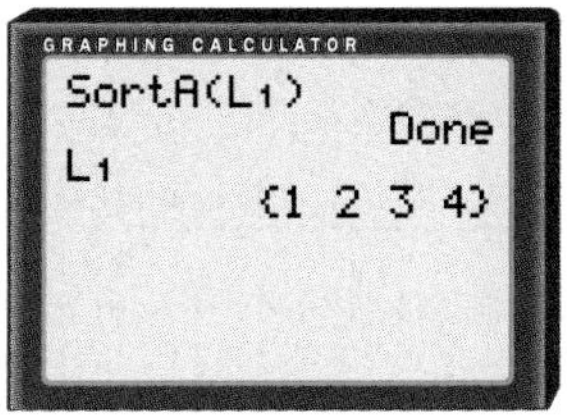 GRAPHING CALCULATOR SortA(L1) 　　　　　Done L1 　　　{1 2 3 4} *Note*: A related function is the SortD(function which sorts a list in descending order.
ShadeNorm(function ShadeNorm(lowerbound, upperbound, mean, standard deviation)	The ShadeNorm(function allows you to shade the area under the probability density graph that a given data point lies between lowerbound and upperbound using a normal distribution with a given mean and standard deviation. The mean and standard deviation are optional. If they are not specified, then the mean is assumed to be zero, and the standard deviation is assumed to be one.

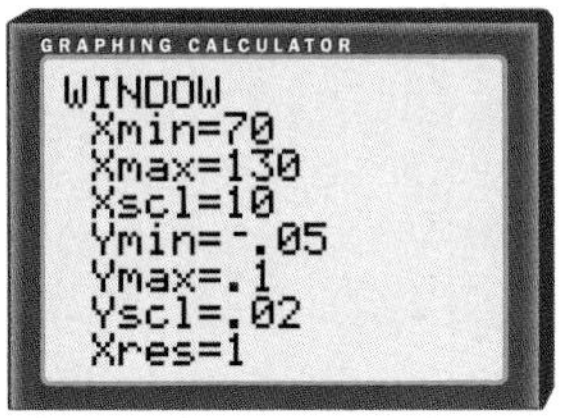

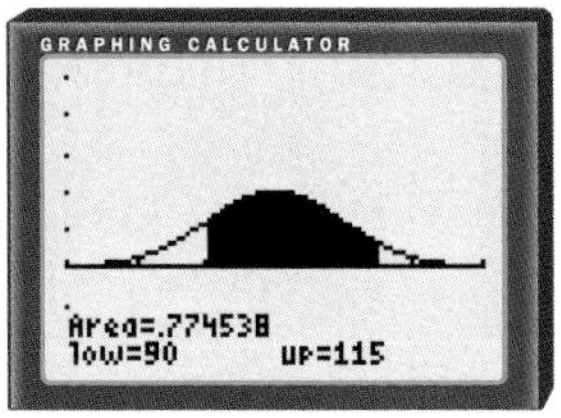

Function or Task	Keystroke(s), Menu, or Screen
| | **Example 1:**<br><br>A particular IQ test has a mean of 100 and a standard deviation of 10. Display and shade the probability density function that a given participant scored between 90 and 115.<br><br>First, adjust your window settings as shown:<br><br>```
GRAPHING CALCULATOR
WINDOW
 Xmin=70
 Xmax=130
 Xscl=10
 Ymin=-.05
 Ymax=.1
 Yscl=.02
 Xres=1
```<br><br>Press 2nd VARS ► to display the **DISTR DRAW** menu. Select 1:ShadeNorm(. Type 90 , 115 , 100 , 10 ) and press ENTER.<br><br>```
GRAPHING CALCULATOR

Area=.774538
low=90      up=115
```<br><br>Notice that the probability represented by the shaded area has also been displayed. |
| **standard deviation**<br><br>stdDev(listname) | The **stdDev(** function located under the **LIST MATH** menu returns the standard deviation of the list specified by **listname**.<br><br>**Example:**<br><br>Select 1:Edit... from the **STAT EDIT** menu to enter the numbers 1, 2, 3, 4, 5, 6, and 7 into L1. Press 2nd MODE to QUIT to the home screen. Press 2nd STAT ► ► to display the **LIST MATH** menu. Select 7:stdDev( and type L1 ). Press ENTER.<br><br>```
GRAPHING CALCULATOR
stdDev(L₁)
      2.160246899
``` |
```

Function or Task	Keystroke(s), Menu, or Screen
STAT PLOT	The plot routines which are used to plot graphs of data entered into the lists of the TI-83/TI-83 Plus are controlled by the **STAT PLOT** screen. This screen is accessed by pressing (2nd) (Y=). The screen contains five options: 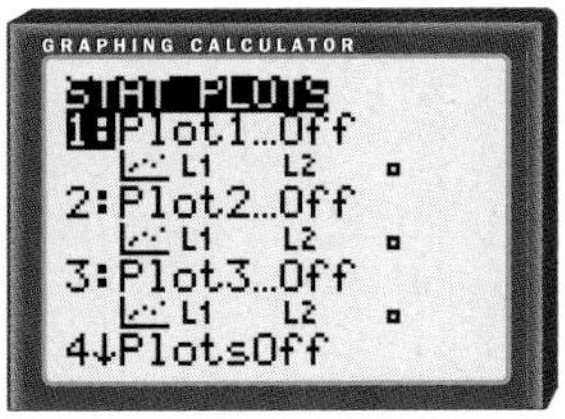

Options 1, 2, and 3 control the three plot routines. Three different statistical plots can be displayed on the graphing screen at one time. Option 4 turns all plots off, and option 5 turns all plots on. If you select a plot, say **Plot1**, then you will see the following screen:

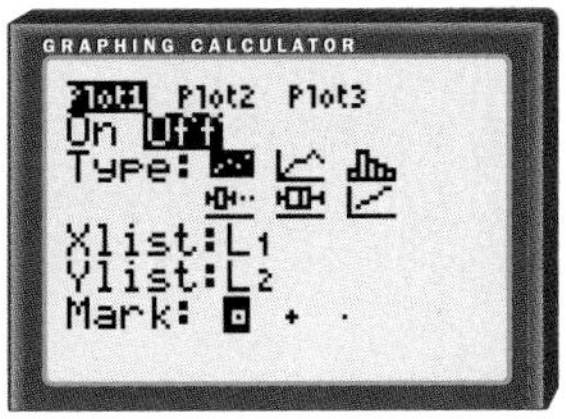

The first line is used to turn **Plot1 On** or **Off**. The second line allows you to select the type of graph you want: scatter plot, *xy*-plot, histogram, modified box plot, box plot, or normal probability plot. The next line or lines let you choose which list or lists will provide the data for the axis or axes. The last line lets you choose one of three symbols to display data points.

| **store matrices** | Store the following matrix in the TI-83/TI-83 Plus:

$$A = \begin{bmatrix} 14 & 10 & 12 \\ 12 & 14 & 10 \\ 8 & 7 & 5 \\ 18 & 15 & 14 \end{bmatrix}$$

On the TI-83 Plus, press (2nd) (x^{-1}) to access the MATRX menu.

On the TI-83, press the (MATRX) key to access the MATRX menu. 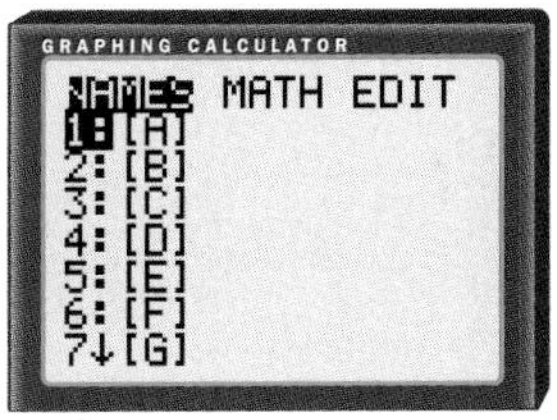 |

Function or Task	Keystroke(s), Menu, or Screen
	Press ▶ ▶ (or ◀) to display the **MATRX EDIT** menu. Select 1:[A], which is matrix [A]. The default dimensions are 1×1 as shown in the following screen:

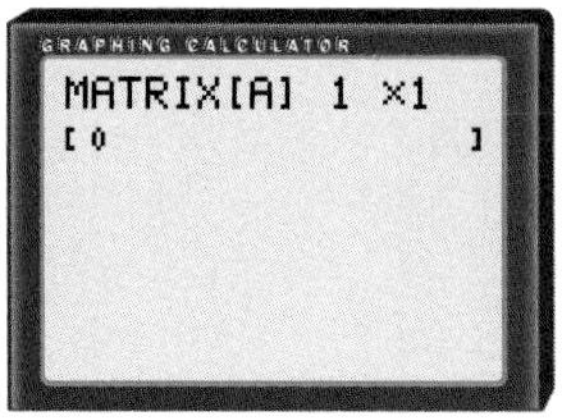

GRAPHING CALCULATOR

MATRIX[A] 1 ×1
[0   ]

Change these dimensions to 4×3. Notice that the matrix enlarges to the required dimensions. Scroll to the first element using the right blue arrow key. Type 14 and press ENTER. Continue to fill in the elements of the matrix by typing each element and pressing ENTER. Once all elements are entered, you will see:

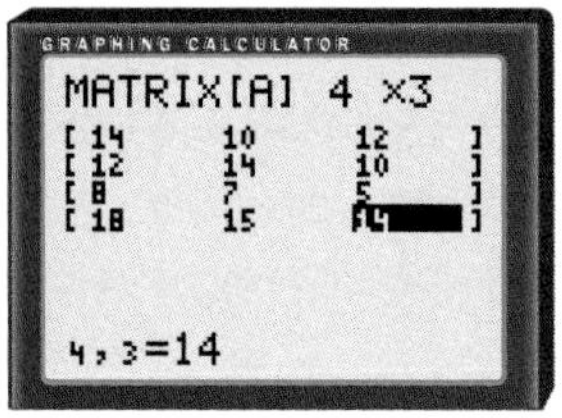

GRAPHING CALCULATOR

MATRIX[A] 4 ×3
[14   10   12   ]
[12   14   10   ]
[8   7   5   ]
[18   15   14   ]

4,3=14

Function or Task	Keystroke(s), Menu, or Screen
sum(function sum(list,start,end)	The **sum(** function is used to find the sum of the elements of a **list** beginning with element **start** and finishing with element **end**. If **start** and **end** are not specified, then the entire list is used. **Example:** Select 1:Edit… from the **STAT EDIT** menu to enter the list of numbers 1, 2, 3, 4, and 5 in L1. Press 2nd STAT ▶ ▶ to display the **LIST MATH** menu. Select 5:sum(and type L1 , 2 , 4). Press ENTER. 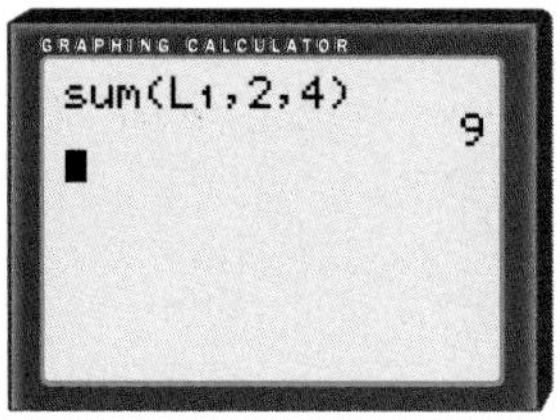

GRAPHING CALCULATOR

sum(L₁,2,4)
                    9
■

Function or Task	Keystroke(s), Menu, or Screen
TRACE instruction	The TRACE instruction allows you to move a cursor along a graph while a readout of the coordinates is displayed as shown below. 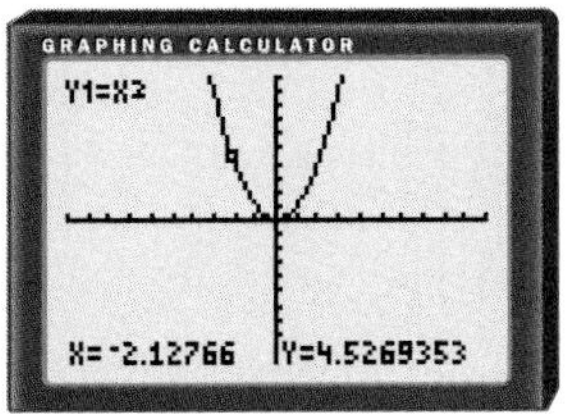 **Example:** Display and trace along the graph of $y = x^2$. Turn off all plots using the STAT PLOT screen. Press the (Y=) key to display the Y= editor. Press the variable key, marked (X,T,θ,n) followed by the (x²) key. To view the graph of $y = x^2$ in the standard viewing window, press (ZOOM) 6. To trace along the graph, press (TRACE). The tracing cursor will appear on the graph. The coordinates of the location of the cursor are displayed at the bottom of the screen. Use the blue left and right arrow keys to move the cursor along the graph.
window settings	The window settings for the current viewing window can be adjusted by pressing the (WINDOW) key. You can set the limits and scales on both the horizontal and the vertical axes. **Example:** Suppose you want to plot the function $y = x^2$ for values of x ranging from –20 to +20. Start by entering the function into the Y= editor. Press (Y=). Clear any existing functions at this time. Move to Y1 using the blue arrow keys if necessary. Press (X,T,θ,n). Press (x²). 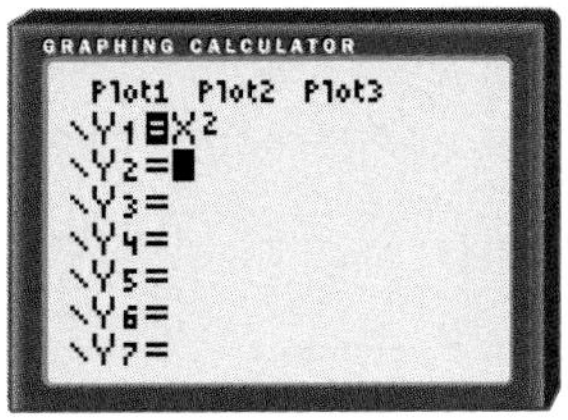

Function or Task	Keystroke(s), Menu, or Screen
	Press [WINDOW]. Set Xmin to –20, Xmax to +20, Xscl to 5, Ymin to 0, Ymax to 400, Yscl to [,] and Xres to 1. Press [GRAPH]. GRAPHING CALCULATOR WINDOW Xmin=-20 Xmax=20 Xscl=5 Ymin=0 Ymax=400 Yscl=50 Xres=1 GRAPHING CALCULATOR Notice where the marks are on the axes. Return to the window settings screen, and experiment with the settings. Check the graph to see the effects of your changes.
Y= editor	The Y= editor is accessed by pressing [Y=]. The Y= editor allows you to enter functions for graphing or tabling purposes. **Example:** To graph $y = x^2$, press the [Y=] key. To obtain the variable X, press [X,T,θ,n]. Press [x^2]. Press [ZOOM] 6 to view the graph of $y = x^2$ in the standard viewing window. GRAPHING CALCULATOR Plot1 Plot2 Plot3 \Y1◻X² \Y2=■ \Y3= \Y4= \Y5= \Y6= \Y7= GRAPHING CALCULATOR You may add as many functions as you have space for. You can select or deselect whether a function is plotted by moving the cursor to the equal sign in the function, and pressing [ENTER] key to toggle selection/deselection.
ZInterval instruction	The ZInterval instruction allows you to find a desired confidence interval for the mean from a finite sample of a distribution whose population mean is not known, but whose standard deviation is known.

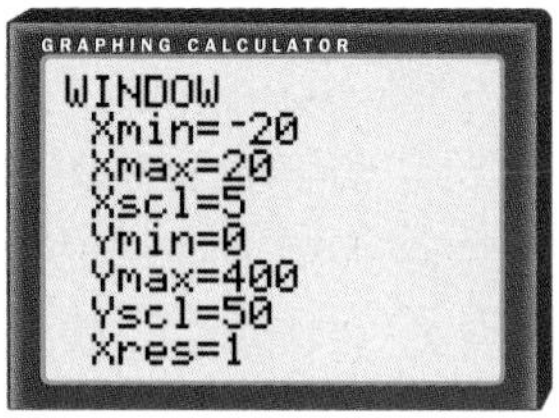

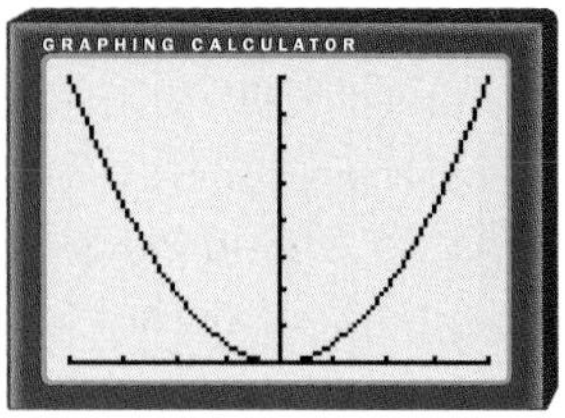

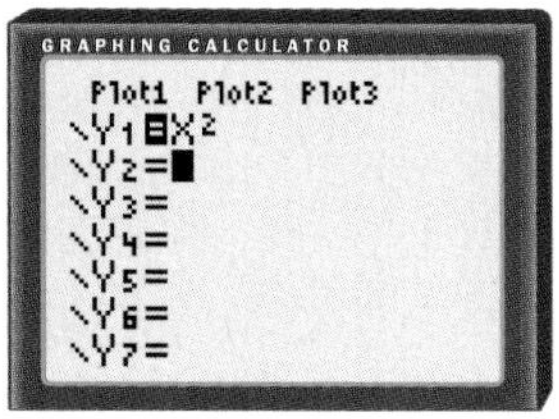

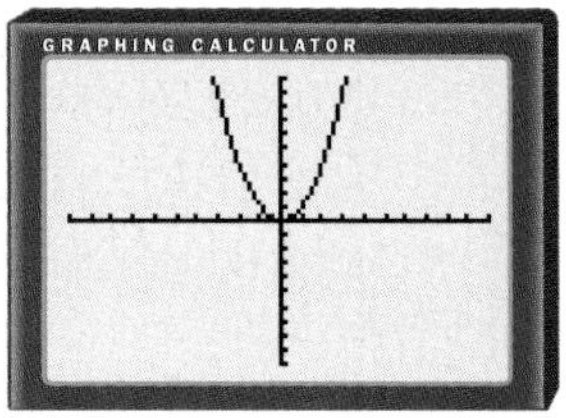

Function or Task	Keystroke(s), Menu, or Screen

Example:

Suppose that a manufacturer knows that the standard deviation for the drying time of latex paints is 10.5 min. A sample of 20 items are painted, and the mean drying time for the sample is found to be 75.4 min. What is the 90% confidence interval for the mean of the population of paint drying times? To determine this, you can use the ZInterval instruction found under the STAT TESTS menu.

Press (STAT) (▸) (▸) to display the STAT TESTS menu. Select 7:ZInterval.

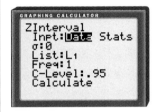

Note that you perform the test with either raw Data or Stats already calculated from a sample. Press (▸) (ENTER) to select the Stats option. Set the remaining parameters as shown:

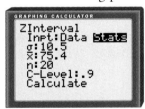

Note that 10.5 is the standard deviation of the distribution, 75.4 is the sample mean, 20 is the number of samples, and 0.9 is the confidence level desired.

Use the arrow keys to scroll down to Calculate and press (ENTER).

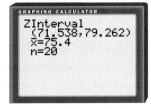

You can be 90% certain that the population mean lies between about 71.5 and 79.3. In this same manner, you can calculate confidence intervals for other confidence levels.

Function or Task	Keystroke(s), Menu, or Screen
	The calculation on page 541 can also be performed using raw **Data**. Press $\boxed{\text{STAT}}$ $\boxed{\blacktriangleright}$ $\boxed{\blacktriangleright}$ to display the **STAT TESTS** menu. Select 7:ZInterval. This time select the **Data** option. Note that the parameters you typed in are still there.

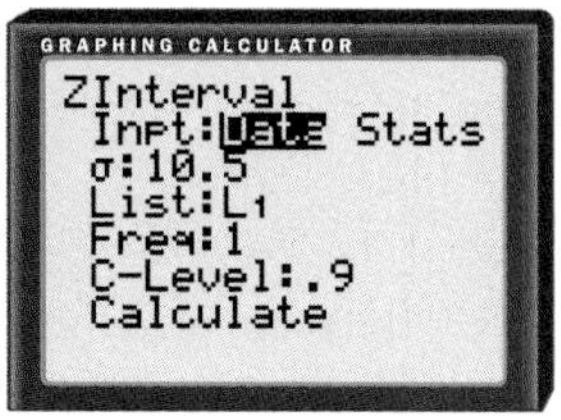

GRAPHING CALCULATOR

```
ZInterval
  Inpt:Data Stats
  σ:10.5
  List:L₁
  Freq:1
  C-Level:.9
  Calculate
```

Note that your sample data must be entered in **L1** (or whatever other list you specify) before attempting to **Calculate** the confidence interval.

Function or Task	Keystroke(s), Menu, or Screen
Z-Test instruction	The **Z-Test instruction** allows you to test the mean from a finite sample of a distribution whose mean is not known, but whose standard deviation is known.

Example:

A manufacturer knows that the standard deviation for the drying time of latex paints is 10.5 min. A sample of 20 items are painted, and the mean drying time is found to be 75.4 min. How confident can the manufacturer be that this would represent the mean drying time of the paint if a larger number of samples were taken?

To determine this, you can use the **Z-Test instruction** located in the **STAT TESTS** menu to determine the probability that the real mean is 80, and that 75.4 is just a statistical variation.

Press $\boxed{\text{STAT}}$ $\boxed{\blacktriangleright}$ $\boxed{\blacktriangleright}$ to display the **STAT TESTS** menu. Select 1:Z-Test to obtain the following screen:

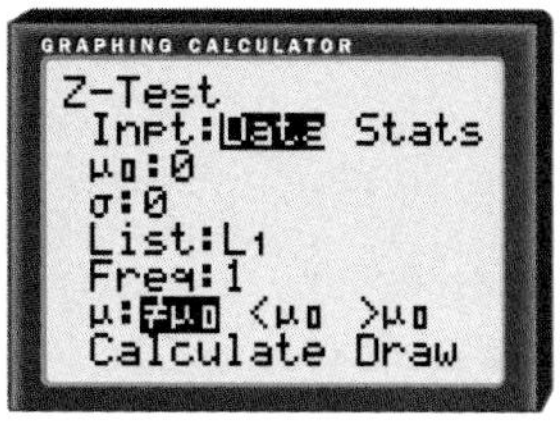

GRAPHING CALCULATOR

```
Z-Test
  Inpt:Data Stats
  μ₀:0
  σ:0
  List:L₁
  Freq:1
  μ:≠μ₀ <μ₀ >μ₀
  Calculate Draw
```

| | You can perform the test with either raw **Data** or **Stats** already calculated from a sample. Use the arrow key, and press (ENTER) to select the **Stats** option. Set the remaining parameters as shown on the right: 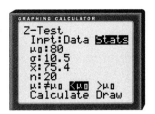 |

Note that 80 is the hypothesized mean, 10.5 is the standard deviation, 75.4 is the sample mean, and 20 is the number of samples.

Use the arrow keys to scroll down to **Calculate** and press (ENTER).

Note that the probability that the real mean is 80 is approximately $p = 0.025$. In the same manner, you can test other values for the mean of the distribution.

You can also display the results graphically. Press (STAT) (▸) (▸) to display the **STAT TESTS** menu. Select **1:Z-Test**. Note that the parameters you typed in are still there. Use the arrow keys to scroll down, and this time select the **Draw** option at the bottom of the screen. Press (ENTER).

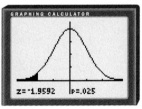

The shaded area represents the probability of getting a sample mean of 75.4 or less if the real mean is 80.

The same test can be performed using raw **Data**. Press (STAT) (▸) (▸) to display the **STAT TESTS** menu. Select **1:Z-Test**. This time select the **Data** option.

Your sample data must be entered in **L1** (or whatever other list you specify) before attempting to **Calculate** or **Draw**.

GRAPHING CALCULATOR

Spreadsheets (Microsoft® Excel and Corel® Quattro® Pro)

Note: The methods provided apply for Microsoft® Excel 2000 and Corel® Quattro® Pro 8 and 9. Methods may vary slightly for other versions.

Function or Task	Keystroke(s), Menu, or Screen

absolute cell referencing	See References:
	    relative referencing
	    absolute referencing
	    mixed referencing

add worksheets **Microsoft® Excel:** Insert/Worksheet **Corel® Quattro® Pro:** Insert/Sheet	**Microsoft® Excel:** If a new worksheet is required, choose Insert/Worksheet. The new worksheet will be inserted before the currently selected worksheet. Simply drag the TAB for a worksheet to move it in the worksheet list. **Corel® Quattro® Pro:** If a new worksheet is required, choose Insert/Sheet. The new worksheet will be inserted before the currently selected worksheet. Simply drag the TAB for a worksheet to move it in the worksheet list.
average **Microsoft® Excel:** =AVERAGE(array) **Corel® Quattro® Pro:** @AVG(array)	**Microsoft® Excel:** The average function is =AVERAGE(array). **Example:** To find the average of 6, 7, 8, 9, and 10, type =AVERAGE(6,7,8,9,10) and press Enter. The result will be 8. To find the average of cells B1 through B10, type =AVERAGE(B1:B10) and press Enter. **Corel® Quattro® Pro:** The average function is @AVG(array). **Example:** To find the average of 6, 7, 8, 9, and 10, type @AVG(6,7,8,9,10) and press Enter. The result will be 8. To find the average of cells B1 through B10, type @AVG(B1..B10) and press Enter.

SPREADSHEETS

Function or Task	Keystroke(s), Menu, or Screen
BINOMDIST function **Microsoft® Excel:** =BINOMDIST(x,n,p,FALSE) **Corel® Quattro® Pro:** @BINOMDIST(x,n,p,FALSE)	The BINOMDIST function returns the binomial distribution probability of an individual term. It returns the probability of getting exactly x successes in n trials of a binomial distribution, where the probability of success on each trial is p. **Example:** Consider the rolling of two dice 20 times. What is the probability of rolling exactly four doubles? In this case, $x = 4$, $n = 20$, and $p = \dfrac{1}{6}$. Hence, BINOMDIST (4,20,1/6,FALSE) will return a value of approximately 0.202.

cell references	See References:
	    relative referencing
	    absolute referencing
	    mixed referencing

Chart feature Insert/Chart...	To make a chart (graph) select the range of x and y data, then choose Insert/Chart.... Be sure to select the column headings too. Step through the Chart Wizard/Expert, supplying the information required. In Corel® Quattro® Pro, you will need to choose an area on the worksheet to put the graph.
combinations function (nCr function) **Microsoft® Excel:** =COMBIN(n,r) **Corel® Quattro® Pro:** @COMB(r,n)	**Microsoft® Excel:** The combinations function is =COMBIN(n,r). **Example:** To find $_{10}C_7$ type =COMBIN(10,7) and press Enter. The result will be 120. **Corel® Quattro® Pro:** The combinations function is @COMB(r,n). **Example:** Notice how the n and r are in counter-intuitive positions. To find $_{10}C_7$ type @COMB(7,10) and press Enter. The result will be 120.

Function or Task	Keystroke(s), Menu, or Screen
CORREL function **Microsoft® Excel:** =CORREL(array1,array2) **Corel® Quattro® Pro:** @CORREL(array1,array2)	The correlation coefficient for two attributes may be calculated using the CORREL function. **Microsoft® Excel:** **Example:** Enter the data as shown below. In cell C1, enter =CORREL(A1:A5,B1:B5). The result should be approximately 0.987. **Corel® Quattro® Pro:** **Example:** Enter the data as shown below. In cell C1, enter @CORREL(A1..A5,B1..B5). The result should be approximately 0.987.

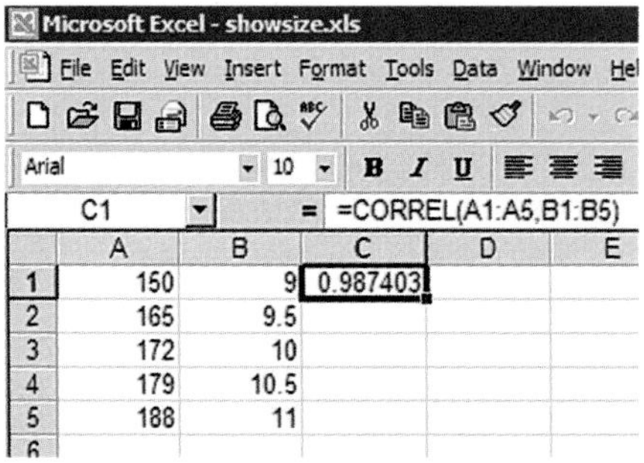

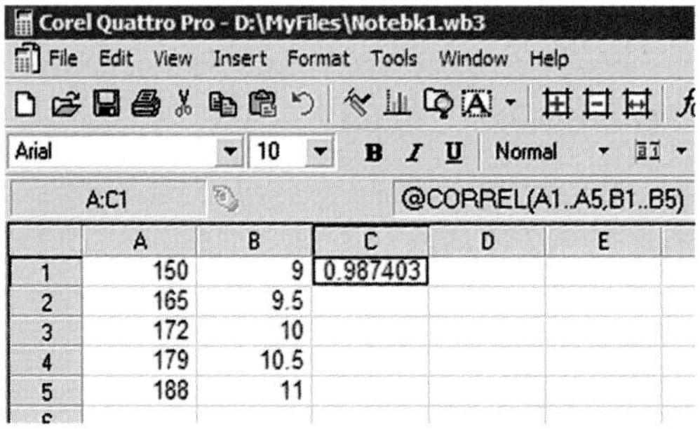

SPREADSHEETS

Function or Task	Keystroke(s), Menu, or Screen
COUNTIF function **Microsoft® Excel:** =COUNTIF(array,value) **Corel® Quattro® Pro:** @COUNTIF(array,value)	The COUNTIF function will count the number of cells in an array that match a value. **Microsoft® Excel:** **Example:** Enter the data as shown below. In cell C1, enter =COUNTIF(B1:B6,"f"). The result should be 4. **Corel® Quattro® Pro:** **Example:** Enter the data as shown below. In cell C1, enter @COUNTIF(B1..B6,"f"). The result should be 4.

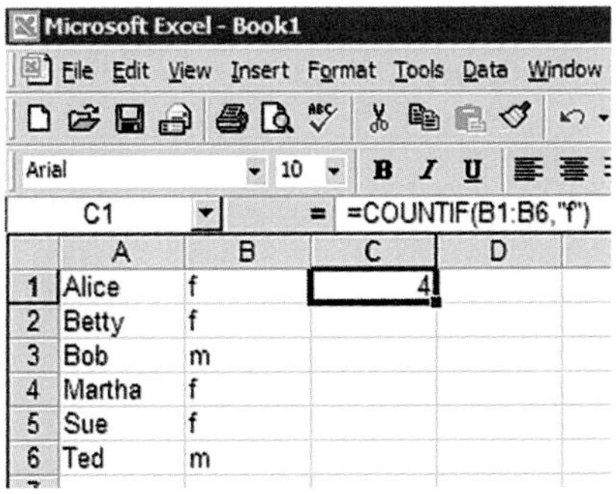

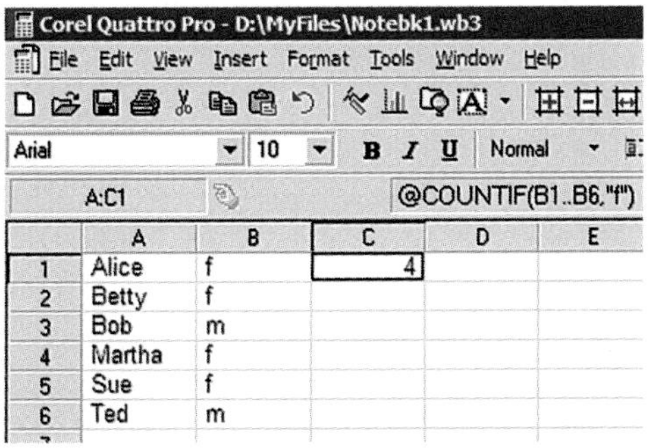

Function or Task	Keystroke(s), Menu, or Screen
FACT(n) function **Microsoft® Excel:** =FACT(n) **Corel® Quattro® Pro:** @FACT(n)	**Microsoft® Excel:** The factorial (!) function is =FACT(n). **Example:** To find 8!, type =FACT(8) and press Enter. The result will be 40 320. *Note*: Microsoft® Excel has a maximum: $170! \doteq 7.3 \times 10^{306}$ **Corel® Quattro® Pro:** The factorial (!) function is @FACT(n). **Example:** To find 8!, type @FACT(8) and press Enter. The result will be 40 320. *Note*: Corel® Quattro® Pro has a maximum: $170! \doteq 7.3 \times 10^{306}$
Fill feature **Microsoft® Excel:** Edit/Fill/Series… **Corel® Quattro® Pro:** Edit/Fill/Fill Series…	Many times you need cells filled with a series of numbers. The series of numbers may be linear or a growth. **Microsoft® Excel:** **Example:** Suppose you need to fill a series of cells with a series, such as 2, 4, 8, … 1 048 576. Enter 1 into cell A1. Now, choose cells A1 through A21 and select Edit/Fill/Series…. Select Growth, enter a step value of 2, and press OK. **Corel® Quattro® Pro:** **Example:** Suppose you need to fill a series of cells with a series, such as 2, 4, 8, … 1 048 576. Enter 1 into cell A1. Now choose cells A1 through A21 and select Edit/Fill/Fill Series…. Enter a starting value of 1, a step value of 2, and a stop value of 1 048 576 (or leave the stop value field blank). Now select Growth and press OK.

Function or Task	Keystroke(s), Menu, or Screen

filtered search

Microsoft® Excel:
Data/Filter/Auto Filter

Corel® Quattro® Pro:
Tools/QuickFilter

Quite often it is necessary to display on the screen cells whose value meet a certain criteria.

Microsoft® Excel:

Example:
Enter and then select the data as shown below:

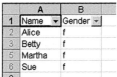

	A	B
1	Name	Gender
2	Alice	f
3	Betty	f
4	Bob	m
5	Martha	f
6	Sue	f
7	Ted	m
8		

Now choose Data/Filter/Auto Filter.

	A	B
1	Name	Gender
2	Alice	f
3	Betty	f
4	Bob	m
5	Martha	f
6	Sue	f
7	Ted	m

Choose the Gender filter by selecting the down arrow beside the word Gender. Choose "f". Now, only the females are displayed.

	A	B
1	Name	Gender
2	Alice	f
3	Betty	f
5	Martha	f
6	Sue	f

Now choose All under the Gender filter. All names are displayed.

To display only the names that start with a letter greater than "C," choose the Name filter. Select Custom…. Fill in the dialog box as follows and click on OK:

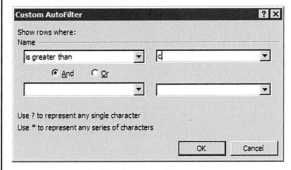

Choose Data/Filter/Auto Filter again to turn off the filtering.

SPREADSHEETS

Function or Task	Keystroke(s), Menu, or Screen
	Corel® Quattro® Pro: **Example:** Enter and then select the data as shown below. 		A	B
1	Name	Gender
2	Alice	f
3	Betty	f
4	Bob	m
5	Martha	f
6	Sue	f
7	Ted	m | Now choose Tools/QuickFilter. |	A	B
1	Name ▼	Gender ▼
2	Alice	f
3	Betty	f
4	Bob	m
5	Martha	f
6	Sue	f
7	Ted	m | Choose the Gender filter by selecting the down arrow beside the word Gender. Choose "f". Now, only the females are displayed. |	A	B
1	Name ▼	Gender ▼
2	Alice	f
3	Betty	f
5	Martha	f
6	Sue	f	 Now choose Show All under the Gender filter. All names are displayed. To display only the names that start with a letter greater than "C," choose the Name filter. Select Custom…. Fill in the dialog box as follows and click on OK: Choose Tools/QuickFilter again to turn off the filtering.

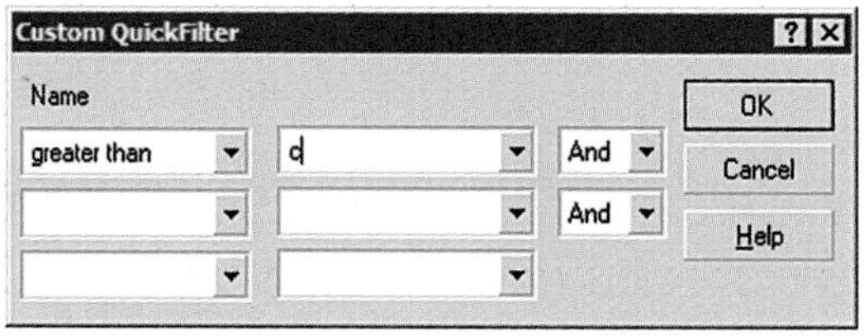

Function or Task	Keystroke(s), Menu, or Screen
Fraction feature **Microsoft® Excel:** Format/Cells…/Fraction **Corel® Quattro® Pro:** Format/Selection… /Numeric Format/Fraction	**Microsoft® Excel:** To display real numbers as fractions, select the cells and then use Format/Cells…/Fraction. Within the dialog box, choose the type of fraction required. **Corel® Quattro® Pro:** To display real numbers as fractions, select the cells and then use Format/Selection…/Numeric Format/Fraction. Within the dialog box, set the denominator required.
INT function **Microsoft® Excel:** =INT(n) **Corel® Quattro® Pro:** @INT(n)	**Microsoft® Excel:** The integer truncation function is =INT(n). **Example:** To convert 8.7 to an integer, type =INT(8.7) and press Enter. The result will be 8. *Note*: The INT function simply removes the decimal portion of the number without rounding. It is recommended that you use the ROUND function if rounding is required. **Corel® Quattro® Pro:** The integer truncation function is @INT(n). **Example:** To convert 8.7 to an integer, type @INT(8.7) and press Enter. The result will be 8. *Note*: The INT function simply removes the decimal portion of the number without rounding. It is recommended that you use the ROUND function if rounding is required.
inverse matrices	See Matrices: inverse
linear regression	See line of best fit

SPREADSHEETS

Function or Task	Keystroke(s), Menu, or Screen
line of best fit	**Microsoft® Excel:** In Microsoft® Excel, set up a table with the data for which you wish to determine the line of best fit. Use the CORREL function to calculate the correlation coefficient. Use the Chart feature to create a scatter plot. Find the line of best fit by selecting Chart/Add Trendline. Check that the default setting is Linear. Select the straight line that appears on your chart, then click Format/Selected Trendline/Options. Check the Display equation on chart box. You can also display r^2. **Corel® Quattro® Pro:** In Corel® Quattro® Pro, set up a table with the data you wish to determine the line of best fit for. Use the CORREL function to calculate the correlation coefficient. Use the Chart feature to create a scatter plot. Find the line of best fit by selecting Tools/Numeric Tools/Regression. Enter the cell ranges for the data, and the program will display regression calculations including the constant (b), the x-coefficient (or slope, a), and r^2.
Matrices: addition and subtraction	**Microsoft® Excel:** Set up your spreadsheet as follows: 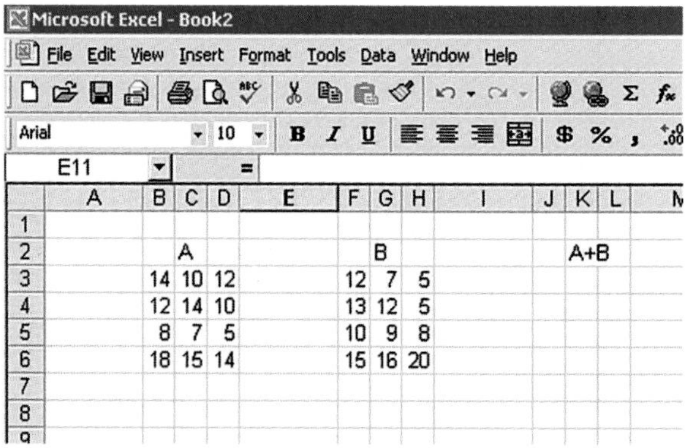

Function or Task	Keystroke(s), Menu, or Screen
	In cell J3, type: =B3+F3 (use "–" for subtraction). Copy this formula across to L3 and then copy this row down to J6 through L6. The result should be as follows: 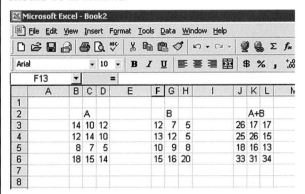 **Corel® Quattro® Pro:** Set up your spreadsheet as follows: 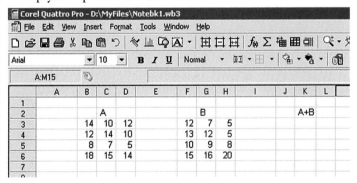 In cell J3, type: +B3+F3 (use "–" for subtraction). Copy this formula across to L3, and then copy this row down to J6 through L6. The result should be as follows: 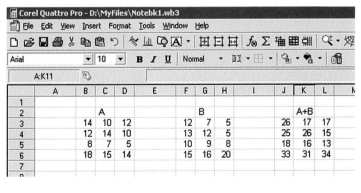

Function or Task	Keystroke(s), Menu, or Screen

Matrices: inverse

Microsoft® Excel:

=INDEX(array, row, col)
=MINVERSE(array)

Corel® Quattro® Pro:

@ARRAY(array)
@MINVERSE(array)

Microsoft® Excel:

In Microsoft® Excel, two functions are required to obtain the inverse of a matrix. The first function is **INDEX** (array, row, col) and the second is **MINVERSE**(array). The MINVERSE function creates a second array that is the inverse of a given array. The INDEX function allows you to specify the row/column of the matrix element to display. That is, you need not display the entire inverse of a matrix, but rather you can choose to display any individual element.

Set up your spreadsheet as follows:

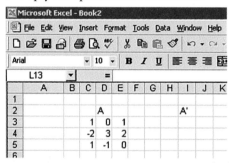

In cell H3, type: =INDEX(MINVERSE($C$3:$E$5),1,1)
In cell I3, type: =INDEX(MINVERSE($C$3:$E$5),1,2)
In cell J3, type: =INDEX(MINVERSE($C$3:$E$5),1,3)

Be sure to use absolute cell referencing as indicated by the "$". Notice how the column number changes as you move from left to right. Copy this row down two more rows. Change the row number for each subsequent row as you did the column number in the examples above. The result should be as follows:

Function or Task	Keystroke(s), Menu, or Screen
	Corel® Quattro® Pro: Corel® Quattro® Pro 8 uses two functions to find the inverse of a matrix, ARRAY(array) and MINVERSE(array). The MINVERSE function creates a second array that is the inverse of a given array. The ARRAY function retrieves the resulting individual row/column elements and displays them in an array. Set up your spreadsheet as follows: 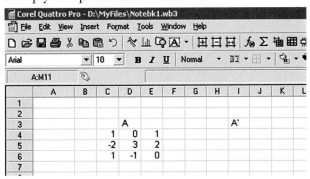 In cell H4, type: @ARRAY(@MINVERSE(C4..E6)) The result should be as follows: 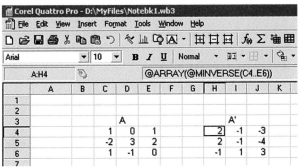 In Corel® Quattro® Pro 9, use Tools/Numeric Tools/Invert instead.

SPREADSHEETS

Function or Task	Keystroke(s), Menu, or Screen

Matrices: multiplication

Microsoft® Excel:

=INDEX(array, row, col)
=MMULT(array1,array2)

Corel® Quattro® Pro:

@ARRAY(array)
@MMULT(array1,array2)

Microsoft® Excel:

In Microsoft® Excel, two functions are required to multiply matrices. The first function is INDEX (array, row, col) and the second is MMULT(array1,array2). The MMULT function creates a third array that is the result of multiplying two matrices. The INDEX function allows you to specify the row/column of the matrix element to display. That is, you need not display the entire inverse of a matrix, but rather you can choose to display any individual element.

Set up your spreadsheet as follows:

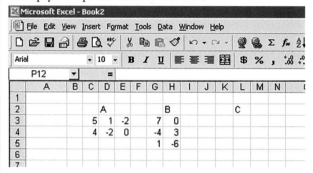

In cell K3, type: =INDEX(MMULT($C$3:$E$4,$G$3:$H$5),1,1)
In cell L3, type: =INDEX(MMULT($C$3:$E$4,$G$3:$H$5),1,2)

Be sure to use absolute cell referencing as indicated by the "$". Notice how the column number changes as you move from left to right. Copy this row down one more row. Change the row number for this subsequent row as you did the column number in the examples above. The result should be as follows:

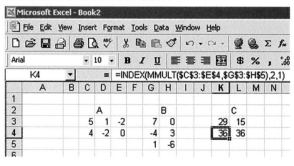

Function or Task	Keystroke(s), Menu, or Screen
	Corel® Quattro® Pro: Corel® Quattro® Pro 8 uses two functions to multiply matrices, ARRAY(array) and MMULT(array1,array2). The MMULT function creates a third array that is the result of multiplying two matrices. The ARRAY function retrieves the resulting individual row/column elements and displays them in an array. Set up your spreadsheet as follows: 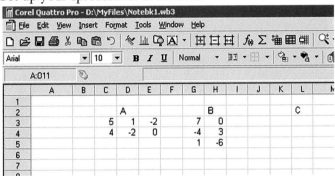 In cell K3, type: @ARRAY(@MMULT(C3..E4,G3..H5)) The result should be as follows: 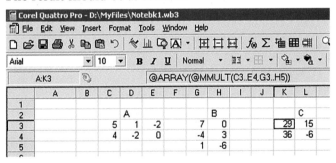 In Corel® Quattro® Pro 9, use Tools/Numeric Tools/Multiply instead.

Matrices: scalar
multiplication

Microsoft® Excel:

Set up your spreadsheet as follows:

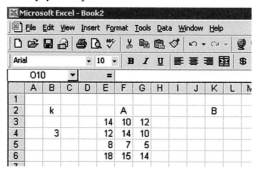

In cell J3, type: =$B$4*E3. Be sure to use absolute cell referencing as indicated by the "$". Copy this formula across to L3. Now copy J3 to L3 down to J6 to L6. The result should be as follows:

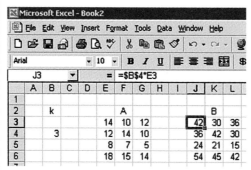

Corel® Quattro® Pro:

Set up your spreadsheet as follows:

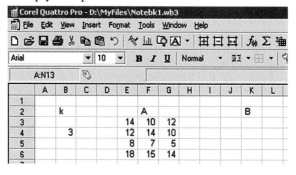

Function or Task	Keystroke(s), Menu, or Screen

In cell J3, type: +$B$4*E3. Be sure to use absolute cell referencing as indicated by the "$". Copy this formula across to L3. Now copy J3 to L3 down to J6 to L6. The result should be as follows:

	A	B	C	D	E	F	G	H	I	J	K	L
1												
2		k				A					B	
3					14	10	12			42	30	36
4		3			12	14	10			36	42	30
5					8	7	5			24	21	15
6					18	15	14			54	45	42

Matrices: storing

Microsoft® Excel:

You store a matrix in Microsoft® Excel as you would any array. Simply enter the matrix (array) into whichever cells you wish to use. A sample matrix (array) is shown:

	A	B	C	D	E
1					
2			A		
3		14	10	12	
4		12	14	10	
5		8	7	5	
6		18	15	14	

Corel® Quattro® Pro:

You store a matrix in Corel® Quattro® Pro as you would with any array. Simply enter the matrix (array) into whichever cells you wish to use. A sample matrix (array) is shown:

	A	B	C	D	E
1					
2			A		
3		14	10	12	
4		12	14	10	
5		8	7	5	
6		18	15	14	
7					

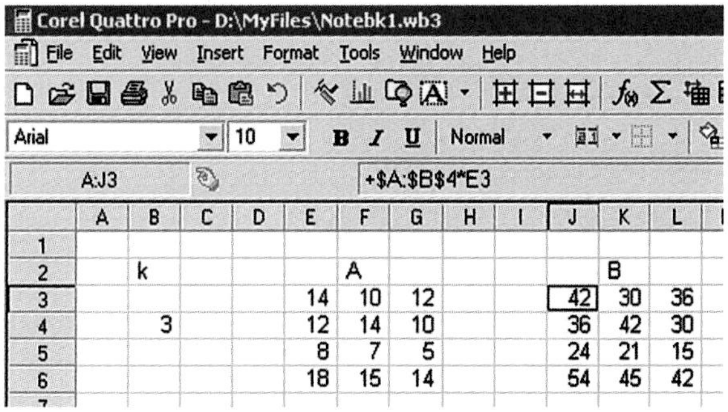

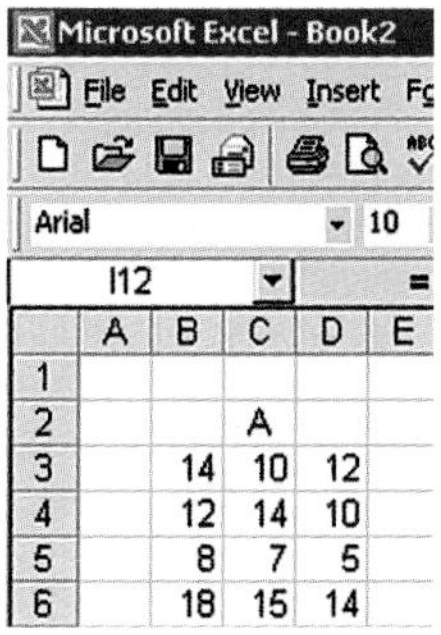

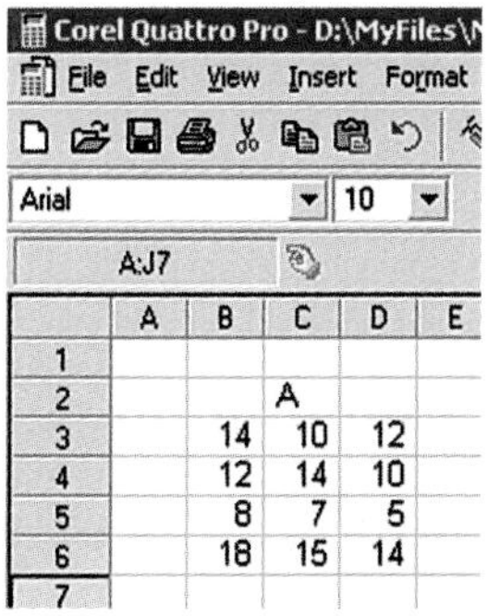

SPREADSHEETS

Function or Task	Keystroke(s), Menu, or Screen

Matrices: transpose

Microsoft® Excel:
Edit/Paste Special...

Corel® Quattro® Pro:
Edit/Paste Special...

Microsoft® Excel:

Set up your spreadsheet as follows:

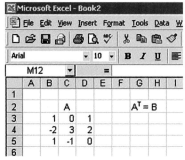

Select and copy the matrix.
Choose the location for the transpose of the matrix.
From the Edit menu choose Paste Special... .
In the Paste Special... dialog box choose Transpose and then OK.
The result should be as follows:

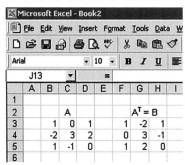

Corel® Quattro® Pro:

Set up your spreadsheet as follows:

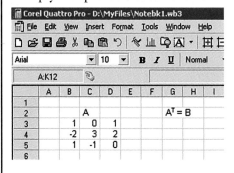

Function or Task	Keystroke(s), Menu, or Screen
	Select and copy the matrix. Choose the location for the transpose of the matrix. From the Edit menu choose Paste Special…. In the Paste Special… dialog box choose Transpose Rows and Columns and then OK. The result should be as follows: 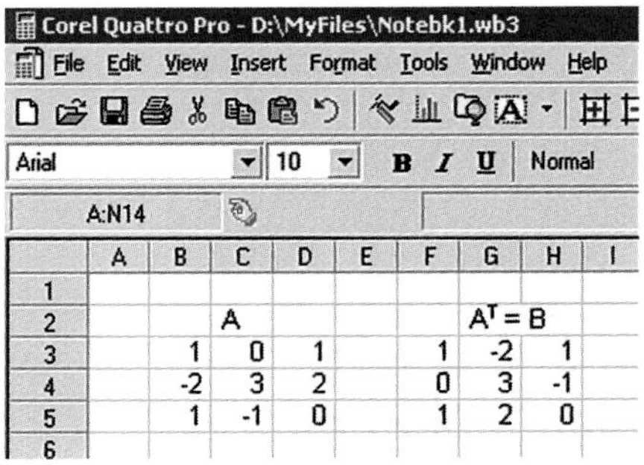 Corel Quattro Pro - D:\MyFiles\Notebk1.wb3 File Edit View Insert Format Tools Window Help Arial 10 **B** *I* U Normal A:N14 |	A	B	C	D	E	F	G	H	I
---	---	---	---	---	---	---	---	---
1									
2			A				A^T = B		
3		1	0	1		1	-2	1	
4		-2	3	2		0	3	-1	
5		1	-1	0		1	2	0	
6										
MAX function **Microsoft® Excel:** =MAX(array) **Corel® Quattro® Pro:** @MAX(array)	**Microsoft® Excel:** The maximum value function is =MAX(array). **Example:** To determine the maximum value in a series of cells such as from A1 to A15, enter =MAX(A1:A15) and press Enter. **Corel® Quattro® Pro:** The maximum value function is @MAX(array). **Example:** To determine the maximum value in a series or matrix such as from cell A1 to A15, enter @MAX(A1..A15) and press Enter.
matrix operations	See Matrices: addition and subtraction Matrices: inverse Matrices: multiplication Matrices: scalar multiplication Matrices: storing Matrices: transpose

SPREADSHEETS

Function or Task	Keystroke(s), Menu, or Screen
mean **Corel® Quattro® Pro:** @MEAN(array) See also average.	**Microsoft® Excel:** See average. **Corel® Quattro® Pro:** The mean value function is @MEAN(array). **Example:** To determine the mean value in a series of cells such as from A1 to A15, enter @MEAN(A1..A15) and press Enter.
median **Microsoft® Excel:** =MEDIAN(array) **Corel® Quattro® Pro:** @MEDIAN(array)	**Microsoft® Excel:** The median function is =MEDIAN(array). **Example:** To find the median of 6, 7, 8, 9, and 10, type =MEDIAN(6,7,8,9,10) and press Enter. The result will be 8. To find the median of cells B1 through B10, type =MEDIAN(B1:B10) and press Enter. **Corel® Quattro® Pro:** The median function is @MEDIAN(array). **Example:** To find the median of 6, 7, 8, 9, and 10, type @MEDIAN(6,7,8,9,10) and press Enter. The result will be 8. To find the median of cells B1 through B10, type @MEDIAN(B1..B10) and press Enter.
mode **Microsoft® Excel:** =MODE(array) **Corel® Quattro® Pro:** @MODE(array)	**Microsoft® Excel:** The mode function is =MODE(array). **Example:** To find the mode of 6, 7, 8, 9, and 10, type =MODE(6,7,8,8,9,10) and press Enter. The result will be 8. To find the mode of cells B1 through B10, type =MODE(B1:B10) and press Enter.

SPREADSHEETS

Function or Task	Keystroke(s), Menu, or Screen
	Corel® Quattro® Pro: The mode function is @MODE(array). **Example:** To find the mode of 6, 7, 8, 9, and 10, type @MODE(6,7,8,8,9,10) and press Enter. The result will be 8. To find the mode of cells B1 through B10, type @MODE(B1..B10) and press Enter.
multiplying matrices	See Matrices: multiplication Matrices: scalar multiplication
NORMDIST function **Microsoft® Excel:** =NORMDIST(boundary, mean, standard deviation,TRUE) **Corel® Quattro® Pro:** @NORMDIST(boundary, mean, standard deviation,1)	**Microsoft® Excel:** The NORMDIST(boundary, mean, standard deviation,TRUE) function allows you to calculate the probability that a given data point lies within a boundary using a normal distribution with a given mean and standard deviation. To calculate the probability that a given data point lies between a *lower boundary* and an *upper boundary*, set up your spreadsheet as follows:

Microsoft Excel - NormalDist.xls

File Edit View Insert Format Tools Data Window

Arial ▾ 10 ▾ **B** *I* U ≡ ≡ ≡

E15 ▾ =

	A	B	C	D
1				
2		lowerbound	90	
3		upperbound	115	
4		mean	100	
5		standard deviation	10	
6				
7				
8		upperbound %		
9		lowerbound %		
10		difference		
11				

In cell C8, type: =NORMDIST(C3,C4,C5,TRUE) to find the *upper boundary* probability.
In cell C9, type: =NORMDIST(C2,C4,C5,TRUE) to find the *lower boundary* probability.
In cell C10, type: =C8-C9 to find the probability that a given data point lies between a *lower boundary* and an *upper boundary*.

SPREADSHEETS

Function or Task	Keystroke(s), Menu, or Screen
	The result should be as follows:

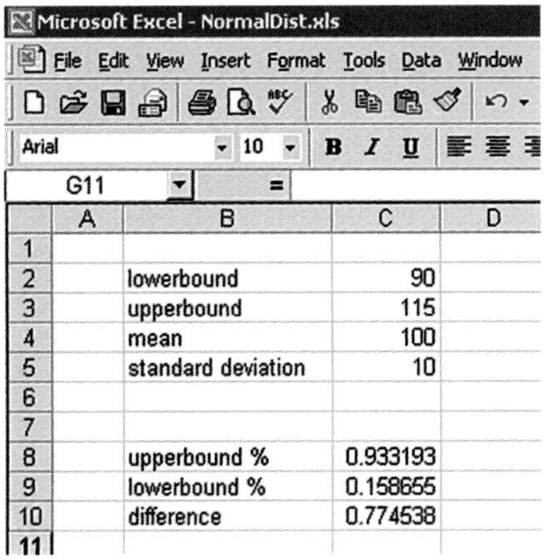

Corel® Quattro® Pro:

The @NORMDIST(boundary, mean, standard deviation,1) function allows you to calculate the probability that a given data point lies within a boundary using a normal distribution with a given mean and standard deviation.

To calculate the probability that a given data point lies between a *lower boundary* and an *upper boundary*, set up your spreadsheet as follows:

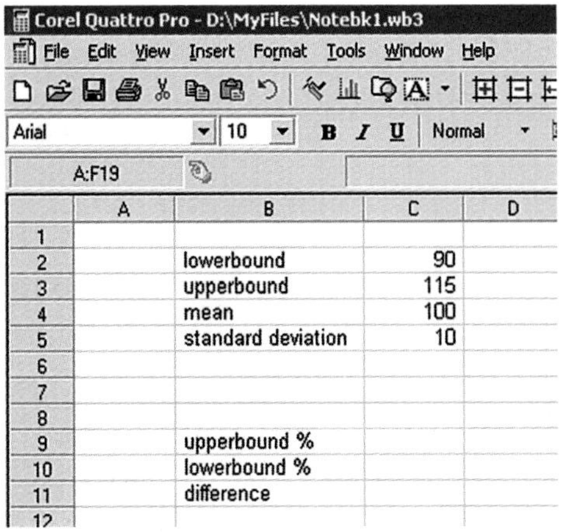

In cell C9, type: @NORMDIST(C3,C4,C5,1) to find the *upper boundary* probability.
In cell C10, type: @NORMDIST(C2,C4,C5,1) to find the *lower boundary* probability.
In cell C11, type: +C9-C10 to find the probability that a given data point lies between a *lower boundary* and an *upper boundary*.

Function or Task	Keystroke(s), Menu, or Screen
	The result should be as follows: 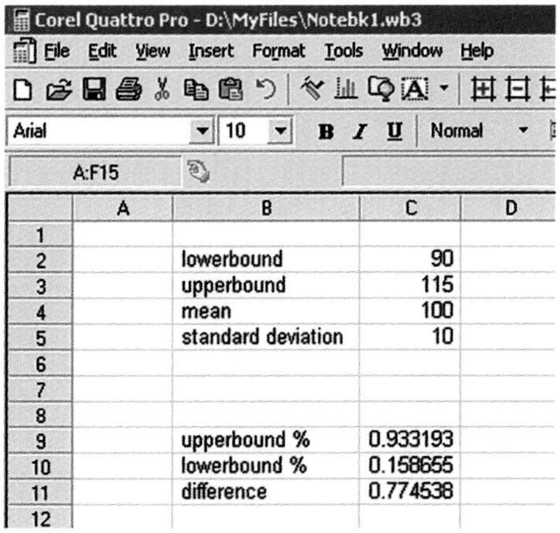 Corel Quattro Pro - D:\MyFiles\Notebk1.wb3 File Edit View Insert Format Tools Window Help Arial 10 B I U Normal A:F15 |	A	B	C	D
---	---	---	---	---
1				
2		lowerbound	90	
3		upperbound	115	
4		mean	100	
5		standard deviation	10	
6				
7				
8				
9		upperbound %	0.933193	
10		lowerbound %	0.158655	
11		difference	0.774538	
12					
permutations function (nPr function) **Microsoft® Excel:** =PERMUT(n,r) **Corel® Quattro® Pro:** @PERMUT(n,r)	**Microsoft® Excel:** The permutations function for Microsoft® Excel is =PERMUT(n,r). **Example:** To find $_{10}P_7$, type =PERMUT(10,7) and press **Enter**. The result will be 604 800. **Corel® Quattro® Pro:** The permutations function for Corel® Quattro® Pro is @PERMUT(*n,r*). Notice that, unlike the combinations function, the *n* and *r* are in intuitive positions. **Example:** To find $_{10}P_7$, type @PERMUT(10,7) and press **Enter**. The result will be 604 800.
RAND function	See random integers random real numbers

Function or Task	Keystroke(s), Menu, or Screen
random integers **Microsoft® Excel:** =lower+round(diff*rand(),0) or =RANDBETWEEN(lower, upper) **Corel® Quattro® Pro:** @RANDBETWEEN(lower, upper)	**Microsoft® Excel:** To generate random integers, use the formula =lower+round(diff*rand(),0). The variable *diff* = *upper* − *lower*. **Example:** To generate a random integer from 6 to 10, type =6+round(4*rand(),0) You can copy this formula to other cells to generate more random integers. *Note:* You can use the RANDBETWEEN(lower,upper) function only if you have installed the Analysis ToolPak. If this function is not available, run the Setup program to install the Analysis ToolPak. After you install the Analysis ToolPak, you must enable it by using the Add-Ins command on the Tools menu. **Corel® Quattro® Pro:** To generate random integers, use the formula @RANDBETWEEN(lower,upper) **Example:** To generate a random integer from 6 to 10, type @RANDBETWEEN(6,10) You can copy the formula to other cells to generate more random integers.
random real numbers **Microsoft® Excel:** =RAND() **Corel® Quattro® Pro:** @RAND	**Microsoft® Excel:** To generate random real numbers, use the formula =RAND(). There is no argument for this function. Simply type the function into any cell. =RAND() will generate a real number from 0 to 1. =6*RAND() will generate a real number from 0 to 6. **Corel® Quattro® Pro:** To generate random real numbers, use the formula @RAND. There is no argument for this function. Simply type the function into any cell. @RAND will generate a real number from 0 to 1. @RAND*6 will generate a real number from 0 to 6.

Function or Task	Keystroke(s), Menu, or Screen
reference data from cells in another worksheet	To reference data that exist in another worksheet or file is very similar to accessing data on a single worksheet. **Microsoft® Excel:** **Example:** Suppose that you have two worksheets. Set cell B2 in Sheet 2 equal to cell A1 in Sheet 1. Select cell B2 on Sheet 2 (where the data are going). Press "=". Select Sheet 1 and then cell A1 and press Enter. Cell B2 on Sheet 2 will now be equal to cell A1 on Sheet 1. The formula in cell B2 on Sheet 2, will be =Sheet1!A1. **Corel® Quattro® Pro:** **Example:** Suppose that you have two worksheets. Set cell B2 in Sheet B equal to cell A1 in Sheet A. Select cell B2 on Sheet B (where the data are going). Press "+". Select Sheet A and then cell A1 and press Enter. Cell B2 on Sheet B will now be equal to cell A1 on Sheet A. The formula in cell B2 on Sheet B will be +A:A1.
References: relative referencing absolute referencing mixed referencing	In spreadsheets there are two types of referencing. *Relative referencing*: These are references to cells that are relative to the position of the formula. *Absolute referencing*: These are references to cells that always refer to a cell's specific location. Depending on the task you wish to perform in a spreadsheet, you can use either type of referencing. If a dollar sign precedes the letter or the number, such as $B$7, the column or row reference is absolute. *Note*: Relative cell references automatically adjust when you copy them, while absolute cell references always point at the same cell. *Mixed referencing*: If you copy the following reference down a column, B$7, then the resulting formulas will always refer to cell B7, because the row has an absolute reference. But, if you copy the formula across a row then the resulting formulas will reference row 7 absolutely and will reference column B relatively.

Function or Task	Keystroke(s), Menu, or Screen
relative cell referencing	See References: 　relative referencing 　absolute referencing 　mixed referencing
ROUND function **Microsoft® Excel:** =ROUND(n,d) **Corel® Quattro® Pro:** @ROUND(n,d)	**Microsoft® Excel:** The rounding function is =ROUND(n,d). **Example:** To round 8.787 to the nearest tenth, type =ROUND(8.787,1) and press Enter. The result will be 8.8. **Corel® Quattro® Pro:** The rounding function is @ROUND(n,d). **Example:** To round 8.787 to the nearest tenth, type @ROUND(8.787,1) and press Enter. The result will be 8.8.
search	The search function is accessed by choosing Find... from the Edit menu or by pressing CTRL-F.
Sort feature **Microsoft® Excel:** Data/Sort.. **Corel® Quattro® Pro:** Tools/Sort..	**Microsoft® Excel:** The sort feature is available through the Data menu. Select the range of cells you wish to sort, then choose Data/Sort.... **Corel® Quattro® Pro:** The sort feature is available through the Tools menu. Select the range of cells you wish to sort then choose Tools/Sort....

Function or Task	Keystroke(s), Menu, or Screen
standard deviation **Microsoft® Excel:** *Population* =STDEVP(array) *Sample* =STDEV(array) **Corel® Quattro® Pro:** *Population* @STD(array) *Sample* @STDS(array)	**Microsoft® Excel:** The standard deviation function is =STDEV(array). **Example:** Determine the standard deviation of a sample listed in cells from A1 to A15. Enter =STDEV(A1:A15) and press Enter. **Corel® Quattro® Pro:** The standard deviation function is @STDS(array). **Example:** Determine the standard deviation of a sample listed in cells from A1 to A15. Enter @STDS(A1..A15) and press Enter.
SUM function **Microsoft® Excel:** =SUM(array) **Corel® Quattro® Pro:** @SUM(array)	**Microsoft® Excel:** **Example:** Determine the sum of a series of cells such as from A1 to A15. Enter =SUM(A1:A15) and press Enter. **Corel® Quattro® Pro:** **Example:** Determine the sum of a series of cells such as from A1 to A15. Enter @SUM(A1..A15) and press Enter.
Variance **Microsoft® Excel:** *Population* =VARP(array) *Sample* =VAR(array) **Corel® Quattro® Pro:** *Population* @VAR(array) *Sample* @VARS(array)	**Microsoft® Excel:** The variance function is =VAR(array). **Example:** To find the sample variance of 6, 7, 8, 9, and 10, type =VAR(6,7,8,9,10) and press Enter. The result will be 1.142857. To find the sample variance of cells B1 through B10, type =VAR(B1:B10) and press Enter. **Corel® Quattro® Pro:** The variance function is @VAR(array). **Example:** To find the sample variance of 6, 7, 8, 9, and 10, type @VARS(6,7,8,9,10) and press Enter. The result will be 1.142857. To find the sample variance of cells B1 through B10, type @VARS(B1..B10) and press Enter.

SPREADSHEETS

Fathom™

Function or Task	Keystroke(s), Menu, or Screen
binomialCumulative() **function** binomialCumulative(x,n,p, min,max)	The binomialCumulative() function is found under the Functions/Distributions/Binomial menu. This function returns the probability of getting x or fewer successes in n trials of a binomial distribution where the probability of success on each trial is p. Normally, x will take on values from 0 to n. However, if min and max are specified, then the value of x will go from min to max in steps of (max − min)/n. **Example:** Consider the rolling of two dice 20 times. What is the probability of rolling up to four doubles? In this case, $x = 4$, $n = 20$ and $p = \dfrac{1}{6}$. Hence, binomialCumulative(4,20,1/6) will return a value of approximately 0.769. You can use this function in a **case table** to create a table of cumulative probabilities as shown in the following screen: 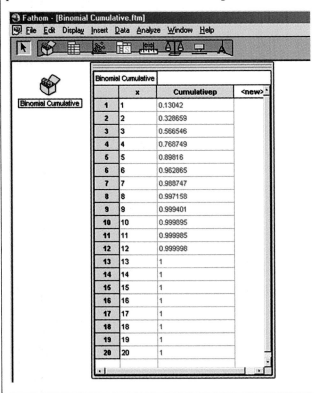

Function or Task	Keystroke(s), Menu, or Screen
binomialProbability() **function** binomialProbability(x,n,p, min,max)	The binomialProbability() function is found under the Functions/Distributions/Binomial menu. It returns the probability of getting exactly x successes in n trials of a binomial distribution, where the probability of success on each trial is p. **Example:** Consider 20 rolls of 2 dice. What is the probability of rolling exactly four doubles? In this case, $x = 4$, $n = 20$, and $p = \dfrac{1}{6}$. Hence, binomialProbability(4,20,1/6) will return a value of approximately 0.202. You can use this function in a **case table** to create a table of probabilities as shown in the following screen: 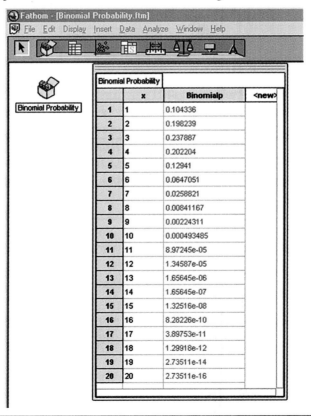

FATHOM™

Function or Task	Keystroke(s), Menu, or Screen
correlation coefficient	The correlation coefficient for two attributes may be calculated using the correlation function under the Functions/Statistical/Two Attributes menu.

Example:

Create a **collection** and **case table** for Height versus ShoeSize, as shown in the screen shot below. Double click on the **collection** box to open the **inspector,** and select the Measures tab. Create a <new> measure called Correlation. Right-click under the Formula column in the Correlation row, and select Edit Formula. Select the correlation function under the Functions/Statistical/Two Attributes menu, and type Height,ShoeSize between the brackets. Click OK. Note that you get a correlation coefficient of about 0.987 between Height and ShoeSize.

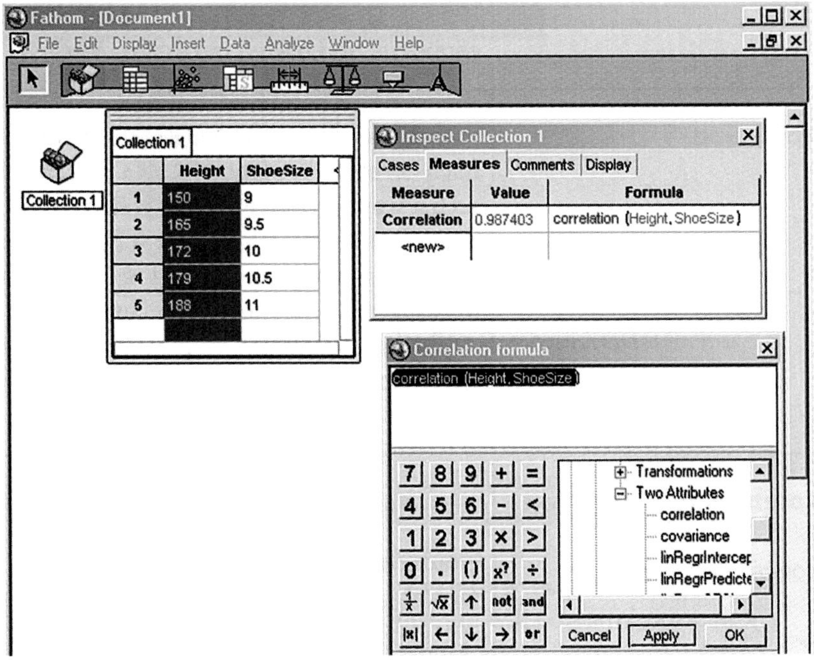

FATHOM™

Function or Task	Keystroke(s), Menu, or Screen
binomialProbability() **function** binomialProbability(x,n,p, min,max)	The binomialProbability() function is found under the Functions/Distributions/Binomial menu. It returns the probability of getting exactly x successes in n trials of a binomial distribution, where the probability of success on each trial is p. **Example:** Consider 20 rolls of 2 dice. What is the probability of rolling exactly four doubles? In this case, $x = 4$, $n = 20$, and $p = \frac{1}{6}$. Hence, binomialProbability(4,20,1/6) will return a value of approximately 0.202. You can use this function in a **case table** to create a table of probabilities as shown in the following screen:

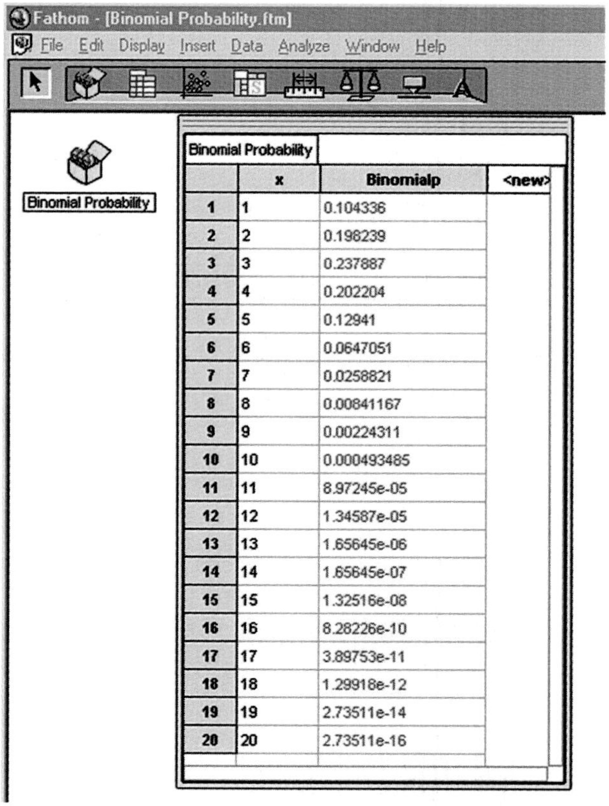

Fathom - [Binomial Probability.ftm]

File Edit Display Insert Data Analyze Window Help

Binomial Probability

	x	Binomialp	<new>
1	1	0.104336	
2	2	0.198239	
3	3	0.237887	
4	4	0.202204	
5	5	0.12941	
6	6	0.0647051	
7	7	0.0258821	
8	8	0.00841167	
9	9	0.00224311	
10	10	0.000493485	
11	11	8.97245e-05	
12	12	1.34587e-05	
13	13	1.65645e-06	
14	14	1.65645e-07	
15	15	1.32516e-08	
16	16	8.28226e-10	
17	17	3.89753e-11	
18	18	1.29918e-12	
19	19	2.73511e-14	
20	20	2.73511e-16	

FATHOM™

Function or Task	Keystroke(s), Menu, or Screen
caseIndex function	The caseIndex function located under Functions/Special menu is like the "row number" in a spreadsheet. **Example:** Suppose that you want an attribute which runs from 1 to 20, perhaps to be used in the calculation of another attribute. Start Fathom and open a new document if necessary. Drag a **collection** box to the workspace. Drag a **case table** to the workspace. Double-click on \<new\> and rename it Index. Right-click on the **case table**, select New Cases and type in 20. Notice that the Index attribute is now numbered from 1 to 20, as shown in the following screen: 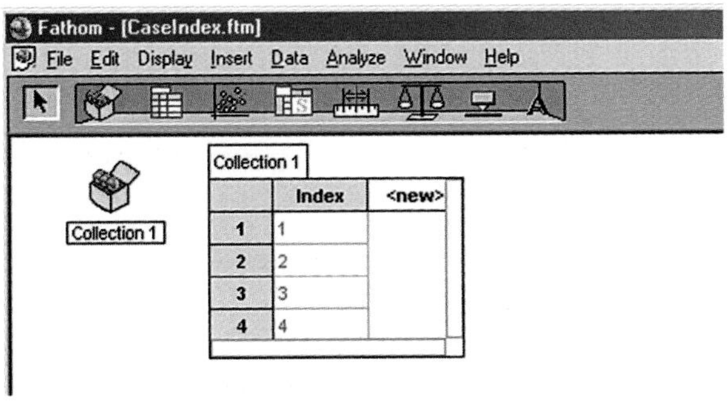
case table	Fathom keeps data for a particular **collection** in a **case table** which is linked to that **collection**. Run Fathom and open a new document, if necessary. Drag a **collection** box to the workspace. Note that it is called Collection 1. Now, drag a case table to the workspace. Notice that it is linked to Collection 1. Now that you have a **case table**, you can specify the attributes you want to keep track of. Suppose that this collection will contain information about the CDs that you own. One attribute might be the title of the CD. Double-click on \<new\>, and type in Title. Under Title, you can type the CD titles that you own. ▼

Function or Task	Keystroke(s), Menu, or Screen
	Your screen should look like this: 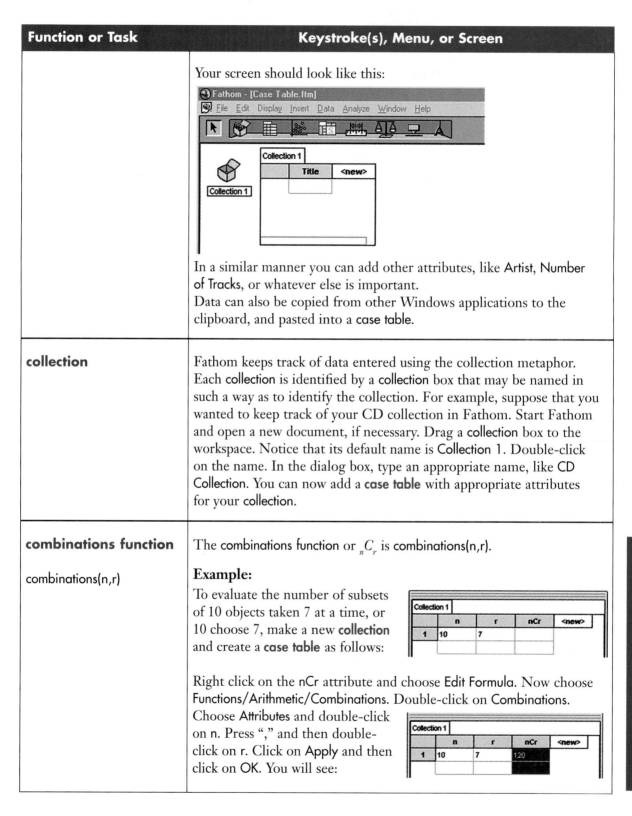 In a similar manner you can add other attributes, like Artist, Number of Tracks, or whatever else is important. Data can also be copied from other Windows applications to the clipboard, and pasted into a **case table**.
collection	Fathom keeps track of data entered using the collection metaphor. Each **collection** is identified by a **collection** box that may be named in such a way as to identify the collection. For example, suppose that you wanted to keep track of your CD collection in Fathom. Start Fathom and open a new document, if necessary. Drag a **collection** box to the workspace. Notice that its default name is **Collection 1**. Double-click on the name. In the dialog box, type an appropriate name, like **CD Collection**. You can now add a **case table** with appropriate attributes for your **collection**.
combinations function combinations(n,r)	The **combinations** function or $_nC_r$ is combinations(n,r). **Example:** To evaluate the number of subsets of 10 objects taken 7 at a time, or 10 choose 7, make a new **collection** and create a **case table** as follows: Right click on the nCr attribute and choose **Edit Formula**. Now choose **Functions/Arithmetic/Combinations**. Double-click on **Combinations**. Choose **Attributes** and double-click on n. Press "," and then double-click on r. Click on **Apply** and then click on **OK**. You will see:

FATHOM™

Function or Task	Keystroke(s), Menu, or Screen
correlation coefficient	The correlation coefficient for two attributes may be calculated using the correlation function under the Functions/Statistical/Two Attributes menu.

Example:

Create a **collection** and **case table** for Height versus ShoeSize, as shown in the screen shot below. Double click on the **collection** box to open the **inspector,** and select the Measures tab. Create a <new> measure called Correlation. Right-click under the Formula column in the Correlation row, and select Edit Formula. Select the correlation function under the Functions/Statistical/Two Attributes menu, and type Height,ShoeSize between the brackets. Click OK. Note that you get a correlation coefficient of about 0.987 between Height and ShoeSize.

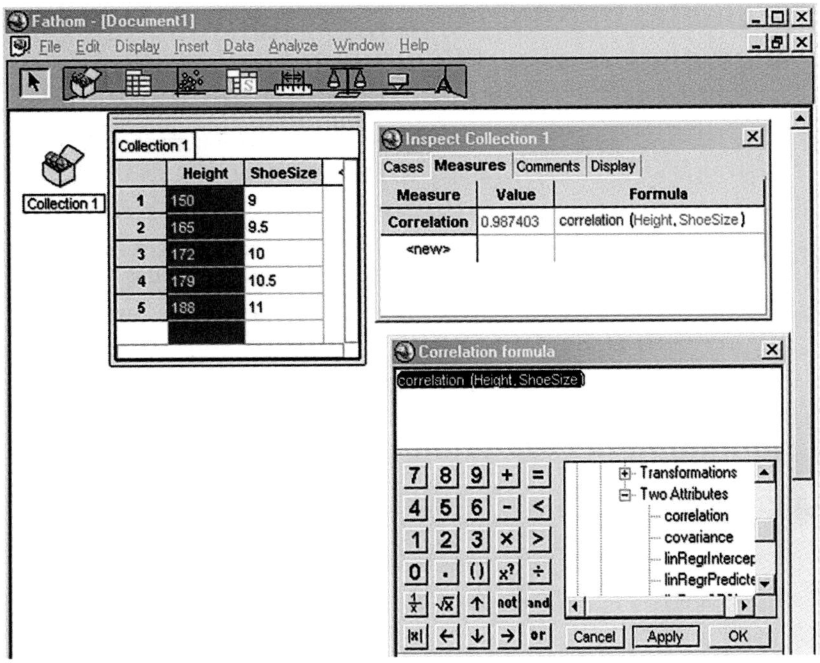

FATHOM™

Function or Task	Keystroke(s), Menu, or Screen
count function	When given a list of data for an attribute, Fathom™ can count how many times a specific condition occurs using the count function under Functions/Statistical/One Attribute menu. For example, suppose that you have a **collection** of data on the students in your school, and you would like to count how many have an entry of 16 under the attribute Age. To see how this function works, run Fathom™ and open a new document if necessary. Drag a **collection** box and then a **case table** to the workspace. Rename the <new> attribute to Age, and enter ages of 14, 15, 16, 15, 17, 16, 15, 18, 16, and 17. Double-click on the **collection** box to open the **inspector**. Select the Measures tab, and rename <new> to Age16. Right-click in the Formula column in the Age16 row, and select the count function under Functions/Statistical/One Attribute menu. Between the brackets, type in the condition Age=16. Click on OK. Notice that the value changes to 3, the number of occurrences of age 16, as shown in the following screen: 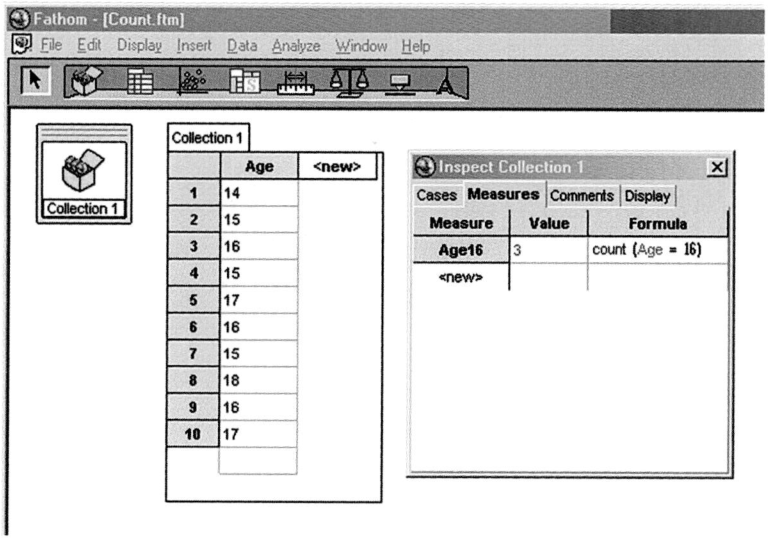

FATHOM™

Function or Task	Keystroke(s), Menu, or Screen
filter	A **filter** may be added to an object in order to select only a subset of the data that is of interest. **Example:** Create a **collection** and a **case table** as shown in the screen shot below: 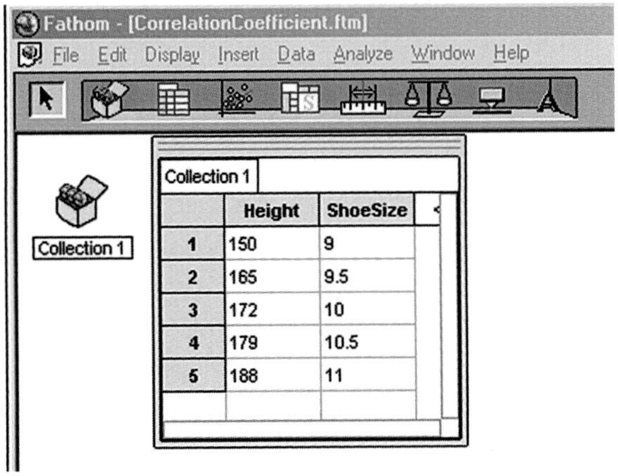 Suppose that you want to show only shoe sizes greater than 10. Click on the **case table** to select it, and select Add Filter from the Data menu. Then, type in the condition ShoeSize>10 and click on OK. You will see: 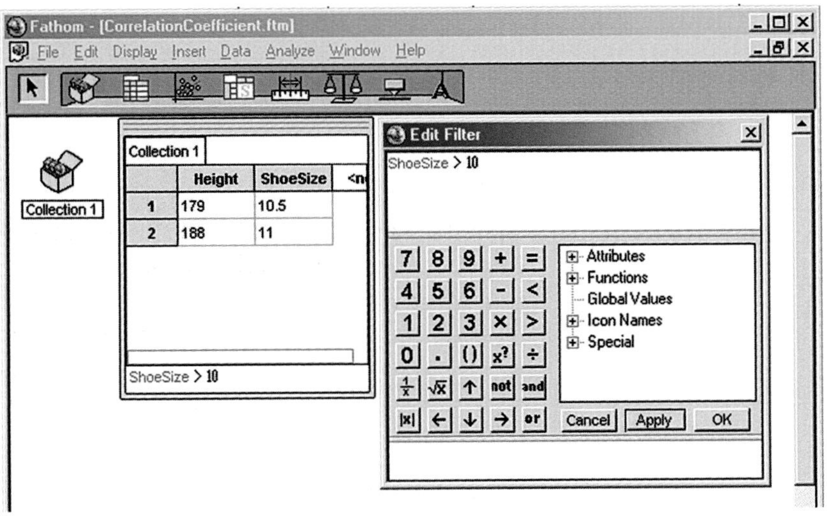

FATHOM™

Function or Task	Keystroke(s), Menu, or Screen
graph icon	The **graph icon** is the third selection on the shelf, to the right of the **collection** box icon and the **case table** icon, as shown in the screenshot below: 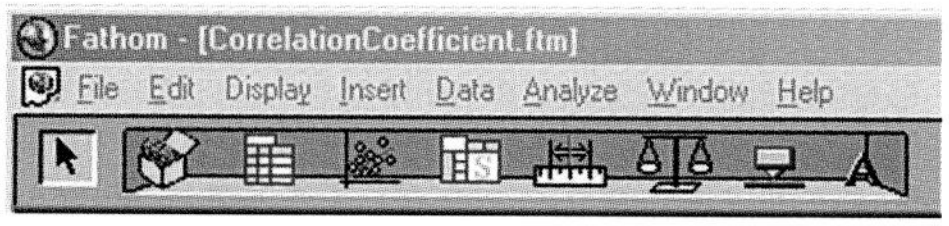 Fathom - [CorrelationCoefficient.ftm] File Edit Display Insert Data Analyze Window Help You can create a new graph by clicking on the **graph icon**, holding the left mouse button down, and dragging it onto the workspace. You can then drag attributes from a **case table** to the axes of the graph.
inspector	The **inspector** for a **collection** can be opened by double-clicking on the **collection** box. A new window will appear, with several panes. The first pane is the **Cases** pane, as shown in the following screen: 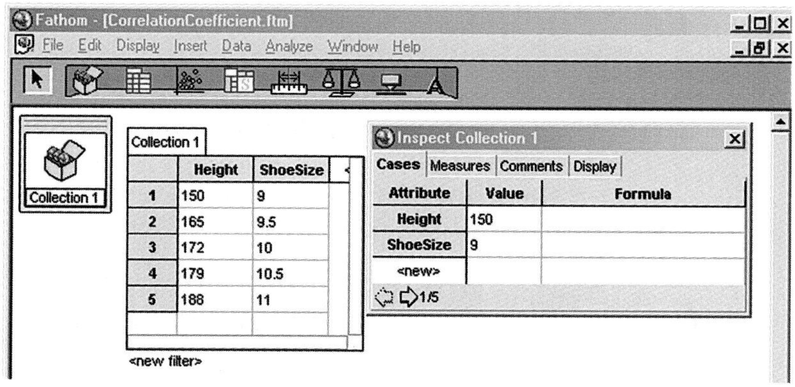 This pane allows you to inspect the **collection** case by case, and is particularly useful for collections that have many attributes. The second pane is the **Measures** pane, which allows you to define measures for the **collection**, such as the **mean**, and **correlation coefficient**, some of which are described elsewhere in this appendix. 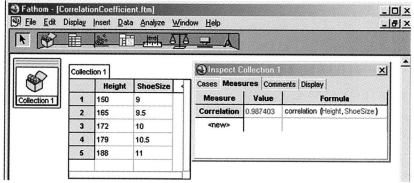

Function or Task	Keystroke(s), Menu, or Screen
	The third pane is the Comments pane, which allows you to add comments relevant to the collection, as shown below:
	The last pane is the Display pane, which allows you to control how the data for a collection appears on the screen. More detail on how to program this feature is available in the *Fathom™ Reference Manual*.
interquartile range iqr(attribute)	The interquartile range function found under the Functions/Statistical/One Attribute menu is used to calculate the interquartile range for an attribute. **Example:** Create a collection and case table as shown in the following screen:
	Double-click on the collection box to open the inspector for the collection, and select the Measures tab. Rename <new> to InterQuartile. Right-click on the Formula column for the InterQuartile measure, and select Edit Formula. Select the iqr() function under the Functions/Statistical/One Attribute menu, and type Marks between the brackets. You will see the interquartile range calculated under Value, as in the screen shown above.

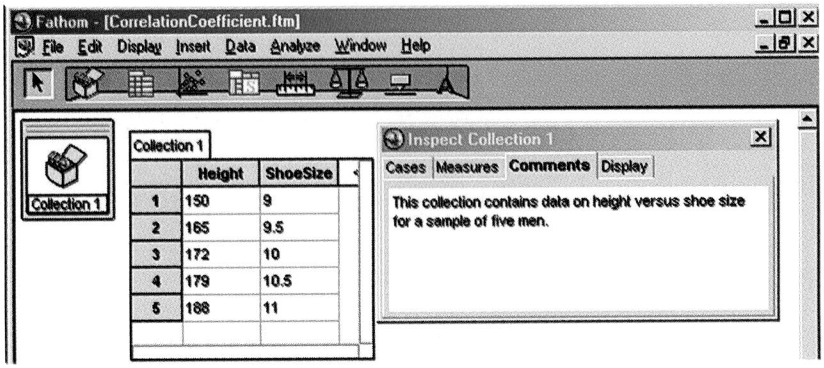

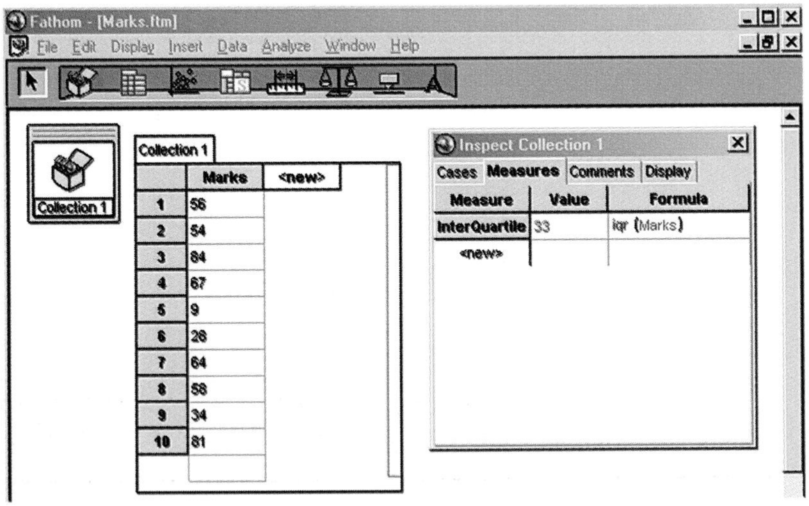

FATHOM™

Function or Task	Keystroke(s), Menu, or Screen
linear regression	Fathom™ generates a **linear regression** line and values when it graphs data. Set up a new **collection** as follows: 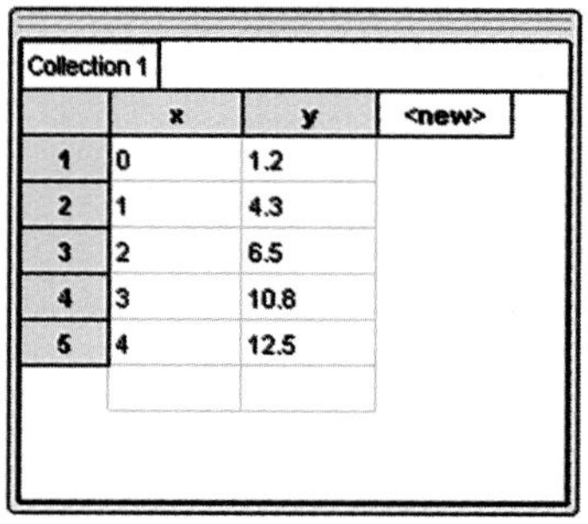 Graph the data as follows: 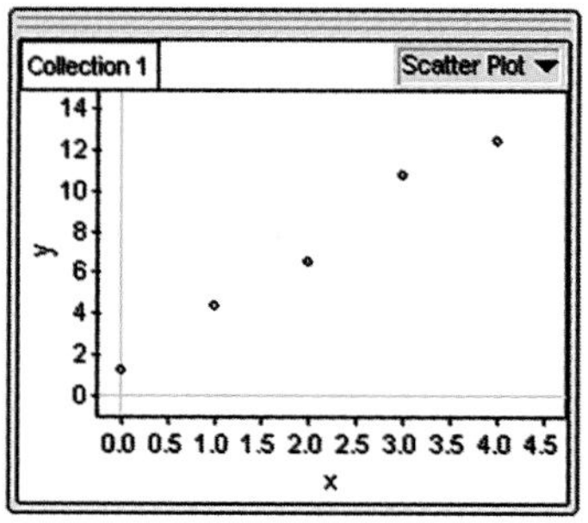 Right-click on the graph and choose **Least-Squares Line**. 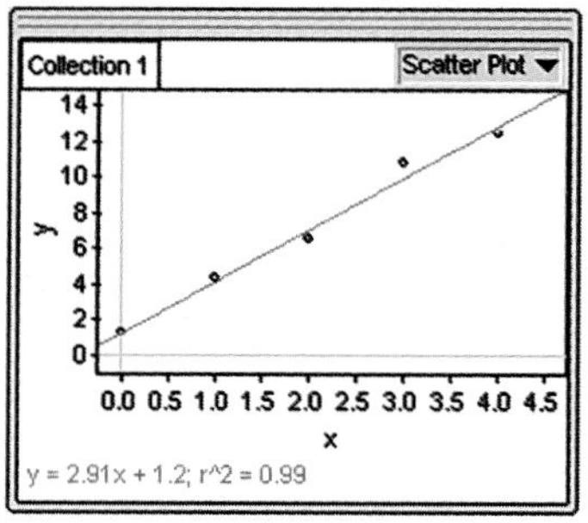 The graph displays an equation of best fit with $a = 2.91$ and $b = 1.2$. The graph also shows the coefficient of determination, r^2.
line of best fit	See **linear regression**.

Function or Task	Keystroke(s), Menu, or Screen
mean	The mean function found under the Functions/Statistical/One Attribute menu is used to calculate the mean of an attribute.

Example:

Create a **collection** and **case table** as shown in the following screen:

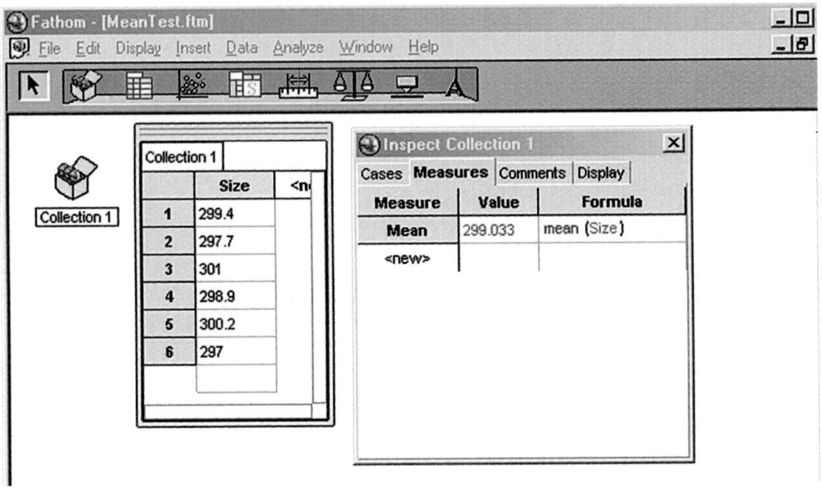

Double-click on the **collection** box to open the **inspector** for the collection, and select the Measures tab. Rename <new> to Mean. Right-click on the Formula column for the Mean measure, and select Edit Formula. Select the **mean()** function under the Functions/Statistical/One Attribute menu, and type Size between the brackets. You will see the mean calculated under Value, as in the screen shown above.

FATHOM™

Function or Task	Keystroke(s), Menu, or Screen
median	The median function found under the Functions/Statistical/One Attribute menu is used to calculate the median for an attribute. **Example:** Create a **collection** and **case table**, as shown in the screen shot below: 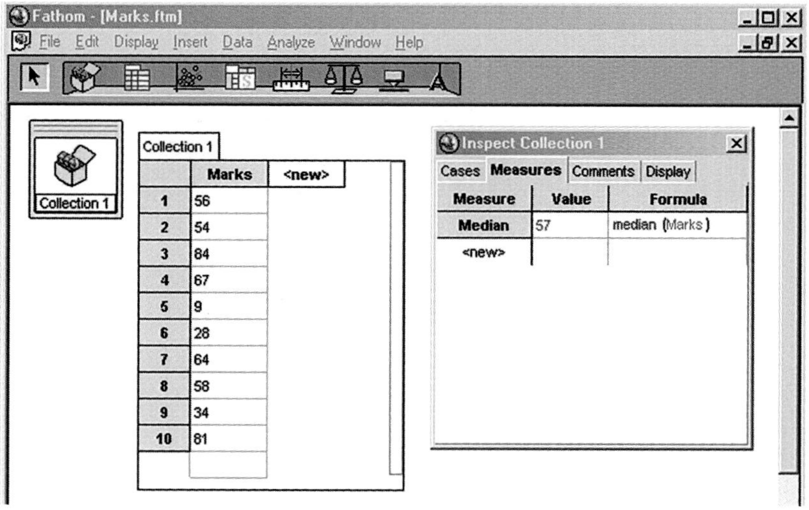 Double-click on the **collection** box to open the **inspector** for the **collection**, and select the Measures tab. Rename \<new\> to Median. Right-click on the Formula column for the Median measure, and select Edit Formula. Select the median() function under the Functions/Statistical/One Attribute menu, and type Marks between the brackets. You will see the median calculated under Value, as in the screen shown above.
mode	There is no **mode** function in Fathom™. You can find the mode for an attribute a by following the procedure for the **median** and then using the formula given below. The derivation of this formula is beyond the scope of this appendix. mean(a, rank(a) - uniqueRank(a) = max(rank(a) - uniqueRank(a))) Another way to find the mode is to **sort** the attribute a and then scroll down the column looking for values that occur multiple times.

FATHOM™

Function or Task	Keystroke(s), Menu, or Screen
non-linear regression	Fathom does not have built-in non-linear regression functions. However, you can plot functions on top of a scatter plot and even control parameters using sliders, in an attempt to match a function to the scatter plot.

Example:
Create a **collection**, **case table**, and graph as shown in the screen shot below:

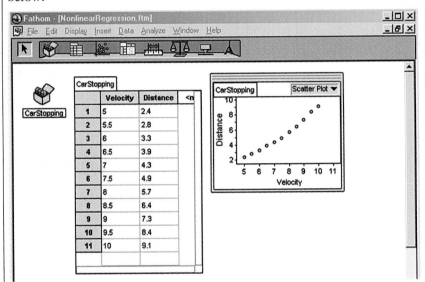

It looks like the relation might be quadratic. To try a fit, select the graph, and then select Plot Function from the Graph menu. You can now enter a formula for Distance as a function of Velocity. You might type in .09×Velocity^2. If this formula does not work, you can try another coefficient. However, a more convenient way to do it is to use a **Slider**. Drag a **Slider** from the shelf to the workspace, and rename it a. Select Plot Function from the Graph menu, and enter the formula aVelocity^2.

 |

Function or Task	Keystroke(s), Menu, or Screen
	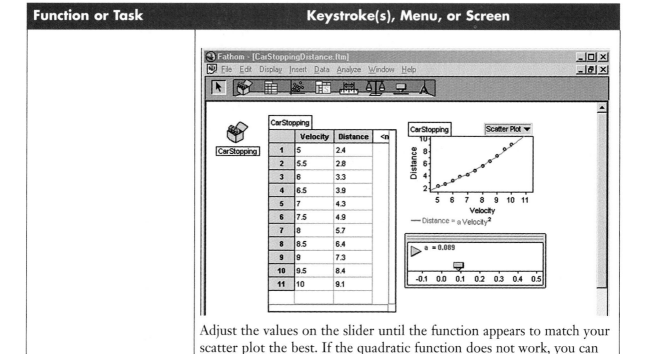
Adjust the values on the slider until the function appears to match your scatter plot the best. If the quadratic function does not work, you can try other functions.
normalCumulative function normalCumulative (x, mean, standard deviation)	The normalCumulative function located under the Functions/Distributions/Normal menu allows you to calculate the probability that a given data point is less than x using a normal distribution with a given mean and standard deviation. **Example:** Suppose that a particular model of tire has a lifetime with a mean of 64 000 km and a standard deviation of 8000 km. What is the probability that a tire will wear out at 60 000 km or less? Create a **collection** and a **case table** as shown in the following screen: 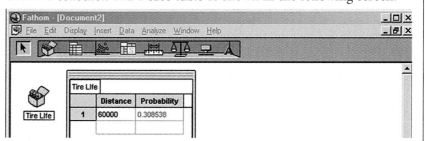 Right-click on the Probability attribute, and select the normalCumulative function under the Function/Distributions/Normal menu. Type Distance,64000,8000 between the brackets. You will get a probability of approximately 0.309.

Function or Task	Keystroke(s), Menu, or Screen
normalQuantile function normalQuantile(p, mean, standard deviation)	The normalQuantile function located under the Functions/Distributions/Normal menu is the opposite of the normalCumulative function. Given a probability p, it allows you to calculate the value of x using a normal distribution with a given mean and standard deviation such that the probability that a given data point is less than or equal to x is p. **Example:** Suppose that a particular model of tire has a lifetime with a mean of 64 000 km and a standard deviation of 8000 km. What distance will see 25% of the tires wear out? Create a **collection** and a **case table** as shown in the following screen:  Right-click on the Distance attribute, and select the normalQuantile function under the Functions/Distributions/Normal menu. Type Probability,64000,8000 between the brackets. You will get a distance of approximately 58 604 km.
quartiles	The quartile functions in Fathom™ are Q1 and Q3, found under the Functions/Statistical/One Attribute menu. **Example:** Create a **collection** and **case table** as shown in the following screen: 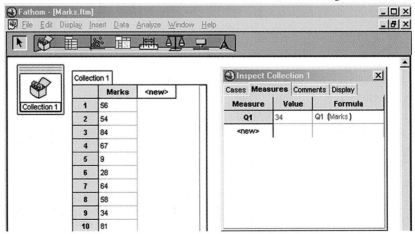

Function or Task	Keystroke(s), Menu, or Screen
	Double-click on the **collection** box to open the **inspector** for the **collection**, and select the Measures tab. Rename <new> to Q1. Right-click on the Formula column for the Median measure, and select Edit Formula. Select the Q1 function under the Functions/Statistical/One Attribute menu, and type Marks between the brackets. You will see the first quartile calculated under Value, as shown in the screen shown above. You can calculate Q3 in a similar manner.
random function random()	Fathom™ has 17 different random functions. random( ) will generate a random real number from 0 to 1. **Example:** Generate ten random numbers from 0 to 1. Open a new **collection**. Create a new **case table**. Double-click on the <new> attribute and rename it Random. To add ten new cases, right-click on the Random attribute and select New Cases.... Type in 10 and press Enter. Right-click on the Random attribute and choose Edit Formula. Choose Functions. Double-click on random. Choose OK. You now have 10 random real numbers between 0 and 1. 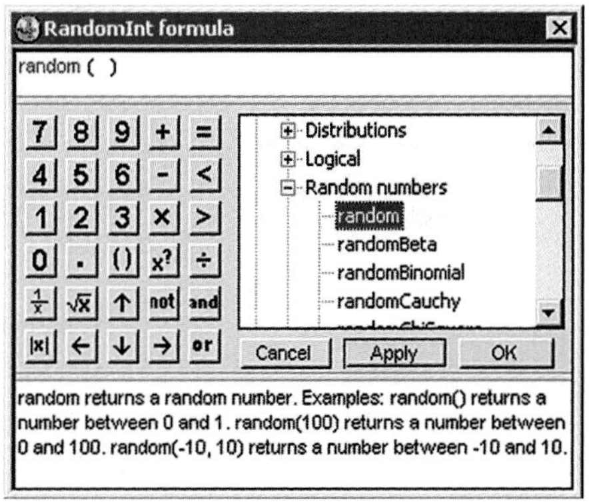
randomInteger function randomInteger(lower, upper)	The randomInteger function will generate random integers from lower to upper. **Example:** Generate 20 random numbers from 6 to 10. Open a new **collection**. Create a new **case table**. Double-click on the <new> attribute and rename it RandomInt.

Function or Task	Keystroke(s), Menu, or Screen
	To add 20 new cases, right-click on the **RandomInt** attribute and select **New Cases...**. Type in 20 and press **Enter**. To generate the random numbers, right-click on the **RandomInt** attribute and choose **Edit Formula**. 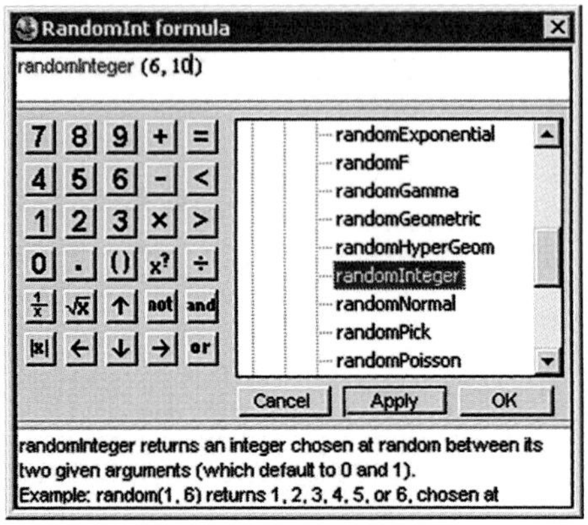 Now choose **Functions**. Double-click on **randomInteger()**. Type 6,10 between the brackets. Choose **OK**. You now have 20 random numbers between 6 and 10.
randomNormal function randomNormal(mean, standard deviation)	The randomNormal function will generate random numbers from a normal distribution with a given **mean** and a given **standard deviation**. **Example:** Generate 20 random numbers from a normal distribution with a mean of 100 and a standard deviation of 10. Open a new **collection**. Create a new **case table**. Double-click on the **<new>** attribute and rename it **RandomNorm**. To add 20 new cases, right-click on the **RandomNorm** attribute and select **New Cases...**. Type in 20 and press **Enter**. To generate the random numbers, right-click on the **RandomNorm** attribute and choose **Edit Formula**.

FATHOM™

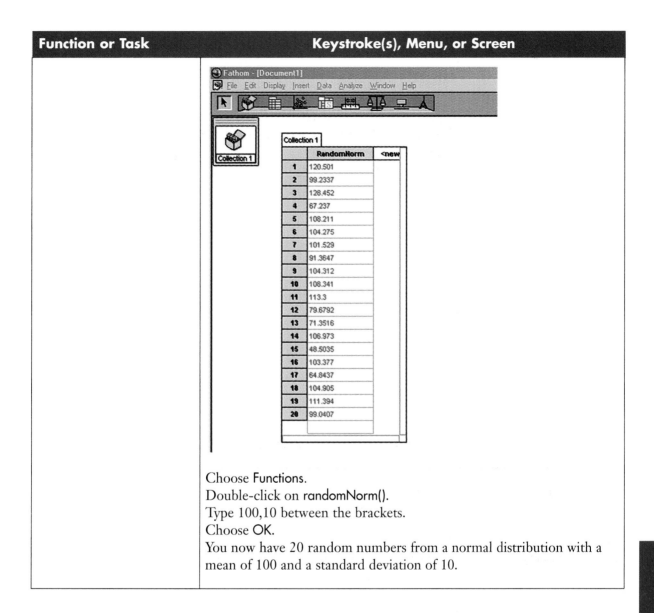

Choose Functions.

Double-click on randomNorm().

Type 100,10 between the brackets.

Choose OK.

You now have 20 random numbers from a normal distribution with a mean of 100 and a standard deviation of 10.

Function or Task	Keystroke(s), Menu, or Screen
scatter plot	You can draw a **scatter plot** by dragging attributes from a **case table** to a graph. **Example:** Create a **collection** and a **case table** as shown in the following screen: 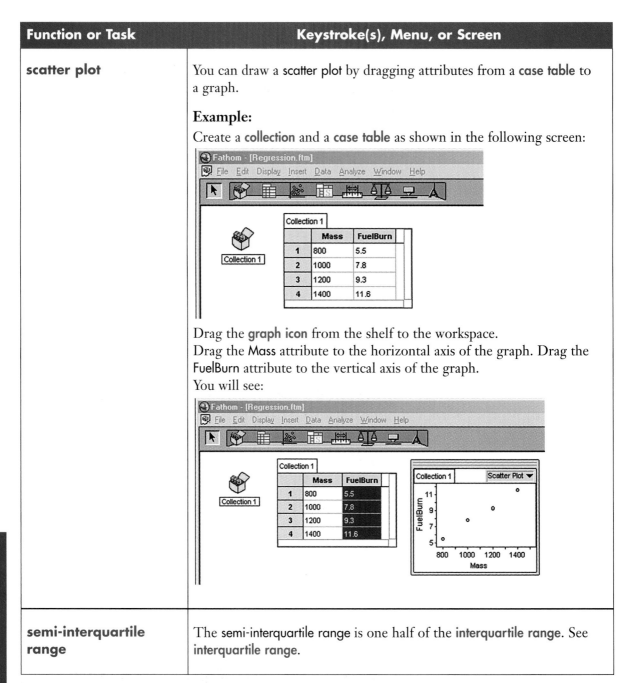 Drag the **graph icon** from the shelf to the workspace. Drag the Mass attribute to the horizontal axis of the graph. Drag the FuelBurn attribute to the vertical axis of the graph. You will see:
semi-interquartile range	The semi-interquartile range is one half of the **interquartile range**. See **interquartile range**.

Function or Task	Keystroke(s), Menu, or Screen
sort	You can sort the entries in a **case table** using the Sort Ascending and Sort Descending functions under the Data menu. **Example:** Create a **collection** and **case table** as shown in the following screen: 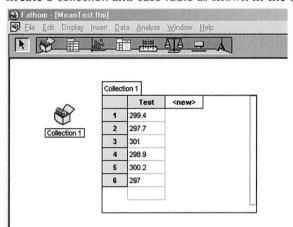 Click on the Test attribute to select the attribute column. Click on the Data menu, and select Sort Ascending. 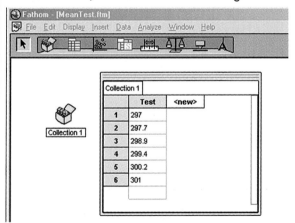 If you want the data sorted in descending order, then select Sort Descending from the Data menu.

FATHOM™

standard deviation ***Population*** popstdDev() ***Sample*** stdDev()	The popstdDev() and stdDev() functions found under the Functions/Statistical/One Attribute menu are used to calculate the standard deviation of an attribute. **Example:** Create a **collection** and **case table** as shown in the following screen: 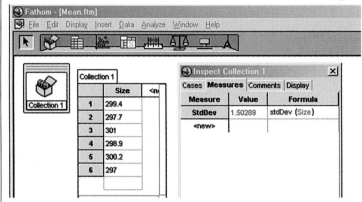 Double-click on the **collection** box to open the **inspector** for the **collection**, and select the Measures tab. Rename \<new> to StdDev. Right-click on the Formula column for the StdDev measure, and select Edit Formula. Select the stdDev() function under the Functions/Statistical/One Attribute menu, and type Size between the brackets. You will see the sample standard deviation calculated under Value.
sum function	The sum function under the Functions/Statistical/One Attribute menu can be used to find the sum of the entries under an attribute. **Example:** Create a **collection** and **case table** as shown below: 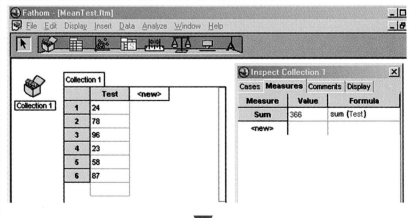

FATHOM™

Function or Task	Keystroke(s), Menu, or Screen
	Double-click on the **collection** box to open the **inspector**, and rename <new> to Sum. Right-click the Formula column at the Sum row and select Edit Formula. Select the sum function under the Functions/Statistical/One Attribute menu and type Test between the brackets. You will see the value of 366 under the Value column.
uniqueRank() function	The uniqueRank() function under the Functions/Statistical/Transformations menu is used to rank the entries in an attribute column of a **case table**. **Example:** Create a **collection** and a **case table** as shown in the following screen: 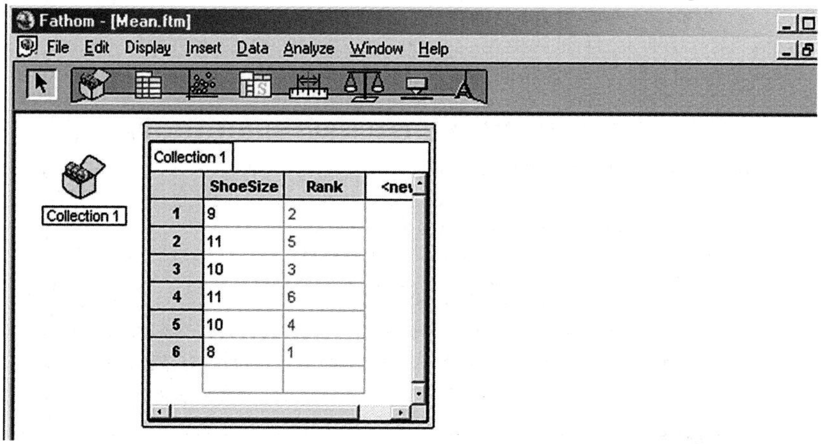 Right-click on the Rank attribute, and select Edit Formula. Select uniqueRank() from the Functions/Statistical/Transformations menu and type ShoeSize between the brackets. The Rank attribute will now show the ranking of each entry under the ShoeSize attribute. *Note*: If the ShoeSize attribute is sorted before applying the uniqueRank() function, then the ranks will be in order.

FATHOM™

Function or Task	Keystroke(s), Menu, or Screen
variance ***Population*** popVariance() ***Sample*** Variance()	The popVariance and variance functions located under the Functions/Statistical/One Attribute menu compute the square of the **standard deviation**. **Example:** Create a **collection** and **case table** as shown in the following screen:

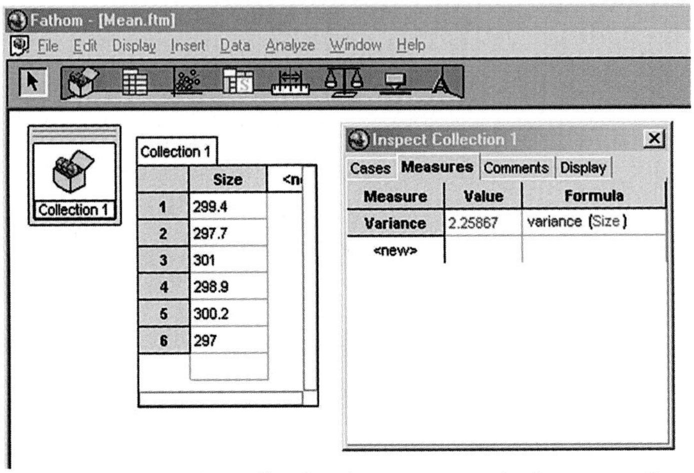

Double-click on the **collection** box to open the **inspector** for the **collection**, and select the Measures tab. Rename <new> to Variance. Right-click on the Formula column for the Variance measure, and select Edit Formula. Select the **variance function** under the Functions/Statistical/One Attribute menu, and type Size between the brackets. You will see the sample variance calculated under Value.

Function or Task	Keystroke(s), Menu, or Screen
zScore function also known as sampleZscore	The zScore function under the Functions/Statistical/Transformations menu calculates how many sample standard deviations a value is from the mean. For example, if a sample has a mean of 100 and a standard deviation of 10, then a value of 120 would have a zScore of 2. **Example:** Create a **collection** and a **case table** as shown in the following screen: 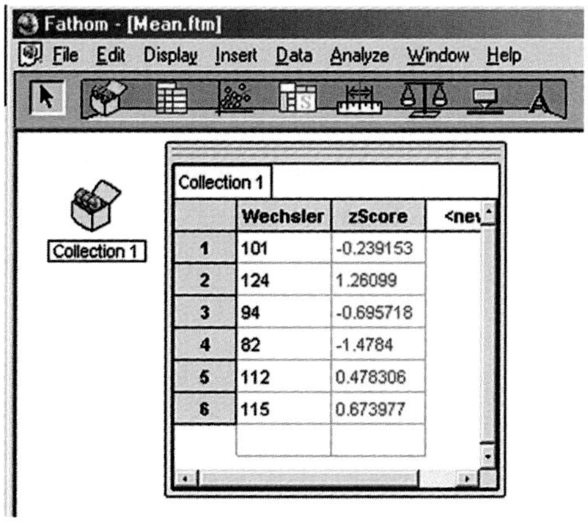 Right-click on the **zScore** attribute, and select **Edit Formula**. Select zScore() from the Functions/Statistical/Transformations menu and type **Wechsler** between the brackets. The **zScore** for each entry in the **Wechsler** attribute is now displayed in the **zScore** attribute column.

FATHOM™

Appendix C Research Skills

What is the difference between a research assignment and a report?

A report generally explains or describes something, whereas a research assignment is an analysis of facts, ideas, data, and expert opinions that usually support a thesis statement.

What is a research assignment?

- A research assignment presents the results of an investigation into a particular topic, and usually involves the following:
  - a thesis statement that is supported by research
  - research of existing facts, ideas, data, and expert opinions
  - your own creative thinking, analysis, and expression of ideas
  - a formal presentation of your findings—in essay, oral and/or technological form
- A research assignment is *not*:
  - a simple analysis of your own thoughts
  - a synopsis of what you have read about your topic
  - a rewriting of someone else's argument with a few other sources added

Developing the Topic and Hypothesis

1. Select a topic of interest to you.
   - Analyse your assignment by highlighting key words and phrases, such as argue, analyse, evaluate, discuss, compare, and contrast.
   - Explore possible topics that are interesting to you.
   - Brainstorm ideas with others, including your teacher.
   - Ensure the topic fulfills the guidelines of the assignment.

2. Are sufficient resources available for your topic?
   - Read about your subject in encyclopedias, textbooks, journals, and on the Internet.
   - Evaluate the quality, quantity, and reliability of the resources.

3. Can the topic be narrowed down to a non-trivial, manageable hypothesis?
   - Think of questions to assist you in defining sub-topics that you will explore.
   - Narrow down the topic to a research question, such as: "What are the effects of ...?" This will provide the focus of your research. Brainstorming is an effective technique for narrowing down your topic.
   - Write a tentative hypothesis statement that summarizes your point of view, is a specific declaration of your main idea, and is a statement (not a question) reflecting your position on a particular topic or issue. Your hypothesis will be formalized as you are actually researching and writing your assignment.

Accessing Resources

1. Explore a variety of different resources.
   - Encyclopedias (print, CD-ROM or Internet-based), the Internet, and textbooks are good initial sources while developing your topic ideas.
   - Scientific journals and expert publications provide opinions and data.
   - The Internet contains a wide variety of information in varying degrees of detail and accuracy. Web sites, newsgroups, and discussion groups may all have valuable information.

2. Use effective search techniques.
   - Identify important concepts for your search.
   - Use key words from your topic:
     ◦ Search for general topic words first.
     ◦ Use synonyms and variations of your keywords.
   - Narrow your search by using Boolean operators, such as "OR", "AND", "NOT" or "+", by placing a phrase in quotation marks or brackets, or by changing the order of the key words in the phrase.
   - Because each search engine on the Internet does not have all web sites comprehensively indexed, you should use a variety of search engines.
     ◦ General search engines keep their own database of web sites. Some examples are:
       - Yahoo (*www.yahoo.ca*)
       - Lycos (*www.lycos.com*)
       - Google (*www.google.com*)
       - AllTheWeb (*www.alltheweb.com*)
       - AltaVista (*www.altavista.com*)
       - Northern Light (*www.northernlight.com*)
       - Dogpile (*www.dogpile.com*)
       - Search Canada (*www.searchcanada.ca*)
       - Canadian Links (*www.canadian-links.com*)
     ◦ Meta search engines access databases kept by general search engines. Some examples are:
       - Metacrawler (*www.metacrawler.com*)
       - Ixquick (*www.ixquick.com*)
       - Ask Jeeves (*www.askjeeves.com*)
     ◦ Use the search engine's online help page. It will provide information on features that are specific to that search engine.
     ◦ Web sites will often have their own internal search engine. Use it to refine your search within the web site.
   - Access online library databases, such as Electric Library Canada, a database of books, newspapers, magazines, and television and radio transcripts. Your local public library may also have online access to its database for people with a valid library card.
   - Use a tracking chart to keep a record of your searches. Record the success or failure of each search. Include the search engine and key words you used, and any web sites you visited (even the non-useful ones).

Evaluating Your Sources

1. Check to make sure your sources are up to date. The date of publication is always cited on all web sites, textbooks, journals, encyclopedias, etc.

2. Is your source reliable and reputable?
   - Government statistical department web sites (e.g., Statcan) are reliable.
   - University resources (web sites and university presses) are generally reliable.
   - Consumer and corporate sources may be biased toward their own position on an issue. Be aware that only data that are favourable to their own side of a discussion may have been published.
   - Is the author a recognized expert in the field? This may be initially difficult to assess. Reliable experts' opinions in the related field of study are regularly cited in other publications.

3. Consider these questions when you are evaluating a source.
   - Is your sample reliable?
   - Is the sample large enough? Small samples may not give reliable results.
   - Is the sample random? Is the sample biased?

4. Take notes and keep records of all your sources.
   - Make a printout of any sources you require. Again, make sure you keep an accurate record of the source.
   - Create and use a tracking chart. Write the complete details of your source, even if you feel you may not use it, as you may change your mind later.
     - If your source is a print source, include all the appropriate information that would go into a bibliography, plus the page number and the Dewey Decimal number.
     - If your source is an online source, include the URL, the title of the web site, the date you accessed the site, any source used as a reference by the web site, as well as any other pertinent information to make accessing the site easier in the future.
   - Take notes on a separate sheet for each source. At the top of the page, write a brief source identification referring to a detailed entry on the tracking chart. Write down information that may support your hypothesis statement or research question: background data, expert opinions, opposing viewpoints, etc. Paraphrase, where possible, and cite your sources. Write in quotation marks any information that you copy verbatim.

Citing Sources

1. Citing your sources.
   - All sources used for ideas, information, data, or opinions must be cited in your assignment. Detailed information regarding these sources must be included. See your teacher for a preferred format such as endnotes, footnotes, or in-text citations.

Plagiarism

1. What is plagiarism?
   - Plagiarism is using or copying another person's ideas, information, data, opinions, or words without acknowledging that it has been taken from the original source. Even if you paraphrase someone else's ideas or words, you must acknowledge the source that you are paraphrasing in the form of footnotes, endnotes, and in the bibliography.

2. What are the consequences of plagiarism?
   - At universities, plagiarism will generally result in failure in the course, and may include the penalties of probation or possibly expulsion from the college or university. Consult your school evaluation policy for the consequences of plagiarism.
   - A permanent blemish will appear on the student's academic record.

3. How can you avoid plagiarism?
   - Do not hand in another student's work as your own.
   - Be familiar with your teacher's or school's system and style of documentation.
   - When taking notes from any source, record the detailed bibliographic information.
   - When you are writing down a direct quote, make sure you use quotation marks.
   - Keep your own ideas separate from those of your sources. When you are making notes, write your name or initials beside your own opinions.
   - Let your own ideas, not those of your sources, steer you in writing your assignment.
   - Never copy material word-for-word and place it into your essay or work unless you clearly indicate that the material is a direct quotation and give full details of the source.

Appendix D Oral Presentation Skills

The following guidelines will help you prepare the oral presentation that is part of your culminating project for this course. These guidelines will take you, step by step, through the preparation and delivery processes of your presentation.

Know Your Purpose and Audience

Purpose

- Your purpose can often be found in the assignment you have been given. Ask yourself these questions:
  - What are you being asked to do?
  - What guidelines are given in the assignment?
  - How will your topic answer the requirements of the assignment?
  - What is your goal in this presentation?

- Your purpose is affected by the method of communication you are using. It is not enough just to read your written report; you must adapt your investigation, data, and conclusions to the circumstances of an oral presentation.

- Your goal is to give your audience a clear understanding of your findings, so your presentation has to be firmly based on your research, data, and analysis of the data. When presenting your research assignment, your content should be objective and informative.

- Select a topic that will meet the requirements of the assignment, fulfill your purpose, and be comprehensible for your listeners to follow in the allotted time given for the presentation. Select the parts of your investigation and report what you think will best serve your purpose.

- You will need a clearly defined central idea, or focus, and an appropriate selection of specific and concrete details to support the central idea. The focus needs to be maintained all the way through your planning and delivery. Lack of focus is one of the most frequent causes of weak oral presentations.

- Have a strong command of your subject. Remember that your audience is going to have a chance to ask you questions.

Audience

- Keep your audience's level of knowledge about your topic in mind while you plan your presentation. They will not be as knowledgeable as you about your topic, and they will not have collected the data or completed the analysis.

- Explain points so that the average listener can understand and see the significance of your problem, your investigation, your data, and your suggested solution.

- Explain anything they are not likely to know or need to be reminded about.

- Create a structure that will enable learners to understand your conclusions and how you arrived at them. Keep your examples and evidence clear, and ensure they are within your audience's understanding of the subject being investigated.

Planning Your Presentation

- Lack of planning and organizational structure is a major reason for poor oral presentations. The more carefully you prepare, plan, and organize your presentation, the better it will be. Take the time to create a clear, understandable, tight, and efficient outline for your presentation.

- Use brainstorming, tree diagrams, mind maps, or any other planning strategy to help organize your thoughts.

- Make sure that you have a clear, single focus.
  - What is your main idea?
  - What evidence do you want to show for your main idea, your investigation, and the solution that you are suggesting for this problem?

- Ensure that the structure of your argument is clear and logical and that your presentation will make that structure understandable to your listeners.

- Include all parts of your presentation in your outline.
  - Write down your main message and focus, and then construct an opening, a body, and a conclusion that support that main message.
  - Decide where the divisions will be and plan how you will let your listeners know about those divisions.

- Decide which organizational strategies will help your audience understand your main idea and your supporting arguments for that idea.
  - How will you take them through your argument (i.e., chronologically, or by comparison, description, analysis, or problem-solution organization)?
  - Your organizational strategy and your structure will all depend on your topic, your purpose, and your audience.

- Include all necessary definitions and descriptions to help your audience understand the significance of your findings.

- Anticipate what questions your audience might ask about your subject and your evidence, and build the answers to those likely questions into your presentation. This will eliminate any confusion in the audience and will help you stay in control of the presentation.

- Determine which visuals you will need and how they can be created to help illustrate points and data in your presentation.

- Rehearse your presentation to ensure it is within the proper length of time.
  - Remember to allocate five to ten minutes for questions and answers at the end of your presentation.
  - In case you finish early, have some optional items ready that would enhance your presentation, but that are not critical to include.

Planning the Introduction

- Plan an effective and interesting opening that explains your main message and catches your audience's interest.
  - Open your presentation with a summary of your investigation and your written report.
  - Define your problem clearly in the beginning so that your listeners know precisely what you are going to be talking about.

- Keep your introduction brief, but also include enough information so your audience knows what the problem is, how it was approached, and what evidence/rationale is going to be offered.

- Make your key ideas very clear in the introduction—these are the points that your audience has to understand.

- Keep your introduction focused so that the presentation is easy to follow.
  - Give your audience a clear statement of your objective and your key points. This statement is your opportunity to focus your audience and to give them a reliable roadmap to follow during the rest of your presentation. Your listeners have to know where you are going to take them, otherwise they may get lost in the details of your investigation.

- Offer any background information that your audience will need to understand the significance of your problem and your solution. Keep your background comments brief so that they do not take over the presentation.

- Explain any limitations of your study and any assumptions that you made.

- As you move from the introduction to the body of your presentation, use a transition. Let your listeners know that you are moving into the evidence for your conclusions.

Planning the Body

- The body of your presentation should include:
  - a description of the criteria used in your investigation
  - a discussion of your findings with relevant factual details
  - a description of your method and the steps you took in your investigation, and the process that you followed
  - an analysis of your data and the methods that you used to perform that analysis
  - the implications of your results and an explanation for your recommendation or conclusion

- The body of your presentation should help your audience understand the rationale for your investigation and the reasons for your conclusions.
  - Build a clear structure for this part of your presentation, and let your audience know the structure. Do not leave it up to your audience to figure out the structure or to draw their own conclusions.
  - Offer your audience a full development of your key ideas and show how your evidence clarifies and justifies your conclusions. Be sure to relate each key idea back to the focus of your presentation.
  - Carefully select details and evidence because you likely will not be able to include all the evidence that you had in your written report.
  - Your examples, evidence, data analysis, and details should all support your main idea and the conclusion of your investigation.
  - Be sure that your evidence is accurate, pertinent, and complete within the boundaries of your subject and your purpose.
  - Use transitions to help people stay on track, to show the logical steps of your argument, and to bridge ideas.
    - You may use something as straightforward as telling them that first you will examine evidence A, then evidence B, and then evidence C.
    - Periodically summarize what you have already discussed and introduce what you are going to say next. This helps keep your listeners on track.

Planning the Conclusion

- The conclusion of your presentation should offer your audience a clear summary of your main ideas and focus of your presentation. Briefly review your main idea and summarize your evidence.
  - You may want to leave the audience with something interesting to think about. Be careful about adding entirely new information in your conclusion, however, as you want to ensure that you leave your listeners with a clear idea of your main message.

- Do not rush in an effort to quickly finish. Planning and practice will help you resist the temptation to hurry your conclusion when you are speaking. Remember that a good presentation deserves a good conclusion.

Creating and Organizing Your Visuals

- Be sure that your visuals are pertinent, readable, and illustrative. They should help your audience understand your data and your evidence. The careful selection and presentation of visuals can dramatically enhance the impact of your message.

- Visuals should be easily seen and read by everyone in the room.
  - Graphs should be clear and uncluttered.
  - Font size should be large enough to be read from the back of the room.
  - Text slides or overheads should be brief and to the point.
  - Mathematical terminology and notations should be clear and correct.

- Visuals should illustrate particular key points or data results of your study. Choose specific, concrete, and significant details and evidence that will support your main focus and purpose.

- Plan how the visuals will fit into your presentation. Note on your outline where you will introduce each visual and how you will explain its purpose. Do not put up visuals and leave your audience to wonder what purpose they are serving.

- If you are using overheads or computer presentation media, be sure you know how to operate the technology and check to be sure it is working on the day of your presentation.

- If you are including handouts, keep them simple and carefully tied to your topic so that they will be helpful rather than be a distraction. If you are including additional background material in your handouts, ask your audience to refrain from reading them during your presentation, so as not to distract their attention.

- Do not forget that most classrooms have a chalkboard, whiteboard, or flipchart as an option for visuals. These tools can be useful for showing the separate steps of a process or demonstrating a mathematical procedure. Be careful when using these types of tools; if you spend too much time writing with your back to your audience, you will lose their attention.

Practising and Rehearsing Your Presentation

- Practise your presentation using cue cards, outlines, or other presentation aids. Practice will make you a better speaker and will make it easier for you to get your message across.

- You can use a tape recorder to listen to how you sound and to see how long your presentation will take. Using a tape recorder can also help you to listen for and eliminate the "ums" and "ahs" that can be distracting to listeners.

- Rehearse your presentation before a live audience, and ask for feedback. Use your visuals during this rehearsal. Tell your listeners beforehand what kind of feedback you are looking for.
  - Can they follow the logic of your structure and understand the focus of your presentation? If they are having trouble, then you probably need to do some revisions.
  - Where are the gaps? What needs to be more fully explained?

Delivering Your Presentation

Follow Your Plan

- As you deliver your presentation, follow the outline that you developed during your planning stage.

- Refer to your outline and use it as your roadmap. Explain the structure of your presentation so your listeners will know exactly where you are taking them.

- Stay focused on your main message, purpose, and audience throughout the presentation.

- Support your main message with appropriate evidence using factual details and relevant examples.

- Use your visual aids effectively and follow your plan when introducing and explaining them. Be sure to give your audience time to read or view the visuals.

Speak Clearly

- Speak loudly and slowly enough so that everyone in the room can hear. Remember that your voice has to be heard as there is no opportunity for your audience to "re-hear" your important points. And once you lose your audience's understanding it is very hard to get it back again.

- Resist the temptation to rush because of nervousness. Practice will help you avoid feeling too nervous.

- Try to avoid using "ums" or "ahs" and unnecessary gaps. The occasional brief pause, however, can be an effective way to emphasize a key point.

- Address everyone in the audience and make eye contact.

- Use body language and gestures appropriately. Try to be natural and relaxed.

Watch Your Language

- Use clear and correct language that is appropriate for the purpose and audience of your presentation. Be sure that your mathematical terminology is correct and consistent throughout your presentation.

- Remember that in a speaking-listening communication situation it is up to you to avoid misunderstandings—you are the speaker so your meaning should be clear at all times.

- Use terms correctly. Be very precise in your word choices, and practise the pronunciation of any difficult words.

- Consider writing difficult words or technical terms on a blackboard, a flip chart, a transparency, or a presentation software slide.

Consider Your Audience

- Watch the reactions of your audience. You may have to adapt content and delivery to your audience's needs. You may need to slow down or to further explain some unfamiliar points if you see that your audience is not following your presentation or has found some point difficult to understand.

- Keep your listeners on track. They will be watching for the main points in your presentation. Make it easy for them to follow and understand those main points.

- Speak to the people in your audience. If you read your presentation to them, you will lose your connection with the audience, and their attention. Use your cue cards or an outline to ensure that you do not just read.

- Bring along your written report, and have it handy in case you need it to answer students' questions.

Answering Questions From the Audience

- Focus your attention on the question and the questioner so that you can respond clearly and directly.

- Remind yourself that when people ask you questions related to something you have said in your presentation, it means that they have identified with your ideas. They have incorporated your presentation into their own minds long enough to consider it and to ask for more information.

- Treat questions from the audience as an opportunity for you to gain some fresh insights into your presentation and to clarify anything that your audience has not understood.

- Practise being an effective, active listener as you listen to the questions. Watch for the key ideas in the question. If you do not understand the question you may want to paraphrase—put the speaker's words in your own words—to be sure the question you have heard is what the speaker intended to ask.

- Try not to let questions disrupt the flow of your presentation. If the question is directly relevant to the particular topic being discussed and can be easily answered without digressing from the topic, then answer it immediately. However, if the question is on a topic that will be addressed later in the presentation, or if it is only peripherally related to the focus of your presentation, then acknowledge the question, but politely defer it to the appropriate time in the presentation or afterward.

- Answer all questions courteously, thoughtfully, and completely. This will demonstrate your flexibility and understanding of your subject.

Areas Under the Normal Distribution Curve

The table lists the shaded area for different values of z.
The area under the entire curve is 1.

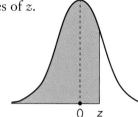

z	0.00	0.01	0.02	0.03	0.04	0.05	0.06	0.07	0.08	0.09
−2.9	0.0019	0.0018	0.0018	0.0017	0.0016	0.0016	0.0015	0.0015	0.0014	0.0014
−2.8	0.0026	0.0025	0.0024	0.0023	0.0023	0.0022	0.0021	0.0021	0.0020	0.0019
−2.7	0.0035	0.0034	0.0033	0.0032	0.0031	0.0030	0.0029	0.0028	0.0027	0.0026
−2.6	0.0047	0.0045	0.0044	0.0043	0.0041	0.0040	0.0039	0.0038	0.0037	0.0036
−2.5	0.0062	0.0060	0.0059	0.0057	0.0055	0.0054	0.0052	0.0051	0.0049	0.0048
−2.4	0.0082	0.0080	0.0078	0.0075	0.0073	0.0071	0.0069	0.0068	0.0066	0.0064
−2.3	0.0107	0.0104	0.0102	0.0099	0.0096	0.0094	0.0091	0.0089	0.0087	0.0084
−2.2	0.0139	0.0136	0.0132	0.0129	0.0125	0.0122	0.0119	0.0116	0.0113	0.0110
−2.1	0.0179	0.0174	0.0170	0.0166	0.0162	0.0158	0.0154	0.0150	0.0146	0.0143
−2.0	0.0228	0.0222	0.0217	0.0212	0.0207	0.0202	0.0197	0.0192	0.0188	0.0183
−1.9	0.0287	0.0281	0.0274	0.0268	0.0262	0.0256	0.0250	0.0244	0.0239	0.0233
−1.8	0.0359	0.0351	0.0344	0.0336	0.0329	0.0322	0.0314	0.0307	0.0301	0.0294
−1.7	0.0446	0.0436	0.0427	0.0418	0.0409	0.0401	0.0392	0.0384	0.0375	0.0367
−1.6	0.0548	0.0537	0.0526	0.0516	0.0505	0.0495	0.0485	0.0475	0.0465	0.0455
−1.5	0.0668	0.0655	0.0643	0.0630	0.0618	0.0606	0.0594	0.0582	0.0571	0.0559
−1.4	0.0808	0.0793	0.0778	0.0764	0.0749	0.0735	0.0721	0.0708	0.0694	0.0681
−1.3	0.0968	0.0951	0.0934	0.0918	0.0901	0.0885	0.0869	0.0853	0.0838	0.0823
−1.2	0.1151	0.1131	0.1112	0.1093	0.1075	0.1056	0.1038	0.1020	0.1003	0.0985
−1.1	0.1357	0.1335	0.1314	0.1292	0.1271	0.1251	0.1230	0.1210	0.1190	0.1170
−1.0	0.1587	0.1562	0.1539	0.1515	0.1492	0.1469	0.1446	0.1423	0.1401	0.1379
−0.9	0.1841	0.1814	0.1788	0.1762	0.1736	0.1711	0.1685	0.1660	0.1635	0.1611
−0.8	0.2119	0.2090	0.2061	0.2033	0.2005	0.1977	0.1949	0.1922	0.1894	0.1867
−0.7	0.2420	0.2389	0.2358	0.2327	0.2296	0.2266	0.2236	0.2206	0.2177	0.2148
−0.6	0.2743	0.2709	0.2676	0.2643	0.2611	0.2578	0.2546	0.2514	0.2483	0.2451
−0.5	0.3085	0.3050	0.3015	0.2981	0.2946	0.2912	0.2877	0.2843	0.2810	0.2776
−0.4	0.3446	0.3409	0.3372	0.3336	0.3300	0.3264	0.3228	0.3192	0.3156	0.3121
−0.3	0.3821	0.3783	0.3745	0.3707	0.3669	0.3632	0.3594	0.3557	0.3520	0.3483
−0.2	0.4207	0.4168	0.4129	0.4090	0.4052	0.4013	0.3974	0.3936	0.3897	0.3859
−0.1	0.4602	0.4562	0.4522	0.4483	0.4443	0.4404	0.4364	0.4325	0.4286	0.4247
0.0	0.5000	0.4960	0.4920	0.4880	0.4840	0.4801	0.4761	0.4721	0.4681	0.4641

z	0.00	0.01	0.02	0.03	0.04	0.05	0.06	0.07	0.08	0.09
0.0	0.5000	0.5040	0.5080	0.5120	0.5160	0.5199	0.5239	0.5279	0.5319	0.5359
0.1	0.5398	0.5438	0.5478	0.5517	0.5557	0.5596	0.5636	0.5675	0.5714	0.5753
0.2	0.5793	0.5832	0.5871	0.5910	0.5948	0.5987	0.6026	0.6064	0.6103	0.6141
0.3	0.6179	0.6217	0.6255	0.6293	0.6331	0.6368	0.6406	0.6443	0.6480	0.6517
0.4	0.6554	0.6591	0.6628	0.6664	0.6700	0.6736	0.6772	0.6808	0.6844	0.6879
0.5	0.6915	0.6950	0.6985	0.7019	0.7054	0.7088	0.7123	0.7157	0.7190	0.7224
0.6	0.7257	0.7291	0.7324	0.7357	0.7389	0.7422	0.7454	0.7486	0.7517	0.7549
0.7	0.7580	0.7611	0.7642	0.7673	0.7704	0.7734	0.7764	0.7794	0.7823	0.7852
0.8	0.7881	0.7910	0.7939	0.7967	0.7995	0.8023	0.8051	0.8078	0.8106	0.8133
0.9	0.8159	0.8186	0.8212	0.8238	0.8264	0.8289	0.8315	0.8340	0.8365	0.8389
1.0	0.8413	0.8438	0.8461	0.8485	0.8508	0.8531	0.8554	0.8577	0.8599	0.8621
1.1	0.8643	0.8665	0.8686	0.8708	0.8729	0.8749	0.8770	0.8790	0.8810	0.8830
1.2	0.8849	0.8869	0.8888	0.8907	0.8925	0.8944	0.8962	0.8980	0.8997	0.9015
1.3	0.9032	0.9049	0.9066	0.9082	0.9099	0.9115	0.9131	0.9147	0.9162	0.9177
1.4	0.9192	0.9207	0.9222	0.9236	0.9251	0.9265	0.9279	0.9292	0.9306	0.9319
1.5	0.9332	0.9345	0.9357	0.9370	0.9382	0.9394	0.9406	0.9418	0.9429	0.9441
1.6	0.9452	0.9463	0.9474	0.9484	0.9495	0.9505	0.9515	0.9525	0.9535	0.9545
1.7	0.9554	0.9564	0.9573	0.9582	0.9591	0.9599	0.9608	0.9616	0.9625	0.9633
1.8	0.9641	0.9649	0.9656	0.9664	0.9671	0.9678	0.9686	0.9693	0.9699	0.9706
1.9	0.9713	0.9719	0.9726	0.9732	0.9738	0.9744	0.9750	0.9756	0.9761	0.9767
2.0	0.9772	0.9778	0.9783	0.9788	0.9793	0.9798	0.9803	0.9808	0.9812	0.9817
2.1	0.9821	0.9826	0.9830	0.9834	0.9838	0.9842	0.9846	0.9850	0.9854	0.9857
2.2	0.9861	0.9864	0.9868	0.9871	0.9875	0.9878	0.9881	0.9884	0.9887	0.9890
2.3	0.9893	0.9896	0.9898	0.9901	0.9904	0.9906	0.9909	0.9911	0.9913	0.9916
2.4	0.9918	0.9920	0.9922	0.9925	0.9927	0.9929	0.9931	0.9932	0.9934	0.9936
2.5	0.9938	0.9940	0.9941	0.9943	0.9945	0.9946	0.9948	0.9949	0.9951	0.9952
2.6	0.9953	0.9955	0.9956	0.9957	0.9959	0.9960	0.9961	0.9962	0.9963	0.9964
2.7	0.9965	0.9966	0.9967	0.9968	0.9969	0.9970	0.9971	0.9972	0.9973	0.9974
2.8	0.9974	0.9975	0.9976	0.9977	0.9977	0.9978	0.9979	0.9979	0.9980	0.9981
2.9	0.9981	0.9982	0.9982	0.9983	0.9984	0.9984	0.9985	0.9985	0.9986	0.9986

ANSWERS

For answers containing diagrams or graphs, refer to the Student e-book.

CHAPTER 1

Review of Prerequisite Skills, pp. 4–5

1. a) −26 **b)** 35 **c)** 1 **d)** 3
2. a) 4 **b)** 3 **c)** 44 **d)** 0 **e)** 30 **f)** 7
3. a) 5 **b)** −2 **c)** 20 **d)** 1 **e)** ±5
f) 5 **g)** −5 **h)** 9
7. a) A specific cell is referred to with a letter for the column and a number for the row. For example, cell A1.
b) In Microsoft® Excel, specify the first and last cells in the range, with a colon between them. For example, A1:A10. In Corel® Quattro® Pro, specify the first and last cells in the range, with two periods between them. For example, A1..A10.
c) Move the cursor to the cell containing the data. Select the Copy command from the Edit menu. Move the cursor to the cell into which you want to copy the data. Select the Paste command from the Edit menu.
d) Move the cursor to the cell containing the data. Point the mouse at the border of the cell, so that the mouse cursor appears as an arrow. Left click the mouse and hold. Drag the data to the new cell, and release the mouse.
e) Move the mouse to where the column letters are displayed near the top of the screen. Point the mouse directly at the border between the column you want to expand and the column to the immediate right. Left click the mouse and hold. Drag to the right until the column is the desired width, and release.
f) Move the cursor to any cell in the column to the immediate right of where you wish to add the new column. Select Columns from the Insert menu.
g) In Microsoft® Excel, the symbol = must precede a mathematical expression. In Corel® Quattro® Pro, the symbol + must precede a mathematical expression.
8. $\triangle ABC \sim \triangle HJG$ (SSS)
9. a) Subtract 3 to get subsequent terms; 56, 53, 50.
b) Divide by 2 to get subsequent terms; 12.5, 6.25, 3.125.
c) Divide by −2 to get subsequent terms; $\frac{1}{16}, -\frac{1}{32}, \frac{1}{64}$.
d) The number of factors of a increases by 1 and the number of factors of b doubles, to get subsequent terms: $aaaaa$, $bbbbbbbbbbbbbbbb$, $aaaaa$.
10. a) 4:1
11. a) $x^2 - 2x + 1$ **b)** $2x^2 - 7x - 4$ **c)** $-5x^2 + 10xy$
d) $3x^3 - 6x^2y + 3xy^2$ **e)** $9x^3 - 9x^2y$ **f)** $ac - ad + bc - bd$
12. a) 0.25 **b)** 0.46 **c)** $0.\overline{6}$ **d)** 11.5 **e)** $0.\overline{857142}$
f) $0.7\overline{3}$

13. a) 46% **b)** 80% **c)** $3.\overline{3}\%$ **d)** 225% **e)** 137.5%

Section 1.1, pp. 10–13

Practise

1. a) iterative **b)** iterative **c)** non-iterative
d) iterative **e)** non-iterative **f)** iterative
3. The order of winners may vary.

Apply, Solve, Communicate

4. Answers will vary.
5. The tracing shows irregularities according to the pulse rate. Sketches may vary.
6. a) Select an item to begin the process. Compare the ranking of this item with the ranking of a second item. The item with the higher ranking is now ranked first. Now, compare the next item's ranking with the first ranked item. The higher ranking of these two is now ranked first. Continue in this fashion with subsequent items to find the item with the highest rank. Once this item is found, work with the remaining items to find the item with the second highest ranking by selecting one of the remaining items to begin the process as outlined above. Continue in this fashion until there are no remaining items.
b) Compare the first two data, then the second and third, and then the third and fourth, and so on, interchanging the order of the data in any pair where the second item is ranked higher. Repeat this process until no data are interchanged.
7. $1 - \left(\dfrac{8}{9}\right)^n$
8. a) 48 **b)** 192 **c)** $3 \times 4^{n-1}$
9. 10, 88, 7714, 59 505 532
10. a) i) Begin with a segment 1 unit long. Branch off at 90° with two segments each one half the length of the previous branch. Repeat the process.
ii) Begin with a segment 1 unit long. Branch off with 3 segments, each of which are one third as long and separated by 60°. Repeat the process.
iii) Begin with a vertical segment 1 unit long. Branch off the top to the left at 120° with a 1-unit segment. Continue upwards on the first branch with a 1-unit segment. Branch off the top right at 120° with a 1-unit segment. Iteration: Go up one, branch left and repeat everything that is one step below. Go up one and branch left. Go up one, branch right and repeat what is on the left, two steps below. Repeat.
b) i) 5 units **ii)** 3 units **iii)** 64 units
c) no **d)** Answers may vary.
11. a) Answers may vary. **b)** Answers may vary.
c) Answers may vary.

12. a) $0, 1, \dfrac{1}{2}, \dfrac{1}{\sqrt{2}}, \dfrac{1}{\sqrt[\sqrt{2}]{2}}, \dfrac{1}{\sqrt[\sqrt[\sqrt{2}]{2}]{2}}, \ldots, \dfrac{1}{\sqrt[\sqrt[\sqrt{2}]{2}]{2}}$

b) $256, 16, 4, 2, \sqrt{2}, \sqrt{\sqrt{2}}, \ldots, \sqrt{\ldots\sqrt{2}}$

c) $2, \dfrac{1}{2}, 2, \dfrac{1}{2}, \ldots$

14. Answers may vary.

15. $\dfrac{\cos\theta}{1 - \cos\theta}$

16. Answers may vary.

17. a) Set up a table with two columns. Write the number in base 10 in the first column. If 2 divides the number with no remainder, write 0 in the second column. If 2 divides the number with remainder 1, write 1 in the second column. Below the first number in the first column, write the value of the first number divided by 2, rounding down, if necessary. If 2 divides the second number in the first column with no remainder, write 0 opposite it in the second column. Otherwise, write a 1. Continue this process until the number in the first column is a 0. Then stop. Read the numbers in the second column, from bottom to top. That is the original number's representation in base 2. For example, converting 23 in base 10 to a number in base 2 would result in the following table:

23	1
11	1
5	1
2	0
1	1
0	

Thus, 23 is 10111 in base 2.

b) i) 10000 **ii)** 10101 **iii)** 100101 **iv)** 10000010

c) i) 10 **ii)** 32 **iii)** 58 **iv)** 511

Section 1.2, pp. 22–23

Practise

1. a) Student marks: 30, 43, 56, 56, 65, 72, 74, 76, 80, 81, 88, 92, 99; Appointment times: 30, 32, 38, 40, 40, 41, 45, 45, 45, 60

b) Student marks: 70.2; Appointment times: 41.6

c) Student marks: 74; Appointment times: 40.5

d) Student marks: 56; Appointment times: 45

2. a) Microsoft® Excel: SUM(A1:A9);
Corel® Quattro® Pro: @SUM(A1..A9)

b) Microsoft® Excel: MAX(F3:K3);
Corel® Quattro® Pro: @MAX(F3..K3)

c) Microsoft® Excel: MIN(A1:K4);
Corel® Quattro® Pro: @MIN(A1..K4)

d) Microsoft® Excel: SUM(A2,B5,C7,D9);
Corel® Quattro® Pro: @SUM(A2,B5,C7,D9)

e) Microsoft® Excel: AVERAGE(F5:M5), MEDIAN(F5:M5), MODE(F5:M5);
Corel® Quattro Pro®: @AVG(F5..M5), @MEDIAN(F5..M5), @MODE(F5..M5)

f) Microsoft® Excel: SQRT(A3);
Corel® Quattro® Pro: @SQRT(A3)

g) Microsoft® Excel: B6^3;
Corel® Quattro® Pro: B6^3

h) Microsoft® Excel: ROUND(D2,4);
Corel® Quattro® Pro: @ROUND(D2,4)

i) Microsoft® Excel: COUNT(D3:M9);
Corel® Quattro® Pro: @COUNT(D3..M9)

j) Microsoft® Excel: PRODUCT(A1,B3,C5:C10);
Corel® Quattro® Pro: A1*B3*C5*C6*C7*C8*C9*C10 or @MULT (A1, B3, C5..C10)

k) Microsoft® Excel: PI;
Corel® Quattro® Pro: @PI

Apply, Solve, Communicate

5. a) To a maximum of 4 decimal places: 0, 1, 0.5, 0.7071, 0.6125, 0.6540, 0.6355, 0.6437, 0.6401, 0.6417

b) To a maximum of 4 decimal places: 256, 16, 4, 2, 1.4142, 1.1892, 1.0905, 1.0443, 1.0219, 1.0109

c) 2, 0.5, 2, 0.5, 2, 0.5, 2, 0.5, 2, 0.5

6. a), b), c)

Name	Protein	Fat	Sugars	Starch	Fibre	Other	TOTALS
Alphabits	2.4	1.1	12.0	12.0	0.9	1.6	30.0
Bran Flakes	4.4	1.2	6.3	4.7	11.0	2.4	30.0
Cheerios	4.0	2.3	0.8	18.7	2.2	2.0	30.0
Crispix	2.2	0.3	3.2	22.0	0.5	1.8	30.0
Froot Loops	1.3	0.8	14.0	12.0	0.5	1.4	30.0
Frosted Flakes	1.4	0.2	12.0	15.0	0.5	0.9	30.0
Just Right	2.2	0.8	6.6	17.0	1.4	2.0	30.0
Lucky Charms	2.1	1.0	13.0	11.0	1.4	1.5	30.0
Nuts 'n Crunch	2.3	1.6	7.1	16.5	0.7	1.8	30.0
Rice Krispies	2.1	0.4	2.9	22.0	0.3	2.3	30.0
Shreddies	2.9	0.6	5.0	16.0	3.5	2.0	30.0
Special K	5.1	0.4	2.5	20.0	0.4	1.6	30.0
Sugar Crisp	2.0	0.7	14.0	11.0	1.1	1.2	30.0
Trix	0.9	1.6	13.0	12.0	1.1	1.4	30.0
AVERAGES	2.52	0.93	8.03	14.99	1.82	1.71	
MAXIMUM	5.1	2.3	14.0	22.0	11.0	2.4	
MINIMUM	0.9	0.2	0.8	4.7	0.3	0.9	

d)

Name	Protein	Fat	Sugars	Starch	Fibre	Other	TOTALS
Bran Flakes	4.4	1.2	6.3	4.7	11.0	2.4	30.0
Shreddies	2.9	0.6	5.0	16.0	3.5	2.0	30.0
Cheerios	4.0	2.3	0.8	18.7	2.2	2.0	30.0
Just Right	2.2	0.8	6.6	17.0	1.4	2.0	30.0
Lucky Charms	2.1	1.0	13.0	11.0	1.4	1.5	30.0
Sugar Crisp	2.0	0.7	14.0	11.0	1.1	1.2	30.0
Trix	0.9	1.6	13.0	12.0	1.1	1.4	30.0
Alphabits	2.4	1.1	12.0	12.0	0.9	1.6	30.0
Nuts 'n Crunch	2.3	1.6	7.1	16.5	0.7	1.8	30.0
Crispix	2.2	0.3	3.2	22.0	0.5	1.8	30.0
Frosted Flakes	1.4	0.2	12.0	15.0	0.5	0.9	30.0
Froot Loops	1.3	0.8	14.0	12.0	0.5	1.4	30.0
Special K	5.1	0.4	2.5	20.0	0.4	1.6	30.0
Rice Krispies	2.1	0.4	2.9	22.0	0.3	2.3	30.0
AVERAGES	2.52	0.93	8.03	14.99	1.82	1.71	
MAXIMUM	5.1	2.3	14.0	22.0	11.0	2.4	
MINIMUM	0.9	0.2	0.8	4.7	0.3	0.9	

7. a) 7.998   **b)** 424 488

8. Look for a menu item such as "Freeze Panes."

9. Enter 1 in cell A1, and the formula A1+1 in cell A2. Copy this formula to the nth row in column A. Enter 1 in cell B1, and formula B1*A2 in cell B2. Copy this formula to the nth row in column B. Cell Bn will contain the value $n \times (n - 1) \times \ldots \times 3 \times 2 \times 1$.

Section 1.3, pp. 31–32

Practise

1. a), b), c), d), and f) would be considered databases.

Apply, Solve, Communicate

Answers to questions 2 to 8 will vary.

Section 1.4, p. 40

Practise

1. a) Answers may vary. For example, on the TI-83 Plus: randInt(1,25,100).
  b) Answers may vary. For example, on the TI-83 Plus: randInt(0,40,24)–20.
2. a) Microsoft® Excel: 1 + 24*RAND().
  Corel® Quattro® Pro: 1 + 24*@RAND.
  Copy this formula to fill 100 cells.
  b) Microsoft® Excel: 1 + ROUND(24*RAND(),0).
  Corel® Quattro® Pro: @RANDBETWEEN(1,25).
  Copy this formula to fill 100 cells.
  c) Microsoft® Excel: –40 + ROUND(80*RAND(),0).
  Corel® Quattro® Pro: @RANDBETWEEN(–40,40).
  Copy this formula to fill 16 cells.
  d) COUNTIF(C10:V40,42.5)

Apply, Solve, Communicate

3. Answers will vary.
4. Answers will vary.
5. a) Answers will vary.
  b) Random integers between 0 and 9 could be generated by recording the last digit of phone numbers.
6. Answers may vary.
7. Answers will vary.
9. Answers will vary.

Section 1.5, pp. 49–52

Practise

1. a) i) A: 2, B: 2, C: 3, D: 2, E: 3   **ii)** traceable
  b) i) P: 4, Q: 5, R: 3, S: 4, T: 5, U: 3   **ii)** not traceable
3. a) 3   **b)** 4

Apply, Solve, Communicate

4. The given map requires only 2 colours.
6. b) Answers may vary. Time Slot 1: English; Time Slot 2: Geography, Geometry; Time Slot 3: Calculus, History; Time Slot 4: Physics, Music; Time Slot 5: French, Mathematics of Data Management
7. a) no   **b)** yes
8. a) A: 4, B: 2, C: 3, D: 3   **b)** 12   **c)** 6
  d) The sum of the degrees divided by 2 equals the number of edges.
9. b) Answers will vary.
10. a) no
11. no
12. inside
13. a) North Bay to Kitchener, Hamilton to Windsor; both links could be backed up by linking Windsor to Thunder Bay.
  b) Charlottetown to Halifax, Halifax to Montréal, Montréal to Toronto, Vancouver to Edmonton, Edmonton to Winnipeg, Winnipeg to Saskatoon; all links could be backed up by linking Saskatoon to Vancouver and Toronto to Charlottetown.
14. a) yes   **b)** yes   **c)** 285 km
15. a) Thunder Bay to Sudbury to North Bay to Kitchener to Windsor
  b) Hamilton to Ottawa to North Bay to Sudbury
  c) Answers may vary.
16. $1856
19. not possible
20. a) yes   **b)** no
21. no
22. no
23. b) three   **c)** Answers may vary.

Section 1.6, pp. 60–62

Practise

1. a) 2×3 **b)** 1×3 **c)** 4×3

2. a) i) 6   **ii)** −4   **iii)** 2
   b) i) a_{31} **ii)** a_{33} **iii)** a_{42}
3. a) MATRICES ARE FUN   **b)** I LOVE MATH
   c) WOW THIS IS FUN
4. Answers may vary.

   a) $[2 \ -1]$, $[1 \ 2 \ 0]$, $\begin{bmatrix} 1 \\ 5 \end{bmatrix}$, $\begin{bmatrix} 1 \\ 0 \\ -3 \end{bmatrix}$

   b) 1×2, 1×3, 2×1, 3×1
5. Answers may vary. Examples are given.

   a) $\begin{bmatrix} 1 & 2 \\ 3 & -1 \end{bmatrix}$, $\begin{bmatrix} 1 & 0 & 0 \\ 0 & 1 & 0 \\ 1 & 0 & 1 \end{bmatrix}$ **b)** 2×2, 3×3

6. a) $\begin{bmatrix} 2 & 3 & 4 & 5 \\ 3 & 4 & 5 & 6 \\ 4 & 5 & 6 & 7 \end{bmatrix}$ **b)** $\begin{bmatrix} 3 & 2 & 3 & 4 \\ 2 & 3 & 6 & 8 \\ 3 & 6 & 3 & 12 \\ 4 & 8 & 12 & 3 \end{bmatrix}$

7. a) $w = -2$, $x = 3$, $y = 5$, $z = 2$
   b) $w = 2$, $x = \pm 3$, $y = 2$, $z = -5$

8. a) $\begin{bmatrix} 5 & 3 \\ -3 & 10 \\ 13 & 2 \\ -5 & -4 \end{bmatrix}$ **b)** $\begin{bmatrix} 5 & 3 \\ -3 & 10 \\ 13 & 2 \\ -5 & -4 \end{bmatrix}$ **c)** not possible

   d) $\begin{bmatrix} 6 & -3 \\ 9 & 27 \\ 15 & 0 \\ -12 & 3 \end{bmatrix}$ **e)** $\begin{bmatrix} -\frac{3}{2} & -2 \\ 3 & \frac{1}{2} \\ -4 & -1 \\ \frac{1}{2} & \frac{5}{2} \end{bmatrix}$ **f)** $\begin{bmatrix} 2 & 10 \\ -18 & -16 \\ 6 & 4 \\ 6 & -12 \end{bmatrix}$

   g) $\begin{bmatrix} 0 & -11 \\ 21 & 25 \\ -1 & -4 \\ -10 & 13 \end{bmatrix}$

9. a) Both sides equal $\begin{bmatrix} 8 & -7 \\ 3 & 2 \\ 5 & 2 \end{bmatrix}$.   **b)** Both sides equal $\begin{bmatrix} 10 & -4 \\ 11 & -4 \\ 9 & 3 \end{bmatrix}$.

   c) Both sides equal $\begin{bmatrix} 40 & -35 \\ 15 & 10 \\ 25 & 10 \end{bmatrix}$.

10. $w = 11$, $x = 8$, $y = 3$, $z = 9$

11. a) $\begin{bmatrix} 4 & -2 & 6 \\ -6 & 3 & -10 \end{bmatrix}$ **b)** 2

Apply, Solve, Communicate

12. a)

	Thunder Bay	Sault Ste. Marie	North Bay	Ottawa	Toronto	
	0	710	1135	1500	1365	Thunder Bay
	710	0	425	790	655	Sault Ste. Marie
	1135	425	0	365	350	North Bay
	1500	790	365	0	400	Ottawa
	1365	655	350	400	0	Toronto

b) same result as in part a)
c) equal
13. a)

	U.S.A.	U.K.	Germany	France	Sweden	
	67	21	20	12	4	physics
	43	25	27	7	4	chemistry
	78	24	16	7	7	physiology/medicine
	10	8	7	12	7	literature
	18	13	4	9	5	peace
	25	7	1	1	2	economic sciences

b) U.S.A.: 241; U.K.: 98; Germany: 75; France: 48; Sweden: 29

14. a)

Field of Study	1997 Males	1997 Females
Social Sciences	28 421	38 244
Education	8 036	19 771
Humanities	8 034	13 339
Health	3 460	9 613
Engineering	10 125	2 643
Agriculture	4 780	6 995
Mathematics	6 749	2 989
Fine & Applied Arts	1 706	3 500
Arts & Sciences	1 730	3 802

Field of Study	1998 Males	1998 Females
Social Sciences	27 993	39 026
Education	7 565	18 391
Humanities	7 589	13 227
Health	3 514	9 144
Engineering	10 121	2 709
Agriculture	4 779	7 430
Mathematics	6 876	3 116
Fine & Applied Arts	1 735	3 521
Arts & Sciences	1 777	3 563

b)

Field of Study	Males
Social Sciences	56 414
Education	15 601
Humanities	15 623
Health	6 974
Engineering	20 246
Agriculture	9 559
Mathematics	13 625
Fine & Applied Arts	3 441
Arts & Sciences	3 507

c)

Field of Study	Females
Social Sciences	77 270
Education	38 162
Humanities	26 566
Health	18 757
Engineering	5 352
Agriculture	14 425
Mathematics	6 105
Fine & Applied Arts	7 021
Arts & Sciences	7 365

d)

Field of Study	Average for Females
Social Sciences	38 635
Education	19 081
Humanities	13 283
Health	9 378.5
Engineering	2 676
Agriculture	7 212.5
Mathematics	3 052.5
Fine & Applied Arts	3 510.5
Arts & Sciences	3 682.5

b)

Age Group	Total
0–4	1 777 330
5–9	2 044 418
10–14	2 049 140
15–19	2 071 614
20–24	2 081 186
25–29	2 109 770
30–34	2 283 166
35–39	2 695 561
40–44	2 611 243
45–49	2 319 848
50–54	2 045 093
55–59	1 555 248
60–64	1 256 573
65–69	1 136 889
70–74	998 277
75–79	804 364
80–84	494 406
85–89	282 415
90+	133 546

15. a)

Matrix 1		Matrix 2	
Age Group	Males	Age Group	Females
0–4	911 028	0–4	866 302
5–9	1 048 247	5–9	996 171
10–14	1 051 525	10–14	997 615
15–19	1 063 983	15–19	1 007 631
20–24	1 063 620	20–24	1 017 566
25–29	1 067 870	25–29	1 041 900
30–34	1 154 071	30–34	1 129 095
35–39	1 359 796	35–39	1 335 765
40–44	1 306 705	40–44	1 304 538
45–49	1 157 288	45–49	1 162 560
50–54	1 019 061	50–54	1 026 032
55–59	769 591	55–59	785 657
60–64	614 659	60–64	641 914
65–69	546 454	65–69	590 435
70–74	454 269	70–74	544 008
75–79	333 670	75–79	470 694
80–84	184 658	80–84	309 748
85–89	91 455	85–89	190 960
90+	34 959	90+	98 587

c)

Age Group	Total
0–4	1 803 990
5–9	2 073 084
10–14	2 079 877
15–19	2 102 680
20–24	2 112 404
25–29	2 141 417
30–34	2 317 413
35–39	2 735 994
40–44	2 650 412
45–49	2 354 646
50–54	2 075 769
55–59	1 578 577
60–64	1 275 422
65–69	1 153 942
70–74	1 013 251
75–79	816 439
80–84	501 822
85–89	286 651
90+	135 549

16. a)

$$\begin{array}{c} \\ \end{array}\begin{array}{cccccccc} \text{K} & \text{L} & \text{M} & \text{NF} & \text{O} & \text{S} & \text{T} & \text{W} \end{array}$$

$$\begin{bmatrix} 0 & 0 & 1 & 0 & 1 & 0 & 1 & 0 \\ 0 & 0 & 0 & 0 & 0 & 0 & 1 & 1 \\ 1 & 0 & 0 & 0 & 1 & 0 & 1 & 0 \\ 0 & 0 & 0 & 0 & 1 & 0 & 1 & 0 \\ 1 & 0 & 1 & 0 & 0 & 0 & 1 & 0 \\ 0 & 0 & 0 & 0 & 0 & 0 & 1 & 0 \\ 1 & 1 & 1 & 1 & 1 & 1 & 0 & 1 \\ 0 & 1 & 0 & 0 & 0 & 0 & 1 & 0 \end{bmatrix} \begin{array}{l} \text{Kingston} \\ \text{London} \\ \text{Montréal} \\ \text{Niagara Falls} \\ \text{Ottawa} \\ \text{Sudbury} \\ \text{Toronto} \\ \text{Windsor} \end{array}$$

b) no direct connection between Niagara Falls and Montréal

c) no direct connection between Montréal and Niagara Falls

d) a direct connection would go both ways

e) the number of direct connections to Kingston

f) the number of direct connections from Kingston

g) direct connections go both ways

Section 1.7, pp. 74–77

Practise

1. a) $\begin{bmatrix} 12 & 47 \\ 21 & -7 \end{bmatrix}$ **b)** $\begin{bmatrix} -13 & -27 \\ -31 & 18 \end{bmatrix}$ **c)** $\begin{bmatrix} -59 & 18 \\ -14 & -63 \end{bmatrix}$

d) not possible **e)** $\begin{bmatrix} 18 & 20 & 4 \\ -16 & -17 & 4 \end{bmatrix}$ **f)** $\begin{bmatrix} -13 \\ 14 \\ -31 \end{bmatrix}$

g) $\begin{bmatrix} -15 & -26 & 1 \\ -7 & -11 & 5 \end{bmatrix}$

4. a) Both sides equal $\begin{bmatrix} 20 & 5 \\ 10 & -5 \end{bmatrix}$.

b) Both sides equal $\begin{bmatrix} 15 & 95 \\ 8 & 32 \end{bmatrix}$.

c) $AB = \begin{bmatrix} 15 & 20 \\ 8 & 8 \end{bmatrix}$, $BA = \begin{bmatrix} 23 & -4 \\ -10 & 0 \end{bmatrix}$

5. a) $\begin{bmatrix} 2 & 0.5 \\ -1 & 0 \end{bmatrix}$ **b)** no inverse **c)** $\begin{bmatrix} \frac{1}{3} & 0 \\ 2 & 1 \end{bmatrix}$

d) $\begin{bmatrix} -1 & 1.5 \\ 2 & -2.5 \end{bmatrix}$ **e)** no inverse

6. a) $\begin{bmatrix} 0.6 & -0.2 & -2 \\ 0 & 0 & -1 \\ 0.4 & 0.2 & -1 \end{bmatrix}$

b) $\begin{bmatrix} 0.0851 & 0.4255 & 0.1064 \\ -0.0638 & -0.3191 & 0.1702 \\ 0.2340 & 0.1702 & 0.426 \end{bmatrix}$

c) $\begin{bmatrix} \frac{1}{6} & -\frac{1}{12} & -\frac{1}{4} & \frac{1}{6} \\ -\frac{1}{3} & \frac{1}{6} & \frac{1}{2} & -\frac{1}{3} \\ \frac{1}{3} & \frac{1}{3} & 0 & -\frac{2}{3} \\ \frac{1}{6} & \frac{5}{12} & \frac{1}{4} & \frac{1}{6} \end{bmatrix}$

Apply, Solve, Communicate

7. a) $A^{-1} = \begin{bmatrix} 2.5 & 2 \\ 1 & 1 \end{bmatrix}$, $(A^{-1})^{-1} = \dfrac{1}{2.5 - 2}\begin{bmatrix} 1 & -2 \\ -1 & 2.5 \end{bmatrix}$

$$= 2\begin{bmatrix} 1 & -2 \\ -1 & 2.5 \end{bmatrix}$$

$$= \begin{bmatrix} 2 & -4 \\ -2 & 5 \end{bmatrix}$$

$$= A$$

b) Both sides equal $\begin{bmatrix} -3 & -1 \\ 2.5 & 1 \end{bmatrix}$.

c) Both sides equal $\begin{bmatrix} 2.5 & 1 \\ 2 & 1 \end{bmatrix}$.

8. Downtown store: $6180, Northern store: $4370, Southern store: $4600

9. a)

$$A = \begin{array}{c}\\\\\\\\\\\\\\\\\\\end{array}\begin{array}{cccccccc} \text{Mon} & \text{Tues} & \text{Wed} & \text{Thurs} & \text{Fri} & \text{Sat} & \text{Sun} \end{array}$$

$$A = \begin{bmatrix} 0 & 8 & 0 & 8 & 8 & 0 & 0 \\ 4 & 4 & 0 & 0 & 6.5 & 4 & 4 \\ 0 & 4 & 4 & 4 & 4 & 8 & 8 \\ 0 & 3 & 3 & 3 & 3 & 8 & 0 \\ 8 & 8 & 8 & 0 & 0 & 0 & 0 \\ 0 & 0 & 3 & 5 & 5 & 8 & 0 \\ 3 & 3 & 3 & 3 & 3 & 0 & 0 \\ 8 & 8 & 8 & 8 & 8 & 0 & 0 \\ 8 & 0 & 0 & 8 & 8 & 8 & 8 \\ 3 & 4.5 & 4 & 3 & 5 & 0 & 0 \end{bmatrix} \begin{array}{l} \text{Chris} \\ \text{Lee} \\ \text{Jagjeet} \\ \text{Pierre} \\ \text{Ming} \\ \text{Bobby} \\ \text{Nile} \\ \text{Louis} \\ \text{Glenda} \\ \text{Imran} \end{array}$$

b) $B = \begin{bmatrix} 7.00 \\ 6.75 \\ 7.75 \\ 6.75 \\ 11.00 \\ 8.00 \\ 7.00 \\ 12.00 \\ 13.00 \\ 7.75 \end{bmatrix} \begin{array}{l} \text{Chris} \\ \text{Lee} \\ \text{Jagjeet} \\ \text{Pierre} \\ \text{Ming} \\ \text{Bobby} \\ \text{Nile} \\ \text{Louis} \\ \text{Glenda} \\ \text{Imran} \end{array}$

c) Chris: $168.00; Lee: $151.88; Jagjeet: $248.00; Pierre: $135.00; Ming: $352.00; Bobby: $168.00; Nicole: $105.00; Louis: $480.00; Glenda: $520.00; Imran: $151.13

d) $2479.01

10. Create a 10×1 matrix to represent the percent of males for each sport and multiply by the number of males. Do a similar scalar multiplication for females.

$$\begin{array}{cc} \text{Males} & \text{Females} \end{array}$$

$$\begin{bmatrix} 1\,325\,007 \\ 1\,432\,440 \\ 954\,960 \\ 429\,732 \\ 549\,102 \\ 393\,921 \\ 549\,102 \\ 429\,732 \\ 346\,173 \\ 358\,110 \end{bmatrix} \begin{bmatrix} 480\,597 \\ 61\,615 \\ 382\,013 \\ 690\,088 \\ 234\,137 \\ 345\,044 \\ 184\,845 \\ 221\,814 \\ 320\,398 \\ 246\,460 \end{bmatrix} \begin{array}{l} \text{Golf} \\ \text{Ice Hockey} \\ \text{Baseball} \\ \text{Swimming} \\ \text{Basketball} \\ \text{Volleyball} \\ \text{Soccer} \\ \text{Tennis} \\ \text{Skiing} \\ \text{Cycling} \end{array}$$

11. a)

Total Cloth (m²)   Total Labour ($)

$$\begin{bmatrix} 1080 \\ 1800 \\ 3750 \\ 4000 \end{bmatrix} \qquad \begin{bmatrix} 10\ 200 \\ 13\ 500 \\ 25\ 000 \\ 22\ 000 \end{bmatrix} \begin{matrix} \text{small} \\ \text{med} \\ \text{LG} \\ \text{XLG} \end{matrix}$$

b) box of 100

$$\begin{bmatrix} \$9\ 064 \\ \$9\ 662 \\ \$10\ 386 \\ \$11\ 645 \\ \$13\ 030 \end{bmatrix} \begin{matrix} \text{xsmall} \\ \text{small} \\ \text{med} \\ \text{LG} \\ \text{XLG} \end{matrix}$$

c) $611 599

12. a) −50, −42, 86, 82, 13, −67, −11, 117, −22, −58, 58, 98, −4, −15, 40, 33, −67, −76, 117, 128

b) 23, −65, −13, 115, −27, 39, 65, −37, −14, 30, 50, −22, −40, 23, 68, −5, −46, −62, 86, 114

13. a) LUNCH TODAY AT JOES

b) HOW ARE YOU TODAY

14. Answers will vary.

15. a)

	A	G	P	S	SI	T	
	0	1	0	0	1	1	Administration
	1	0	0	0	0	1	Guidance
	0	0	0	1	0	1	Parents
	0	0	1	0	0	1	Students
	1	0	0	0	0	0	Superintendent
	1	1	1	1	0	0	Teachers

b) If $a_{ij} = 1$, there is open communication between i and j; if $a_{ij} = 0$, there is no open communication between i and j.

c) the number of open links of communication between parents and others

d)

	A	G	P	S	SI	T	
	3	1	1	1	0	1	Administration
	1	2	1	1	1	1	Guidance
	1	1	2	1	0	1	Parents
	1	1	1	2	0	1	Students
	0	1	0	0	1	1	Superintendent
	1	1	1	1	1	4	Teachers

e) 1; Parents-Teachers-Administration

f)

	A	G	P	S	SI	T	
	3	2	1	1	1	2	Administration
	2	2	1	1	1	2	Guidance
	1	1	2	2	0	2	Parents
	1	1	2	2	0	2	Students
	1	1	0	0	1	1	Superintendent
	2	2	2	2	1	4	Teachers

Each a_{ij} represents the number of lines of communication with no intermediary links and exactly one intermediary link between i and j.

16. a) 2; Honey Bridge-Green Bridge-Connecting Bridge, Honey Bridge-Blacksmith Bridge-Connecting Bridge

b) 3 **c)** yes

17. a) 14 **b)** 18

18. Answers will vary.

19. Rewrite the system $ax_1 + bx_2 = c$, $dx_1 + ex_2 = f$ as a matrix product $\begin{bmatrix} a & b \\ d & e \end{bmatrix} \begin{bmatrix} x_1 \\ x_2 \end{bmatrix} = \begin{bmatrix} c \\ f \end{bmatrix}$. Then, multiply both sides on the left by the inverse of the matrix $\begin{bmatrix} a & b \\ d & e \end{bmatrix}$.

20. Answers will vary.

21. Answers will vary.

Review of Key Concepts, pp. 78–81

1.1 The Iterative Process

1. b) 30

2. a) Beginning with the 3-word MATH triangle, place two copies of this triangle on each subsequent row; the copy to the extreme left has the T in the top word of the triangle centred under the first M of the previous line and the copy to the extreme right has the M in the top word of the triangle centred under the last T of the previous line.

c) The repeated word MATH forms a pyramid.

3. b) 16 cm² **c)** 64 cm²

4. Answers may vary.

1.2 Data Management Software

5. Answers may vary; for example, Microsoft® Excel: spreadsheet functions; Fathom™: graphing; Microsoft® Access: sorting.

6. a) 0 **b)** 30 **c)** 12

7. Corel® Quattro® Pro: enter A:A3 in cell B2 in Sheet B and copy this formula through cell B9; Microsoft® Excel: enter Sheet1!A3 in cell B2 in Sheet2 and copy this formula through cell B9.

8. Use a filter to select the even Celsius degrees.

1.3 Databases

Answers to questions 9 to 12 will vary.

1.4 Simulations

13. Answers may vary.

14. a) randInt(18,65)

b) Microsoft® Excel: 18 + ROUND(47*RAND(),0). Corel® Quattro® Pro: @RANDBETWEEN(18,65).

15. Remove 2 cards from a standard deck of 52 playing cards. To simulate a repeated brainteaser, draw 5 cards, one at a time with replacement, and reshuffle after every draw. If the same card appears more than once, this will count as a success—getting two chocolate bars with the same brainteaser. This experiment must be repeated many times to obtain an approximation to the probability of the actual event.

16. For each of 500 simulations, draw 5 random numbers from 1 to 50—if any two are the same, this counts as a success—getting two chocolate bars with the same brainteaser.

1.5 Graph Theory

17. a) 3 **b)** 3

18. a) i) yes **ii)** yes **iii)** yes

b) i) yes **ii)** no **iii)** no

c) i) yes **ii)** no **iii)** no

20. Time Slot 1: *Gone With the Wind*, *The Amazon Queen*; Time Slot 2: *Curse of the Mummy*, *West Side Story*, *Ben Hur*; Time Slot 3: *Citizen Kane*, *Jane Eyre*

21. a)

$$\begin{array}{cccccc} A & D & M & P & S & T \\ \end{array}$$

0	0	0	1	0	1	Afra
0	0	1	0	1	1	Deqa
0	1	0	1	1	0	Mai
1	0	1	0	1	0	Priya
0	1	1	1	0	1	Sarah
1	1	0	0	1	0	Tanya

b) no **c)** Sarah **d)** Afra

1.6 Modelling With Matrices

22. a) 4×3

b) i) -8 **ii)** 5 **iii)** -2

c) i) a_{23} **ii)** a_{42} **iii)** a_{12}

23.
$$\begin{bmatrix} 1 & 2 & 3 \\ 2 & 4 & 6 \\ 3 & 6 & 9 \\ 4 & 8 & 12 \end{bmatrix}$$

24. a) $\begin{bmatrix} 9 & 3 & -5 \\ -12 & 9 & 5 \end{bmatrix}$ **b)** not possible **c)** not possible

d) $\begin{bmatrix} 12 & 9 \\ -3 & 21 \\ 18 & 6 \end{bmatrix}$ **e)** $\begin{bmatrix} -3 & -\dfrac{1}{2} & 2 \\ \dfrac{5}{2} & -\dfrac{9}{2} & 0 \end{bmatrix}$ **f)** $\begin{bmatrix} 36 & 3 \\ 6 & 33 \\ 24 & 21 \end{bmatrix}$

g) $\begin{bmatrix} 11 & -9 \\ 5 & 4 \\ 1 & 10 \end{bmatrix}$ **h)** not possible

25. a)

$$\begin{bmatrix} 15 \\ 17 \\ 4 \\ 15 \\ 8 \\ 12 \\ 10 \end{bmatrix}, \begin{bmatrix} 10 \\ 3 \\ 15 \\ 20 \\ 12 \\ 5 \\ 15 \end{bmatrix}, \begin{bmatrix} 17 \\ 13 \\ 17 \\ 12 \\ 12 \\ 16 \\ 23 \end{bmatrix} \begin{array}{l} \text{basketballs} \\ \text{volleyballs} \\ \text{footballs} \\ \text{baseballs} \\ \text{soccer balls} \\ \text{pk tennis balls} \\ \text{pk golf balls} \end{array}$$

b) $\begin{bmatrix} 8 \\ 7 \\ 2 \\ 23 \\ 8 \\ 1 \\ 2 \end{bmatrix}$ basketballs, volleyballs, footballs, baseballs, soccer balls, pk tennis balls, pk golf balls

c) $\begin{bmatrix} 6 \\ 6 \\ 2 \\ 18 \\ 6 \\ 1 \\ 2 \end{bmatrix}$ basketballs, volleyballs, footballs, baseballs, soccer balls, pk tennis balls, pk golf balls

26. Answers will vary.

1.7 Problem Solving With Matrices

27. a) $\begin{bmatrix} -21 & 35 \\ -21 & 21 \end{bmatrix}$ **b)** $\begin{bmatrix} 4 & 5 \\ -62 & -4 \end{bmatrix}$ **c)** $\begin{bmatrix} -14 & 35 \\ 42 & -21 \end{bmatrix}$

d) not possible **e)** $\begin{bmatrix} 26 & 20 & 13 \\ -14 & 4 & 30 \\ -59 & -46 & 61 \end{bmatrix}$

28. a) $A^t = \begin{bmatrix} 1 & 8 \\ 5 & -2 \end{bmatrix}$, $B^t = \begin{bmatrix} 0 & 6 \\ 4 & -1 \end{bmatrix}$

b) Both sides equal $\begin{bmatrix} 30 & -12 \\ -1 & 34 \end{bmatrix}$.

29. a) XYZ: $7350, YZX: $4080, ZXY: $12 400

b) Find the column sum of the product matrix $\begin{bmatrix} 7\,350 \\ 4\,080 \\ 12\,400 \end{bmatrix}$.

30. Answers may vary.

31. a) $A \times A^{-1} = I$

c) $\begin{bmatrix} \dfrac{3}{2} & -\dfrac{5}{2} \\ -1 & 2 \end{bmatrix}$

32. a)

$$A = \begin{array}{c} \begin{array}{ccccc} B & LF & P & SF & S \end{array} \\ \begin{bmatrix} 0 & 0 & 0 & 0 & 0 \\ 1 & 0 & 0 & 0 & 1 \\ 0 & 1 & 0 & 1 & 1 \\ 1 & 1 & 0 & 0 & 1 \\ 1 & 0 & 0 & 0 & 0 \end{bmatrix} \begin{array}{l} \text{Bacteria} \\ \text{Large Fish} \\ \text{Plants} \\ \text{Small Fish} \\ \text{Snails} \end{array} \end{array}$$

b) $\begin{bmatrix} 0 & 0 & 0 & 0 & 0 \\ 1 & 0 & 0 & 0 & 0 \\ 3 & 1 & 0 & 0 & 2 \\ 2 & 0 & 0 & 0 & 1 \\ 0 & 0 & 0 & 0 & 0 \end{bmatrix}$ **c)** 2

d) $\begin{bmatrix} 0 & 0 & 0 & 0 & 0 \\ 2 & 0 & 0 & 0 & 1 \\ 3 & 2 & 0 & 1 & 3 \\ 3 & 1 & 0 & 0 & 2 \\ 1 & 0 & 0 & 0 & 0 \end{bmatrix}$ Each a_{ij} represents the number of routes from i to j with at most one intermediary link.

e) $\begin{bmatrix} 0 & 0 & 0 & 0 & 0 \\ 0 & 0 & 0 & 0 & 0 \\ 3 & 0 & 0 & 0 & 1 \\ 1 & 0 & 0 & 0 & 0 \\ 0 & 0 & 0 & 0 & 0 \end{bmatrix}$

f) Plants-Small Fish-Large Fish-Bacteria; Plants-Small Fish-Snails-Bacteria; Plants-Large Fish-Snails-Bacteria

Chapter Test, pp. 82–83

1. a) Answers will vary. **b)** Answers will vary.

2. $0, \dfrac{1}{2}, \dfrac{2}{5}, \dfrac{5}{12}, \dfrac{12}{29}, \dfrac{29}{70}, \ldots$ After the first term, the numerator plus the denominator of the previous term added to the numerator of the following term gives the denominator of the following term.

3. b) 6

4. a) C1+C2+C3+C4+C5+C6+C7+C8 or SUM(C1:C8)

b) MIN(A5:G5) **c)** (5−SQRT(6))/(10+15)

Month	Balance	Payment	Interest	Principal	New Balance
February	1000.00	88.88	5.00	83.88	916.12
March	916.12	88.88	4.58	84.30	831.82
April	831.82	88.88	4.16	84.72	747.10
May	747.10	88.88	3.74	85.14	661.96
June	661.96	88.88	3.31	85.57	576.38
July	576.38	88.88	2.88	86.00	490.39
August	490.39	88.88	2.45	86.43	403.96
September	403.96	88.88	2.02	86.86	317.10
October	317.10	88.88	1.59	87.29	229.80
November	229.80	88.88	1.15	87.73	142.07
December	142.07	88.88	0.71	88.17	53.90
January	53.90	88.88	0.27	88.61	-34.71

In the spreadsheet, cell D2 contains the formula 0.005*B2, cell E2 contains the formula C2–D2, cell F2 contains the formula B2–E2, and cell B3 contains the formula F2. These cells are copied down to 12 months.

b) A final payment of $88.88 overpays the loan by $34.71 and so the final payment is $88.88 – $34.71 = $54.17.

c) Highlight the balance column, and from the Insert menu, select Chart. From this menu, select line graph.

6. Answers may vary.

7. Use the TI-83 Plus: randInt(1, 50); use a spreadsheet function such as INT(RAND()*51); write the integers from 1 to 50 on equal-sized slips of paper and draw slips at random from a hat, replacing the slip after each draw.

8. b) 4 colours

9. a) No; more than two vertices have an odd degree.
b) Pinkford-Brownhill-Whiteford-Redville-Blueton-Greenside-Blacktown-Orangeton-Pinkford

10. Yes; there are exactly two vertices with an odd degree (in the associated network diagram)

11. a) 4×3 **b)** 9   **c)** a_{12}
d) No; the inner dimensions do not match.

12. a) $\begin{bmatrix} 10 & 9 \\ 19 & -5 \end{bmatrix}$ **b)** not possible   **c)** not possible

d) $\begin{bmatrix} 22 & 70 \\ 20 & 40 \\ -11 & -35 \end{bmatrix}$ **e)** $\begin{bmatrix} 8 & 5 & -4 \\ -2 & 0 & 1 \end{bmatrix}$

13. $4175

CHAPTER 2

Review of Prerequisite Skills, p. 90

1. a) $79   **b)** $16.99   **c)** $479   **d)** $64.69
2. a) $13.50/h   **b)** $0.83
3. 30%
4. $188.89
5. a) mean: 25.8, median: 26, mode: 26
   b) mean: 21, median: 21, mode: no mode
   c) mean: 20.3, median: 18, mode: 10, 18
   d) mean: 43.2, median: 41, mode: 70
   e) mean: 242.2, median: 207.5, mode: no mode
   f) mean: 33.2, median: 33.5, mode: 32
6. a) approximately $1.44   **b)** 1997
   c) yearly increases in price
   d) 10.4% **e)** domain: {1996–2001}, range: {1.44–1.59}

Section 2.1, pp. 101–103

Practise

1. a) Some intervals have common endpoints; a 38-year-old could be placed in either of two intervals.
   b) The intervals 81–85 and 86–90 are omitted.
2. a) bar graph   **b)** histogram   **c)** bar graph
   d) histogram
3. b) 19.4%   **c)** Answers will vary.
   d) Less than 50% of the respondents order red meat.
4. Answers will vary.

Apply, Solve, Communicate

5. a) 53   **b)** size: 5; number 11

c)

Score	Tally	Frequency
39.5–44.5	III	3
44.5–49.5	I	1
49.5–54.5	II	2
54.5–59.5	IIII	4
59.5–64.5	II	2
64.5–69.5	II	2
69.5–74.5	II	2
74.5–79.5	HHT	5
79.5–84.5	II	2
84.5–89.5	I	1
89.5–94.5	III	3

g) Frequency polygon: shows the changes in frequency from one interval to the next. Relative frequency polygon: shows the changes in frequency relative to the total number of scores. Cumulative frequency: shows the rate of change of frequency from one interval to the next and the total number of scores.

6. d) Discrepancies are due to chance variation.

7. a) ungrouped

b)

Number of Cups	Tally	Frequency
0	II	2
1	LHT III	8
2	LHT I	6
3	III	3
4	I	1

d) Answers may vary.

8. a) grouped

b)

Purchase Amount ($)	Tally	Frequency
5.00–14.99	II	2
15.00–24.99	III	3
25.00–34.99	LHT	5
35.00–44.99	IIII	4
45.00–54.99	IIII	4
55.00–64.99	III	3
65.00–74.99	III	3
75.00–84.99	II	2
85.00–94.99	II	2
95.00–104.99		0
105.00–114.99	II	2

d) Answers will vary.

9. a)

Speed (km/h)	Tally	Frequency
61–65		0
66–70	IIII	4
71–75	LHT III	8
76–80	LHT II	7
81–85	II	2
86–90	I	1
91–95	I	1
96–100	I	1

d) 12   **e)** 5

10. a)

Salary	Tally	Frequency
0 – 499 999	II	2
500 000 – 999 999	LHT LHT LHT II	17
1 000 000 – 1 499 999	II	2
1 500 000 – 1 999 999	I	1
2 000 000 – 2 499 999		0
2 500 000 – 2 999 999		0
3 000 000 – 3 499 999	I	1
3 500 000 – 3 999 999		0
4 000 000 – 4 499 999	I	1

11. a) 1

b) Relative frequencies are a percent of the total, and so their sum is 100%, or 1.

12.

Score	Frequency
29.5–39.5	1
39.5–49.5	2
49.5–59.5	4
59.5–69.5	8
69.5–79.5	5
79.5–89.5	3
89.5–99.5	2

13. The data does support the theory since fewer than 50% of those surveyed purchased 2 or more of the band's CDs. However, this evidence is far from conclusive.

14. a) 2, 4, 6, 7, 8, 9, 10, 11, 12, 13, 14
e) chance variation

15. a) 54.5–64.5; this interval has the steepest line segment.
b) 8

16. Answers may vary.

Section 2.2, pp. 109–112

Practise

1. a) 11.5   **b)** 11.5% **c) i)** 8.14 **ii)** $57.80

Apply, Solve, Communicate

2. a) to have a collection of items that are representative of purchases of typical Canadians
b) No, the changes in all prices are used to compute an average.

3. a) 1996 **b)** 0   **c)** 1999–2001   **d)** no

4. a) an increase in prices and a decrease in purchasing power
b) They give a measure of the increase in prices.
c) In general, an increase in one of these indices would coincide with an increase in the other.
d) Housing will have a significant weighting in the CPI.

5. a) scale on vertical axis; period of time over which the graph ranges; scale on horizontal axis

b) Trends may be observed over a longer period of time. A period of time over which the index has doubled is shown.

c) Detailed month-by-month variations are not exhibited. Year-by-year growth rates are not easily calculated.

d) 1980

e) The average retail price has doubled from 1980 to 1992.

6. a) The mean annual change is lower for each graph.

b) Energy costs have increased more rapidly than the average cost of other goods and services.

c) the percent change in the CPI due to energy costs alone

d) about 8%

e) significantly lower

7. b)

	Means Values Per Game		
Season	Goals	Assists	Points
1	0.15	0.20	0.35
2	0.16	0.24	0.40
3	0.25	0.33	0.58
4	0.24	0.46	0.70
5	0.34	0.44	0.78

d) The graph of the actual number of points produces the most steeply increasing curve and so is likely the one that the agent would use during negotiations. The manager will likely use the graphs of means.

8. b) Overall, there is a downward trend in the annual change; however, a year in which there is a decrease is frequently followed by a year in which there is an increase, and vice-versa.

9. Answers may vary.

10. Answers may vary.

11. Answers may vary.

12. Answers may vary.

13. a) Ontario and Alberta

b) No; the four larger circles have areas roughly proportional to the emissions they represent, but the area of the circle representing 199 100 is not $\frac{199\ 100}{638} \doteq 312$ times the area of the circle representing 638.

c) Advantage: easy visual comparisons can be made. Disadvantage: precise values are difficult to determine.

d) Prince Edward Island, New Brunswick, Nova Scotia.

e) Answers may vary.

15. b) Answers will vary.

16. Answers will vary.

17. Answers will vary.

Section 2.3, pp. 117–118

Practise

1. a) students in the particular school

b) grade 10 students in the particular school

c) you

d) all those who have listened to the Beatles' music

e) all those who have tried the new remedy

2. a) voluntary-response sample   **b)** stratified sample

c) convenience sample   **d)** systematic sample

e) multi-stage sample   **f)** cluster sample

3. a) stratified sample

b) simple random sample, convenience sample

c) voluntary-response sample

Apply, Solve, Communicate

4. sample $\frac{12}{37} \doteq 32\%$ of the members of each group; 4 children, 3 teens, 5 adults

5. Obtain a list of the students in the school. Calculate $\frac{n}{20}$ to the nearest integer, where n is the number of names on the list. Choose a name at random from the first $\frac{n}{20}$ names. Then, select every $\frac{n}{20}$th name.

6. Answers may vary.

7. a) cluster sample

b) No, not every member of the community had an equal chance of being surveyed.

c) To the same extent as the member of the community centre was randomly selected.

8. a) convenience sample

b) i) yes   **ii)** no

c) In part ii), a simple random sample should be used.

9. a) voluntary-response sample

b) yes

c) not if the host claims that the callers are representative of the population

10. Answers will vary.

11. Answers will vary.

12. Answers will vary.

Section 2.4, pp. 123–124

Practise

1. a) sampling bias   **b)** response bias

c) measurement bias—loaded question

d) sampling bias—the cluster may not be representative

Apply, Solve, Communicate

2. a) A random sample of area residents should be taken.

b) The responses could be returned anonymously.

c) Remove the preamble to the question and the phrase "forward-thinking."

d) A random sample of area residents should be taken.

3. Answers may vary.

a) Which party will you vote for in the next federal election?

b) same as a)

c) Do you think first-year calculus is an easy or difficult course?

d) Who is your favourite male movie star?

e) Do you think that fighting should be eliminated from professional hockey?

4. Answers may vary.

6. Not statistically valid; the sample may contain response bias and also sampling bias. The radio station may have a conservative audience.

7. Answers may vary.

8. Answers may vary.

Section 2.5, pp. 133–135

Practise

1. a) 2.58, 2.4, 2.4   **b)** 14.6, 14.5, 14 and 18

2. Answers may vary.

   a) 1, 2, 3, 4, 5, 6, 7, 8   **b)** 1, 2, 3, 4, 6, 7, 8, 9

   c) 1, 1, 1, 2, 3, 4, 4, 4   **d)** 1, 2, 3, 4, 4, 5, 6, 7

Apply, Solve, Communicate

3. 82

4. a) mode   **b)** mean or median

   c) median   **d)** mean

5. Enzo

6. The two outliers lower the mean mark for all the students by 3.3, as opposed to lowering the mean mark for Class B by 5.7. This is due to the much larger size of the group of all students—the additional higher marks reduce the effect of the outliers.

7. a) 77.95%   **b)** 74%

8. a) Paulo: 3.75; Janet: 4.25; Jamie: 4

   b) Paulo: 3.625; Janet: 4; Jamie: 4.125

   c) Jamie

9. a) mean: 8, median: 7.5, mode: 7

   b) mode   **c)** 10

10. a) 90   **b)** $\dfrac{x_1 + \ldots + x_{15}}{15} = 6 \Rightarrow x_1 + \ldots + x_{15} = 90$

11. a) $45 300   **b)** $45 000

   c) mean: $500, median: $0

12. a) 34.5, 32.5, 48   **b)** mean or median

   c)

Age	Tally	Frequency
16.5–22.5	LHT III	8
22.5–28.5	LHT	5
28.5–34.5	III	3
34.5–40.5	II	2
40.5–46.5	IIII	4
46.5–52.5	LHT III	8

   f) 34.1, 31.5, 19.5 and 49.5

13. a) 16.5–22.5 and 46.5–52.5

   b) No; there is more than one model interval.

14. a) none   **b)** none

15. a) 67.4¢/L

   b) find the average of x when xy is a constant, for example finding the average speed for travelling a given distance

16. a) 2.38%

   b) finding the average of percents, ratios, indexes, or growth rates

Section 2.6, pp. 148–150

Practise

1. a) 7.1, 2.2, 5.0   **b)** $27.42, $10.88, 118.35

2. a) median: 6; Q1: 4.5; Q3: 10; interquartile range: 5.5; semi-interquartile range: 2.75

   b) median: 71.5; Q1: 60.5; Q3: 79.5; interquartile range: 19; semi-interquartile range: 9.5

3. a) first quartile   **b)** second quartile   **c)** second quartile

4. a) 25th   **b)** 50th   **c)** 75th

5. −0.49, −1.23, 1.48, 0, 0.25 if data are a sample

Apply, Solve, Communicate

6. a) 79.4, 57.53, 3309.7   **b)** 68, 52, 26

   d) 260

7. a) 35.8, 35.0, 5.9 if OR nurses are considered as a sample; 35.8, 33.6, 5.8 if they are considered as a population

   b) 36.5, 3, 1.5

8. a) The standard deviation will be smaller and the left-hand whisker will be shorter.

   b) new standard deviation: 3.4

9. a) 804 674, 6.475×10^{77}, 150 000, 75 000

   c) −0.87

   d) After calculating the new average salary and standard deviation, the z-score will be −0.037.

10. Chi-Yan seems to have better control over her drives (a lower standard deviation in their distances). If their putting abilities are essentially equal, then Chi-Yan is more likely to have a better score in a round of golf.

11. There must be either $4n + 2$ or $4n + 3$ (where $n \geq 1$) data points in the distribution.

12. a) 7, 8, 8, 8, 8, 9; these could be the sizes of the first six pairs of shoes sold in a day at a shoe store.

13. No, the standard deviation cannot be much less than half of the semi-interquartile range. For example, {0, 0, 0, 0, 5, 5, 5, 5, 5, 5, 5, 10, 10, 10, 10} has $s = 3.8$ and a semi-interquartile range of 5.

14. a) i) 68   **ii)** 90

   b) 12th to 14th   **c)** 86.5, 78

16. $\sum(x - \bar{x}) = \sum x - \sum \bar{x}$

$$\bar{x} = \frac{1}{n}\sum x, \text{ so } \sum \bar{x} = n\bar{x}$$

$$\sum x = n\bar{x}$$

$$\sum(x - \bar{x}) = \sum x - \sum \bar{x}$$
$$= n\bar{x} - n\bar{x}$$
$$= 0$$

17. a) By definition, $s = \sqrt{\dfrac{\Sigma(x - \bar{x})^2}{n - 1}}$.

Then, $\Sigma(x - \bar{x})^2 = \Sigma(x^2 - 2x\bar{x} + (\bar{x})^2)$

$= \Sigma x^2 - 2\bar{x}\Sigma x + \Sigma(\bar{x})^2$

$= \Sigma x^2 - 2n(\bar{x})^2 + n(\bar{x})^2$

$= \Sigma x^2 - n(\bar{x})^2$

$= \Sigma x^2 - \dfrac{(n\bar{x})^2}{n}$

$= n(\Sigma x^2) - (\Sigma x)^2$

b) The individual deviations do not need to be calculated, round-off error from calculation of mean is avoided, and the quantities in this formula are supplied by a graphing calculator.

18. a) midrange: 65.5; interquartile range: 27.5

b) Both measures are half of a certain range of data; the midrange is affected by outliers, whereas the interquartile range is not.

19. a) mean deviation: 17.6, standard deviation: 22.9

b) Both are measures of the deviations from the mean; the standard deviation emphasizes the larger differences by squaring.

Review of Key Concepts, pp. 151–153

Section 2.1

1. a)

Sales	Tally	Frequency
3	I	1
4	I	1
5	IIII	4
6	LHT	5
7	III	3
8	I	1
9	I	1

2. a)

Mass(g)	Tally	Frequency
230.5–240.5	II	2
240.5–250.5	I	1
250.5–260.5	I	1
260.5–270.5	III	3
270.5–280.5	LHT	5
280.5–290.5	II	2
290.5–300.5	III	3
300.5–310.5	II	2
310.5–320.5	III	3
320.5–330.5	I	1
330.5–340.5	I	1
340.5–350.5	I	1

3. a) categorical data

Section 2.2

4. a) times-series/index   **b)** fuel oil and other fuel

b) 1992 is the base year

d) i) coffee and tea **ii)** rent

e) Rent is controlled in some provinces.
Coffee and tea prices are weather dependent.

5. a) i) about \$8.80   **ii)** about \$7.05

b) about \$620   **c)** about \$1.74

Section 2.3

6. a) With a systematic sample, members are selected from the population at regular intervals whereas a stratified sample divides the population into groups according to characteristics and selects members from each group appropriately.

b) A convenience sample would be appropriate when respondents who are representative of the population are easily accessible—for example, sampling of classmates to determine a choice of field trip.

c) A voluntary-response sample is convenient and inexpensive since respondents submit information themselves. A disadvantage is the possibility of bias.

7. a) multi-stage sample

b) The classes from which students would be selected would be chosen at random; then, within these classes students could be chosen systematically from a class list.

8. Answers will vary.

Section 2.4

9. a) measurement bias; method of data collection (asking a leading question) is poor and the group of children may not be representative and so the sampling technique may also be poor.

b) response bias; method of data collection is poor.

c) response bias; method of data collection is poor.

d) measurement bias; the sample (convenience) is not necessarily representative of the population and the method of data collection (asking a leading question) is poor.

10. a) The sample of children should be representative of the population under consideration—a systematic sample of this population could be chosen. The question should be reworded to eliminate phrases such as "junk food."

b) The teacher could collect the data in such a way as to make respondents anonymous—this may remove the response bias.

c) The musician should play before a more random collection of people—perhaps at a coffee shop—to see how well the song is received.

d) A convenience sample may still be appropriate, but not necessarily at a public library. The potential respondents should first be asked if they are familiar with the work of Carol Shields and, if so, they may then be asked their opinion on her work. Questions with phrases such as "critically acclaimed" should not be used.

Section 2.5

11. a) 5.9, 6, 6
   b) Each measure describes these data equally well.
12. a) mean: 285; median: 285
   b) mean: 288.6; median: 285
   c) The results in part a) are estimates.
13. a) median
   b) No; the distribution of ages is skewed.
14. a) 90.2   **b)** 88.9
   c) Students entering an engineering program must be proficient in sciences and mathematics courses which are given higher weightings.
15. Answers may vary. For example, the mode is useful when determining the "average" size such as for hats.

Section 2.6

16. a) standard deviation: 1.5; interquartile range: 2; semi-interquartile range: 1
   c) no
17. a) The interquartile range is not necessarily symmetric about the median.
   b) Add these differences and divide by 2.
18. a) i) 267.5, 310   **ii)** 242.5, 267.5, 310, 332.5
   b) The first quartile coincides with the 25th percentile and that the third quartile coincides with the 75th percentile. Rounding may produce slightly different values.
19. a) i) −1   **ii)** 2.33   **iii)** 0   **iv)** −2.53
   b) i) 115   **ii)** 70   **iii)** 122.5   **iv)** 82
20. a) 68.1, 68; almost equal
   b) standard deviation: 13.9; interquartile range: 30
   c) standard deviation; the interquartile range is too large.

Chapter Test, pp. 154–155

1. a)

Score	Tally	Frequency
44.5–49.5	II	2
49.5–54.5	I	1
54.5–59.5	IIII	4
59.5–64.5	HHT	5
64.5–69.5	II	2
69.5–74.5	III	3
74.5–79.5		0
79.5–84.5		0
84.5–89.5		0
89.5–94.5	I	1
94.5–99.5	II	2

2. a) mean: 66.06, median: 63.5, mode: 70
   b) standard deviation: 14.52, variance: 210.68
   c) 13, 6.5
3. b) 92, 97, 98
   c) There is no right whisker, so all the data above Q3 must be outliers.

4. The median is the best descriptor. The mean and mode are too high.
5. a) two   **b)** 97, 98
6. Steven, with a weighted score of 4.
7. a) stratified sample
   b) cluster sample
   c) voluntary-response sample
   d) simple random sample
   e) convenience sample
8. Obtain a list of the names of the 52 children. Randomly select one of the first 6 names on the list and then choose every 6th name from that point on.
9. a) i) measurement bias—loaded question; may lead to many negative responses
   ii) response bias—many false responses may be given
   iii) sampling bias—the sample includes only business executives who are more likely to favour the channel
   b) i) Reword the question and omit the references to "run around" and "get into trouble."
   ii) Require the audience members to respond anonymously.
   iii) A sample of the relevant population should be selected randomly.
10. The company is stating that all of its funds have returns below the first quartile of comparable funds.

CHAPTER 3

Review of Prerequisite Skills, p. 158

2. i) a) $y = -1.16x + 21.17$ **b)** $y = 0.73x + 1.16$
   ii) a) x-intercept: 18.3, y-intercept: 21.17
   b) x-intercept: −1.6, y-intercept: 1.16
   iii) a) 13.05   **b)** 6.3
3. a) slope: 3, y-intercept: −4   **b)** slope: −2, y-intercept: 6
   c) slope: 2, y-intercept: $-\dfrac{7}{6}$
4. a) y-intercept: 0, x-intercept: 0
   b) y-intercept: −6, x-intercepts: −6, 1
   c) y-intercept: 2, x-intercepts: −0.67, 1
5. a) i) base: 3, coefficient: 0.5
   ii) base: 2, coefficient: 1
   iii) base: 0.5, coefficient: 100
   c) i) x has a large negative value
   ii) x has a large negative value
   iii) x becomes very large
6. a) 36   **b)** 55
7. a) 6   **b)** 5
9. a) Every member of the population has an equal chance of being selected and the selection of any particular individual does not affect the chances of any other individual being chosen.

b) Scan the population sequentially and select members at regular intervals. The sampling interval size is determined by dividing the population size by the sample size.

c) a data point or observation that lies a long way from the main body of data

10. a) a measurement technique or sampling method that systematically decreases or increases the variable it is measuring

b) surveys that might include sensitive questions

c) surveys that ask leading questions

d) Answers may vary.

Section 3.1, pp. 168–170

Practise

1. Answers may vary.

a) strong positive

b) moderate positive

c) weak negative

d) zero

e) weak positive

2. a) independent: cholesterol level; dependent: heart disease

b) independent: practice; dependent: success rate

c) independent: amount of fertilizer; dependent: height

d) independent: level of education; dependent: income

e) independent: running speed; dependent: pulse rate

Apply, Solve, Communicate

3. a) moderate positive   **b)** moderate to strong negative

c) hours watching TV   **d)** 0.755, −0.878; yes

4. a) Rogers method: strong positive linear correlation, 0.961 Laing System: weak negative linear correlation, −0.224

b) The scatter plot will be reflected in the line $y = x$; the correlation coefficient will remain unchanged.

c) Rogers method: $r = 0.961$; Laing System: $r = -0.224$

5. a) moderate negative linear correlation

b) −0.61   **c)** yes

d) The point (1, 2) is an outlier.

e) Karrie may have worked only part of last year.

6. b) −0.050; weak negative linear correlation

c) no relationship between the two variables

7. b) moderate to strong positive linear correlation

c) 0.765

8. Answers may vary.

9. Answers may vary.

10. a) yes, for certain ranges of temperatures

11. Answers may vary.

12. a) Interchanging the order of multiplication in the formula for covariance does not affect the covariance; the variables you consider to be dependent and independent do not affect the spread in the values of these variables. Then, interchanging the independent and dependent variables, but retaining variable names, the correlation coefficient is calculated as $\dfrac{s_{YX}}{s_Y \times s_X}$, which is the same as $\dfrac{s_{XY}}{s_X \times s_Y}$ or r.

13. Answers may vary.

14. Answers may vary.

Section 3.2, pp. 180–183

Practise

1. a) (92, 18)   **b)** (4, 106)

2. a) a: $y = 0.61x + 17.2$ b: $y = -39.9x + 392.9$

b) a: $y = 1.12x + 3.2$ b: $y = -98.7x + 794.4$

3. $y = 1.17x - 22.5$

Apply, Solve, Communicate

5. a) moderate positive linear correlation

b) $y = 0.442x + 5.51$ **c)** 78.4 kg   **d)** 166 cm

e) The correlation between height and weight is not perfect.

6. a) moderate positive linear correlation

b) 0.568; $y = 1.67x - 20.42$ **c)** (17, 15)

d) 0.759; $y = 0.92x - 9.61$

e) Yes; the correlation is greater.

f) Answers will vary.

g) yes

7. b) $y = -162.92x + 4113.1$ **c)** 1018   **d)** $16

8. a) (4.2, 1.5) is an outlier.   **b)** 0.447; $y = 1.71x + 2.6$

c) 0.996; $y = 2.34x + 2.42$

d) b: 0.700; $y = 1.55x - 0.97$

e) b: 0.681; $y = 2.63x - 2.15$

f) Answers will vary.

g) Answers will vary.

9. a) $y = 0.6x + 45.9$; 0.765   **b)** 196

c) (192, 140); possibly due to a recession

d) a: $y = 0.67x + 35.2$; 0.957   b: 203

e) Answers will vary.

10. a) moderate to strong positive linear correlation

b) 0.865; $y = 1.28x + 19.31$

c) No; the correlation is not perfect.

e) It appears that the wolf population depends on the rabbit population.

11. b) $y = 0.627x + 29.6$

14. No; consider the points (1, 2), (2, 3), (3, 5), (4, 5), (5, 6), (6, 6), (7, 8) (8, 9). All points except (3, 5) and (6, 6) lie on the line $y = x + 1$. The point (3, 5) is one unit above the line and the point (6, 6) is one unit below the line. Performing a linear regression gives the line of best fit as $y = 0.93x + 1.32$ with correlation 0.976.

15. a) $a = \dfrac{n(\sum yx) - (\sum y)(\sum x)}{n(\sum y^2) - (\sum y)^2}$, $b = \bar{x} - a\bar{y}$

b) no

c) no

Section 3.3, pp. 191–194

Practise

1. a) iii   **b)** i   **c)** iv   **d)** ii

2. a) $y = -1.43x^2 - 2.46x + 4.64$; $r^2 = 0.966$

b) $y = 0.89x^3 + 3.68x^2 + 3.29x + 1.01$; $r^2 = 0.959$

c) $y = 1.60(1.77)^x$ or $y = 1.60(e^{0.57x})$; $r^2 = 0.979$

Apply, Solve, Communicate

3. b) $r = 0.99$; $y = 6.48x - 6.53$ **c)** $r^2 = 0.9996$; $y = 1.475x^{2.018}$

  d) Power regression fits the data a bit more closely.

  e) i) 10.9 m² **ii)** 4.4 m

  f) If the trees' outward growth is proportional to their upward growth, the areas will be proportional to the squares of the heights.

4. b) $y = 1.05x^{0.75}$

  c) The curve fits these data well; $r^2 = 0.999$

  d) i) 8 kJ/day **ii)** 314 kJ/day

5. b) $y = 97.4(0.38)^x$ or $y = 97.4(e^{-0.98x})$

  c) The curve fits these data well; $r^2 = 0.999$

  d) 0.73 h or approximately 44 min

6. b) $y = 0.157x^2 - 1.10x + 28.3$; $r^2 = 0.83$

7. a) power level $= 2.02 \times 10^{-3}d^{-2}$, where d is the distance to the transmitter in kilometres

  b) i) 2.02 μW/m² **ii)** 126 μW/m² **iii)** 0.81 μW/m²

8. a) $y = \dfrac{1647}{1 + 46.9e^{-0.39x}}$

  d) The growth rate levels off due to environmental constraints.

9. a) $y = 99.8(1.3)^x$ billion or $y = 99.8e^{0.264x}$ where $x =$ years since 1995

  b) after 10.7 years

  c) the average number of the crop-destroying insects that each arachnid will kill

  d) population $= 99.8(1.3)^x$ billion $- k \times (100$ million $\times 2^x)$, where k is the number of insects destroyed by each arachnid

10. For linear regression, the coefficient of determination is equal to the square of the linear correlation constant.

11. a) i) A: 3, B: 4

  ii) A: $y = 0.3125x^3 - 4.625x^2 + 20x - 19$; $r^2 = 1$
  B: $y = -0.042x^4 + 1.153x^3 - 10.49x^2 + 35.19x - 30.98$; $r^2 = 1$

Section 3.4, pp. 199–201

Practise

1. a) cause-and-effect relationship: alcohol consumption impairs driving ability

  b) common-cause factor: achievement in physics and calculus requires similar skills

  c) reverse cause-and-effect relationship: better job performance leads to increases in pay

  d) accidental relationship: no causal relationship between the variables

  e) presumed relationship: seems logical that a student who has obtained a number of scholarships would be attractive as an employee, but there are many other qualities that employers seek

  f) cause-and-effect relationship: coffee consumption keeps people awake

  g) reverse cause-and-effect relationship: higher number of medals won at Olympic games encourages investors to fund athletic programs

2. a) overall athletic ability

  b) overall disregard of safe driving practices

  c) a large income

Apply, Solve, Communicate

3. Traffic accidents cause traffic congestion.

4. a) the teachers of the class

  b) no

  c) Conduct further trials giving teachers time to become comfortable with the new teaching method and examine the results of classes where the same teacher has used the different methods to minimize the effect of this possible extraneous variable.

5. The overall increase in computer use is likely to have caused parallel increases in the fortunes of the companies.

6. a) Accidental relationship is likely.

7. a) Accidental relationship is likely.

  b) cause-and-effect relationship

8. Answers may vary.

9. In a double-blind study, neither the participants in a trial nor the investigators are aware of which intervention the participants are given. This method may prevent performance bias by the participants and may prevent detection bias by the investigators.

10. a) no **b)** a decrease in the number of graduates hired

11. b) $y = 1.06x - 46\,092$; 0.973

  c) Answers may vary; for example, economy.

  d) Answers may vary.

12. Answers may vary.

13. Hawthorne effect: an increase in worker productivity produced by the psychological stimulus of being singled out and made to feel important. Placebo effect: the measurable, observable, or felt improvement in health not attributable to treatment.

14. a) Create groups consisting entirely of members of the same gender.

  b) Compare results of same-gender groups and mixed-gender groups.

Section 3.5, pp. 209–211

Apply, Solve, Communicate

1. Poor attendance could be a result of mathematics anxiety.

2. income

3. a) Yes, there is a strong negative correlation, $r = 0.9$.

  b) The sample may not be representative, and is much too small in any event.

  c) using a sufficiently large random sample

4. The results are skewed in favour of customer satisfaction since only satisfied customers return.

5. a) The line of best fit is $y = -2.1x + 84.9$; for values of x greater or equal to 17, the value of y is less than 50.

  b) The outlier (20, 30) skews the regression due to the small sample size.

  c) Investigate the possibility of removing the outlier; consider a larger random sample.

6. a) –$120 million   **b)** no
  c) The cubic model has no logical relationship to the situation and gives inaccurate predictions when extrapolated beyond the data.
  d) Gina could obtain data for the years before 1992 to see if the trends are clearer over a longer period. Salaries that tend to have the same percent increase every year will fit an exponential model. Non-linear models may work better with the years elapsed since 1992 as the independent variable in place of the date.
7. Answers may vary.
8. b) 1995   **c)** Answers may vary.
  e) $y = 0.011x + 0.447$
  f) in 1998, when productivity began increasing
9. Answers may vary.
10. Answers may vary.

Review of Key Concepts, pp. 212–213

Section 3.1

1. a)   i) moderate positive linear correlation
    ii) strong negative linear correlation
    iii) no relationship
  b)   i) approximately 0.66
    ii) approximately –0.93
    iii) approximately 0.03
2. a) moderate negative linear correlation
  b) –0.651   **c)** Answers may vary.

Section 3.2

3. $y = -1.20x + 87.9$
4. b) 0.740; $y = 0.87x + 36.15$ **c)** (177, 235)
  d) 0.966; $y = 0.94x + 15.57$ **e)** 254, 251; 3 points

Section 3.3

5. b) $y = -5.24x^2 + 10.1x + 0.095$; $r^2 = 0.9997$
  c) 4.9 m **d)** 1.9 s
  e) r^2 is very close to 1, but the data are only seven measurements with limited precision.
6. a) $y = \dfrac{450}{1 + 44e^{-0.29x}} - 10$
  b) Answers may vary.
  c) accelerated, drove at constant velocity, slowed to a stop

Section 3.4

7. a) An external variable causes two variables to change in the same way, for example, sunny weather could boost sales of both sunscreen and ice-cream cones.
  b) The dependent and independent variables are reversed in the process of establishing causality, for example, the lower the golf score, the more hours a golfer will spend practising.
  c) A variable distinct from the dependent and independent variables that influences either the dependent or independent variable, for example, in trying to find a correlation between traffic congestion and traffic accidents, the time of day is an extraneous variable.

8. a) Results from experimental and control groups are compared to minimize the effects of extraneous variables. The independent variable is varied for the experimental group but not for the control group.
  b) A control group may be required to reduce the placebo effect.
9. a) In an accidental relationship, a correlation exists without any causal relationship whereas in a logical or presumed relationship, a correlation seems logical although no causal relationship is apparent.
  b) accidental relationship: a correlation between an NFL team from a certain conference winning the super bowl and the state of the market; logical relationship: a correlation between musical and mathematical ability
10. common-cause factors
11. Answers may vary.

Section 3.5

12. Answers may vary.
13. a) A hidden variable is an extraneous variable that is not easily detected.
  b) The presence of a hidden variable may be detected by analysing any disjoint clusters of data separately.

Chapter Test, pp. 214–215

1. a) A correlation with value –1; the dependent variable decreases at a constant rate as the independent variable increases.
  b) The design of experiments and analysis of data to establish the presence of relationships between variables.
  c) A data point or observation that lies a long way from the main body of data.
  d) A variable distinct from the dependent and independent variables that influences either the dependent or independent variable.
  e) An extraneous variable that is not easily detected.
2. a) –0.8   **b)** 1   **c)** 0.3   **d)** 0.6   **e)** –1
3. a) strong positive linear correlation
  b) 0.935   **c)** $y = 0.517x + 1.93$ **d)** 8.1
4. b) $y_1 = 1.79x + 57.9$; $r = 0.618$
  c) (2, 38)
  d) $y_2 = 0.894x + 66$; $r = 0.819$
5. y_2, assuming at most 9 h of study
6. Answers may vary.
7. b) $y = 0.011x^2 - 0.167x + 0.855$; $r^2 = 0.9995$
  c) $y = 0.854(0.812)^x$ or $y = 0.854e^{-0.208x}$; $r^2 = 0.9997$
  d) Both fit the data very well.
  e) quadratic: 0.286 lumens, exponential: 0.107 lumens
  f) exponential; the quadratic curve begins to increase:
8. Use control groups.
9. The more money a person invests, the higher their income will be.

CUMULATIVE REVIEW: CHAPTERS 1 TO 3, pp. 218–220

1. a) $\begin{bmatrix} 2 & -6 \\ -10 & -8 \\ -8 & -2 \end{bmatrix}$ **b)** not possible

c) not possible   **d)** $\begin{bmatrix} 59 & 12 \\ -60 & 11 \end{bmatrix}$

e) not possible   **f)** $\begin{bmatrix} \dfrac{4}{37} & \dfrac{1}{37} \\ \dfrac{5}{37} & \dfrac{8}{37} \end{bmatrix}$

2. a) Place 1 in the centre. Place 2 one square to the right of 1. Place the numbers beginning with 3 in a counterclockwise circle around 1. When you reach 9 move one square to the right and place 10. Continue in a counterclockwise circle as before.

b)

37	36	35	34	33	32	31
38	17	16	15	14	13	30
39	18	5	4	3	12	29
40	19	6	1	2	11	28
41	20	7	8	9	10	27
42	21	22	23	24	25	26
43	44	45	46	47	48	49

3. a) no   **b)** yes   **c)** no   **d)** maybe

4. a) voluntary response   **b)** simple random sample
     c) observational study

5. a) strong negative   **b)** moderate positive
     c) strong positive   **b)** weak positive

6. Answers may vary.
     a) either a common-cause factor or a presumed relationship
     b) common-cause factor   **c)** accidental relationship
     d) cause-and-effect relationship

7. Answers may vary.

8. a) i) no   ii) yes   iii) yes
     b) i) yes   ii) yes   iii) no

9. Answers may vary.

10. a) Buffalo   **b)** Buffalo

11. a) 0, 0, 0, 0, 0, 0, 0, 0, 0, 0, 0, 1, 1, 1, 1, 1, 1, 1, 2, 2, 2, 2, 2,
     3, 3, 3, 3, 3, 4, 4, 5, 5, 6, 7, 7, 7, 8, 9, 10, 10, 14, 15
     b) 3.4

c)

Class Interval	Tally	Frequency
−0.5–2.5	HHT HHT HHT HHT III	23
2.5–5.5	HHT IIII	9
5.5–8.5	HHT	5
8.5–11.5	III	3
11.5–14.5	I	1
14.5–17.5	I	1

12. a)

Class Interval	Tally	Frequency
15.9–25.9	II	2
25.9–35.9	HHT	5
35.9–45.9	HHT II	7
45.9–55.9	HHT	5
55.9–65.9	HHT II	7
65.9–75.9	IIII	4
75.9–85.9	III	3
85.9–95.9	II	2
95.9–105.9		0
105.9–115.9	III	3
115.9–125.9	I	1
125.9–135.9	I	1

c) 60%

13. a) response bias   **b)** response bias
     c) sampling bias   **d)** measurement bias

14. a) 198.6, 193.5, 260; the mean or median is the most descriptive; the mode is the least descriptive.
     b) 40.4, 170, 222, 52
     c) The standard deviation is a measure of the spread of the scores about the mean; 25% of the scores are below the 1st quartile, 170, and 75% are above; 75% of the scores are below the 3rd quartile, 222, and 25% are above; 50% of the scores fall within the interquartile range.
     d) 193.5
     e) No; this person is in the 75th percentile.

15. a) independent: practice hours; dependent: batting average
     b) strong positive correlation
     c) 0.95; $y = 0.0098x + 0.0321$
     d) i) 0.190   ii) 0.160   iii) 0.375

16. Answers will vary.

17. b) predictions may vary; $6.3 million
     c) factors may vary: the economy, health of the employee, change of employer, design of 2003 cars

d)

Year	Index
1997	100
1998	137.5
1999	200
2000	287.5
2001	437.5
2002	587.5

1997 = 100

e) Answers may vary.

18. a) 100 **b)** 1980 **c)** 1980–1990
  d) 5.2%

CHAPTER 4

Review of Prerequisite Skills, p. 224

1. 8 ways
2. b) 12
3. a) 7 **b)** 28 **c)** 84
  d) 1, 3, 6, 10, ... ; $t_n = \dfrac{n(n+1)}{2}$
4. a) 15 **b)** 45 **c)** 90 **d)** 150
5. a) 96 **b)** 72 **c)** 8
6. a) 11 **b)** 209.1 **c)** 120 **d)** 2450 **e)** 605
7. a) $\dfrac{x}{2} - \dfrac{y}{2} + 1$ **b)** $x^2 + 4x + 4$ **c)** $x^2 + 2$
  d) $x^2 - 4x + 3$ **e)** $\dfrac{3y+2}{x}$

Section 4.1, pp. 229–231

Practise

1. 12
2. 5
3. 16
4. 8
5. a) 12 **b)** 6 **c)** 9
6. Use the product rule; for each of the 10 books there are 4 choices of a pen, and so there are $10 \times 4 = 40$ choices in total.

Apply, Solve, Communicate

7. b) 18
8. 7776
9. 15
10. 64
11. 28
12. 32

13. a) Canada: 17 576 000; U.S.A.: 100 000 **b)** 17 476 000
14. 160
15. 12
16. a) 12 **b)** 26 **c)** 15 **d)** 14
  e) Multiplicative counting principle in parts a) and b).
    Additive counting principle in parts c) and d).
17. a) 90 **b)** 56
18. a) 17 576 000 **b)** 456 976 000
19. 72
20. a) 30 **b)** 6
22. 1620
23. 23 816
24. a) 60 840
  b) Answers may vary. The answer given in part a) assumes that any string of three numbers taken from the list 0, 1, ..., 39 is a possible combination, provided no successive numbers are the same.
  c) 59 319
  d) If the first number can also be dialled counter clockwise from 0, the number of possible combinations increases to 118 638, assuming the remaining numbers are dialled clockwise.
  e) 2 372 760, with assumptions similar to those in part b)
25. a) i) 2 **ii)** 9 **iii)** 22
  b) No; the number of subsequent moves depends on the particular previous move.

Section 4.2, pp. 239–240

Practise

1. a) 6! **b)** 8! **c)** 3! **d)** 9!
2. a) 210 **b)** 110 **c)** 168 **d)** 5 405 400
  e) 592 620 **f)** 30 270 240
3. a) $_6P_3$ **b)** $_9P_4$ **c)** $_{20}P_4$ **d)** $_{101}P_5$ **e)** $_{76}P_7$
4. a) 5040 **b)** 43 680 **c)** 20 **d)** 3024 **e)** 5040

Apply, Solve, Communicate

6. a) 720 **b)** 24 **c)** 11 880
7. a) 12! **b)** 9! **c)** $(n+5)!$
8. The sequence $t_n = n!$ can be given by the recursion formula $t_1 = 1$, $t_n = n \cdot t_{n-1}$. A recursion formula is an example of an iterative process.
9. a) 5040 **b)** 720 **c)** 120 **d)** 1440
10. 479 001 599
11. 79 833 600
12. 201 600
13. 479 001 600
14. 6840
15. a) 311 875 200 **b)** 5.74×10^{16}
  c) 7 893 600 **d)** 24
16. a) 9 **b)** 30 240
17. 479 001 600
18. a) 5040 **b)** 4320
19. No; there are 39 916 800 possible orders.
22. 120
23. 2^{97}

24. 3 628 800

Section 4.3, pp. 245–246

Practise

1. a) 2 of the letter *a*, 2 of the letter *m*, and 2 of the letter *t*.
  b) the notebook of the same colour
  c) the food items of the same kind
  d) Thomas and Richard
2. a) 20 160   **b)** 907 200   **c)** 420   **d)** 9 081 072 000
3. 10
4. a) 720   **b)** 120   **c)** 180   **d)** 30

Apply, Solve, Communicate

5. 56
6. 210
7. 19 380
8. a) SUYFS: 60, YATTS: 60, SPEEXO: 360, HAREMM: 360
  b) Answers may vary.
9. 2520
10. a) 720   **b)** 20   **c)** 90
11. 11 732 745 024
12. 1260
13. 15
15. 4032
16. 20
17. a) $\dfrac{165!}{72! \times 36! \times 57!}$ **b)** Estimates will vary.
18. 504
19. 190 590 400

Section 4.4, pp. 251–253

Practise

2. a) $t_{8,3}$ **b)** $t_{52,41}$ **c)** $t_{17,11}$ **d)** $t_{n-1,\,r-1}$
3. a) 4096   **b)** 1 048 576   **c)** 33 554 432   **d)** 2^{n-1}
4. a) 8   **b)** 11   **c)** 14   **d)** 16

Apply, Solve, Communicate

5. a)

Row	Sum/Difference	Result
0	1	1
1	$1 - 1$	0
2	$1 - 2 + 1$	0
3	$1 - 3 + 3 - 1$	0
4	$1 - 4 + 6 - 4 + 1$	0
5	$1 - 5 + 10 - 10 + 5 - 1$	0
6	$1 - 6 + 15 - 20 + 15 - 6 + 1$	0
⋮	⋮	⋮

  b) 0   **c)** 0
6. a) $t_{2n,\,n}$ **b)** *n* odd: 0, *n* even: $(-1)^{0.5n} t_{n,\,0.5n}$
7. a) The digits in 11^n are the same as the digits in row *n* of Pascal's triangle.
  b) Power: $11^5 = 161\ 051$. Add the digits in the 5th row of

Pascal's triangle as follows:
1 5 10 10 5 1
1 6 1 0 5 1
  c) Power: $11^6 = 1\ 771\ 561$. Add the digits in the 6th row of Pascal's triangle as follows:
1 6 15 20 15 6 1
1 7 7 1 5 6 1
Power: $11^7 = 19\ 487\ 171$. Add the digits in the 7th row of Pascal's triangle as follows:
1 7 21 35 35 21 7 1
1 9 4 8 7 1 7 1
8. a) 2, 5, 9   **b)** $t_{n-3,\,1} + t_{n-2,\,2}$ **c)** 14, 20
9. The terms of Pascal's triangle in prime-numbered rows are divisible by the row-number (with the exceptions of the leading and final 1).
10. a) 0, 1, 3, 7, 15, ... Yes, the rows numbered $2^n - 1$ for $n \ge 0$ contain only odd numbers.
  b) no   **c)** odd
11. a) $t_{n+2,\,n-1}$ **b)** 364
12. a) $1 + 4 + \ldots + n^2 = t_{n+1,\,n-2} + t_{n+2,\,n-1}$
  b) 385
13. a) $\dfrac{n^2 + n + 2}{2}$ **b)** 121
15. Answers will vary.
16. a) $3t_{n+1,\,n-1}$ **b)** 165
17. Answers may vary.
18. c) Answers will vary.
19. a) The top term is 1, and the terms on the extreme left and right of row *n* are generated by increasing the denominator of the extreme left and right terms of row $n - 1$ by 1. Every other term is obtained by subtracting the term to its immediate right from the term immediately above and to the right. Alternatively, each term is obtained by subtracting the term to its immediate left from the term immediately above and to the left. The values will be the same in each method.
  b) $\dfrac{1}{7}, \dfrac{1}{42}, \dfrac{1}{105}, \dfrac{1}{140}, \dfrac{1}{105}, \dfrac{1}{42}, \dfrac{1}{7}, \dfrac{1}{8}, \dfrac{1}{56}, \dfrac{1}{168},$
    $\dfrac{1}{280}, \dfrac{1}{280}, \dfrac{1}{168}, \dfrac{1}{56}, \dfrac{1}{8}$
  c) Answers may vary.

Section 4.5, pp. 256–259

Practise

2. a) 128   **b)** 672   **c)** 80
3. 1, 15, 105, 455, 1365, 3003, 5005, 6435, 6435; 1, 17, 136, 680, 2380, 6188, 12 376, 19 448, 24 310

Apply, Solve, Communicate

4. a) 126   **b)** 270   **c)** 136
5. 56
6. No; there are only 126 different routes.
7. a) 32   **b)** 308
8. a) 20   **b)** the square labelled with 3: 54 routes
9. a) 4   **b)** 4   **c)** 12   **d)** 16   **e)** yes

10. 25

11. a) either 3 or 4; either 1 or 6

12. b) Conjectures will vary.   **c)** Conjectures will vary.

13. $t_{n,r} = t_{n-3,r-3} + 3t_{n-3,r-2} + 3t_{n-3,r-1} + 3t_{n-3,r}$

15. $\dfrac{(m+n)!}{m!\,n!}$

16. 269

17. 30.8 L

18. yes

19. 320

20. Answers may vary.

Review of Key Concepts, p. 260

4.1 Organized Counting

2. 336

3. 480

4. 48

4.2 Factorials and Permutations

5. $0, 1, 2, 3$

6. 60

7. 1320

4.3 Permutations With Some Identical Items

8. 4200

9. a) 5040 **b)** 1260 **c)** 630

10. $900\ 900$

4.4 Pascal's Triangle

11.

```
            1
          1   1
        1   2   1
      1   3   3   1
    1   4   6   4   1
```

12. 128

13. Answers may vary.

4.5 Applying Pascal's Method

14. Pascal's method is iterative since previous terms are used to determine subsequent terms.

15. 204

Chapter Test, p. 261

1. a) 16

2. a) $\dfrac{15!}{9!} = 3\ 603\ 600$ **b)** $\dfrac{6!}{4!} = 30$

   c) $\dfrac{7!}{4!} = 210$ **d)** $\dfrac{9!}{0!} = 362\ 880$

   e) $\dfrac{7!}{7!} = 1$

3. a) i) 27 **ii)** 39

   b) For codes of the same length the multiplicative counting principle applies by multiplying the number of pulses possible for each position. The additive counting principle applies by adding the numbers of codes of various lengths since these are mutually exclusive cases.

4. a) 840 **b)** 480 **c)** 360

5. 6

6. a) 12 **b)** $226\ 800$ **c)** $362\ 880$

7. a) $t_{n,n-2}$ **b)** 15

CHAPTER 5

Review of Prerequisite Skills, p. 264

1. a) $40\ 320$ **b)** 336 **c)** 552 **d)** 144

2. a) 120 **b)** 90 **c)** 12 **d)** 210

3. a) $30\ 240$ **b)** 240 **c)** $3\ 628\ 800$ **d)** $6\ 720$

4. a) 1.96×10^{39} **b)** 2.69×10^{23}

   c) 8.84×10^{30} **d)** 2.13×10^{35}

5. $35\ 152$

6. a) 60 **b)** $226\ 800$ **c)** 1320

7. a) 1 **b)** $-64x^3$ **c)** $576\ 240x^4y^2$ **d)** $\dfrac{21}{x^4}$

   e) $432x^6y^5$ **f)** $\dfrac{3}{2}x^2y^3$ **g)** $-75x^5y^3$ **h)** $-8x^3y^3$

8. a) $x^2 - 10x + 25$ **b)** $25x^2 - 10xy + y^2$

   c) $x^4 + 10x^2 + 25$ **d)** $x^3 - 7x^2 - 5x + 75$

   e) $x^4 - 2x^2y + y^2$ **f)** $4x^2 + 12x + 9$

   g) $x^3 - 10x^2 + 32x - 32$ **h)** $4x^4 + 12x^2y + 9y^2$

   i) $4x^3 - 4x^2 - 7x - 2$ **j)** $x^3 - 3x^2y + 4y^3$

9. a) $\displaystyle\sum_{n=0}^{4} 2^n$ **b)** $\displaystyle\sum_{n=1}^{5} nx^n$ **c)** $\displaystyle\sum_{n=2}^{\infty} \dfrac{1}{n}$

10. a) $4 + 6 + 8 + 10$ **b)** $x + \dfrac{x^2}{2!} + \dfrac{x^3}{3!} + \dfrac{x^4}{4!}$

   c) $2^1 + 1^2 + 2^2 + 2^2 + 2^3 + 3^2 + 2^4 + 4^2 + 2^5 + 5^2 = 117$

Section 5.1, pp. 270–272

Practise

1. a) i) apple   **ii)** apple, orange, pear, banana
   iii) apple, orange, pear, banana
   iv) apple, banana
   iv) apple, orange, pear, banana

   b) i) 5 **ii)** 4 **iii)** 4 **iv)** 4 **v)** 3

   c) i) apple, pear; orange, pear; apple, orange
   ii) apple, banana
   iii) apple, orange; apple, pear; apple, banana; orange, pear; orange, banana; pear, banana

2. a) 82 **b)** 10 **c)** 56 **d)** 9 **e)** 6 **f)** 50

Apply, Solve, Communicate

3. 18

4. a) 95% **b)** 5% **c)** 45%

   d) 50% **e)** 35%

5. 5

6. 10

7. b) There are many consistent sets of information. If x represents the number of people that did not read books, newspapers, or magazines, then x can have any value between 0 and 15, and the number who read both newspapers and magazines is $25 + x$; the number who read newspapers only is $35 - x$, and the number who read magazines only is $15 - x$.

8. a) 5   **b)** 40

9. a) 22   **b)** 20   **c)** 19   **d)** 20

10. a) $n(A \cup B \cup C) = n(A) + n(B) + n(C) - n(A \cap B) - n(B \cap C)$
$- n(C \cap A) + n(A \cap B \cap C)$

b) $n(A \cup B \cup C \cup D) = n(A) + n(B) + n(C) + n(D) - n(A \cap B)$
$- n(B \cap C) - n(C \cap D) - n(D \cap A) - n(A \cap C) - n(B \cap D)$
$+ n(A \cap B \cap C) + n(B \cap C \cap D) + n(C \cap D \cap A) +$
$n(D \cap A \cap B) - n(A \cap B \cap C \cap D)$

c) $(A_1 \cup A_2 \cup \ldots \cup A_n) = \sum_{i=1}^{n} A_i - \sum_{\substack{i,j \\ i<j}}^{n} n(A_i \cap A_j)$
$+ \sum_{\substack{i,j,k \\ i<j<k}}^{n} n(A_i \cap A_j \cap A_k) - \ldots + (-1)^{n+1} n(A_1 \cap A_2 \cap \ldots \cap A_n)$

Section 5.2, pp. 279–281

Practise

1. a) 210   **b)** 435   **c)** 8568
d) 560   **e)** 3876   **f)** 53 130

2. a) 11, 11, equal   **b)** 55, 55, equal   **c)** 165, 165, equal

Apply, Solve, Communicate

3. 105

4. 270 725

5. 165

6. 74 613

7. 33 649

8. a) 2 220 075

9. a) 55   **b)** 165   **c)** 990
d) 6; the number of ways to order the three people in each combination of part b) is $_3P_3 = 3!$ or 6.

10.

Number of Cards Chosen	Combinations	Number of Cards Chosen	Combinations
0	1	27	4.77551E+14
1	52	28	4.26385E+14
2	1326	29	3.5287E+14
3	22100	30	2.70534E+14
4	270725	31	1.91992E+14
5	2598960	32	1.25995E+14
6	20358520	33	7.63604E+13
7	133784560	34	4.2672E+13
8	752538150	35	2.19456E+13
9	3679075400	36	1.03632E+13
10	15820024220	37	4.48138E+12
11	60403728840	38	1.76897E+12
12	2.06379E+11	39	6.35014E+11
13	6.35014E+11	40	2.06379E+11
14	1.76897E+12	41	60403728840
15	4.48138E+12	42	15820024220
16	1.03632E+13	43	3679075400
17	2.19456E+13	44	752538150
18	4.2672E+13	45	133784560
19	7.63604E+13	46	20358520
20	1.25995E+14	47	2598960
21	1.91992E+14	48	270725
22	2.70534E+14	49	22100
23	3.5287E+14	50	1326
24	4.26385E+14	51	52
25	4.77551E+14	52	1
26	4.95919E+14		

11. 26 460

12. 1 051 050

13. a) 65 780   **b)** 4950   **c)** 15 015
d) 25 025   **e)** 3465

14. a) 125 970   **b)** 9450   **c)** 203 490
d) 257 754

15. 42 504

16. a) 330   **b)** 35

17. 14 000

18. 190

20. a) 13!; the number of ways to arrange the 13 cards
b) $_{13}C_5 \times _{13}C_2 \times _{13}C_3 \times _{13}C_3$ **c)** $_{13}C_5 \times _{39}C_8$
d) 8 211 173 256, 79 181 063 676

21. a) 2 743 372 800   **b)** 2 472 422 400   **c)** 8 892 185 702 400

22. a) 330   **b)** 150   **c)** 5   **d)** 15

Section 5.3, pp. 286–288

Practise

1. 15

2. 11

3. a) 4   **b)** 16   **c)** 128

4. 255

5. a) combinations; $_{12}C_3$ **b)** permutations; $_{12}P_3$
c) combinations; $_3C_1 \times _4C_1 \times _2C_1$ **d)** permutations; $_4P_2$

Apply, Solve, Communicate

6. a) 3003   **b)** 2926

7. a) 287 **b)** 216

8. a) 593 775 **b)** 2925 **c)** 590 850

9. a) 792 **b)** 35 772 **c)** 21 252

10. a) 28 561 **b)** 2 023 203 **c)** 844 272

11. a) 48 **b)** 36 **c)** 12

12. 252

13. 2047

15. a) 1365 **b)** 364 **c)** 1001

16. a) i) 5 **ii)** 9

17. 19

18. 9240

19. a) 7.15×10^{18} **b)** 3 628 800 **c)** 256

21. a) DEVEL: 8, VEENT: 8, PAPNYS: 18, SIFOSY: 18

b) Because some choices could be identical.

22. 150

23. a) 6 **b)** 7 **c)** 6

24. 23

Section 5.4, pp. 293–295

Practise

1. a) $_{16}C_{10} + {}_{16}C_{11}$ **b)** $_{42}C_{35} + {}_{42}C_{36}$ **c)** $_{n}C_{r} + {}_{n}C_{r+1}$

d) $_{33}C_{5}$ **e)** $_{16}C_{10}$ **f)** $_{n+1}C_{r+1}$

g) $_{17}C_{8}$ **h)** $_{23}C_{8}$ **i)** $_{n-1}C_{r}$

2. a) 4 **b)** 2 **c)** 5

3. a) 13 **b)** 6 **c)** 21

4. a) $_{11}C_{2}$, 55 **b)** $_{11}C_{0}$, 1 **c)** $_{11}C_{5}$, 462

Apply, Solve, Communicate

5. a) 512 **b)** 0 **c)** 32 768 **d)** 2^{n}

6. 14

7. a) i) $2 \times {}_{2n-1}C_{n}$

ii) $(-1)^{\frac{n}{2}} \times 2 \times {}_{n-1}C_{\frac{n}{2}}$ for n even, 0 for n odd.

iii) $_{n-1}C_{2} - 1$

b) i) 155 117 520 **ii)** 924 **iii)** 77

8. 9

9. a) $x^7 + 7x^6y + 21x^5y^2 + 35x^4y^3 + 35x^3y^4 + 21x^2y^5 + 7xy^6 + y^7$

b) $64x^6 + 576x^5y + 2160x^4y^2 + 4320x^3y^3 + 4860x^2y^4 + 2916xy^5 + 729y^6$

c) $32x^5 - 400x^4y + 2000x^3y^2 - 5000x^2y^3 + 6250xy^4 - 3125y^5$

d) $x^8 + 20x^6 + 150x^4 + 500x^2 + 625$

e) $2187a^{14} + 20\ 412a^{12}c + 81\ 648a^{10}c^2 + 181\ 440a^8c^3 + 241\ 920a^6c^4 + 193\ 536a^4c^5 + 86\ 016a^2c^6 + 16\ 384c^7$

f) $-38\ 880c^{10} + 64\ 800c^8p - 43\ 200c^6p^2 + 14\ 400c^4p^3 - 2400c^2p^4 + 160p^5$

10. a) $59\ 049x^{10} + 196\ 830x^9y + 295\ 245x^8y^2 + 262\ 440x^7y^3 + 153\ 090x^6y^4$

b) $59\ 049x^{10} - 196\ 830x^9y + 295\ 245x^8y^2 - 262\ 440x^7y^3 + 153\ 090x^6y^4$

c) The terms are identical, except the coefficients of the terms involving odd powers are negative in $(3x - y)^{10}$.

11. a) $x^{10} - 5x^7 + 10x^4 - 10x + 5x^{-2} - x^{-5}$

b) $16y^4 + 96y + 216y^{-2} + 216y^{-5} + 81y^{-8}$

c) $64x^{12} + 192x^{10.5} + 240x^9 + 160x^{7.5} + 60x^6 + 12x^{4.5} + x^3$

d) $5k^5m^{-2} + 10k^5m^{-4} + 10k^5m^{-6} + 5k^5m^{-8} + k^5m^{-10} + k^5$

e) $y^{3.5} - 14y^{2.5} + 84y^{1.5} - 280y^{0.5} + 560y^{-0.5} - 672y^{-1.5} + 448y^{-2.5} - 128y^{-3.5}$

f) $162m^8 - 432m^{5.5} + 432m^3 - 192m^{0.5} + 32m^{-2}$

12. a) $(x + y)^6$ **b)** $(y^3 + 2)^4$ **c)** $(3a - b)^5$

13. a) $0.5^5(1 + 1)^5 = 1$ **b)** $(0.7 + 0.3)^7 = 1$

c) $(7 - 1)^9 = 6^9$ or 10 077 696

14. a) $x^4 + 8x^2 + 32x^{-2} + 16x^{-4} + 24$

b) $\left(x + \dfrac{2}{x}\right)^4 = \left(\dfrac{1}{x}(x^2 + 2)\right)^4 = \dfrac{1}{x^4}(x^2 + 2)^4$

15. a) $15\ 625x^6 + 56\ 250x^5y + 84\ 375x^4y^2 + 67\ 500x^3y^3 + 30\ 375x^2y^4 + 7290xy^5 + 729y^6$

b) $59\ 049x^{10} - 131\ 220x^8y^2 + 116\ 640x^6y^4 - 51\ 840x^4y^6 + 11\ 520x^2y^8 - 1024y^{10}$

16. a) 5.153 92 after 6 terms **b)** 7

17. a) 210 **b)** 1023

19. a) $16x^5, 208x^4, 1080x^3$

b) $4096x^7, -11\ 264x^6, 8704x^5$

c) $x^{36}, -45x^{34}, 12x^{33}$

20. a) $x^2 + 2xy + 2xz + y^2 + 2yz + z^2$

b) $x^3 + 3x^2y + 3x^2z + 3xy^2 + 6xyz + 3xz^2 + y^3 + 3y^2z + 3yz^2 + z^3$

c) $x^4 + 4x^3y + 4x^3z + 6x^2y^2 + 12x^2yz + 6x^2z^2 + 4xy^3 + 12xy^2z + 12xyz^2 + 4xz^3 + y^4 + 4y^3z + 6y^2z^2 + 4yz^3 + z^4$

d) $\displaystyle\sum_{\substack{i,j,k=0 \\ i+j+k=n}}^{n} \dfrac{n!}{i!j!k!} x^iy^jz^k$

e) $x^5 + 5x^4y + 5x^4z + 10x^3y^2 + 20x^3yz + 10x^3z^2 + 10x^2y^3 + 30x^2y^2z + 30x^2yz^2 + 10x^2z^3 + 5xy^4 + 20xy^3z + 30xy^2z^2 + 20xyz^3 + 5xz^4 + y^5 + 5y^4z + 10y^3z^2 + 10y^2z^3 + 5yz^4 + z^5$

21. a) $B^5 + 5B^4G + 10B^3G^2 + 10B^2G^3 + 5BG^4 + G^5$

b) The coefficient of G^k is the number of ways of having k girls in a family of 5 children.

c) 10 **d)** 5

22. a) 32 **b)** 10 **c)** 31

Review of Key Concepts, pp. 296–297

5.1 Organized Counting With Venn Diagrams

1. a) R2, R3, R5, R6, R7, R8 **b)** R7, R8

c) R5, R8 **d)** all regions

2. a) $n(A \text{ or } B) = n(A) + n(B) - n(A \text{ and } B)$

b) Because the elements counted by the last term have been counted twice in previous terms.

c) Answers will vary.

3. a) colour television, computers, and no dishwashers; colour television, dishwashers, and no computers; dishwashers, computers, and no colour televisions; colour television and no computers or no dishwashers; computers and no colour televisions or no dishwashers; dishwashers and no colour television or no computers; no colour televisions, no computers, and no dishwashers

b) 33%, 19%, 1%, 14%, 1%, 1%, 1%, respectively

c) Answers will vary.

5.2 Combinations

4. a) 95 548 245 **b)** 1 037 158 320 **c)** 1 081 575

d) 10 272 278 170 **e)** 45 **f)** 105

g) 5   **h)** 25   **i)** 1365
j) 53 130   **k)** 12 870   **l)** 27 405
5. 9900
6. 4 134 297 024
7. Order is important.

5.3 Problem Solving With Combinations

8. 512
9. 31
10. 210
11. 256
12. 756 756
13. 459

5.4 The Binomial Theorem

14. a) 6   **b)** 3
15. a) $x^8 + 8x^7y + 28x^6y^2 + 56x^5y^3 + 70x^4y^4 + 56x^3y^5 + 28x^2y^6 + 8xy^7 + y^8$
  b) $4096x^6 - 6144x^5y + 3840x^4y^2 - 1280x^3y^3 + 240x^2y^4 - 24xy^5 + y^6$
  c) $16x^4 + 160x^3y + 600x^2y^2 + 1000xy^3 + 625y^4$
  d) $16\,807x^5 + 36\,015x^4 + 30\,870x^3 - 13\,230x^2 + 2835x - 243$
16. a) $x^6 + 6x^5y + 15x^4y^2 + 20x^3y^3 + 15x^2y^4 + 6xy^5 + y^6$
  b) $1296x^4 - 4320x^3y + 5400x^2y^2 - 3000xy^3 + 625y^4$
  c) $3125x^5 + 6250x^4y + 5000x^3y^2 + 2000x^2y^3 + 400xy^4 + 32y^5$
  d) $729x^6 - 2916x^5 + 4860x^4 - 4320x^3 + 2160x^2 - 576x + 64$
17. a) $128x^7$, $2240x^6y$, $16\,800x^5y^2$
  b) $4096x^6$, $-6144x^5y$, $3840x^4y^2$
18. The $(r + 1)$st term in the expansion is given by $t_{r+1} = {}_6C_r(2x)^{n-r}(-3y)^r$. Evaluate each term from $r = 0$ to $r = 6$ and sum them.
19. $32x^5$
20. 43 750
21. $a = 1$, $x = 0.5$
22. $y^{24} - 24y^{20} + 240y^{16} - 1280y^{12} + 3840y^8 - 6144y^4 + 4096$
23. $(4x^2 - 3)^5$

Chapter Test, pp. 298–299

1. a) 1   **b)** 52   **c)** 220   **d)** 40 225 345 056
2. a) ${}_{11}C_8$ **b)** ${}_{22}C_{15}$
3. a) $81x^4 - 432x^3 + 864x^2 - 768x + 256$
  b) $128x^7 + 1344x^6y + 6048x^5y^2 + 15\,120x^4y^3 + 22\,680x^3y^4 + 20\,412x^2y^5 + 10\,206xy^6 + 2187y^7$
4. a) $32\,768x^5 - 61\,440x^4 + 46\,080x^3 - 17\,280x^2 + 3240x - 243$
  b) $64x^6 - 960x^5y + 6000x^4y^2 - 20\,000x^3y^3 + 37\,500x^2y^4 + 37\,500xy^5 + 15\,625y^6$
5. a) 1001   **b)** 35   **c)** 441   **d)** 966
6. a) 1140   **b)** 6840   **c)** no; order is important in part b).
7. a) 118   **b)** 1
8. a) 35   **b)** 34
9. a) 32 767 **b)** 216
10. $x = -0.1$, $n = 9$
11. $64x^6 - 576x^5 + 2160x^4 - 4320x^3 + 4860x^2 - 2916x + 729$
12. a) 66   **b)** 239 500 800

CHAPTER 6

Review of Prerequisite Skills, pp. 302–303

1. a) 35%   **b)** 4%   **c)** 95%   **d)** 0.8%   **e)** 8.5%   **f)** 37.5%
2. a) 0.15   **b)** 0.03   **c)** 0.85   **d)** 0.065   **e)** 0.265   **f)** 0.752
3. a) $\dfrac{3}{25}$ **b)** $\dfrac{7}{20}$ **c)** $\dfrac{67}{100}$ **d)** $\dfrac{1}{25}$ **e)** $\dfrac{1}{200}$ **f)** $\dfrac{49}{50}$
4. a) 25%   **b)** 86.7% **c)** 78.6% **d)** 70%   **e)** 44.4% **f)** 65%
7. 12 ways, assuming that Benoit will wear an article of each type.
8. 6.33966×10^9
9. a) i) 2, 4, 6, 8, 10   **ii)** 1, 3, 5, 7, 9
    iii) 4, 6, 8, 9, 10   **iv)** 1, 4, 9
10. a) To count the elements of a set, each element must be counted exactly once. If an element is included in the count more than once, it must be excluded the appropriate number of times. For example, to count the elements of the set A or B, each element of A and B is counted (included) twice in the sum $n(A) + n(B)$ and so must be excluded once.
Thus, $n(A \text{ or } B) = n(A) + n(B) - n(A \text{ and } B)$.
  b) 22
11. a) 720   **b)** 1   **c)** 240   **d)** 220   **e)** 9900 **f)** 143
12. a) 60   **b)** 7   **c)** 30   **d)** 362 880 **e)** 100   **f)** 9900
13. 259 459 200
14. 800
15. a) 20   **b)** 4   **c)** 1   **d)** 1   **e)** 315   **f)** 100
  g) 190   **h)** 190
16. 84
17. a) 6188 **c)** 2200
18. square matrices: a), d); row matrix: b); column matrix: f)
19. a) [0.505 0.495]
  b) Not possible; the number of columns of B does not equal the number of rows of A.
  c) $\begin{bmatrix} 0.49 & 0.51 \\ 0.4675 & 0.5325 \end{bmatrix}$
  d) $\begin{bmatrix} 0.4765 & 0.5235 \\ 0.479875 & 0.520125 \end{bmatrix}$
  e) Not possible; A is not a square matrix. **f)** [0.58]

Section 6.1, pp. 312–313

Practise

1. a) $\dfrac{1}{2}$ **b)** $\dfrac{1}{4}$ **c)** $\dfrac{7}{8}$ **d)** $\dfrac{2}{9}$ **e)** $\dfrac{29}{36}$ **f)** $\dfrac{3}{13}$
2. a) 100% **b)** 0%   **c)** Answers may vary.
  d) Answers may vary.
3. a) i) (2, 3), (2, 4), (3, 2), (3, 3), (3, 4), (4, 2), (4, 3), (4, 4)
    ii) no outcomes
    iii) (2, 2)
  b) Player B
4. a) 0.625 **b)** 0.281 25 **c)** 0.8

Apply, Solve, Communicate

5. b) $\dfrac{3}{8}$ **c)** $\dfrac{1}{16}$

d) Assuming that a win or loss by either team is equally likely.

6. $\dfrac{2}{5}$

7. a) Player B has a winning probability of $\dfrac{11}{18}$ and so has the advantage.

b) Player B has a winning probability of $\dfrac{5}{9}$ and so has the advantage.

c) Player A wins if 1, 2, or 3 are thrown. Player B wins if 4, 5, or 6 are thrown.

8. a) $\dfrac{13}{24}$ **b)** 5

9. Answers will vary.

10. b) 0.169

d) by multiplying the branch probabilities

11. a) Trimble 36%, Yakamoto 23%, Audette 41%

b) Answers may vary.

c) In part a), votes for Jonsson are not counted. In part b), voters who intended to vote for Jonsson can now cast their votes for one of the remaining three candidates.

12. a) Answers may vary; for example, always choose B or C.

b) Answers may vary; for example, alternately choose B and C.

Section 6.2, pp. 318–319

Practise

1. a) 2:3   **b)** $\dfrac{3}{5}$

2. $\dfrac{1}{20}$

3. a) 1:35   **b)** 5:13   **c)** 5:7   **d)** 0

Apply, Solve, Communicate

4. a) 1:6   **b)** 1:7   **c)** 5:3   **d)** 1:25   **e)** 4:5

5. a) 6:5   **b)** 8:3

6. $\dfrac{7808}{19\ 683}$

7. a) 7:41   **b)** 11:13

8. odds against $A = \dfrac{P(A')}{P(A)} = \dfrac{1}{\frac{P(A)}{P(A')}} = \dfrac{1}{\text{odds in favour of } A}$

9. 3:7

10. a) 10:3   **b)** 210:11

12. a) 3:7   **b)** 2:3   **c)** 1:1

13. No; the associated probabilities of a win, a loss, and a tie do not total 1.

14. approximately 43:7

15. $\dfrac{P(A)}{P(A')} = \dfrac{h}{k}$; $kP(A) = hP(A')$; $kP(A) = h(1 - P(A))$;

$(h + k)P(A) = h$; $P(A) = \dfrac{h}{h + k}$

Section 6.3, pp. 324–326

Practise

1. $\dfrac{1}{3}$

2. 0.074

3. a) $\dfrac{5}{14}$ **b)** $\dfrac{25}{28}$

4. $\dfrac{2}{7}$

Apply, Solve, Communicate

5. a) $\dfrac{7}{102}$ **b)** $\dfrac{4}{17}$

6. $\dfrac{7}{30}$

7. a) $\dfrac{1}{28}$ **b)** $\dfrac{1}{56}$

c) Names are drawn at random.

8. a) $\dfrac{3}{10}$ **b)** $\dfrac{1}{120}$

c) Assuming that the friends arrive individually.

9. a) $\dfrac{14}{969}$ **b)** $\dfrac{1001}{4845}$

10. a) i) $\dfrac{1}{540\ 000}$ **ii)** $\dfrac{1}{54\ 000}$ **iii)** $\dfrac{1}{5400}$

b) 27 000   **c)** 270 000

11. $\dfrac{1}{5}$

12. a) i) $\dfrac{1}{209}$ **ii)** $\dfrac{170}{209}$

b) $\dfrac{1}{42}$

14. a) i) $\dfrac{3}{490}$ **ii)** $\dfrac{25}{784}$

b)–e) Answers may vary.

15. $\dfrac{35}{128}$

16. 7

Section 6.4, pp. 334–335

Practise

1. a) i) dependent   **ii)** independent   **iii)** independent
   iv) dependent   **v)** independent   **vi)** dependent

2. 0.06

3. $\dfrac{1}{1296}$

4. a) 84.55%   **b)** 0.55%   **c)** 99.45%

Apply, Solve, Communicate

5. a) Biff

6. 0.51

7. $\dfrac{121}{300}$

9. 0.2

10. a) 59.5%   **b)** 9:191

c) More time spent studying one subject means less time spent studying the other.

11. Assuming independence, the probability is $\dfrac{1}{12^5}$.

12. $\dfrac{13}{80}$

13. a) $\dfrac{33}{182}$ b) Answers may vary.   c) Answers may vary.

14. 1.3%

15. a) Each of the n tosses is independent and has a probability of $\dfrac{1}{2}$ of getting a head.

b) $\dfrac{127}{128}$

16. $\dfrac{64}{729}$

17. 0.48

18. 0.2187

Section 6.5, pp. 340–343

Practise

1. a) non-mutually exclusive   b) non-mutually exclusive
c) mutually exclusive   d) mutually exclusive
e) non-mutually exclusive   f) non-mutually exclusive

2. a) $\dfrac{1}{9}$ b) $\dfrac{1}{3}$ c) $\dfrac{4}{9}$

3. a) 7%   b) 81%   c) 74%   d) 26%

Apply, Solve, Communicate

4. a) i) $\dfrac{124}{365}$ ii) $\dfrac{92}{365}$

5. a) $\dfrac{13}{55}$ b) $\dfrac{4}{11}$ c) $\dfrac{1}{2}$ d) $\dfrac{3}{22}$

6. a) 0.55   b) 0.148   c) 0.948

7. a) $\dfrac{3}{28}$ b) $\dfrac{4}{7}$ c) $\dfrac{5}{14}$ d) $\dfrac{15}{28}$

8. b) A and B are mutually exclusive.

9. Answers will vary.

10. $\dfrac{37}{120}$; mutually exclusive since bald patches relate to does only.

11. a) 0.232   b) 0.107

13. b) $\dfrac{7}{8}$ c) $\dfrac{83}{120}$

14. 7:33

15. a) $\dfrac{3}{25}$ b) no

16. a) i) 0.418   ii) 0.058   iii) 0.040
iv) 0.960   v) 0.436
b) 0.090   c) 0.697   d) not independent

Section 6.6, pp. 353–356

Practise

1. a) Not a probability vector since the components do not sum to 1.

b) A probability vector.
c) Not a probability vector, since it is not a row vector.
d) Not a probability vector since there is a negative component.
e) A probability vector.

2. a) Not a transition matrix since the components of the 3rd row do not sum to 1.
b) A transition matrix.
c) Not a transition matrix since it is not square.

3. a) Initially, ZapShot has 67% of the market for digital cameras. E-pics has 33% of the market.
b) If the customer's initial purchase was with ZapShot, there is a 60% chance that the customer returns to ZapShot, and a 40% chance that the customer switches to E-pics. If the customer's initial purchase was with E-pics, there is a 50% chance that the customer returns to E-pics, and a 50% chance that the customer switches to ZapShot.

Apply, Solve, Communicate

4. a) Answers may vary.   b) $\begin{bmatrix} \dfrac{5}{9} & \dfrac{4}{9} \end{bmatrix}$

c) E-pics
d) Although ZapShot has a greater chance of drawing customers back than E-pics, the initial market share for ZapShot in question 3 was too high to sustain.

5. a) Regular since all are non-zero entries.
b) Not regular since all powers have entries equal to 0.
c) Regular since the second power has all non-zero entries.

6. a) $\begin{array}{cc} \text{W} & \text{L} \end{array}$
$\begin{bmatrix} 0.7 & 0.3 \\ 0.4 & 0.6 \end{bmatrix} \begin{array}{c} \text{W} \\ \text{L} \end{array}$
b) 0.52
c) $\begin{bmatrix} \dfrac{4}{7} & \dfrac{3}{7} \end{bmatrix}$; in the long run the team will win $\dfrac{4}{7}$ of their games.

7. a) [0.5 0.5]
b) $\begin{array}{cc} \text{RP} & \text{BP} \end{array}$
$\begin{bmatrix} 0.65 & 0.35 \\ 0.75 & 0.25 \end{bmatrix} \begin{array}{c} \text{RP} \\ \text{BP} \end{array}$
c) [0.7 0.3], [0.68 0.32]
d) approximately 0.68

8. a) $\begin{array}{ccc} \text{S} & \text{C} & \text{R} \end{array}$
$\begin{bmatrix} 0.5 & 0.3 & 0.2 \\ 0.25 & 0.35 & 0.4 \\ 0.2 & 0.35 & 0.45 \end{bmatrix} \begin{array}{c} \text{S} \\ \text{C} \\ \text{R} \end{array}$
b) 0.305   c) 0.310
d) Assuming that the transition matrix is seasonally invariant.

9. a) 0.25   b) 0.5415
c) Yes; approximately 67% of the time the value of the stock will rise or remain unchanged.

10. a) [0.7 0.3]
b) $\begin{array}{cc} \text{D} & \text{B} \end{array}$
$\begin{bmatrix} 0.8 & 0.2 \\ 0.35 & 0.65 \end{bmatrix} \begin{array}{c} \text{D} \\ \text{B} \end{array}$
c) 0.364

12. a) Answers may vary. **b)** $\begin{bmatrix} \frac{5}{9} & \frac{4}{9} \end{bmatrix}$

c) $P = \begin{bmatrix} 0.6 & 0.4 \\ 0.4 & 0.6 \end{bmatrix} \begin{matrix} C \\ S \end{matrix}$; $S^{(n)} = \begin{bmatrix} \frac{1}{2} & \frac{1}{2} \end{bmatrix}$

$\quad\quad\quad\ \ \begin{matrix} C & S \end{matrix}$

13. a) not regular; steady state vector: $\begin{bmatrix} \frac{1}{2} & \frac{1}{2} \end{bmatrix}$

b) regular; steady state vector: $\begin{bmatrix} \frac{1}{3} & \frac{2}{3} \end{bmatrix}$

c) not regular; steady state vector: $\begin{bmatrix} 1 & 0 \end{bmatrix}$

14. a) $\begin{bmatrix} 0.4711 & 0.3471 & 0.1818 \\ 0.4711 & 0.3471 & 0.1818 \\ 0.4711 & 0.3471 & 0.1818 \end{bmatrix}$

b) $[0.4711\ 0.3471\ 0.1818]$

c) A regular Markov chain will reach the same steady state regardless of the initial probability vector.

15. $m = 0.7$, $n = 0.3$

Review of Key Concepts, pp. 357–359

6.1 Basic Probability Concepts

1. a) $\frac{7}{20}$ **b)** $\frac{3}{4}$

2. a) $\frac{1}{4}$ **b)** $\frac{1}{6}$ **c)** chance variation

3. a) Answers may vary; normally low.
b) Answers may vary; normally high.

6.2 Odds

4. 1:7
5. a) 7:13 **b)** 1:1 **c)** 17:3

6.3 Probability Using Counting Techniques

6. b) no **c)** $\frac{14}{285}$ **d)** $\frac{11}{57}$

7. $\frac{1}{72}$

8. a) $\frac{1}{12}$ **b)** $\frac{1}{120}$ **c)** $\frac{1}{720}$

6.4 Dependent and Independent Events

9. a) independent **b)** dependent
c) dependent **d)** independent
10. a) 0.3 **b)** subjective
11. a) $\frac{1}{2}$ **b)** $\frac{1}{3}$ **c)** $\frac{1}{2}$ **d)** 0.40
12. 0.025

6.5 Mutually Exclusive Events

13. a) mutually exclusive **b)** mutually exclusive
c) non-mutually exclusive **d)** non-mutually exclusive
14. a) $\frac{5}{26}$

15. b) $\frac{6}{13}$

16. a) 11:4 **b)** 2:13
17. a) 0.45 **b)** 0.87

6.6 Applying Matrices to Probability Problems

18. a) $[1\ 0]$
b) $\begin{bmatrix} 0.5 & 0.5 \\ 0.35 & 0.65 \end{bmatrix} \begin{matrix} S \\ N \end{matrix}$

$\quad\ \ \begin{matrix} S & N \end{matrix}$

c) 0.5, 0.425, 0.4118 **d)** 0.4118 **e)** 0.4118
19. a) $\begin{bmatrix} 0.7 & 0.2 & 0.1 \\ 0.25 & 0.6 & 0.15 \\ 0.3 & 0.05 & 0.65 \end{bmatrix} \begin{matrix} BP \\ S \\ GW \end{matrix}$

$\quad\quad \begin{matrix} BP & S & GW \end{matrix}$

b) $[0.4775\ 0.2703\ 0.2523]$
c) approximately 25%

Chapter Test, pp. 360–361

1. b) $\frac{1}{8}$ for all three tosses coming up heads; $\frac{1}{2}$ for a single toss coming up heads; $\frac{7}{8}$ for at least one toss in three coming up heads

2. a) $\frac{2}{3}$ **b)** 4:5

3. $\left(\frac{1}{2}\right)^{10} = \frac{1}{1024}$

4. 0.05

5. a) i) $\frac{19}{30}$ **ii)** $\frac{11}{30}$

b) that nobody works at two jobs

6. 1:3

7. b) i) 0.46 **ii)** 0.54

8. a) $\frac{1}{10}$ **b)** $\frac{1}{60}$

9. a) $\frac{7}{99}$ **b)** $\frac{35}{66}$ **c)** $\frac{13}{66}$ **d)** $\frac{2}{33}$

10. a) i) $[1\ 0]$
ii) $\begin{bmatrix} 0.07 & 0.93 \\ 0.004 & 0.996 \end{bmatrix} \begin{matrix} D \\ ND \end{matrix}$

$\quad\quad \begin{matrix} D & ND \end{matrix}$

iii) 0.0049
b) Regular; all entries of the transition matrix are non-zero.

CUMULATIVE REVIEW: CHAPTERS 4 TO 6, p. 364

1. a) 5040 **b)** 7 **c)** 7 **d)** 5040
e) 21 **f)** 21
2. a) $243x^5 - 810x^4y + 1080x^3y^2 - 720x^2y^3 + 240xy^4 - 32y^5$
b) $2(a - b)^4$
3. a) 19 770 609 664 **b)** 19 461 693 888
4. 399 168 000

5. $\frac{7}{10}$

6. $\frac{4}{9}$

7. a) 24 b) 16

8. a) $\frac{1}{14}$ b) $\frac{3}{7}$

9. $\frac{1}{20}$

10. 0.073

11. a) 0.18 b) 0.0324

12. 0.64

13. a) Sasha: 40%; Pedro: 60%
   b) 47.5%
   c) Sasha: 53, Pedro: 47

CHAPTER 7

Review of Prerequisite Skills, p. 368

1. a) $\frac{2}{81}$ b) $\frac{9}{64}$ c) $\frac{36}{625}$ d) 6.28 e) 0.5904

2. a) $\sum_{k=1}^{12} t_k$ b) $\sum_{k=0}^{9} k_9C_k$ c) $\sum_{k=1}^{6} \frac{k+1}{k+2}$ d) $\frac{1}{6}\sum_{k=0}^{5} a_k$

3. a) $1 + 4 + 9 + 16 + 25 + 36 = 91$
   b) $b_0 + b_1 + b_2 + b_3 + b_4 + b_5 + b_6 + b_7 + b_8 + b_9 + b_{10} + b_{11} + b_{12} + b_{13} + b_{14}$
   c) $_7C_0 + _7C_1 + _7C_2 + _7C_3 + _7C_4 + _7C_5 + _7C_6 + _7C_7 = 128$
   d) $0.7 + (0.3)(0.7) + (0.3)^2(0.7) + (0.3)^3(0.7) + (0.3)^4(0.7) + (0.3)^5(0.7) + (0.3)^6(0.7) + (0.3)^7(0.7) + (0.3)^8(0.7) = 1 - 0.3^9$

4. a) $x^6 + 6x^5y + 15x^4y^2 + 20x^3y^3 + 15x^2y^4 + 6xy^5 + y^6$
   b) $0.4^4 + 4(0.4)^3(0.6) + 6(0.4)^2(0.6)^2 + 4(0.4)(0.6)^3 + 0.6^4 = 1$

   c) $\left(\frac{1}{3}\right)^5 + 5\left(\frac{1}{3}\right)^4\left(\frac{2}{3}\right) + 10\left(\frac{1}{3}\right)^3\left(\frac{2}{3}\right)^2 + 10\left(\frac{1}{3}\right)^2\left(\frac{2}{3}\right)^3$
   $+ 5\left(\frac{1}{3}\right)\left(\frac{2}{3}\right)^4 + \left(\frac{2}{3}\right)^5 = 1$

   d) $p^n + _nC_1p^{n-1}q + _nC_2p^{n-2}q^2 + \dots + _nC_nq^n = \sum_{k=0}^{n} _nC_kp^{n-k}q^k$

5. a) $\frac{1}{6}$ b) $\frac{1}{18}$ c) $\frac{5}{9}$ d) $\frac{5}{36}$ e) $\frac{1}{6}$

6. $\frac{1}{16}$

7. a) $\frac{1}{676}$ b) $\frac{150}{169}$

Section 7.1, pp. 374–377

Practise

1. a) discrete b) continuous c) continuous
   d) discrete e) continuous f) discrete

2. a) uniform
   b) not uniform; not everyone has an equal probability of being selected.
   c) uniform d) uniform
   e) not uniform; some values will have a higher probability of being selected.
   f) uniform
   g) not uniform; some totals have a higher probability than other totals.

3. a) 10.75 b) 2 775 250 c) $\frac{29}{10}$

4. a) $\frac{1}{2}$ b) 4.5

Apply, Solve, Communicate

5. a) uniform b) $\frac{1}{10\ 000}$ c) $\frac{1}{1000}$

6. a) b) 7 c)

x	P(x)
2	$\frac{1}{36}$
3	$\frac{2}{36}$
4	$\frac{3}{36}$
5	$\frac{4}{36}$
6	$\frac{5}{36}$
7	$\frac{6}{36}$
8	$\frac{5}{36}$
9	$\frac{4}{36}$
10	$\frac{3}{36}$
11	$\frac{2}{36}$
12	$\frac{1}{36}$

x	P(x)
3	$\frac{1}{216}$
4	$\frac{3}{216}$
5	$\frac{6}{216}$
6	$\frac{10}{216}$
7	$\frac{15}{216}$
8	$\frac{21}{216}$
9	$\frac{25}{216}$
10	$\frac{27}{216}$
11	$\frac{27}{216}$
12	$\frac{25}{216}$
13	$\frac{21}{216}$
14	$\frac{15}{216}$
15	$\frac{10}{216}$
16	$\frac{6}{216}$
17	$\frac{3}{216}$
18	$\frac{1}{216}$

expected sum: 10.5

7. tetrahedron: 2.5, cube: 3.5, octahedron: 4.5, dodecahedron: 6.5, icosahedron: 10.5

8. a) $\dfrac{7}{2\ 000\ 000}$ **b)** \$1.48

9. a) -0.5 **b)** No; the expected value is not 0.

10. \$1.05

11.

x	P(x)
0	0.25
1	0.5
2	0.25

expected number: 1

12. 26 cm^2

13. a) Answers may vary.   **b)** 2

14. a) Answers may vary.

b) Fatima has the greater probability of having two sons.

15. a) i) $\dfrac{1}{343}$ **ii)** $\dfrac{24}{343}$ **iii)** $\dfrac{180}{343}$ **iv)** $\dfrac{18}{343}$ **v)** $\dfrac{120}{343}$

c) no

17. b)

x	P(x)
0	$\dfrac{21}{66}$
1	$\dfrac{35}{66}$
2	$\dfrac{10}{66}$

c) $\dfrac{14}{99}$ **d)** 1.7

18. a) i) 5.3% **ii)** 5.3% **iii)** 5.3% **iv)** 5.3% **v)** 5.3%

b) Answers may vary.

c) The house advantage decreases.

19. inner region: $\dfrac{4}{25}$, middle region: $\dfrac{5}{25}$, outside region: $\dfrac{16}{25}$

20. Answers may vary.

a) expectation: 2.625   **b)** expectation: 1.75

21. equal

Section 7.2, pp. 385–387

Practise

1. a) binomial

b) not binomial; there are different probabilities of a success.

c) not binomial; there is no fixed probability of success.

d) binomial

2. a)

x	P(x)
0	0.327 68
1	0.4096
2	0.2048
3	0.0512
4	0.0064
5	0.000 32

b)

x	P(x)
0	0.003 906 25
1	0.031 25
2	0.109 375
3	0.218 75
4	0.273 43 75
5	0.218 75
6	0.109 375
7	0.031 25
8	0.003 906 25

Apply, Solve, Communicate

3. a)

x	P(x)
0	0.735 09
1	0.232 13
2	0.030 54
3	0.002 14
4	8.4609×10^{-5}
5	1.7813×10^{-6}
6	1.5625×10^{-7}

b) 0.3

4. Answers may vary.

5. 0.2003

6. a) Answers may vary.   **b)** 0.9358   **c)** 1.5

7. a) 0.8939   **b)** 5.2

8. a) 0.0319   **b)** 2   **c)** Answers may vary.

9. Answers may vary.

10. 0

11. $12\dfrac{2}{3}$%

12. a) i) $(p + q)^6 = p^6 + 6p^5q + 15p^4q^2 + 20p^3q^3 + 15p^2q^4 + 6pq^5 + q^6$

ii) $(0.2 + 0.8)^5 = 0.2^5 + 5(0.2)^4(0.8) + 10(0.2)^3(0.8)^2 + 10(0.2)^2(0.8)^3 + 5(0.2)(0.8)^4 + (0.8)^5$

b) Each term is of the form of the probability of a certain number of successes for a binomial random variable.

14. a) Poisson approximation: 1.2276×10^{-8}, binomial distribution: 9.7936×10^{-9}

b) Poisson approximation: 5.3559×10^{-15}, binomial distribution: 4.4357×10^{-15}

c) Poisson approximation: 5.2337×10^{-7}, binomial distribution: 3.9805×10^{-7}

15. Answers may vary. Approximately 96% of the time the number of heads flipped by a fair coin in 20 flips will be between 6 and 14 inclusive.

16. a) $\dfrac{n!}{n_1!n_2!n_3!}\,p^{n_1}q^{n_2}r^{n_3}$; where $n_1 + n_2 + n_3 = n$

b) $\dfrac{10!}{2!4!4!}\left(\dfrac{1}{6}\right)^2\left(\dfrac{1}{6}\right)^4\left(\dfrac{4}{6}\right)^4 \doteq 0.0133$

17. There is a 79% chance that at least one model will be damaged.

Section 7.3, pp. 394–396

Practise

1. a) geometric
  b) not geometric; the number of trials is fixed.
  c) not geometric; the random variable for the waiting time is not discrete.
  d) geometric
  e) not geometric

2. a)

x	P(x)
0	0.2
1	0.16
2	0.128
3	0.1024
4	0.081 92
5	0.065 536

b)

x	P(x)
0	0.5
1	0.25
2	0.125
3	0.625
4	0.031 25
5	0.015 625

Apply, Solve, Communicate

3. a) 0.0700   **b)** 11 rolls
4. a) 0.1001   **b)** 3 flights
5. No; the expected waiting time to roll a sum of 2 is 35 tries.
6. $\dfrac{16}{81}$
7. a) 0.0573   **b)** 0.5578   **c)** 11.5 tests
8. a) 0.0314   **b)** 9 hang-ups
9. a) 0.1296   **b)** 1.5 deliveries
10. a) 0.0426   **b)** 1.9 people
11. a) 0.0188   **b)** 0.0776   **c)** 49 cards
12. a) 0.0176   **b)** 13.3 shots
13. a) 0.0728   **b)** 65.7 chips
14. a) Answers may vary.   **b)** 4.5
15. a) Answers may vary.   **b)** 18.15
16. $P(x) = q^{x-1}p,\ E(x) = \dfrac{1}{p}$
17. The MTBF is the expected value of a continuous distribution that measures time before the failure of a component. The probability of the drive failing in any one-

year period cannot be calculated knowing only the mean of the distribution—the distribution itself must be known. Probability of failure changes with time.

18. $4p^2q^3$
19. a) i) $\dfrac{1}{20}$ **ii)** $\dfrac{1}{20}$ **iii)** $\dfrac{1}{20}$ **iv)** 0.45
  b) 95
  c) The trials are not indpendent.

Section 7.4, pp. 404–405

Practise

1. a) hypergeometric
  b) not hypergeometric; this is a geometric distribution.
  c) not hypergeometric; this is a binomial distribution.
  d) not hypergeometric; this is a binomial distribution.
  e) hypergeometric   **f)** hypergeometric
  g) hypergeometric

2. a)

x	P(x)
0	0.05
1	0.45
2	0.45
3	0.05

b)

x	P(x)
0	0.017 857 143
1	0.267 857 143
2	0.535 714 286
3	0.178 571 429

Apply, Solve, Communicate

3. a) $\dfrac{14}{33}$ **b)** $\dfrac{7}{6}$
4. a) 2167   **b)** chance variation in survey results
5. a) Answers may vary.   **b)** 0.2173
6. a) Answers may vary.   **b)** 8.3
7. a) 0.6339   **b)** 1.25
8. a) 0.3777   **b)** 1.6
9. a) 0.0707   **b)** 2.1
10. a) 0.2668   **b)** 0.9886   **c)** 5.3
11. a) 0.1758   **b)** 1.8
12. a) 0.0152   **b)** 0.3818   **c)** 0.1818
  d) Red: 0.73, Black: 1.82, Green: 1.45
14. a) 4.34×10^{-8} **b)** 2.17×10^{-7} **c)** 0.842
15. If n is very large, non-replacement of successes will not change the ratio of successes to the population a great deal, and a binomial distribution will be a good approximation to the hypergeometric distribution.
16. $\dfrac{63}{256}$
17. 2.4 older scientists versus 2 younger scientists.

Review of Key Concepts, pp. 406–407

7.1 Probability Distributions

1. A uniform distribution is one in which each outcome has an equal probability. For example, when rolling a fair die, each of the six faces has an equal chance of facing upward.

2. a) forward 3.5 squares

b) Since the expected move is not 0, the game is not fair.

3. −\$4.54

4. 57 m²

5. 8-sided

7.2 Binomial Distributions

6. A binomial distribution is one in which there are a certain number of independent trials and two possible outcomes for each trial, with the probability of these outcomes being constant in each trial. For example, counting the number of times a 1 is rolled in 10 rolls of a fair die.

7. a) 0.0009   **b)** 2

8.

x	P(x)
0	0.401 877 572
1	0.401 877 572
2	0.160 751 029
3	0.032 150 206
4	0.003 215 021
5	0.000 128 601

9. a) 36.8%   **b)** 6.2%

10. a) $252 \times \dfrac{19^5}{20^{10}}$ or 0.000 0609   **b)** 0.5

11. a)

x	P(x)	xP(x)
0	$\left(\dfrac{4}{5}\right)^{15}$	0
1	$15 \times \dfrac{4^{14}}{5^{15}}$	$15 \times \dfrac{4^{14}}{5^{15}}$
2	$105 \times \dfrac{4^{13}}{5^{15}}$	$210 \times \dfrac{4^{13}}{5^{15}}$
3	$455 \times \dfrac{4^{12}}{5^{15}}$	$1365 \times \dfrac{4^{12}}{5^{15}}$
4	$1365 \times \dfrac{4^{11}}{5^{15}}$	$5460 \times \dfrac{4^{12}}{5^{15}}$
5	$3003 \times \dfrac{4^{10}}{5^{15}}$	$15015 \times \dfrac{4^{10}}{5^{15}}$
6	$5005 \times \dfrac{4^{9}}{5^{15}}$	$30030 \times \dfrac{4^{9}}{5^{15}}$
7	$6435 \times \dfrac{4^{8}}{5^{15}}$	$45045 \times \dfrac{4^{8}}{5^{15}}$
8	$6435 \times \dfrac{4^{7}}{5^{15}}$	$51480 \times \dfrac{4^{7}}{5^{15}}$
9	$5005 \times \dfrac{4^{6}}{5^{15}}$	$45045 \times \dfrac{4^{6}}{5^{15}}$
10	$3003 \times \dfrac{4^{5}}{5^{15}}$	$30030 \times \dfrac{4^{5}}{5^{15}}$
11	$1365 \times \dfrac{4^{4}}{5^{15}}$	$15015 \times \dfrac{4^{4}}{5^{15}}$
12	$455 \times \dfrac{4^{3}}{5^{15}}$	$5460 \times \dfrac{4^{3}}{5^{15}}$
13	$105 \times \dfrac{4^{2}}{5^{15}}$	$1365 \times \dfrac{4^{2}}{5^{15}}$
14	$15 \times \dfrac{4}{5^{15}}$	$210 \times \dfrac{4}{5^{15}}$
15	$\dfrac{1}{5^{15}}$	$15 \times \dfrac{1}{5^{15}}$

7.3 Geometric Distributions

12. A geometric distribution is one in which there are independent trials that have two possible outcomes, with the probability of these outcomes being constant in each trial. The number of trials before a success is of interest. For example, counting the number of rolls of a fair die before a 1 is rolled.

13. 5

14. a) 0.0213   **b)** 0.1356   **c)** 40.7 boards

15. a) $\left(\dfrac{24}{25}\right)^5\left(\dfrac{1}{25}\right)$ **b)** 24 trials before a 10 appears

16.

x	P(x)	xP(x)
1	$\dfrac{1}{36}$	$\dfrac{1}{36}$
2	$\dfrac{35}{36^2}$	$2 \times \dfrac{35}{36^2}$
3	$\dfrac{35^2}{36^3}$	$3 \times \dfrac{35^2}{36^3}$
4	$\dfrac{35^3}{36^4}$	$4 \times \dfrac{35^3}{36^4}$
5	$\dfrac{35^4}{36^5}$	$5 \times \dfrac{35^4}{36^5}$
6	$\dfrac{35^5}{36^6}$	$6 \times \dfrac{35^5}{36^6}$
7	$\dfrac{35^6}{36^7}$	$7 \times \dfrac{35^6}{36^7}$
8	$\dfrac{35^7}{36^8}$	$8 \times \dfrac{35^7}{36^8}$
9	$\dfrac{35^8}{36^9}$	$9 \times \dfrac{35^8}{36^9}$
10	$\dfrac{35^9}{36^{10}}$	$10 \times \dfrac{35^9}{36^{10}}$

7.4 Hypergeometric Distributions

17. A hypergeometric distribution is one in which there are a certain number of dependent trials, each having two possible outcomes. The number of successes is of interest. For example, counting the number of clubs dealt in a hand of 5 cards from a standard deck of 52 cards.

18. a)

x	P(x)
0	0.006 993 007
1	0.093 240 093
2	0.326 340 326
3	0.391 608 392
4	0.163 170 163
5	0.018 648 019

b) 2.7

19.

x	P(x)	xP(x)
0	$\dfrac{{}_{16}C_0 \times {}_{36}C_7}{{}_{52}C_7} = 0.062\ 40$	0
1	$\dfrac{{}_{16}C_1 \times {}_{36}C_6}{{}_{52}C_7} = 0.232\ 95$	0.239 25
2	$\dfrac{{}_{16}C_2 \times {}_{36}C_5}{{}_{52}C_7} = 0.338\ 15$	0.676 30
3	$\dfrac{{}_{16}C_3 \times {}_{36}C_4}{{}_{52}C_7} = 0.246\ 57$	0.739 70
4	$\dfrac{{}_{16}C_4 \times {}_{36}C_3}{{}_{52}C_7} = 0.097\ 13$	0.388 53
5	$\dfrac{{}_{16}C_5 \times {}_{36}C_2}{{}_{52}C_7} = 0.020\ 57$	0.102 85
6	$\dfrac{{}_{16}C_6 \times {}_{36}C_1}{{}_{52}C_7} = 0.002\ 15$	0.012 93
7	$\dfrac{{}_{16}C_7 \times {}_{36}C_0}{{}_{52}C_7} = 0.000\ 09$	0.000 60

20. 153

21. a) Simulations will vary. **b)** $P(x) = \dfrac{{}_{32}C_x \times {}_{50}C_{32-x}}{{}_{98}C_{32}}$

Chapter Test, pp. 408–409

1. a) hypergeometric   **b)** geometric
   c) hypergeometric   **d)** binomial
   e) uniform   **f)** binomial
   g) hypergeometric   **h)** geometric

2. –$1.86

3. a) 0.2743   **b)** 0.2189   **c)** 0.96   **d)** 1.92

4. 34.56 cm²

5. a)

Number of Prizes	Probability
0	0.430 467
1	0.382 638
2	0.148 803
3	0.033 067
4	0.004 593
5	0.000 408
6	2.27E -05
7	7.2E-07
8	1E-08

b) 50

6. a) 0.1314   **b)** 35 rolls

7. Answers may vary.

8. a) 0.469   **b)** 0.019 73   **c)** 2.5

9. a) $\dfrac{1}{6^4}$ **b)** 13.7

10. Answers may vary.

11. 43

CHAPTER 8

Note: *For some normal probability questions, answers calculated in a spreadsheet will differ slightly from those done on a graphing calculator.*

Review of Prerequisite Skills, pp. 412–413

1. b) 1   **c) i)** $y = 10^{-\frac{x}{2}}$ **ii)** $y = 10^{-2x}$
2. b) 150.4, 27.2
3. a) 21.9, 23, 24
   c) 4.2
4. mean: 24.9; median: 22; modes: 12, 14, 22; standard deviation: 14.8
5. mean: 77.3; median: 80; modes: 89, 90; standard deviation: 16.7
7. c) The curve $y = \sqrt{x^3}$ is almost a perfect fit.
   d) The data satisfies $T = \sqrt{R^3}$, where T is the time for one revolution and R is the mean distance from the sun.
8. a) −1.05   **b)** 2.49   **c)** 0.13
   d) 25.59 to 51.01   **e)** 12.88 to 63.72   **f)** 21.78 to 54.82
9. Gavin's z-score is 2.28 while Patricia's is only 2.10.
10. a) 0.0683   **b)** 0.0922
   c) 0.9761   **d)** 10.2
11. a) 0.047   **b)** 0.957   **c)** 2.75

Section 8.1, pp. 419–421

Practise

1. a) iii   **b)** ii   **c)** i

Apply, Solve, Communicate

2. a) about 76   **b)** Answers may vary.
3. c) Estimates may vary.
   d) mean: 176.9; standard deviation: 27
4. b) positively skewed
   c) Lower proportion of older workers; lower accident rate among more experienced workers.
   d) 420; average
5. Answers will vary.
6. b) $\frac{5}{6}$ **c)** $\frac{1}{2}$
7. b) Yes; waiting times follow an exponential distribution.
   c) 5.9   **d)** $y = 0.17e^{-0.17x}$
8. a) 1, ∞   **b)** 0.78, both estimates are low by about 10%.
   c) The estimate from the simulation is low due to random chance. The slope of the exponential curve becomes less steep as x increases, so the midpoint of the intervals underestimates the area under the curve.
   d) Increase the length of time being simulated to a week or more. Estimate the area under the exponential curve using narrower interval widths, such as tenths or hundredths of a minute.
9. a) 0.58   **b)** 0.24
10. c) 20.05, 7.23
11. b) bell curve
   c) so that the total area under the curve is 1

Section 8.2, pp. 430–431

Practise

1. 0.16, 0.16, 0.32, 0.18

Apply, Solve, Communicate

2. 0.8944
3. 13.33%
4. a) 88th percentile   **b)** 6th percentile
   c) 99th percentile
5. Ty Cobb: 4.15, Ted Williams: 4.26, George Brett: 4.07 Using z-scores, Williams is top ranked, followed by Cobb and then Brett.
6. 0.766%
7. 5.02%
8. a) 2.21%   **b)** 45.77%   **c)** 11.6%
9. a) 62.76%   **b)** 14.60%
10. a) 0.0316   **b)** 145.5 cm to 168.5 cm
11. a) $y = \dfrac{10(x - 55)}{13} + 70$, where x is the old mark and y is the new mark.
   b) 89   **c)** 58
12. between 20 cm and 43 cm
13. 4.35 min

Section 8.3, pp. 438–441

Apply, Solve, Communicate

1. a) 110.3, 10.7   **b)** 0.8315
2. a) \$32 461.20, \$3375.77   **b)** 0.26
3. a) \$67.37, \$17.00   **b)** 0.6677   **c)** 0.1534
   d) With repeated sampling, the mean of the sample means will more closely approximate the true population mean and may yield more accurate results.
4. b) no
   c) Canada: 125.5, 25.9; Ontario: 128.3, 34.4
   d) yes
5. Answers depend on current data.
6. Answers depend on current data.
7. Answers depend on current data.
8. a) 43.9, 10.866   **b)** 0.408
9. 367
10. a) no   **b)** no
   c) Yes, the data could be part of a larger set that is normally distributed.
11. 0.0742
13. a) Compound the gain from year to year.
   b) 5-year mean 78.22%, standard deviation 26.52%. Ratio of standard deviation to return is markedly less for the 5-year returns: fluctuations in the annual returns tend to cancel out over longer periods.
   c) 1-year mean 20.05, standard deviation 7.23; 5-year mean 101.08, standard deviation 27.29. Again, ratio of s to $\bar{x}$ is lower for the longer period.

Technology Extension, pp. 442–444

Normal Probability Plots

1. 20 min, 5.71 min
2. yes
3. a) 0.5   **b)** 76.9%

Section 8.4, pp. 449–450

Practise

1. a) reasonable   **b)** not reasonable since $np < 5$
c) reasonable   **d)** reasonable

2.

Sample Size, n	Probablility of Success, p	μ	σ	Probablility
60	0.4	24	3.79	$P(X < 22) = 0.2550$
200	0.7	140	6.48	$P(X < 160) = 0.9999$
75	0.6	45	4.24	$P(X > 50 = 0.0974$
250	0.2	50	6.32	$P(X > 48) = 0.5937$
1000	0.8	800	12.65	$P(780 < X < 840) = 0.9375$
90	0.65	58.5	4.52	$P(52 < X < 62) = 0.6539$
100	0.36	36	4.80	$P(X = 40) = 0.0587$
3000	0.52	1560	27.36	$P(X = 1650) = 0.000\,0653$

Apply, Solve, Communicate

3. 0.150
4. 0.129
5. 0.69
6. Normal approx: 0.056; Binomial Dist.: 0.0541
7. Normal approx: 0.080; Binomial Dist.: 0.0775
8. a) i) 0.9273   **ii)** 0.8948   **b)** 6.7
9. a) i) 0.9507   **ii)** 0.0432
b) i) 0.91   **ii)** 0.0433
10. a) 0.0312   **b)** 0.0507
11. a) i) 0.209   **ii)** 0.209
b) i) 0.213   **ii)** 0.207
c) very closely
13. a) 32   **b)** 26.4, 37.6

Section 8.5, pp. 457–458

Practise

1. 20, 0.86; 12, 0.8; 5, 0.33; 40, 0.8; 8.4, 0.39; 17.6, 1.11; 73.9, 1.35
2. a) accept H_0 **b)** accept H_0 **c)** accept H_0
d) accept H_0 **e)** reject H_0

Apply, Solve, Communicate

3. $P(z > 11.79) < 10\%$; the machine needs to be adjusted.
4. No, $P(z < -1.23) > 10\%$; that only 31 agreed in the survey could have been due to chance variation.
5. $P(z \geq 0.35) > 0.01$; recommend acceptance—there is not enough evidence to reject the claim that the drug causes serious side effects in less than 0.01% of the population.
6. $\alpha = 5\%$: $P(z \leq -1.09) > 5\%$; there is not enough evidence to reject the claim that the product holds 28% of the market.
7. $P(z \leq -3.30) < 0.05\%$; the evidence refutes the company's claim even for $\alpha = 1\%$.
8. The standard deviation in loan amounts. A significance level of 1% would be appropriate due to the small sample size.
9. 50%
10. The advertising campaign was probably not a success.
11. Larger samples are likely to give more accurate statistics but will be more time-consuming and costly. Repeated sampling verifies the measurement process but will not detect bias.
13. Whether to accept your hypothesis for the value of μ.

Section 8.6, pp. 464–467

Practise

1. a) $14.2 < \mu < 15.8$ **b)** $27.7 < \mu < 32.3$
c) $5.6 < \mu < 7.2$ **d)** $29.3 < \mu < 31.1$
e) $40.4 < \mu < 43.2$ **f)** $3.9 < \mu < 4.6$
2. a) There is a 95% probability that between 39% and 45% of high school graduates expect to earn over $100 000 per year by the time they retire.
b) There is a 90% probability that between 43% and 53% of decided voters will vote for the incumbent in the next election.
c) There is a 73% probability that between 24% and 32% of teenagers will purchase the latest CD by the rock band.

Apply, Solve, Communicate

3. $24.48 < \mu < 25.52$
4. $15.8 < \mu < 19.4$
5. $0.192 < p < 0.368$
6. $0.119 < p < 0.121$
7. $61.0 < \mu < 71.4$
8. $13.7\% < \mu < 18.3\%$
9. $18.3\% \pm 4.4\%$
10. 68
11. 44
12. 936
13. Answers may vary.
14. Answers may vary.
15. a) $-0.111, 0.355$ **b)** 0.62
c) yes   **d)** $-0.296 < \mu < 0.074$
16. 106
17. Answers may vary.
20. 1690

Review of Key Concepts, pp. 468–471

8.1 Continuous Probability Distributions

1. a) 0.6   **b)** 0.28
2. a) the exponential distribution   **b)** 0.55
3. b) approximately symmetric and bell-shaped
   c) mean: 8.8, standard deviation: 8.07

8.2 Properties of the Normal Distribution

4. a) 0.0856   **b)** 0.6486   **c)** 0.5695
5. a) 0.0228   **b)** 65.68%   **c)** 35.3
6. a) 0.2119   **b)** 69.15%   **c)** about 1158

8.3 Normal Sampling and Modelling

7. a) A histogram shows that these data could be normally distributed.
   b) 35.4, 10.5   **c)** using a normal distribution: 0.695
8. a) 2.304, 0.069   **b)** 21.67%
9. a) 75.7, 4.35
   b) The standard deviation of the class average scores equals the standard deviation of the individual scores divided by the square root of the class size.
   c) 31
10. 0.503

8.4 Normal Approximation to the Binomial Distribution

11. a) 0.176
   b) Yes; 110 is almost five standard deviations above expectation.
12. 0.8291

8.5 Repeated Sampling and Hypothesis Testing

13. $P(z \geq 5.3) < 0.10$; it takes longer than 0.9 h on average to paint a car.
14. The coach is correct at this significance level.
15. The evidence does not support the claim that the advertising campaign was a success at a 20% significance level.

8.6 Confidence Intervals

16. $5.2 < \mu < 5.6$
17. $0.304 < p < 0.416$
18. a) $1515 < \mu < 3185$
   b) The confidence interval includes less than half of the data because they are not normally distributed.
19. $45.5 < \mu < 394.8$ (in thousands)
20. 358
21. 210
22. a) $0.382 < p < 0.658$ **b)** Answers may vary.

Chapter Test, pp. 472–473

1. Answers may vary.
   a) test mark distribution for a large class
   b) driving time between Toronto and Hamilton
   c) waiting times for service at a bank machine

   d) the distribution of shoe sizes of adult males and females
2. a) 0.588   **b)** 9.12%   **c)** about 24
3. Currently about 18% of the cans contain the exceptional amounts. A lower standard deviation would be better.
4. a) 9.88, 3.48   **b)** 0.486
5. a) 0   **b)** 41%
   c) 9% have less than 20% cashews.
   d) "At least 19% cashews in every can" or "average 22% cashews."
6. 0.6807
7. a) 0.178 assuming defect rate is 0.03
   b) If the defect rate is 3%, there is a 17.8% chance that 65 or more defective cars will be produced during a shift. If more samples yield similar numbers of defective cars, changes may be required. Otherwise, the high number of defects may have been just a chance variation.
8. $0.352 < p < 0.527$
9. a) 384
   b) Yes, the poll indicates that the politician has a 78% chance of losing the election if he faces a single opponent.
   c) The politician should try to convince some undecided voters to vote for him.

CUMULATIVE REVIEW: CHAPTERS 7 AND 8, p. 476

1. –68.8¢
2. a) 37%   **b)** 15.9%
3.

Number of Heads, x	Probability, P(x)
0	0.007 8125
1	0.054 6875
2	0.164 0625
3	0.273 4375
4	0.273 4375
5	0.164 0625
6	0.054 6875
7	0.007 8125

4. a) 0.0158   **b)** 0.9810   **c)** 8.25
5. 0.38%
6. a) Answers will vary.   **b)** 0.2616   **c)** 5.45
   d) Statistical fluctuations could cause noticeable differences unless the number of simulated trials is quite large.
7. 504
8. a) $1.58 < \mu < 1.76$
   b) Wolves are normally distributed.

COURSE REVIEW, pp. 477–479

1. a) 2:3　**b)** 3:2

2. Answers may vary.

3. a) E, F, G, and H

　b)　i) no　　　　**ii)** no　　　　**iii)** yes

4. 15

5.

Day of Month	Probability
1	$\frac{2}{61}$
2	$\frac{2}{61}$
3	$\frac{2}{61}$
4	$\frac{2}{61}$
5	$\frac{2}{61}$
6	$\frac{2}{61}$
7	$\frac{2}{61}$
8	$\frac{2}{61}$
9	$\frac{2}{61}$
10	$\frac{2}{61}$
11	$\frac{2}{61}$
12	$\frac{2}{61}$
13	$\frac{2}{61}$
14	$\frac{2}{61}$
15	$\frac{2}{61}$
16	$\frac{2}{61}$
17	$\frac{2}{61}$
18	$\frac{2}{61}$
19	$\frac{2}{61}$
20	$\frac{2}{61}$
21	$\frac{2}{61}$
22	$\frac{2}{61}$
23	$\frac{2}{61}$
24	$\frac{2}{61}$
25	$\frac{2}{61}$
26	$\frac{2}{61}$
27	$\frac{2}{61}$
28	$\frac{2}{61}$
29	$\frac{2}{61}$
30	$\frac{11}{366}$
31	$\frac{7}{366}$

6. a) Independent trials in which there are only two possible outcomes.

　b) Unlike a hypergeometric distribution, trials in a geometric distribution are independent.

7. a) 3　**b)** 6

　c) No, with $2 coins and quarters, there are only three choices.

　d) 12

8. Answers may vary.

　a) sampling students from each of the grades 9, 10, 11, and 12 in numbers according to the proportion each grade comprises of the total; ensures representation from all grades in the proper proportions

　b) sampling persons from randomly-chosen households in Ontario for their provincial political preference; each person has an equal opportunity to be a part of the sample

　c) to sample 6 students from a class of 30, choose a random number from 1 to 5 and, from a class list, choose this student and every 6th student thereafter; easy to obtain the required percentage of respondents

9. a) 10 879 286 400　　　　**b)** about 20 719.3 years

10. a) 6.7, 6.5

11. a) weak negative correlation

　b) -0.46; $y = -0.10x + 9.66$

　c) (46, 11); -0.98; $y = -0.068x + 7.04$;

　d) the faster the speed, the fewer the errors

12. No; both are dairy products and so have a common-cause relationship.

13. a) 600　**b)** random sample of all households in Ontario

14. a) 64　**b)** 729

15. 1 632 960

16. 256 851 595

17. a) 120　**b)** 14　　**c)** 8　　**d)** 51　　**e)** 50

18. 17 153 136

19. a) $\dfrac{x^5}{243} + \dfrac{5x^4y}{27} + \dfrac{10x^3y^2}{3} + 30x^2y^3 + 135xy^4 + 243y^5$

　b) $\dbinom{n+1}{2}$

20. $\dfrac{1}{5}$

21. a) 0.438 866　　　**b)** $3.\overline{3}$

22. $\dfrac{19}{24}$

23. a) yes　　　　　**b)** no

24. a) 0.080　　**b)** 0.594　　　　**c)** 5.06

25. a) UltraChicken: 0.2807; Churrasqueira Champion: 0.5439; Mac's Chicken: 0.1754

　b) that transition matrix does not change

26. a) 0.014　　　**b)** 0.88　　　**c)** 46.6

27. Leading questions offer information that may then be used by the respondent, when it otherwise would not have been. Loaded questions are intended to directly influence a response by supplying information in a certain way to obtain the desired result.

28. a) 0.069　　　**b)** 13.4

29. Answers will vary.

APPENDIX A

Evaluating expressions, p. 496

1. a) -31 **b)** 2 **c)** 271 **d)** -422 **e)** 48

f) -227 **g)** -42 **h)** -699 **i)** -36 **j)** $-\dfrac{14}{81}$

k) $\dfrac{1}{20}$ **l)** 4900

Exponent laws, p. 496

1. a) $1024x^{10}$ **b)** $9x^8y^{10}$ **c)** $1250x^9y^7$ **d)** $\dfrac{3}{400}$

e) a^7b **f)** $-9x^4y^4$ **g)** $\dfrac{1536y^3}{x^2}$ **h)** $-50x^{15}y^7$

Fractions, percents, decimals, p. 497

1. a) $75\%,\ 0.75$ **b)** $50\%,\ 0.5$ **c)** $840\%,\ 8.4$

d) $\dfrac{17}{50},\ 0.34$ **e)** $\dfrac{3}{10\,000},\ 0.0003$ **f)** $\dfrac{7}{125},\ 0.056$

g) $\dfrac{9}{20},\ 45\%$ **h)** $\dfrac{3}{100},\ 3\%$ **i)** $2\dfrac{17}{25},\ 268\%$

2. 14.3%

3. $\$23$

Graphing data, p. 498

1. b) Initially sales are high, but then they begin to fall off as the number of appearances increases.

c) Graphs that do not exhibit changes over time would not be appropriate, for example, a circle graph.

2. a) 176

c) 0: 20%, 1: 14%, 2: 32%, 3: 15%, 4: 10%, 5: 7%, 6: 2%

e) A line graph would not be appropriate.

Graphing quadratic equations, p. 500

1. a) y-intercept: 0, x-intercept: 0

b) y-intercept: 0, x-intercept: 0

c) y-intercept: 5, x-intercepts: none

d) $y = 3(x + 1)^2 - 8$, y-intercept: -5, x-intercepts: 0.6, -2.6

e) $y = -2(x - 2)^2 + 18$, y-intercept: 10, x-intercepts: -1, 5

f) $y = 5\left(x - \dfrac{4}{5}\right)^2 - \dfrac{6}{5}$, y-intercept: 2, x-intercepts: 0.3, 1.3

g) $y = (x + 3)^2 - 18$, y-intercept: -9, x-intercepts: -7.2, 1.2

h) $y = -\left(x - \dfrac{5}{2}\right)^2 + \dfrac{37}{4}$, y-intercept: 3, x-intercepts: -0.5, 5.5

i) $y = 4\left(x + \dfrac{3}{4}\right)^2 - \dfrac{5}{4}$, y-intercept: 1, x-intercepts: -1.3, -0.2

Mean, median, mode, p. 501

1. a) 16.7, 17, 18 **b)** 82.1, 79, 79

Number patterns, p. 502

1. a) subtract 3; 6, 3, 0

b) multiply by 3: 108, 324, 972

c) perfect squares; 16, 25, 36

d) add 2; 8, 10, 12

e) divide by -3: $-\dfrac{1}{3}, \dfrac{1}{9}, -\dfrac{1}{27}$

f) divide by -2: $-10, 5, -\dfrac{5}{2}$

g) multiply by -1: 1, -1, 1

h) subtract 10: $-15, -25, -35$

i) powers of 2 or multiply by 2: 32, 64, 128

j) multiply by $\dfrac{2}{3}$: $\dfrac{2}{9}, \dfrac{4}{27}, \dfrac{8}{81}$

k) multiply p by q^x: $pqqqq, pqqqqq, pqqqqqq$

l) append twice the previous number of b's: $abbbbbbbb$, $abbbbbbbbbbbbbbbb$, $abbbbbbbbbbbbbbbbbbbbbbbbbbbbbbbb$

m) add alternately to lower right and upper right

n) add next larger square base

Order of operations, p. 503

1. a) 33 **b)** 195 **c)** 108 **d)** 64.4

e) 0.5 **f)** $1\dfrac{1}{2}$ **g)** $\dfrac{419}{270}$ **h)** $\dfrac{9}{8}$

Ratio of areas, p. 503

1. a) $1:4$ **b)** $1:9$ **c)** $4:1$ **d)** $125:1$

Scatter plots, p. 504

1. b) As altitude increases, temperature decreases.

d) -6.5; the rate of change of temperature with respect to altitude

e) $t = -6.5a + 11.25$ **f)** $-66.8°$

2. b) As floor space increases, heating cost increases.

d) 9.68; the rate of change of heating cost with respect to floor space

e) $c = 9.68s - 295.5$ **f)** $\$3092.50$

Sigma notation, p. 505

1. a) $\displaystyle\sum_{i=1}^{6} 2^i$ **b)** $\displaystyle\sum_{i=3}^{15} i$ **c)** $\displaystyle\sum_{i=1}^{7} \dfrac{i}{i+1}$ **d)** $\displaystyle\sum_{i=5}^{11} ix$

2. a) $3 + 6 + 9 + 12 + 15 + 18 + 21 + 24 + 27 + 30$

b) $5.2 + 5.4 + 5.8 + 5.16 + 5.32 + 5.64 + 5.128 + 5.256$

c) $7 + 9 + 11 + 13 + 15 + 17 + 19 + 21 + 23 + 25 + 27 + 29$

d) $\dfrac{4}{9} + \dfrac{5}{16} + \dfrac{6}{25} + \dfrac{7}{36}$

Similar triangles, p. 505

1. $\triangle QPR \sim \triangle FGH$ because
$QP{:}FG = QR{:}FH = PR{:}GH = 2{:}1$

Simplifying expressions, p. 506

1. a) $x^2 + 4xy + 4y^2$ **b)** $4x^2 - 20x + 25$

c) $x^4 + 2x^2y + y^2$ **d)** $25x^2 - 20xy + 4y^2$

e) $12x^3 + 28x^2 + 19x + 4$ **f)** $75y^3 + 40y^2 - 68y + 16$

g) $k^3 - 2k^2 - 15k + 36$ **h)** $125m^3 - 15mn^2 + 2n^3$

i) $-64x^3 - 128x^2y - 20xy^2 + 50y^3$

Solving equations, p. 506

1. a) -4 **b)** -6 **c)** -3 **d)** 18 **e)** $\dfrac{15}{2}$ **f)** 1.575

g) -11 **h)** -4 **i)** ± 6 **j)** ± 12 **k)** 2 **l)** 6

Substituting into equations, p. 507

1. a) -3

b) 27

c) -19

d) $9x^4 - 36x^2 + 38$ **e)** 11

f) $3x^2 + 12x + 6$

g) 966

h) $6x^2 - 11$

i) $27x^4 - 54x^2 + 102$

j) 69

k) $x^4 + 4x^2 + 6$

l) $10\ 406$

m) 7

n) 7797

2. no

GLOSSARY

A

absolute cell referencing A spreadsheet feature that blocks automatic adjustment of cell references when formulas are moved or copied. References preceded by a dollar sign—$A$1, for example—are left unchanged.

accidental relationship A correlation between two variables that happens by random chance.

action plan A logical sequence of specific steps for completing a project or testing a hypothesis.

addition rule for mutually exclusive events The principle relating the probabilities of events that cannot occur at the same time. For example, if events A and B are mutually exclusive, then $P(A \text{ or } B) = P(A) + P(B)$.

addition rule for non-mutually exclusive events The principle relating the probabilities of events that can occur at the same time. For example, if events A and B are not mutually exclusive, then $P(A \text{ or } B) = P(A) + P(B) - P(A \text{ and } B)$.

additive counting principle (rule of sum) The principle that, if one mutually exclusive action can occur in m ways, a second in n ways, a third in p ways, and so on, then there are $m + n + p$ … ways in which one of these actions can occur.

adjacent vertices Vertices that are connected by an edge.

algorithm A procedure or set of rules for solving a problem.

arrow diagram A diagram that uses lines and arrows to illustrate the sequence of steps in a process.

B

balloting process A sampling process in which the instrument is an anonymous ballot.

bar graph A chart or diagram that represents quantities with horizontal or vertical bars whose lengths are proportional to the quantities.

Bernoulli trials Repeated independent trials measured in terms of successes and failures.

bias Systematic error or undue weighting in a statistical study.

bimodal Having two modes, or "humps." See *mode*.

binomial distribution A distribution having independent trials whose outcomes are either success or failure. The probability of success is unchanged from one trial to the next, and the random variable is the number of successes in a given number of trials.

binomial theorem A theorem giving the expansion of powers of a binomial: For any natural number n, $(a + b)^n = {}_nC_0a^n + {}_nC_1a^{n-1}b + \ldots + {}_nC_ra^{n-r}b^r + \ldots + {}_nC_nb^n$

box-and-whisker plot A graph that summarizes a set of data by representing the first quartile, the median, and the third quartile with a box and, the lowest and highest data with the ends of lines extending from the box.

C

CANSIM and **CANSIM II** The Canadian Socio-economic Information Management System, an extensive database compiled by Statistics Canada. It profiles the Canadian people, economy, and industries. CANSIM II is the updated version of CANSIM.

categorical data Data that can be sorted or divided by type rather than by numerical values.

cause-and-effect relationship A relationship in which a change in an independent variable (X) produces a change in a dependent variable (Y).

cell references A letter and number that indicate the column and row of a cell in a spreadsheet. For example, B3 refers to the third row of column B.

census An official count of an entire population or class of things.

circle graph A graph that represents quantities with segments of a circle that are proportional to the quantities. Circle graphs are also called pie charts.

circuit A path in a network that begins and ends at the same vertex.

class A set of data whose values lie within a given range or interval.

classical probability The probability of an event deduced from analysis of the possible outcomes. Classical probability is also called theoretical or *a priori* probability.

cluster sampling A survey of selected groups within a population. This sampling technique can save time and expense, but may not give reliable results unless the clusters are representative of the population.

coefficient of determination (generalized correlation coefficient) A measure of how closely a curve fits a set of data. The coefficient of determination is denoted by r^2, and can be calculated using the formula $r^2 = \dfrac{\Sigma(y_{est} - \bar{y})^2}{\Sigma(y - \bar{y})^2}$, where $\bar{y}$ is the mean y value and y_{est} is the value estimated by the best-fit curve.

coding matrix A matrix used to encode a message.

column matrix A matrix having only one column.

column sum The sum of the entries in a column of a matrix.

combination A selection from a group of items without regard to order. The number of combinations of r items taken from a set of n items is denoted by $_nC_r$, $C(n, r)$, or $\binom{n}{r}$, and equal to $\dfrac{n!}{r!(n-r)!}$.

combinatorics The branch of mathematics dealing with ideas and methods for counting, especially in complex situations.

common-cause factor An external variable that causes two variables to change in the same way.

common element An element that is in two sets.

complement of an event The set of all outcomes that are not included in an event. The complement of an event A is the event that event A does *not* happen, and is denoted as A' or $\sim A$.

complete network A network that has an edge between every pair of vertices.

compound event An event consisting of two or more events.

conditional probability of an event, $P(B|A)$ The probability that event B occurs, given that event A has already occurred.

confidence interval A range of values that is centred on the sample mean and has a specified probability of including the population mean, μ. For example, there is a 0.9 probability that μ will lie between the upper and lower limits of the 90% confidence interval.

confidence level The probability that a measurement or conclusion is correct. The confidence level is equal to $1 - \alpha$, where α is the *significance level*.

connected network A network having at least one path connecting each pair of vertices.

consumer price index (CPI) A collective measure of the cost of items purchased by a typical family.

continuity correction Treating the values of a discrete variable as continuous intervals in order to use a normal approximation for a binomial distribution.

continuous Involving a variable or data that can have an infinite number of possible values in a given interval. A continuous function or distribution can be graphed as a smooth curve.

control group The group for which the independent variable is held constant in an experiment or statistical study.

convenience sample A sample selected simply because it is easily accessible. Such samples may not be random, so their results are not always reliable.

correlation coefficient A summary statistic that gives a quantified measure of the linear relationship between two variables. Sometimes referred to as the *Pearson product-moment coefficient of correlation*, this coefficient is denoted by r and can be calculated using the formula $r = \dfrac{n\Sigma xy - (\Sigma x)(\Sigma y)}{\sqrt{[n\Sigma x^2 - (\Sigma x)^2][n\Sigma y^2 - (\Sigma y)^2]}}$

correlational research The application of statistical methods to determine the relationship between two variables.

covariance The mean of the products of the deviations of two variables.

cubic function A polynomial function that can be written in the form $y = ax^3 + bx^2 + cx + d$, where a, b, c, and d are numerical coefficients and $a \neq 0$.

cumulative-frequency graph A graph that shows the running total of the frequencies for values of the variable starting from the lowest value.

cumulative probability The probability that a variable is less than a certain value.

curve of best fit The curve that fits closest to the data points in a scatter plot or best represents the relationship between two variables.

D

data Facts or pieces of information. A single fact is a *datum*.

database An organized store of records.

degree The number of edges that begin or end at a vertex in a network; also, the highest power of the variables in an equation or polynomial.

dependent event An event whose outcome depends directly on the outcome of another event.

dependent (or **response**) **variable** A variable whose value is affected by another variable.

deviation The difference in value between a datum and the mean.

dimensions The number of rows and columns in a matrix, usually expressed in the form $m \times n$ with the number of rows listed first.

direct linear correlation A relationship in which one variable increases at a constant rate as the other variable increases, a perfect positive linear correlation.

discrete variable A variable that can take on only certain values within a given range.

disjoint events Events that cannot occur at the same time, mutually exclusive events.

E

edges Line segments in a network.

element of a set An item or member in a set.

empirical probability The number of times that an event occurs in an experiment divided by the number of trials. The empirical probability is also known as the experimental or relative-frequency probability.

entry A number appearing in a matrix.

E-STAT An interactive educational web site hosted by Statistics Canada. This site enables students to access data from the CANSIM database.

expected value, $E(X)$ The predicted mean of all possible outcomes of a probability experiment.

experimental group The group for which the independent variable is changed in an experiment or statistical study.

exponential distribution A continuous probability distribution that predicts waiting times between consecutive events in a random sequence of events.

exponential function A function that can be written in the form $y = ab^x$, where a and b are numerical coefficients.

exponential regression An analytic technique for finding the equation with the form $y = ab^x$ or $y = ae^{kx}$ that best models the relationship between two variables.

extraneous variables Variables that affect or obscure the relationship between an independent and a dependent variable.

extrapolation Estimating variable values beyond the range of the data.

F

factorial, $n!$ A product of sequential natural numbers having the form
$n! = n \times (n-1) \times (n-2) \times \ldots \times 2 \times 1.$
The notation $n!$ is read "n factorial."

fair game A game in which the expectation is 0.

field A location in a database record where specific data are displayed or entered.

first-step probability vector, $S^{(1)}$ The row matrix in a Markov chain that gives the probabilities of being in any state after one trial.

fractal A geometric figure that is created using an iterative process and self-similar shapes.

frequency diagram A diagram, such as a histogram or frequency polygon, that shows the frequencies with which different values of a variable occur.

frequency polygon A plot of frequencies versus variable values with the resulting points joined by line segments.

function A built-in formula in a graphing calculator, spreadsheet, or other software.

fundamental (multiplicative) counting principle The principle that, if a task or process is made up of stages with separate choices, the total number of choices is $m \times n \times p \times \ldots$, where m is the number of choices for the first stage, n is the number of choices for the second stage, p is the number of choices for the third stage, and so on.

G

geometric distribution A distribution having independent trials whose outcomes are either success or failure. The probability of success is unchanged from one trial to the next, and the random variable is the number of trials before a success occurs.

graph In graph theory, a collection of line segments and vertices, which can represent interconnections between places, items, or people; a network.

graph theory A branch of mathematics in which graphs or networks are used to represent relationships and solve problems in many fields.

gross domestic product (GDP) A measure of a country's overall economic output, including all goods and services.

H

hidden variable An extraneous variable that is difficult to recognize.

histogram A bar graph in which the areas of the bars are proportional to the frequencies for various values of the variable.

hypergeometric distribution A distribution having dependent trials whose outcomes are either success or failure. The probability of success changes from one trial to the next, and the random variable is the number of successes.

hypothesis A proposition or thesis that is assumed to be true in order to investigate its validity.

hypothesis test A statistical procedure that uses a sample to determine the probability that a statement is correct.

I

identity matrix A matrix having entries of 1 along the main diagonal and zeros for all other entries, such as $\begin{bmatrix} 1 & 0 & 0 \\ 0 & 1 & 0 \\ 0 & 0 & 1 \end{bmatrix}$.

independent event An event whose probability is not affected by the outcome of another event.

independent (or **explanatory**) **variable** A variable that affects the value of another variable.

index A number relating the value of a variable, or group of variables, to a base level, which is often the value on a particular date.

indirect method A problem-solving technique for finding a quantity by determining another quantity from which the first can be derived. Often the first quantity is difficult to calculate directly.

inflation An increase in the overall price of goods and services.

initial probability vector, $S^{(0)}$ A matrix that represents the probabilities of the initial state of a Markov chain.

initial value A value given for a term in the first step of a recursion formula.

instrument In statistics, any form of data-collection mechanism, such as questionnaires, personal interviews, telephone survey, or direct measurement.

interpolate Estimate a value between two known values.

interquartile range The range of the central half of a set of data when the data are arranged in numerical order.

intersection The set of elements common to two or more sets. The intersection of sets A and B is often written as $A \cap B$.

interval A set of all numbers between two given numbers.

inverse linear correlation A relationship in which one variable increases at a constant rate as the other decreases, a perfect negative linear correlation.

inverse matrix The matrix that produces the identity matrix when multiplied by a given matrix. The inverse of matrix A is written as A^{-1}. Only square matrices can have inverses.

iteration The process of repeating the same procedure over and over.

L

leading question A question which prompts a particular answer.

least-squares fit An analytic technique for determining the line of best fit by minimizing the sum of the squares of the deviations of the data from the line.

line of best fit The straight line that passes closest to the data points on a scatter plot and best represents the relationship between two variables.

linear correlation A relationship in which changes in one variable tend to be proportional to the changes in another.

linear regression An analytic technique for finding the equation with the form $y = ax + b$ that best models the relationship between two variables.

loaded question A question containing information or language intended to influence the respondents' answers.

lurking variable An extraneous variable that is difficult to recognize, a hidden variable.

M

margin of error The width of a confidence interval.

Markov chain A probability model in which the outcome of any trial depends directly on the outcome of the previous trial.

mathematical model A model that describes the relationship between variables in a quantitative fashion.

matrix A rectangular array of numbers used to manage and organize data.

mean The sum of the values in a set of data divided by the number of values.

mean absolute deviation The mean of the absolute values of the *deviations* of a set of data.

measurement bias Bias resulting from a data-collection method that consistently either under- or over-estimates a characteristic of the population.

measures of central tendency The values around which a set of the data tends to cluster.

measures of spread Quantities that indicate how closely a set of data clusters around its central values.

median The middle value of a set of data ranked from highest to lowest. If there is an odd number of data, the median is the midpoint between the two middle values.

member An item or element of a set.

midrange Half of the sum of the highest value and the lowest value in a set of data.

mind map A tool for organizing related topics and generating ideas about them and related sub-topics.

mind web A mind map in which related topics at the same level are joined with dotted lines.

modal interval For grouped data, the interval which contains more data than any other interval.

mode The value in a distribution or set of data that occurs most frequently.

modified box plot A box-and-whisker plot that shows outliers as separate points instead of including them in the whiskers.

multi-stage sampling A sampling technique that uses several levels of random sampling.

mutually exclusive events Events that cannot occur at the same time.

N

negative skew The pulling to the left of the tail in an asymmetric probability distribution.

neighbours In a network, vertices that are joined by an edge.

network In graph theory, a collection of line segments and vertices, which can represent interconnections between places, items, or people.

network matrix A matrix that represents a network.

node A point in a network at which edges end or meet, a vertex.

non-linear regression An analytical technique for finding a curve of best fit for a set of data.

non-mutually exclusive events Events that can occur simultaneously.

non-response bias Bias occurring when particular groups are under-represented in a survey because they choose not to participate.

normal distribution A common continuous probability distribution in which the data are distributed symmetrically and unimodally about the mean.

normal probability plot A graph of the data in a sample versus the z-scores of the corresponding quantiles for a normal distribution. If the plot is approximately linear, the underlying population can be assumed to be normally distributed.

nth-step probability vector, $S^{(n)}$ The row matrix in a Markov chain that gives the probabilities of being in any state after n trials.

null set A set that has no elements.

O

odds against The ratio of the probability that the event will not occur to the probability that it will occur.

odds in favour The ratio of the probability that the event will occur to the probability that it will not occur.

ogive A graph of a cumulative-frequency distribution.

outcome A possible result, a component of an event.

outliers Points in a set of data that are significantly far from the majority of the other data.

P

Pascal's formula A formula relating the combinations in Pascal's triangle: ${}_nC_r = {}_{n-1}C_{r-1} + {}_{n-1}C_r$. This formula is also known as Pascal's theorem.

path A connected sequence of vertices in a network.

percentile A set of values that divides an ordered set of data into 100 intervals, each having the same number of data.

perfect negative linear correlation A relationship in which one variable increases at a constant rate as the other variable decreases. A graph of the one variable versus the other is a straight line with a negative slope.

perfect positive linear correlation A relationship in which one variable increases at a constant rate as the other variable increases. A graph of the one variable versus the other is a straight line with a positive slope.

permutation An arrangement of items in a definite order. The total number of permutations of distinct n items is denoted by ${}_nP_n$ or $P(n, n)$ and is equal to $n!$.

pictograph A chart or diagram that represents quantities with symbols.

planar network A network that can be drawn such that its edges do not cross anywhere except at vertices.

polynomial regression An analytic technique for finding the polynomial equation that best models the relationship between two variables.

population All individuals that belong to a group being studied.

positive skew The pulling to the right of the tail in an asymmetric probability distribution.

power regression An analytic technique for finding the equation with the form $y = ax^b$ that best models the relationship between two variables.

presumed relationship A correlation that does not seem to be accidental even though no cause-and-effect relationship or common-cause factor is apparent.

principle of inclusion and exclusion The principle that the total number of elements in either set A or set B is the number in A plus the number in B minus the number in both A and B:
$n(A \cup B) = n(A) + n(B) - n(A \cap B)$

probability theory A branch of mathematics that deals with chance, random variables, and the likelihood of outcomes.

probability density The probability per unit of a continuous variable.

probability-density function An equation that describes or defines the curve for a probability distribution.

probability distribution The probabilities for all possible outcomes of an experiment, often shown as a graph of probability versus the value of a random variable.

probability experiment A well-defined process consisting of a number of trials in which clearly distinguishable outcomes are observed.

probability of an event A, $P(A)$ A quantified measure of the likelihood that event A will occur. The probability of an event is always a value between 0 and 1.

product rule for dependent events The principle that the probability of event B occurring after event A has occurred is $P(A \text{ and } B) = P(A) \times P(B|A)$, where $P(B|A)$ is the *conditional probability* of event B.

product rule for independent events The principle that the probability of independent events A and B both occurring is $P(A \text{ and } B) = P(A) \times P(B)$.

Q

quadratic function A function that can be written in the form $y = ax^2 + bx + c$, where a, b, and c are numerical coefficients and $a \neq 0$.

quantile One of a set of values that divide a set of data into groups with equal numbers of data; the variable value corresponding to a given cumulative probability. For example, the first quartile has the cumulative probability $P(X \leq x) = 0.25$.

R

random variable A variable that can have any of a set of different values. In statistics, a random variable is often denoted by a capital letter (commonly X or Y), while its individual values are denoted by the corresponding lowercase letter.

randomization A technique that ensures that all members of a population are equally likely to be selected for a sample. Such techniques reduce the likelihood that results will be inappropriately weighted in favour of one particular group within the population.

range The difference between the highest and lowest values in a set of data.

raw data Unprocessed information.

record A set of data that is treated as a unit in a database.

recursion formula A formula for calculating a series of terms, each of which is derived from the preceding terms.

regression analysis An analytic technique for determining the relationship between two variables.

regular Markov chain A Markov chain that always achieves a steady state.

relational database Databases in which different sets of records can be linked and sorted in complex ways based on the data contained in the records.

relative cell referencing A spreadsheet feature that automatically adjusts cell references in formulas when they are moved or copied.

relative frequency The frequency of a value or group of values expressed as a fraction or percent of the whole data set.

repeated sampling A sampling method that uses two or more independent samples from the same population.

residual The difference between the observed value of a variable and the corresponding value predicted by the regression equation.

residual plot A graph of the residuals of a set of data versus the independent variable.

response bias Bias that occurs when participants in a survey deliberately give false or misleading answers.

reverse cause-and-effect relationship A relationship in which the presumed dependent and independent variables are reversed in the process of establishing causality.

row matrix A matrix having only one row.

row sum The sum of the entries in a row of a matrix.

S

sample A group of items or people selected from a population.

sample space, *S* The set of all possible outcomes in a probability experiment.

sampling A process of selecting a group from a population in order to estimate the characteristics of the entire population.

sampling bias Bias resulting from a sampling frame that does not reflect the characteristics of the population.

sampling frame The members of a population that actually have a chance of being selected for a sample.

scalar A quantity having only magnitude (as opposed to a vector, which also has a direction).

scatter plot A graph in which data are plotted with one variable on the *x*-axis and the other on the *y*-axis. The pattern of the resulting points can show the relationship between the two variables.

self-similar shape A shape containing components that have the same geometrical characteristics.

semi-interquartile range One half of the *interquartile range*.

seed value A value given for a term in the first step of a recursion formula, an initial value.

set A group of items.

significance level, α The probability that the result of a hypothesis test will be incorrect.

simple random sample A sample in which every member of a population has an equal and independent chance of being selected.

simulation An experiment, model, or activity that imitates real or hypothetical conditions.

square matrix A matrix with the same number of rows as columns.

standard deviation The square root of the mean of the squares of the deviations of a set of data. The standard deviation is given by the formulas $\sigma = \sqrt{\dfrac{\Sigma(x - \mu)^2}{N}}$ for a population and $s = \sqrt{\dfrac{\Sigma(x - \bar{x})^2}{n - 1}}$ for a sample.

standard normal distribution A normal distribution in which the mean is equal to 0 and the standard deviation is equal to 1.

statistical bias Any factor that favours certain outcomes or responses and hence systematically skews the survey results.

statistical fluctuation The difference between characteristics measured from a sample and those of the entire population. Such differences can be substantial when the sample size is small.

statistical inference A process used to determine whether causality exists between the variables in a set of data.

statistics The gathering, organization, analysis, and presentation of numerical information.

Statistics Canada A federal government department that collects, summarizes, and analyses a broad range of Canadian statistics.

steady-state vector A probability vector in a Markov chain that remains unchanged when multiplied by the transition matrix.

stratified sample A sample in which each stratum or group is represented in the same proportion as it appears in the population.

stratum A group whose members share common characteristics, which may differ from the rest of the population.

subjective probability An estimate of the likelihood of an event based on intuition and experience—an educated guess.

subset A set whose elements are all also elements of another set.

systematic sample A sample selected by listing a population sequentially and choosing members at regular intervals.

T

theoretical probability The probability of an event deduced from analysis of the possible outcomes. Theoretical probability is also called classical or *a priori* probability.

time-series graph A plot of variable values versus time with the adjacent data points joined by line segments.

total variation The sum of the squares of the deviations for a set of data, $\sum(y-\bar{y})^2$.

traceable network A network whose vertices are all connected to at least one other vertex and whose edges can all be travelled exactly once in a continuous path.

transition matrix, *P* A matrix representing the probabilities of moving from any initial state to any new state in a given trial.

transpose matrix A matrix in which the rows and columns have been interchanged so that a_{ij} becomes a_{ji}.

trial A step in a probability experiment in which an outcome is produced and tallied.

triangular numbers The sum of the first n natural numbers: $1 + 2 + \ldots + n$. Triangular numbers correspond to the number of items stacked in a triangular array.

U

uniform probability distribution A probability distribution in which each outcome is equally likely in any single trial.

unimodal Having only one mode, or "hump." See *mode*.

union The set of all elements contained in two or more sets. The union of sets A and B is often written as $A \cup B$.

universal set, *S* The set containing all elements involved in a particular situation.

V

variable A quantity that can have any of a set of values.

variance The mean of the squares of the deviations for a set of data. Variance is denoted by σ^2 for a population and s^2 for a sample.

Venn diagram A pictorial representation of one or more sets, in which each set is represented by a closed curve.

vertex A point in a network at which edges end or meet. Also called a node.

voluntary-response sample A sampling technique in which participation is at the discretion or initiative of the respondent.

W

waiting time or **period** The number of unsuccessful trials or the elapsed time before success occurs. Waiting time is the random variable in geometric and exponential probability distributions.

weighted mean A measure of central tendency that reflects the greater significance of certain data.

Z

zero matrix A matrix in which all entries are zero.

z-score The number of standard deviations from a datum to the mean. The z-score of a datum is given by the formula $z = \dfrac{x - \bar{x}}{s}$.

TECHNOLOGY INDEX

sheet referencing, 20–21
simulation, 36–37
sorting, 18–19
standard deviation, 139

N

Normal distribution, 423, 426–429, 434–435, 438

O

Organizers, 14

R

Referencing
  cells, 15–17
  worksheets, 20–21
Relational databases, 14
Relative cell referencing, 16

S

Schedulers, 14
Searching, 19–20
Simulation, 33–40
  Fathom™, 37–39
  graphing calculator, 35
  spreadsheet, 36–37
Sorting, 18–19
Spreadsheets, 544–569
  adding a worksheet, 20
  binomial distribution, 382
  charting 17–18, 417
  counting, 36–37, 417
  combinations, 277, 399
  correlation coefficients, 166
  data grouping, 95–96
  deviations, 139
  exponential, 417
  factorials, 234
  fill and series, 16–17, 391
  formulas and functions, 14–15, 391, 544–569
  frequency diagrams, 95–97
  linear regression, 175–176, 178
  matrices, 58–59, 66–67, 69
  mean, median, and mode, 128–129, 434
  non-linear regression, 186–189

normal distribution, 435
permutation function, 238
random numbers, 36–37, 417
regression analysis, 175–176
searches, 19–20
standard deviation, 139, 434
SUM function, 15, 391, 400, 418
variance, 139
worksheets, 20–21
Statistical software, 14–15
SUM function, 15, 391, 400, 418

T

Technology Tool Cross-Reference Table, 508–509

V

Variable
  continuous, 91
  discrete, 91

W

Web Connection, 8, 12, 27, 28, 47, 86, 92, 107, 108, 112, 127, 150, 179, 194, 202, 211, 216, 245, 247, 272, 310, 314, 318, 356, 378, 387, 402, 425, 439, 440, 463, 474, 483, 484
Word-processor, 14

INDEX

Repeated sampling, 451–453
  simulating, 451
Research skills, 594–597
  accessing sources, 595
  citing sources, 597
  evaluating sources, 596
Residuals, 172, 179
Response bias, 122
Reverse cause-and-effect, 196–197, 199
Row matrix, 54
Row sum, 53
Rule of sum, 228, 266–267, 337–338

S

Sample, 113–116, 133, 137–138
  bias, 119-122
  central value, 137
  cluster, 116
  convenience, 116
  extrapolating from, 113
  frame, 114
  interval, 114–115
  multi-stage, 116
  vs population, 128–130
  simple random, 114
  size, 203–205, 209, 462–463
  stratified, 115–116
  systematic, 114
  techniques, 203–204, 209
  voluntary-response, 116
Sample space, 305
Scalar, 56
Scatter plots, 159–161, 164–167, 172–173, 175, 177, 188–190, 203, 206, 414, 504–505
Seed value, 9
Self-similar shape, 8
Semi-interquartile range, 141–2
Sets, 265–270
  common elements, 266
  elements, 266
  intersection, 267
  members, 266
  null set, 283
  principle of inclusion and exclusion, 267–270
  subset, 266, 283
  union, 267

Sierpinski triangle, 7–8
Sigma notation, 126, 505
Significance level, 456
Similar triangles, 505–506
Simplifying expressions, 506
Simulation, 33–40, 320, 369, 378, 388, 397, 416–418, 451
Skewed distributions, 415
Solving equations, 506–507
Spread, measures of , 136–147
Square matrix, 54
Standard deviation, 137–140
Standard normal distribution, 426
Statistical fluctuation, 308
Statistics, 202
Statistics Canada, 28–30, 104, 106
Steady state, 349–352
Stratified sample, 115–116
Subset, 266, 283
Substituting into equations, 507
Survey
  bias, 119–122
  control group, 197–199
  experimental group, 197–198
Systematic sample, 114

T

Tetrahedral numbers, 252
Time-series graphs, 105–109, 207–208
Traceable network, 44–45, 48
Transition matrix, 345
Tree diagrams, 226, 507
Trials, 305, 378–379
Triangular numbers, 249–251

U

Uniform distribution, 372–374, 415–416
Unimodal distribution, 415
Union of sets, 267
Universal set, 265–269

V

Variables, 91, 100, 104, 197
  continuous, 91–92, 100, 370–371, 374
  correlation between, 161

dependent, 160, 197
discrete, 91, 100, 370–371, 374
explanatory, 160
extraneous, 197–199, 204–206, 208–209
hidden, 206–209
independent, 160, 177, 197–198
lurking, 206–209
random, 370–371, 374, 389, 394
relationships between, 159, 171–179, 184
response, 160
Variance, 137, 139–140
Venn diagrams, 265–270, 337, 339
Vertex, 41–46, 48
  adjacent, 43, 48
  degree, 43–45, 47–48
Voluntary-response sample, 116

W

Waiting time (waiting period), 388–389, 392–394, 416–418
Weighted mean, 130–131, 133

Z

Z-scores, 146–147, 426, 433, 453

CREDITS

Photographs

iv Statistics Canada, *www.statcan.ca*, January 7, 2002; **v** CORBIS RF/MAGMA PHOTO; **vi** Ian Crysler; **vii** Eyewire; **1** PhotoDisc; **3** Dick Hemingway; **6** CORBIS RF/MAGMA PHOTO; **10** Adapted from *www.ec.gc.ca/water/en/nature/prop/e_prop.html*; **14** CORBIS RF/MAGMA PHOTO; **28** Ed Young/CORBIS/MAGMA PHOTO; **30** Statistics Canada, *www.statcan.ca*, January 7, 2002; **33** Reuters; **41** Toronto Transit Commission; **53** COA/CP Picture Archive; **63** Harry How/Allsport; **86 Top** CORBIS RF/MAGMA PHOTO, **Bottom** *www.comstock.com*; **87** AFP/CORBIS/MAGMA PHOTO; **89** Duomo/CORBIS/MAGMA PHOTO; **91** Dick Hemingway; **104** The Weather Network; **113** Ivy Images; **119** Warren Clements; **125** Gary Sussman/AP Photo/CP Picture Archive; **136** CORBIS RF/MAGMA PHOTO; **157** Rachel Epstein/PhotoEdit; **171** David Tanaka; **184** PAUL A. SOUDERS/CORBIS/MAGMA PHOTO; **195** Melanie Brown/PhotoEdit; **202** The DUPLEX © 1999, Glenn McCoy. Reprinted with permission of UNIVERSAL PRESS SYNDICATE. All rights reserved; **221** CORBIS RF/MAGMA PHOTO; **223** Jeff Greenberg/Valan Photos; **225** Dick Hemingway and John Fowler/Valan Photos; **232** Jeff Greenberg/Valan Photos; **241** CORBIS RF/MAGMA PHOTO; **245** © Tribune Media Services, Inc. All rights reserved. Reprinted with permission; **259** © Patrick Ward/CORBIS/MAGMA PHOTO; **263** Dick Hemingway; **265** Dick Hemingway; **273** Dick Hemingway; **282** Bill Ivy/Ivy Images; **288** © Tribune Media Services, Inc. All rights reserved. Reprinted with permission; **289** Archivo Iconographico, S.A./CORBIS/MAGMA PHOTO; **301** CORBIS RF/MAGMA PHOTO; **304** Wonderfile/Masterfile; **314** *www.comstock.com*; **320** CORBIS/MAGMA PHOTO; **327** CORBIS/MAGMA PHOTO; **336** Zoran Milich/Masterfile; **344** Toronto Stock Exchange; **356** Wonderfile/Masterfile; **365** Mary Ellen McDonald/CORBIS/MAGMA PHOTO; **369** *www.comstock.com*; **378** Owen Franzen/CORBIS/MAGMA PHOTO; **388** Tom Hauck/Allsport; **397** Jeff Codge/Getty Images/The Image Bank; **411** PhotoDisc; **414** Lynn Goldsmith/CORBIS/MAGMA PHOTO; **422** PhotoDisc; **432** Richard Lautens/Toronto Star/CP Picture Archive; **445** CORBIS/MAGMA PHOTO; **451** Ian Crysler; **459** CORBIS/MAGMA PHOTO; **482** Charles E. Rotkin/CORBIS/MAGMA PHOTO; **486** David Tanaka; **488–489 Top** Robert Ressmeyer/CORBIS/MAGMA PHOTO, **Centre** PhotoDisc, **Bottom** J Silver/SuperStock; **491** CORBIS/MAGMA PHOTO; **493** Eyewire

Data

4 Statistics Canada, *www.statcan.ca/english/Pgdb/People/Labour/labor01.htm*, November 7, 2001; **4** Statistics Canada, *www.statcan.ca/english/land/geography/phys08a.htm*, November 7, 2001; **61 question 13** *The Top 10 of Everything*, Russell Ash, Reader's Digest Canada, p. 68, **question 14** Statistics Canada, *www.statcan.ca/English/Pgdb/People/Education/educ21.htm*,

January 7, 2002; **62** Statistics Canada, *www.statcan.ca/english/Pgdb/People/Population/demo10a.htm*, November 7, 2001; **65** Maclean's Guide to Canadian Universities and Colleges 2000; **76** Statistics Canada, *www.statcan.ca/english/Pgdb/People/Population/demo10a.htm*, November 7, 2001; **104–105** Statistics Canada, *www.statcan.ca*, January 7, 2002; **107** Statistics Canada, *www.statcan.ca*, January 7, 2002; **110** Statistics Canada, *www.statcan.ca*, January 7, 2002 and The Demographic Viability of Wolves, Table 1, *www.cpaws-ov.org/images/vucetichpacquet.pdf*; **111–112** Statistics Canada, *www.statcan.ca*, January 7, 2002; **151** Statistics Canada, *www.statcan.ca*, *January 7, 2002*; **183** The Worldwatch Institute, *www.worldwatch.org*; **201** Statistics Canada, *www.statcan.ca*, January 7, 2002 and Bankruptcy branch, Industry Canada; **215** Ontario Road Safety Annual Report, *The Toronto Star*, July 25, 2001, p. A20; **343** Statistics Canada, CANSIM, cross-classified table 00580602; **411** USGS Earthquake Hazards Program, National Earthquake Information Center, *neic.usgs.gov/neis/eqlists/7up.html*; **440** and **466** The Demographic Viability of Wolves, Table 1, *www.cpaws-ov.org/images/vucetichpacquet.pdf*; **477** Statistics Canada, *www.statcan.ca*, January 7, 2002; **485** Ontario Road Safety Annual Report, *The Toronto Star*, July 25, 2001, p. A20.

Statistics Canada Information is used with the permission of the Minister of Industry, as Minister responsible for Statistics Canada. Information on the availability of the wide range of data from Statistics Canada can be obtained from Statistics Canada's regional offices, its World Wide Web site at *www.statcan.ca*, and its toll-free access number 1-800-263-1136.

Illustrations

8 Greg Duhaney; **10** Greg Duhaney; **247** Alana Lai; **367** Greg Duhaney

Screen Captures

Calculator templates, Texas Instruments Incorporated; Corel® Quattro® Pro, Corel and Quattro trademarks or registered trademarks of Corel Corporation or Corel Corporation Limited, reprinted by permission; Microsoft® Excel, reprinted by permission from Microsoft Corporation; Fathom™, Key Curriculum Press.

Technical Art

Tom Dart, Alana Lai, Claire Milne, and Greg Duhaney of First Folio Resource Group, Inc.

Chapter Expectations

© Queen's Printer for Ontario, 2000.

Key Equations

Matrix Operations

Transpose

$A^t_{m \times n} = B_{n \times m}$, where $b_{ij} = a_{ji}$

Scalar Multiplication

$kA = C$, where $c_{ij} = ka_{ij}$

Addition

$A + D = E$, where $e_{ij} = a_{ij} + d_{ij}$

Multiplication

$A_{m \times n} F_{n \times p} = G_{m \times p}$, where $g_{ij} = \sum_{k=1}^{n} a_{ik} f_{kj}$

Inverse

For $H = \begin{bmatrix} a & b \\ c & d \end{bmatrix}$, $H^{-1} = \dfrac{1}{ad - bc} \begin{bmatrix} d & -b \\ -c & a \end{bmatrix}$ if $ad \neq bc$

Statistics of One Variable

	Population	Sample	Weighted Mean
Mean:	$\mu = \dfrac{\Sigma x}{N}$	$\bar{x} = \dfrac{\Sigma x}{n}$	$\bar{x}_w = \dfrac{\Sigma w_i x_i}{\Sigma w_i}$

Variance:

$$\sigma^2 = \frac{\Sigma(x - \mu)^2}{N} \qquad s^2 = \frac{\Sigma(x - \bar{x})^2}{n - 1}$$

Standard Deviation:

$$\sigma = \sqrt{\frac{\Sigma(x - \mu)^2}{N}} \qquad s = \sqrt{\frac{\Sigma(x - \bar{x})^2}{n - 1}}$$

$$= \sqrt{\frac{n\Sigma x^2 - (\Sigma x^2)}{n(n - 1)}}$$

Z-score:

$$z = \frac{x - \mu}{\sigma} \qquad z = \frac{x - \bar{x}}{s}$$

Grouped Data:

$$\mu \doteq \frac{\Sigma f_i m_i}{\Sigma f_i} \qquad \bar{x} \doteq \frac{\Sigma f_i m_i}{\Sigma f_i}, \text{ where } m_i \text{ is midpoint of } i\text{th interval}$$

$$\sigma \doteq \sqrt{\frac{\Sigma f_i (m_i - \mu)^2}{N}} \qquad s \doteq \sqrt{\frac{\Sigma f_i (m_i - \bar{x})^2}{n - 1}}$$

Statistics of Two Variables

Correlation Coefficient

$$r = \frac{s_{XY}}{s_X s_Y}$$

$$= \frac{n\Sigma xy - \Sigma x \Sigma y}{\sqrt{n\Sigma x^2 - (\Sigma x)^2}\sqrt{n\Sigma y^2 - (\Sigma y)^2}}$$

Least Squares Line of Best Fit

$y = ax + b$, where $a = \dfrac{n\Sigma xy - \Sigma x \Sigma y}{n\Sigma x^2 - (\Sigma x)^2}$ and $b = \bar{y} - a\bar{x}$

Coefficient of Determination

$$r^2 = \frac{\Sigma(y_{est} - \bar{y})^2}{\Sigma(y - \bar{y})^2}$$

Permutations and Organized Counting

Factorial: $n! = n \times (n - 1) \times (n - 2) \times \ldots \times 3 \times 2 \times 1$

Permutations

r objects from n different objects: ${}_nP_r = \dfrac{n!}{(n - r)!}$

n objects with some alike: $\dfrac{n!}{a!b!c!\ldots}$

Combinations and the Binomial Theorem

Combinations

r items chosen from n different items: $_nC_r = \dfrac{n!}{(n-r)!r!}$

at least one item chosen from n distinct items: $2^n - 1$

at least one item chosen from several different sets of identical items: $(p+1)(q+1)(r+1)\ldots -1$

Pascal's Formula: $_nC_r = {_{n-1}C_{r-1}} + {_{n-1}C_r}$

Binomial Theorem: $(a+b)^n = \displaystyle\sum_{r=0}^{n} {_nC_r}\, a^{n-r}b^r$

Introduction to Probability

Equally Likely Outcomes: $P(A) = \dfrac{n(A)}{n(S)}$

Complement of A: $P(A') = 1 - P(A)$

Odds: odds in favour of $A = \dfrac{P(A)}{P(A')}$

If odds in favour of $A = \dfrac{h}{k}$, $P(A) = \dfrac{h}{h+k}$

Conditional Probability: $P(A|B) = \dfrac{P(A \text{ and } B)}{P(B)}$

Independent Events: $P(A \text{ and } B) = P(A) \times P(B)$

Dependent Events: $P(A \text{ and } B) = P(A) \times P(B|A)$

Mutually Exclusive Events: $P(A \text{ or } B) = P(A) + P(B)$

Non-Mutually Exclusive Events: $P(A \text{ or } B) = P(A) + P(B) - P(A \text{ and } B)$

Markov Steady State: $S^{(n)} = S^{(n)}P$

Discrete Probability Distributions

Expectation: $E(x) = \displaystyle\sum_{i=1}^{n} x_i P(x_i)$

Discrete Uniform Distribution: $P(x) = \dfrac{1}{n}$

Binomial Distribution: $\qquad P(x) = {_nC_x}\, p^x q^{n-x}$ $\qquad E(x) = np$

Geometric Distribution: $\qquad P(x) = q^x p$ $\qquad E(x) = \dfrac{q}{p}$

Hypergeometric Distribution: $\qquad P(x) = \dfrac{_aC_x \times {_{n-a}C_{r-x}}}{_nC_r}$ $\qquad E(x) = \dfrac{ra}{n}$

Continuous Probability Distributions

Exponential Distribution: $y = ke^{-kx}$, where $k = \dfrac{1}{\mu}$

Normal Distribution: $y = \dfrac{1}{\sigma\sqrt{2\pi}} e^{-\frac{1}{2}\left(\frac{x-\mu}{\sigma}\right)^2}$

Normal Approximation to Binomial Distribution: $\mu = np$ and $\sigma = \sqrt{npq}$ if $np > 5$ and $nq > 5$

Distribution of Sample Means: $\mu_{\bar{x}} = \mu$ and $\sigma_{\bar{x}} = \dfrac{\sigma}{\sqrt{n}}$

Confidence Intervals: $\bar{x} - z_{\frac{\alpha}{2}} \dfrac{\sigma}{\sqrt{n}} < \mu < \bar{x} + z_{\frac{\alpha}{2}} \dfrac{\sigma}{\sqrt{n}}$ $\qquad \hat{p} - z_{\frac{\alpha}{2}} \dfrac{\sqrt{\hat{p}(1-\hat{p})}}{\sqrt{n}} < p < \hat{p} + z_{\frac{\alpha}{2}} \dfrac{\sqrt{\hat{p}(1-\hat{p})}}{\sqrt{n}}$